W9-DFR-310
APR 27 2000

St. Olaf College

MAR 0 4 1999

Science Library

A PRACTICAL GUIDE TO HEAVY TAILS

A PRACTICAL GUIDE TO HEAVY TAILS

Statistical Techniques and Applications

Robert J. Adler
Raisa E. Feldman
Murad S. Taqqu
Editors

Birkhäuser
Boston • Basel • Berlin

QA
274.4
P73
1998

Robert J. Adler
Dept. of Statistics
University of North Carolina
Chapel Hill, NC 27599
and
Faculty of Industrial Engineering
and Management
Technion
Haifa, Israel 32000

Raisa E. Feldman
Dept. of Statistics
University of California
Santa Barbara, CA 93106

Murad S. Taqqu
Dept. of Mathematics
Boston University
Boston, MA 02215

Library of Congress Cataloging-in-Publication Data

A practical guide to heavy tails : statistical techniques and applications /
Robert J. Adler, Raisa E. Feldman, Murad S. Taqqu, editors.
p. cm.
Includes bibliographical references.
ISBN 0-8176-3951-9 (alk. paper). -- ISBN 3-7643-3951-9 (Basel :
alk. paper)
1. Gaussian distribution. 2. Gaussian processes. I. Adler,
Robert J. II. Feldman, Raisa E., 1958- . III. Taqqu, Murad S.
QA274.4.P73 1998 978731
519.2'4--dc21 CIP

AMS Subject Classification: 60E07, 60G10, 60K30, 62G18, 62M10, 6209

Printed on acid-free paper
© 1998 R.J. Adler, R.E. Feldman, M.S. Taqqu

Birkhäuser 

Copyright is not claimed for works of U.S. Government employees.
All rights reserved. No part of this publication may be reproduced, stored in a retrieval system, or transmitted, in any form or by any means, electronic, mechanical, photocopying, recording, or otherwise, without prior permission of the copyright owner.

Authorization to photocopy items for internal or personal use of specific clients is granted by Birkhäuser Boston provided that the appropriate fee is paid directly to Copyright Clearance Center (CCC), 222 Rosewood Drive, Danvers, MA 01923, USA (Telephone: (978) 750-8400), stating the ISBN, the title of the book, and the first and last page numbers of each article copied. The copyright owner's consent does not include copying for general distribution, promotion, new works, or resale. In these cases, specific written permission must first be obtained from the publisher.

ISBN 0-8176-3951-9
ISBN 3-7643-3951-9

Typeset and reformatted from authors' disks
Printed and bound by Hamilton Printing Co., Rensselaer, NY
Printed in the United States of America

9 8 7 6 5 4 3 2 1

36713066

Contents

VII. Numerical Procedures

Preface

Ever since data have been collected, they have fallen into two quite distinct groups: "good data," which meant that their owners knew how to perform the analysis, and "bad data," which were difficult, if not impossible, to handle.

Since the development of modern statistics over half a century ago, good data were typically those whose distribution was amenable to the tools of the theory, tools which invariably assumed the distributions of a first course in statistics, i.e., normal, chi-squared, etc. Bad data, on the other hand, came in many forms; one type tended to jump around too much and involved outliers that contained important information. This, in short, was data with heavy-tailed histograms. Economists and modern financial analysts, for example, have been well aware for almost 30 years that much economic data falls into this "bad" category. Data of this kind, however, arise in a far wider variety of fields than economics, and include statistical physics, automatic signal detection, and telecommunications, to name just three. What made this type of data "bad," however, was nothing intrinsic, but rather the absence of well-developed statistical techniques for its analysis.

Heavy-tailed distributions and processes have been studied for decades by probabilists and mathematical statisticians, with the last decade or so having seen major advances. Many of these are summarized in the 1994 monograph on *Stable Non-Gaussian Random Processes* by Samorodnitsky and Taqqu, which provides a theoretical background to the papers in this volume. The current collection, however, is directed to the general practitioner and is primarily concerned with techniques for data analysis.

Interestingly, despite the lack of a large-scale coordinated effort to develop techniques for the analysis of heavy-tailed data, it turns out that there are really a good number of them, scattered through a variety of disciplines. It was in an attempt to bring together these various disciplines, and to "compare notes," that a small workshop was held in Santa Barbara in December 1995, with ONR support, and it was from the success of that workshop that the current volume grew.

We set about collecting expository papers on applications, data analytic techniques, and models for heavy-tailed distributions and processes. Together with our authors, we worked hard to write in a style easily accessible to readers in different disciplines. We impressed on our contributors the need to keep the elusive "user" in mind, and as a result we believe that the papers in this volume, updated bibliographically for current use, will go a long way in helping the practitioner who encounters heavy-tailed data. They provide tools, examples of different approaches, and a lead into the applied literature.

In this spirit, the volume opens with a section on applications. The two main applications considered are in the areas of computer networking and financial and insurance modelling. Crovella, Taqqu and Bestavros present convincing evidence of the heavy-tailed nature of the size distributions of files sent over the World Wide Web, and discuss the implications of this for network traffic, a topic that is continued in a paper by Willinger, Paxson and Taqqu which discusses related structural modelling problems.

On the economic side, Müller, Dacorogna and Pictet discuss the importance of heavy tails in the analysis of high frequency financial data, and look at the problem of tail decay parameter estimation in this setting, while Mittnik, Rachev and Paolella discuss some general questions of heavy-tailed modelling in financial markets. The problem of risk management, in insurance and other financial settings, is treated in a paper by Bassi, Embrechts and Kafetzaki via the use of quantile information.

The second grouping of papers centers around the problem of time series analysis for heavy-tailed data. Adler, Feldman and Gallagher give a comprehensive introduction to "Box-Jenkins" modelling in the stable setting, including a large number of simulations to indicate what does and what does not work. Calder and Davis describe parameter estimation in the stable time setting, followed by Taqqu and Teverovsky who treat the important problem of estimating long range dependence in finite and infinite variance series. These papers are followed with a thought provoking article by Resnick, which discusses a number of unexpected surprises and problems related to non-linearities and heavy-tailed modelling.

One of the interesting aspects of working with heavy-tailed, infinite variance time series is that many of the techniques used in finite variance series carry through with amazing success, although the technical details (such as the asymptotic sampling distributions of parameter estimates) may change dramatically. This is a recurring theme in all of the papers in this section, and is taken up again by Mikosch, who looks at the behavior of "periodogram" estimates from heavy-tailed data. The section closes with an illuminating article on sampling based Bayesian inference for heavy-tailed time series by Ravishanker and Qiou.

The third section of the volume contains two papers on general parameter estimation problems in the heavy-tailed setting. Pictet, Dacorogna and Müller describe an analysis of tail index estimation through Monte-Carlo simulation of synthetic data, in order to evaluate several tail estimators available in the literature. Ultimately, they recommend a bootstrapped and jacknifed version of the well known Hill estimator. A different approach to tail index estimation is taken by Kogon and Williams, who recommend working with the empirical characteristic function, and who suggest a method of getting around the heavy computational problems usually associated with this approach.

Sections 4–6 focus on specific statistical and modelling problems in which heavy-tailed distributions or processes play a central role. McCulloch considers the general regression problem when the error distribution is stable, while LePage, Podógorski and Ryznar discuss two resampling techniques for multiple linear regression with heavy-tailed errors. One is based on resampling permutations of residuals to the least squares estimates while the second exploits random flip signs. Both techniques are used to develop effective statistical inference for regression in the heavy-tailed setting.

Two more focused papers on signal processing follow. The first, by Tsakalides and Nikias, discusses the "direction of arrival" estimation problem — a classical signal/noise problem — in a setting of stable noise. Their approach is via maximum likelihood estimation, which restricts their model to the Cauchy case, when likelihoods can be explicitly computed via analytic formulas (more on this below). The second paper in this area, by Tshirintzis, presents and analyzes a model for heavy-tailed interference arising from multiple users in communications networks.

Three general types of models are presented by Goldie and Klüppelberg, Rosiński, and Samorodnitsky, who treat, respectively, subexponential distributions, the structure of stationary Lévy-stable processes, and shot noise processes with heavy-tailed shocks. These three papers, taken together, provide a solid insight into the structure of stable processes, and give a good indication of the wealth of models that exist in this area.

The volume closes with four papers related to the numerical aspects of stable distributions, two each by McCulloch and Nolan. There is no question that the development of fast and accurate numerical methods for computing stable densities is one of the main issues facing heavy-tailed modelling today.

Since the introduction of stable models, the impracticability of computing stable densities has been one of the main reasons for the need to develop non-standard statistical techniques in this setting. One could not, for example, employ the all but ubiquitous maximum likelihood techniques of standard (i.e., Gaussian) statistical analysis when there was no practical way of computing a likelihood. With the advent of ever faster computers and new numerical techniques, this possibility is close to being realized; the fact that we have no analytic form for the stable density may soon no longer be a problem.

In his two papers in this closing section, McCulloch discusses the general problem of numerical approximation of the symmetric stable distribution and density, and presents some tables for the maximally skewed case. Nolan discusses approximation, estimation, simulation and identification problems for multivariate stable distributions, and, in a short but important paper on numerical methods, gives us a URL for a battery of useful computer programs.

While reiterating the applied nature of this volume, it is important to note that many of the questions posed in the individual papers will require heavy theoretical analysis to be fully answered. Consequently, although we did not plan it this way, we rather expect that it will make a good source book for theoreticians as well, re-emphasizing once again that the best theory is usually born from an application.

Finally, as editors, we have two sets of acknowledgments to make. The first is to our authors and referees. They all worked very hard to prepare papers that were useful and readable, rather than just "clever," as we are all trained to do nowadays. We take this opportunity to apologize to them for all the rewriting we demanded.

Secondly, we must thank our granting agencies: RA is indebted to the Israel Science Foundation, the US-Israel Binational Science Foundation, the Office of Naval Research, and most recently, the National Science Foundation for support. RF thanks the Office of Naval Research. MT thanks the National Science Foundation.

Robert J. Adler
Raise E. Feldman
Murad S. Taqqu

Contributors

Robert J. Adler
Department of Statistics
University of North Carolina
Chapel Hill, NC 27599-3260
adler@stat.unc.edu

Faculty of Industrial
Engineering and Management
Technion
Haifa, Israel 32000
robert@ieadler.technion.ac.il

Franco Bassi
Department of Mathematics
ETH-Zentrum
CH-8092 Zurich, Switzerland
bassi@math.ethz.ch

Azer Bestavros
Computer Science Department
Boston University
111 Cummington St.
Boston, MA 02215-2411
best@bu.edu

Matt Calder
Department of Statistics
Colorado State University
Fort Collins, CO 80523
calder@stat.colostate.edu

Mark E. Crovella
Computer Science Department
Boston University
111 Cummington St.
Boston, MA 02215-2411

Richard A. Davis
Department of Statistics
Colorado State University
Fort Collins, CO 80523
rdavis@colostate.edu

Michel M. Dacorogna
Olsen & Associates AG
Research Institute for Applied Economics
Seefeldstrasse 2333
CH-8008 Zurich, Switzerland
daco@olsen.ch

Paul Embrechts
ETH-Zentrum
Department of Mathematics
CH-8092 Zurich, Switzerland
embrechts@math.ethz.ch

Raisa E. Feldman
Dept. of Statistics & Appl. Probability
University of California
Santa Barbara, CA 93106
epstein@pstat.ucsb.edu

Colin Gallagher
Dept. of Statistics
and Applied Probability
Santa Barbara, CA 93106
colin@pstat.ucsb.edu

Charles M. Goldie
School of Mathematical Sciences
University of Sussex
Brighton BN1 9QH, England
C.M.Goldie@sussex.ac.uk

Maria Kafetzaki
Department of Mathematics
ETH-Zentrum
CH-8092 Zurich, Switzerland
kafetz@math.ethz.ch

Stephen M. Kogon
MIT Lincoln Laboratory
244 Wood St.
Lexington, MA 02173-9108
kogon@ll.mit.edu

Claudia Klüppelberg
Department of Mathematics
Center for Mathematical Sciences
Munich University of Technology
D-80290 Munich, Germany
cklu@mathematik.uni-mainz.de

Raoul LePage
Department of Statistics
and Probability
Michigan State University
East Lansing, MI 48824
entropy@msu.edu

J. Huston McCulloch
Economics Department
Ohio State University
1945 N. High Street
Columbus, OH 43210
mcculloch.2@osu.edu

Thomas Mikosch
Department of Mathematics
University of Groningen
P.O. Box 800
NL-9700 AV Groningen
T.Mikosch@math.rug.nl

Stefan Mittnik
Institute of Statistics and Econometrics
Christian Albrechts University at Kiel
Olshausenstr. 40-60
D-24098 Kiel, Germany
mittnik@stat-econ.uni-kiel.de

Ulrich A. Müller
Olsen & Associates AG
Research Inst. for Applied Economics
Seefeldstrasse 233
CH-8008 Zurich, Switzerland
ulrichm@olsen.ch

Chrysostomos L. Nikias
Signal and Image Processing Institute
Department of Electrical
Engineering Systems
University of Southern California
Los Angeles, CA 90089-2564
nikias@sipi.usc.edu

John P. Nolan
Department of Mathematics
American University
4400 Massachusetts Avenue, NW
Washington, DC 20016
jpnolan@american.edu

Don B. Panton
Department of Finance
University of Texas at Arlington
Arlington, TX 76019
panton@uta.edu

Marc S. Paolella
Inst. of Statistics and Econometrics
Christian Albrechts University at Kiel
Olshausenstr. 40-60
D-24098 Kiel, Germany
paolella@stat-econ.uni-kiel.de

Vern Paxson
Lawrence Berkeley National Lab
University of California
Berkeley, CA 94720
vern@ee.lbl.gov

Olivier V. Pictet
Olsen & Associates AG
Research Institute
for Applied Economics
Seefeldstrasse 233
CH-8008 Zurich, Switzerland
olivier@olsen.ch

Krzysztof Podgórski
Dept. of Mathematical Sciences
IUPUI
Indianapolis, IN 46202-3216
kpodgorski@math.iupui.edu

Zuqiang Qiou
Palisades Research Inc.
101 Bedford Place
Morganville, NJ 07751
zuqiang@stat.uconn.edu

Svetlozar T. Rachev
Dept. of Statistics & Appl. Probability
University of California
Santa Barbara, CA 93106
rachev@pstat.ucsb.edu

Nalini Ravishanker
Department of Statistics
University of Connecticut
U-120, 196 Auditorium Road
Storrs, CT 06269
ravishan@uconnvm.uconn.edu

Sidney I. Resnick
School of Operations Research
Cornell University
Ithaca, NY 14853
sid@orie.cornell.edu

Jan Rosiński
Department of Mathematics
University of Tennessee
Knoxville, TN 37996-1300
rosinski@math.utk.edu

Michal Ryznar
Institute of Mathematics
Technical University of Wroclaw
50-370 Wroclaw, Poland
ryznar@graf.im.pwr.wroc.pl

Gennady Samorodnitsky
School of Operations Research
and Industrial Engineering
Cornell University
Ithaca, New York 14583
gennady@orie.cornell.edu

Murad S. Taqqu
Department of Mathematics
Boston University
111 Cummington Street
Boston, MA 02215-2411
murad@math.bu.edu

Vadim Teverovsky
Dept. of Mathematics
Boston University
111 Cummington Street
Boston, MA 02215
vt@bu.edu

Panagiotis Tsakalides
Signal & Image Processing Institute
Dept. of Electrical Engineering Systems
University of Southern California
Los Angeles, CA 90089-2564
tsakalid@sipi.usc.edu

George A. Tsihrintzis
Center for Electromagnetics Research
Northeastern University
Boston, MA 02215
geoatsi@sipi.usc.edu

Douglas B. Williams
Center for Signal and Image Processing
School of Electrical and Computer Engineering
Georgia Institute of Technology
Atlanta, GA 30332-0250
dbw@ee.gatech.edu

Walter Willinger
AT & T Laboratories
Information Sciences Research
600 Mountain Ave
Florham Park, NJ 07932
willinger@att.research.com

Alex White
Dept. of Statistics & Probability
Michigan State University
East Lansing, MI 48824
whiteale@pilot.msu.edu

I. Applications

Heavy-Tailed Probability Distributions in the World Wide Web

Mark E. Crovella, Murad S. Taqqu and Azer Bestavros [1] [2] [3]

Abstract

The explosion of the World Wide Web as a medium for information dissemination has made it important to understand its characteristics, in particular the distribution of its file sizes. This paper presents evidence that a number of file size distributions in the Web exhibit heavy tails, including files requested by users, files transmitted through the network, transmission durations of files, and files stored on servers. In addition, we argue that because of the presence of caching in the Web, the size distribution of transmitted files is primarily determined by the distribution of files available on the Web, and is relatively insensitive to the distribution of files requested by users. Finally, we discuss some of the implications of heavy-tailed transmission durations and relate these results to self-similarity in network traffic.

1. Introduction

The World Wide Web was designed and initially developed at the European Laboratory for Particle Physics (then called CERN) as a distribution method for scientific documents. Since its public release in 1992 the World Wide Web ("the Web") has been enthusiastically adopted by commercial, educational, and governmental users as a method of easily organizing and distributing information that is rich in multimedia content: text, graphics, animation, audio, and video. Growth of the Web has been very rapid; since its inception it has doubled in size roughly every nine months. As of early 1996 typical estimates of the number of documents available on the Web are in the range of 50 million [Bra96].

At the end of 1996, the Web generated more data traffic on the Internet than any other application. Consequently, characteristics of the Web have implications for network engineering, capacity planning, and performance evaluation of the Internet. In addition, since the Web is so widely used, a thorough understanding of its characteristics is an important goal in its own right. In this paper we describe a number of empirical

[1]This work was supported in part by NSF grants CCR-9501822, CCR-9308344, NCR-9404931 and DMS-9404093 at Boston University.

[2]*AMS Subject classification:* 60K30, 60G18, 60E07.

[3]*Keywords:* infinite variance, self-similarity, stable distribution, Pareto, Lognormal distribution, Hill estimator, Internet, caching.

characteristics of the Web, concentrating on measurements of probability distributions.

To place our measurements in context, we start by presenting background on how the Web is organized and implemented in the current Internet. The basic infrastructure used by the Web consists of host computers (each functioning as a *client* or a *server* or both) and a connecting network (usually the global Internet). In addition, *caches* may be present at various points in the system, and we describe these and their effects on data traffic over the Internet. We then describe details of our data collection methods, which were based on adding measurement apparatus to the Web clients at our site, and on conducting a survey of a number of Web servers.

Next we show that a number of our measurements of the Web are consistent with the conclusion of heavy-tailed distributions. We first present evidence indicating that when files are transmitted over the network, the transmission durations appear to follow a heavy-tailed distribution. We then show that this effect may be explained by the sizes of the files themselves, as they too exhibit heavy tails. In fact, our data indicates that both the distribution of files requested by users, as well as the distribution of files available on a set of servers, are heavy-tailed.

Thus, our measurements indicate that heavy-tailed distributions appear in a number of related datasets associated with the Web. One of the questions we try to address in this paper is: which of these datasets is the primary cause of heavy-tailed distributions in the Web? That is, are the distributional characteristics of transmission durations mainly dependent on user requests or on the available files? Surprisingly, we show that the heavy-tailed property of transmission times is likely to be caused mainly by the distribution of available files, rather than by the nature of user requests.

We conclude with some observations on the implications of those heavy-tailed distributions for network performance evaluation and engineering. We note that the presence of a heavy tail in the distribution of transmission lengths is a possible cause of network traffic *self-similarity* [LTWW94, WPT98] with scaling parameter $H > 1/2$, which means that traffic shows noticeable bursts at all scales of interest. We further note that self-similarity is a significant factor in the performance of networks like the Internet; in particular, delays in transferring data can be much more severe when traffic shows self-similarity than would be predicted by traditional traffic models.

2. Studying the World Wide Web

In this section we present an overview of how the Web is organized and implemented, of how we took measurements of the Web, and of the methods we used to analyze the data collected.

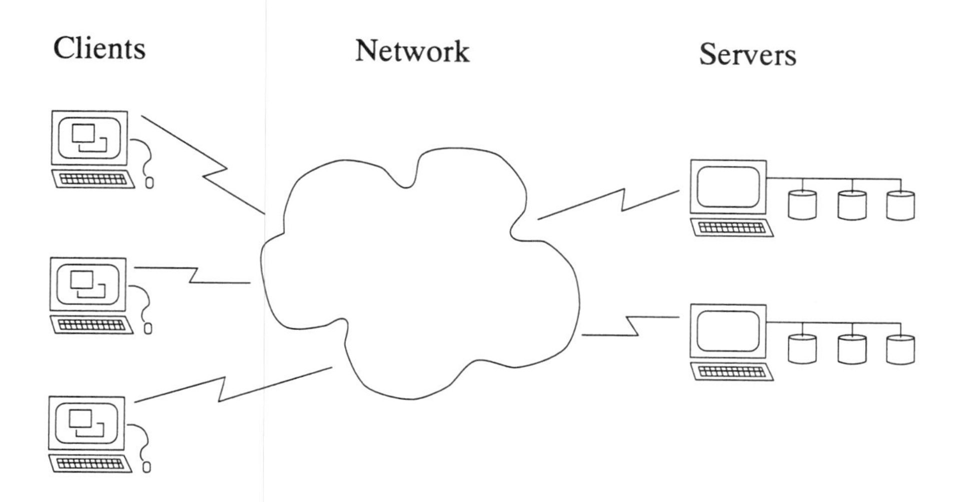

Figure 1: Clients, Network, and Servers in the World Wide Web

2.1 Organization and Implementation of the Web

The remarkable popularity of the Web seems to arise from a combination of its utility and its ease of use. It is useful as a means of publishing and delivering information in a wide variety of formats: raw data, formatted text, graphics, animation, audio, video, and even software. Its ease of use stems from the fact that it hides the details of contacting remote sites on the Internet, transporting data across the network, and formatting, displaying or playing the requested information regardless of the type of particular computers involved.

Information in the Web exists as files on computer systems (*hosts*); each file has a globally unique identifier called a Uniform Resource Locator (URL). It is only necessary to know a file's URL in order to transfer it from wherever it is stored (potentially anywhere in the global Internet) and display it on the user's local computer.

The Web is organized using a client-server model. Each file is stored on a specific host, specified as part of its URL; such hosts are *servers*. When a user requests a file, it is transferred to the user's local host, the *client*. (In fact, a single host can act as both a client and a server.) The software used by the client to retrieve and display files is called a *browser*. Figure 1 shows a schematic view of this model. Clients (which typically have display and input devices) retrieve files from servers (which have storage devices) using the network. From the standpoint of the participants, the specific topology of the network is unknown — hence its representation as an amorphous cloud.

A more detailed view including implementation details is shown in Figure 2. The configuration shown is representative of the connections between a set of hosts at a single

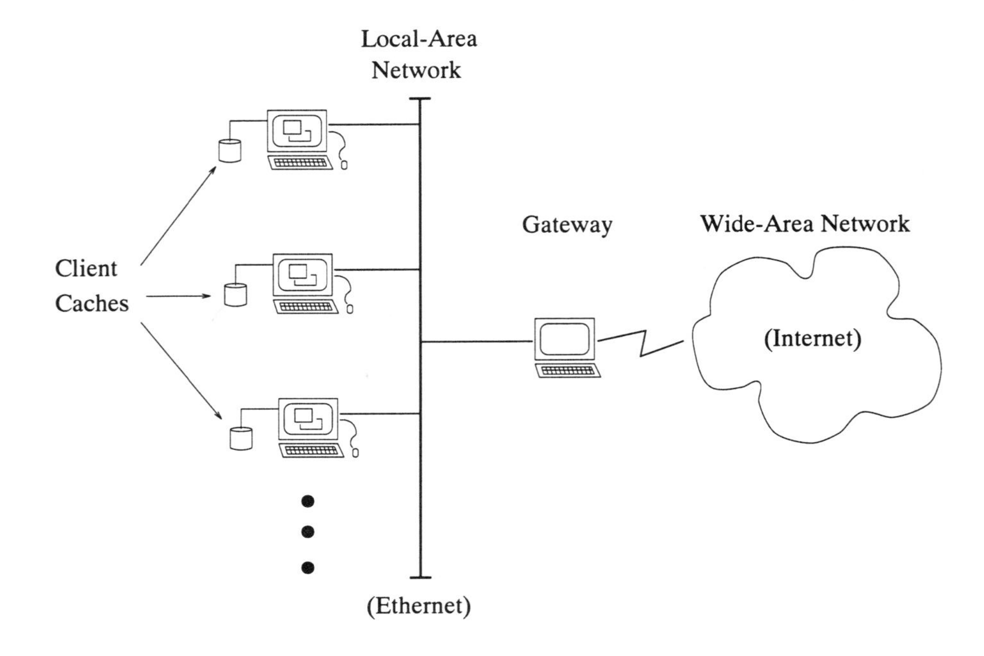

Figure 2: Implementation of Client Connections to the Web

site and the Internet. Clients are interconnected with a Local-Area Network (LAN) which is typically an Ethernet. The LAN is then connected via a special computer (a *gateway* or, equivalently, a *router*) to the Wide-Area Network (WAN), that is, the Internet.

All clients on the LAN employ *caching* to speed up access to WWW files. A cache is a set of copies of WWW files, kept on a local storage device — either main memory or disk. When a request is made, the client's browser software first checks to see if the file being requested can be found in the cache; if so, it need not be retrieved over the Internet but instead can be copied directly from the cache. The difference in response time between service from the cache versus service from the Internet can amount to two to three orders of magnitude; hence, most browsers have from the earliest days of the Web implemented some form of caching.

To implement caching, the browser examines each URL request made by the user. If the request can be served from the cache, it is a *cache hit;* if not, it is a *cache miss.* Whenever a file is retrieved from the network as a result of a cache miss, it then becomes a candidate for subsequent caching. Since the browser can only use a limited amount of storage for implementing the cache, it must decide whether to cache the file, and if so, whether to evict some other file(s) from the cache in order to find space for the new file. Such a set of decisions is called a *cache management policy.*

2.2 Web Data

Measurements of Web activity can be made at a variety of points in the network; in particular, two important measurement points are at the client and at the server. Server measurements are generally easy to obtain because one of the server's roles is to assess its own impact on its host system. As a result, most servers keep detailed records of each access made to them. On the other hand, clients typically peform very little recordkeeping associated with their activities. Unfortunately, it is difficult to use server records to obtain a picture of Web activity on a LAN, since each client on the LAN may visit many different servers over a short time.

Thus, in order to capture all of the Web activity on a LAN, it was necessary to perform measurement on Web browsers. We added measurement apparatus to the browsers in use at Boston University's Computer Science Department. To do this, we modified the Web browser *NCSA Mosaic* [fSA] and installed it for general use. This browser is available in source code form, and permission has been granted for using and modifying the code for research purposes. Most important, at the time of the study (November 1994 through February 1995) Mosaic was the browser preferred by nearly all users at our site. Thus by instrumenting only this program we were able to measure nearly all of the local Web activity. Since that time, other browsers have become more popular than Mosaic; because these (commercial) browsers are not easily instrumented, collecting an equivalent set of data at the current time would be significantly more difficult.

In this paper we will refer to a single execution of Mosaic as a *session,* and the record of all URLs accessed in a session as a *trace.* Each trace is stored in a separate file called a *log.* Each line of a log corresponds to a single URL requested by the user; it contains the machine name, the time stamp when the request was made, the URL, the size of the file in bytes (including the overhead of the HTTP protocol) and the file retrieval time in seconds (reflecting only actual communication time, and not including the intermediate processing performed by Mosaic in a multi-connection transfer). Timestamps are accurate to 10 ms.

To collect our data we installed our instrumented version of Mosaic in the general computing environment at Boston University's Computer Science Department, which consists principally of 37 SparcStation 2 workstations connected in a local network. The data used in this paper was collected during the period 17 January 1995 to 28 February 1995. This data is freely available from Boston University [CBC95] and from the Internet Traffic Archives [Hal]. To our knowledge, these are the first such traces generally available to the research community.

Three of the most important datasets we collected are:

1. The set of *file requests:* this is a record of all requests for URLs made by users. This dataset contains many duplicate requests, which can occur when a user requests

a file more than once, or when more than one user requests the same file. Many such requests result in cache hits, which means that they were satisfied without generating any network traffic.

2. The set of *file transfers:* this consists of all of the cache misses. Each element in this set corresponds to a single instance of a file being transferred over the network. Thus it is a proper subset of the set of file requests. However, despite the action of caching, some files are still transferred more than once over the network, so this dataset still contains some duplicate files.

3. The set of *unique files:* this set contains exactly one entry for each file, regardless of how many times it was requested or transferred. Thus this set is also a proper subset of the previous two.

For each of these datasets, we are primarily concerned with the distribution of object sizes measured in bytes. Descriptive statistics are shown in Table 1.

Sessions 4,700 Users 591			
File Requests	575,775	Bytes Requested	2,713 MB
File Transfers	130,140	Bytes Transferred	1,849 MB
Unique Files	46,830	Unique Bytes	1,088 MB
Transmission Times ↔ File Transfers Available Files 146,400			

Table 1: Summary Statistics for Trace Data Used in This Study

We also collected two additional datasets. Of particular importance for the study of network traffic is the set of *transmission times.* For each item in the set of file transfers, there is a corresponding item in this set which specifies how much time was required to transfer the given file.

Finally, although our results are largely based on client measurements, for comparison purposes we also examine some measurements made at servers. In these measurements our goal was to gain a rough picture of the size distribution of available files present on Web servers. This set is somewhat elusive because it is constantly changing. To obtain an approximate snapshot of available files at the time of the study, we surveyed a subset of Web servers. To do this we started from a list of over 500 known Web servers. From these we selected those which provided freely available usage reports using a software package called *www-stat 1.0* [Reg]; there were 32 such servers. These usage reports provide information sufficient to determine the size distribution of the files present on the server for files that were accessed during the reporting period. We

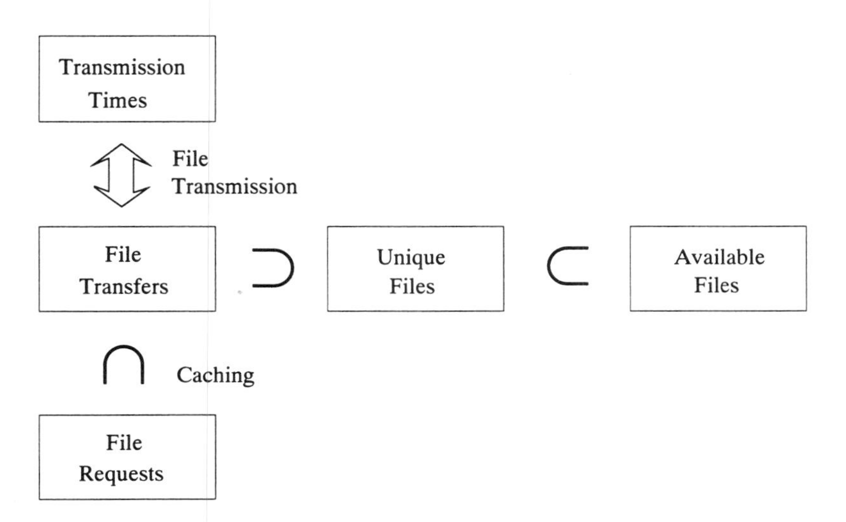

Figure 3: Relationship Between Datasets ($\updownarrow$ indicates a one-to-one relation and $\subset$, a subset relation).

collected the results from all 32 servers into a single dataset, which consists of 146,400 files containing a total of 3,674 MB. We use this dataset as a representative of the set of *available files* on the Web. While the collection method we used does not provide a truly random sample of files available in the Web, it sufficed to assess whether the heavy-tailed property might be present in the set of available files.

The relationship between the datasets considered here is shown in Figure 3. Note that while the set of available files is conceptually a superset of the set of unique files, in the case of our datasets this is not strictly the case because the set of available files was collected independently of the others. We will discuss these relationships in more detail in Section 3.3.

2.3 Estimating Tail Weight in Web Data

The distributions we use in this paper have the property of being *heavy-tailed.* We say here that a random variable X follows a heavy-tailed distribution if

$$P[X > x] \sim x^{-\alpha}, \quad \text{as } x \to \infty, \quad 0 < \alpha < 2.$$

The simplest heavy-tailed distribution is the *Pareto* distribution, with probability mass function

$$p(x) = \alpha k^{\alpha} x^{-\alpha-1}, \quad \alpha, k > 0, \quad x \geq k.$$

and cumulative distribution function

$$F(x) = P[X \leq x] = 1 - (k/x)^{\alpha}.$$

Our results attempt to estimate the values of α for a number of empirically measured distributions. To do so, we use two methods:

1. Log-log *complementary distribution* (CD) plots; and
2. the *Hill* estimator [Hil75].

CD plots show the complementary cumulative distribution $\overline{F}(x) = 1 - F(x) = P[X > x]$ on log-log axes. Plotted in this way, heavy-tailed distributions have the property that

$$\frac{d \log \overline{F}(x)}{d \log x} \sim -\alpha,$$

for large x. In practice we obtain an estimate for α by plotting the CD plot of the dataset and selecting a minimal value x_0 of x above which the plot appears to be linear. Then we select equally-spaced points from among the CD points larger than x_0 and estimate the slope using least-squares regression. Equally-spaced points are used because the point density varies over the range used, and the preponderance of data points for small file sizes would otherwise unduly influence the least-squares regression.

Our second approach to estimating tail weight is using the Hill estimator. The Hill estimator gives an estimate of α as a function of the k largest elements in the data set; it is defined as

$$\mathcal{H}_{k,n} = \frac{1}{k} \sum_{i=1}^{k} (\log X_{(i)} - \log X_{(k+1)})$$

where $X_{(1)} \leq \ldots \leq X_{(n)}$ denote the dataset's order statistics, i.e., the data items arranged according to size. In practice the Hill estimator is plotted against k for small values of k; if the estimator stabilizes to a consistent value this provides an estimate of α.

Although typically the Hill estimator is plotted as a function of k, in many cases it is more informative to plot as a function of $\log k$ [RS96, Res98], which we do in this paper. In addition, we label the ordinate in terms of $\log(P[X > x])$. This is possible because when using the empirical distribution, the probability that X is greater or equal to the kth order statistic $X_{(k)}$ is k/n, and hence the logarithm of that probability corresponds to $\log(k/n) = \log k - \log n$. The advantage of this labeling scheme is that it allows an easy comparison with the CD plot.

Finally, we use a simple test based on the theory of stable distributions, which we call the Limit Distribution (LD) test. We start by aggregating the dataset in question (X_i) over blocks of size m:

$$X_t^{(m)} = \sum_{i=(t-1)m+1}^{tm} X_i.$$

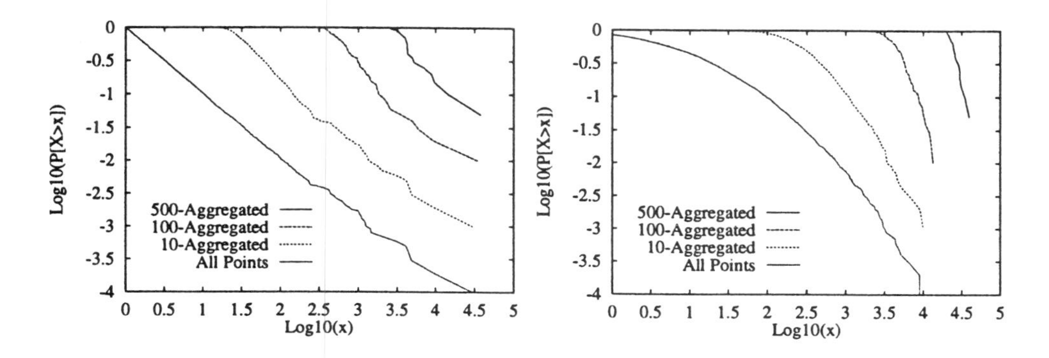

Figure 4: Comparison of LD Test for Pareto (left) and Lognormal (right) Distributions

This process is repeated for a number of large values of m (in our case, for $m = 10, 100$, and 500). If the original dataset follows a distribution that belongs to the domain of attraction of a stable distribution with $\alpha < 2$, then the tails of the m-aggregated datasets will all tend to follow power-law behavior with the same α. If the original dataset follows a finite-variance distribution, then the aggregated datasets will tend to the normal distribution and their tails will decline exponentially.

This difference can be observed on a CD plot. For datasets with finite variance, the slope will increasingly decline as m increases, reflecting the underlying distribution's approximation of a normal distribution. For datasets with infinite variance, the slope will remain roughly constant with increasing m.

An example is shown in Figure 4. The figure shows the LD test for aggregation levels of 10, 100, and 500 as applied to two synthetic datasets. On the left the dataset consists of 10,000 samples from a Pareto distribution with $\alpha = 1.0$. On the right the dataset consists of 10,000 samples from a lognormal distribution with $\mu = 2.0$, $\sigma = 2.0$. These parameters were chosen so as to make the Pareto and lognormal distributions appear approximately similar for $\log_{10}(x)$ in the range 0 to 4. In each plot the original CD plot for the dataset is the lowermost line; the upper lines are the CD plots of the aggregated datasets. Increasing aggregation level increases the average value of the points in the dataset (since the sums are not normalized by the new mean) so greater aggregation levels show up as higher lines in the plot. The figure clearly shows the qualitative difference between finite and infinite variance datasets. The Pareto dataset is characterized by parallel lines, while the lognormal dataset is characterized by lines that seem roughly convergent.

3. WWW Size Distributions

As discussed in Section 2, our results are based on measurements of five datasets:

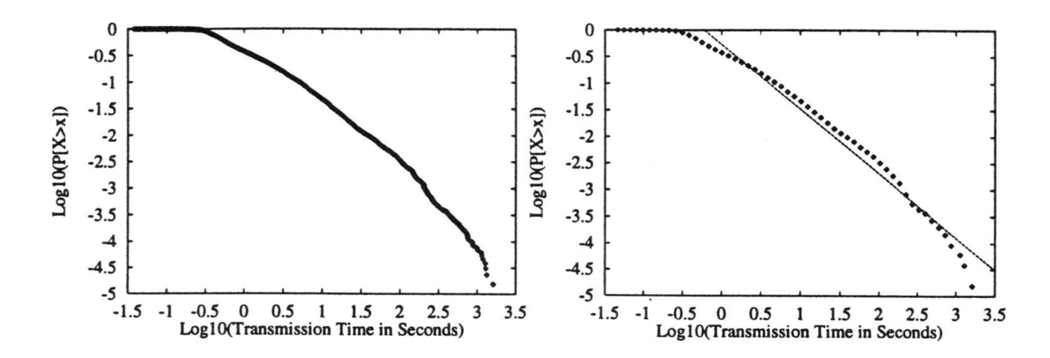

Figure 5: CD of Transmission Times of Web Files

file requests, file transfers, transmission times, unique files, and available files. In this section we present the principal results concerning distributions in these datasets. For each, we show that the corresponding empirical distribution is consistent with heavy-tailed behavior. Also, we describe the differences among these distributions and suggest some causal relationships between them.

3.1 The Distribution of Transmission Times

The set of transmission times has the most direct connection with network engineering. As we will discuss in Section 4, the conclusion of heavy-tailed behavior for this dataset is especially important because it provides support for an explanation of the observed self-similarity of network traffic.

The set of transmission times appears to exhibit heavy-tailed characteristics. Figure 5 (left side) presents the CD plot of the durations of all 130,140 file transfers occurring during the measurement period. The figure shows that for values greater than about $\log_{10}(x) = -0.5$ ($x \approx 0.3$ seconds), the plot is nearly linear — consistent with a power law upper tail. The least squares fit shown on the right side of Figure 5 ($R^2 = 0.98$) has a slope of -1.21, corresponding to an $\hat{\alpha} = 1.21$. The plot appears to have some curvature, although we show below that this distribution appears to be in the domain of attraction of an infinite variance (stable) distribution.

Figure 7 shows the use of the Hill estimator on this dataset. On the left side, the Hill estimator is plotted with a linear scale in the ordinate. Vertical lines are plotted at the 90th, 80th, and 50th percentile of the dataset.

On the right side, the same data is plotted using a log scale in the ordinate. The Hill estimator is hard to interpret for this dataset, and both the Hill plot and the slight curvature in the CD plot suggest two regimes for α in this dataset. For our purposes these plots are strongly suggestive of the presence of heavy-tailed distributions. Similar comments apply to the other datasets given in the following sections.

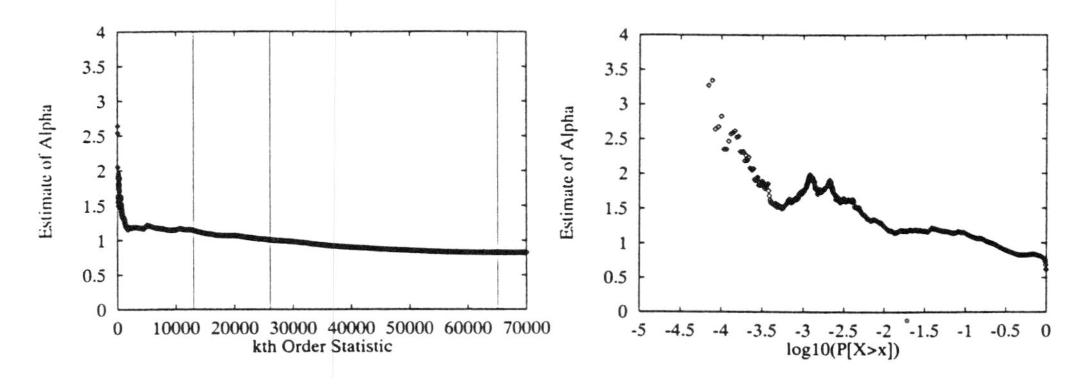

Figure 6: Hill Estimator For Transmission Times of Web Files; Linear Ordinate (left) and Log Scale Ordinate (right)

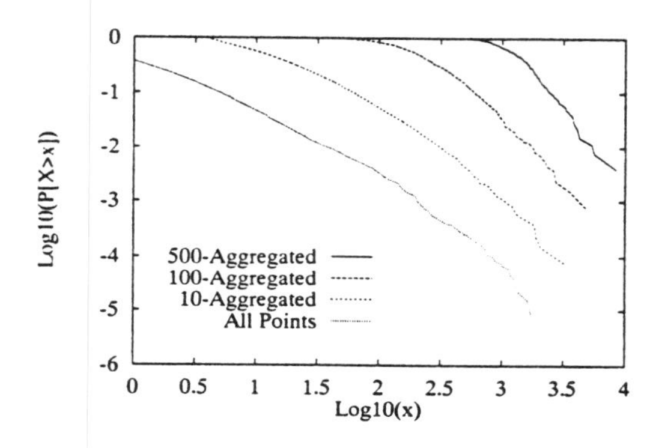

Figure 7: LD Test For Transmission Times of Web Files

To assess whether this dataset is consistent with a conclusion of infinite variance, we use the LD Test described in Section 2.3. The results for our dataset of transmission times is shown in Figure 7, and shows that as we aggregate the dataset, the slope of the tail does not change appreciably; that is, under the LD Test, transmission times behave more like the Pareto distribution (left side of Figure 4) than the Lognormal distribution (right side of Figure 4).

3.2 Why are Transmission Times Heavy-Tailed?

To understand why transmission times are heavy-tailed, we now examine size distributions of Web files themselves. In particular, we present distributions for four datasets: file requests, file transfers, unique files, and available files.

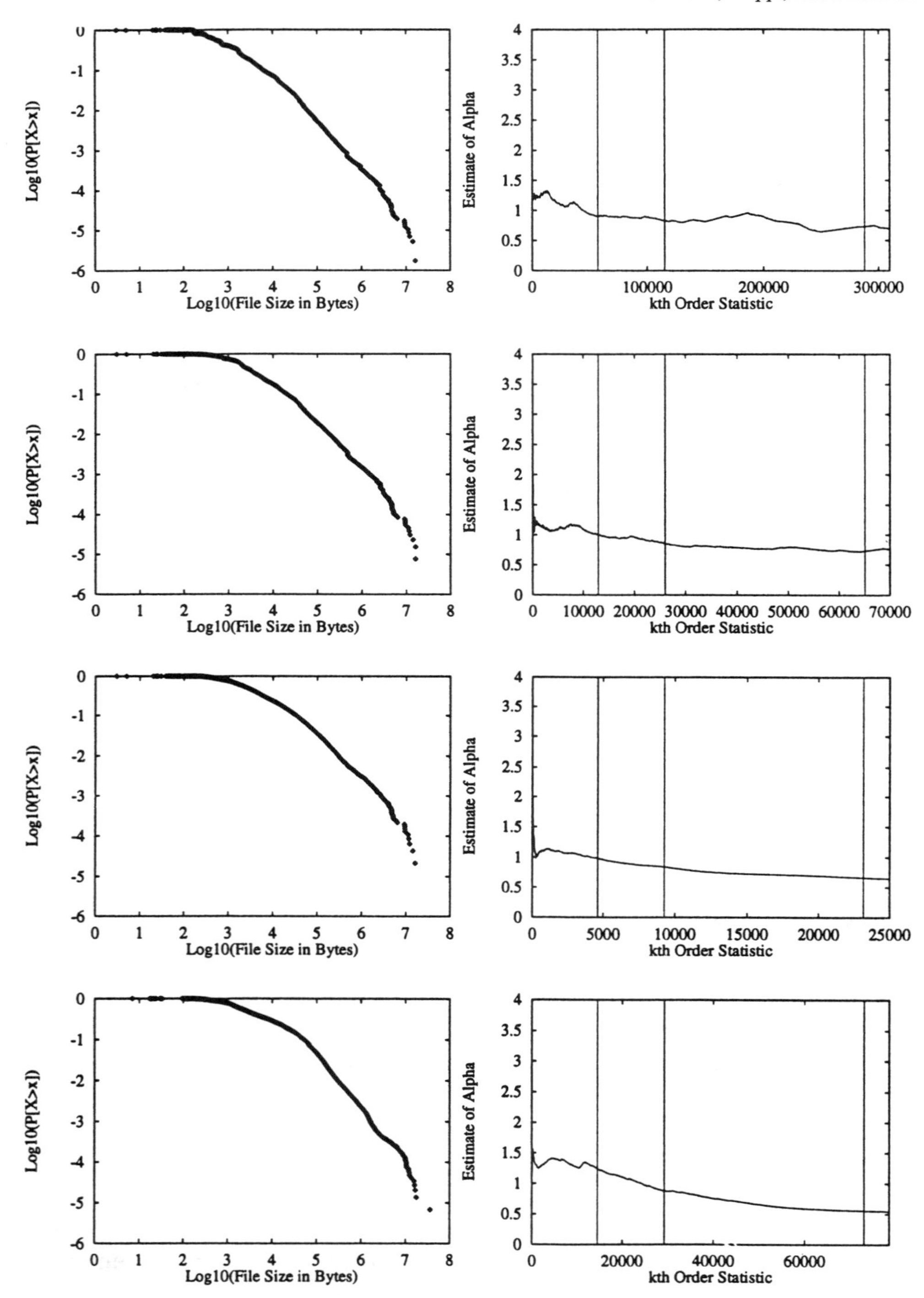

Figure 8: Evidence of Heavy Tails in File Requests (top), File Transfers (second), Unique Files (third), and Available Files (bottom).

First we present evidence that these distributions each show heavy-tailed behavior. Figure 8 shows CD plots and Hill plots for these datasets. Each of the CD plots shows nearly linear behavior over a large range of the random variable — typically about four orders of magnitude. In addition, each of the Hill plots shows reasonably stable behavior over a long range.

At the top of the figure are the plots for file requests; for these 575,775 data points the slope of the CD plot yields an $\hat{\alpha}$ of about 1.16. Next are the plots for file transfers; for these 130,140 items the slope estimate yields $\hat{\alpha}$ of about 1.06. Third are the plots for the unique files; for these 46,830 items the slope estimate is $\hat{\alpha}$ of about 1.05. At the bottom are the plots for the set of available files; for these 146,400 items the slope estimate is $\hat{\alpha}$ of 1.06.

These data show that heavy-tailed distributions seem to be present in each of our datasets of file sizes. In particular, the sizes of file transfers show heavy tails, and so do transmission times.

It is important to note that transmission times depend not only on file sizes but also on the condition of the network at the time. In fact, there does not seem to be a strong sample correlation between file sizes and transmission times; Figure 9 illustrates this point by showing a scatterplot of transmission time versus file size for our datasets. It seems likely, however, that heavy-tailed file sizes are responsible, albeit perhaps in an indirect manner, for heavy-tailed transmission times.

An important question then is why file transfers show a heavy-tailed distribution. On the one hand, it is clear that file requests constitute user "input" to the system. It's natural to assume that file requests therefore might be the primary determiner of heavy-tailed file transfers. If this were the case, then perhaps changes in user behavior might affect the heavy-tailed nature of file transfers, and by implication, the self-similar properties of network traffic.

In fact, in the next subsection we argue that file requests are *not* intrinsically responsible for the heavy-tailed nature of file transfers. Rather, we will argue that the set of file transfers is more closely determined by what files are *available* in the Web than it is by what files are requested. That is, we will contend that the heavy-tailed nature of transmission times is more strongly determined by the size distribution of available files than it is by the size distribution of file requests. To do so we rely on characteristics of the set of unique files.

3.3 The Nature of Unique File Sets and the Action of Infinite Caches

Our argument is based on two important properties of the set of unique files. First, the set of unique files is the limit for the set of transferred files as the cache size grows to infinity. Second, measurements indicate that the set of unique file is approximately

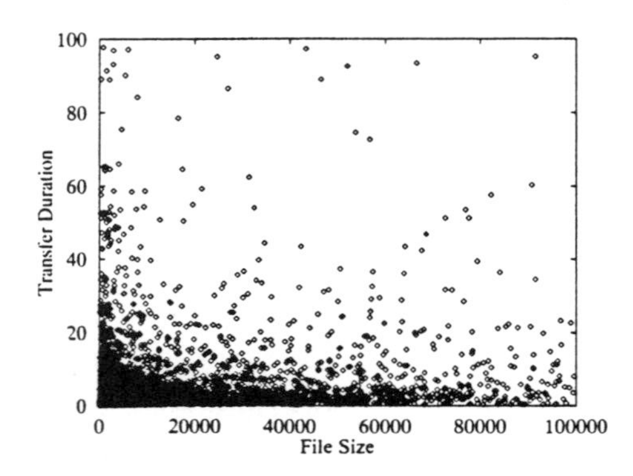

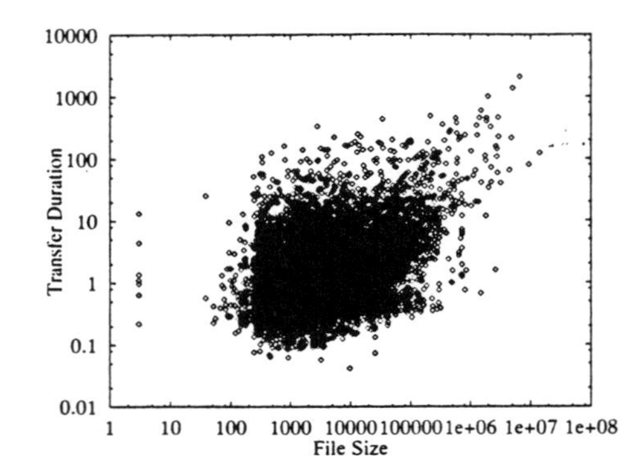

Figure 9: Scatter Plot of File Sizes vs. Transmission Times (left: linear axes; right: logarithmic axes)

distributed like the set of available files. Together these two properties suggest that for systems with large caches, the distribution of the set of transferred files will naturally approximate that of available files — and will be relatively independent of the particular requests made by users. We now discuss and support these points.

First, we explain why the set of unique files is the limit for the set of available files as the cache size grows to infinity. To start, it is helpful to imagine the behavior of a cache of infinite size. Such a cache would never have to evict any file. As a result, the only references that would cause cache misses would be those that happen to be the *first* reference made to the particular file. In our measurements, this set is in fact the set of unique files (whose distribution was shown in the last subsection).

In practice, the goal of any real cache management policy may be described as the attempt to simulate an infinite cache using finite resources. The better a real cache performs, the closer the set of transferred files will be to the set of unique files. In our measurements, it seems that NCSA Mosaic was able to achieve a reasonable approximation of the performance of an infinite cache, despite its finite resources: from Table 1 we can calculate that NCSA Mosaic achieved a 77% hit rate ($1-130140/575775$), while a cache of infinite size (shared by all users) would achieve a 92% hit rate ($1 - 46830/575775$).

The role of the set of unique files as the limit set for the set of file transfers can be seen in our data. In Figure 10 (left side), we show the distributions of file requests, file transfers, and unique files on the same axes. In this figure we can see how the distributions are changing as we progress from requested files to transferred files to unique files. The figure shows that set of file transfers (i.e., cache misses) is *intermediate* in distributional characteristics between the set of file requests and the set of unique files.

The plot on the left side of Figure 10 shows that the median of the set of file transfers is larger than the median of the set of file requests. As it happens, in these datasets, users tended to request small files more often, per file, than large files. The sets of file requests

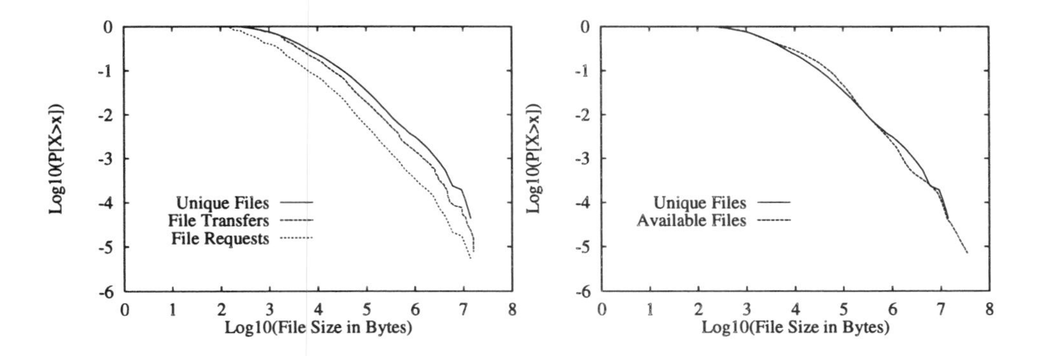

Figure 10: CD plots of the Different Distributions

and file transfers would be identical if there were no caching taking place. Because, in our case, caching is fairly effective, one cannot relate the two datasets. However, *whatever* the preferences of users happened to be — for large files, or small files, or neither — the action of caching is to move the distributional characteristics of the set of file transfers closer to that of the set of unique files. Thus, depending on the effectiveness of caching, the median transfer size may be closer either to the median of the set of file requests, or to that of the set of unique files.

The second property of our unique file set is that it approximates in distribution the set of available files. That is, it appears that the set of unique files can be considered, with respect to sizes, to be a sample from the set of all files available in the Web.

Evidence of this effect is shown on the right side of Figure 10. This figure plots the CD of the sets of unique files and available files on the same axes. While in this case the set of unique files is not a subset of the set of available files mentioned in Table 1, we are considering this set of available files as representative of the entire set of files accessible via the Web. The figure shows that, in contrast to the comparisons on the left side of Figure 10, these two distributions are nearly identical over their entire shared range. This suggest that for our data, the distribution of the set of unique transfers seems approximately that of the set of files available on the Web.

To summarize our argument we show the relations between the important datasets indicated in Figure 3.

First, all the distributions appear to be heavy-tailed. Second, the set of unique files is a subset of the set of file transfers, but this subset relation tends to equality as cache sizes grow large. Third, the set of unique files is also a subset of the set of files available on the Web, but these datasets tend to be approximately equal in distribution. This means that the set of file transfers can be expected in general to be similar to the distribution of files *available* on the Web and, because of caching, relatively insensitive to the *particular requests* made by users.

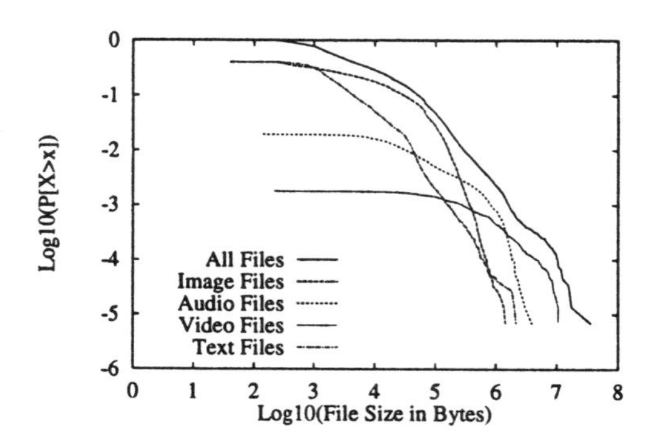

Figure 11: CD of File Sizes of 32 Web Sites

3.4 Why are Available Files in the Web Heavy-Tailed?

Available files in the Web appear heavy-tailed (Figure 8, bottom). A possible explanation might be that the explicit support for multimedia formats may encourage larger file sizes, thereby increasing the tail weight of distribution sizes. While we find that multimedia does increase tail weight to some degree, based on our evidence it does not seem to be the root cause of the heavy tails. This can be seen in the plot shown in Figure 11.

Figure 11 was constructed from the dataset of available files on Web servers, by categorizing all server files into one of seven categories. The categories we used were: *images, audio, video, text, archives, preformatted text,* and *compressed files.* This simple categorization was able to encompass 85% of all files. From this set, the categories *images, audio, video* and *text* accounted for 97%. The cumulative distribution of these four categories, expressed as a fraction of the total set of files, is shown on in Figure 11. In the figure, the upper line is the distribution of all files, which is the same as the plot shown on the right side of Figure 10. The three intermediate lines, from upper to lower, are the components of that distribution attributable to images, audio, and video, respectively. The lowest line is the component attributable to text (HTML) alone.

The figure shows that the effect of adding multimedia files to the set of text files serves to increase the weight of the tail. However, it also suggests that the distribution of text files may itself be heavy-tailed. Using least-squares fitting for the portions of the distributions in which $\log_{10}(x) > 3$, we find that for all files available $\hat{\alpha} = 1.06$ (as previously mentioned) but that for the text files only, $\hat{\alpha} = 1.36$ ($R^2 = 0.98$). The effects of the various multimedia types are also evident from the figure. In the approximate range of 1,000 to 30,000 bytes, tail weight is primarily increased by images. In the approximate range of 30,000 to 300,000 bytes, tail weight is increased mainly by audio files. Beyond 300,000 bytes, tail weight is increased mainly by video files.

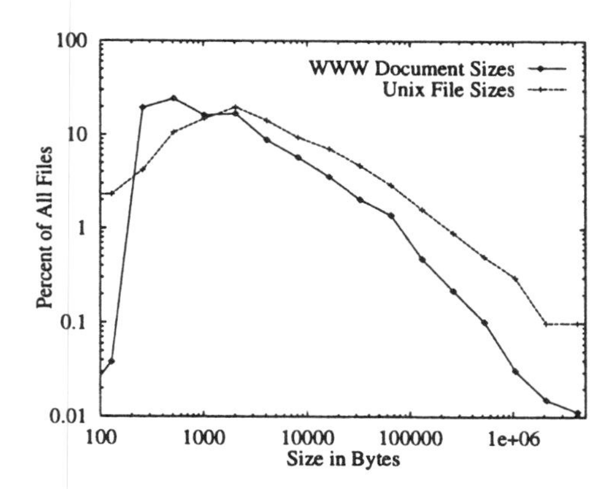

Figure 12: Comparison of Unix and WWW File Sizes

As an another example suggesting that multimedia is not fundamentally responsible for the heavy-tailed nature of available Web files, we compare the distribution of available files in the Web with an overall distribution of files found in a survey of Unix file systems. While there is no truly "typical" Unix file system, an aggregate picture of file sizes on over 1000 different Unix file systems is reported in [Irl94]. In Figure 12 we compare the distribution of available files in the Web with that data. The Figure plots the two histograms on the same log-log scale.

Surprisingly, Figure 12 shows that in our Web data, there is a *stronger* preference for small files than in Unix file systems.[4] The Web favors documents in the 256 to 512 byte range, while Unix files are more common in the 1KB to 4KB range. More importantly, the tail of the distribution of available files in the Web is not nearly as heavy as the tail of the distribution of Unix files. Thus, despite the emphasis on multimedia in the Web, this data suggests that Web file systems may be currently more biased toward small files than are typical Unix file systems.

The fact that file size distributions have very long tails has been noted before, particularly in file system studies [Sat81, Flo86, BHK$^+$91], but without including power-law distributions and measurements of α. In fact, the authors in [PF94] studied a set of transfer sizes that were made using the more general FTP protocol; FTP is used to transfer files but does not include the notions of hypertext or multimedia presentation that have made the Web so popular. They found that the upper tail of the distribution of data bytes in FTP bursts was well suited to a Pareto distribution with $0.9 \leq \alpha \leq 1.1$. Thus our results indicate that with respect to the upper-tail distribution of file sizes, Web traffic does not differ significantly from the more general case of FTP traffic.

[4]However, not shown in the figure is the fact that while there are virtually no Web files smaller than 100 bytes, there are a significant number of Unix files smaller than 100 bytes, including many zero- and one-byte files.

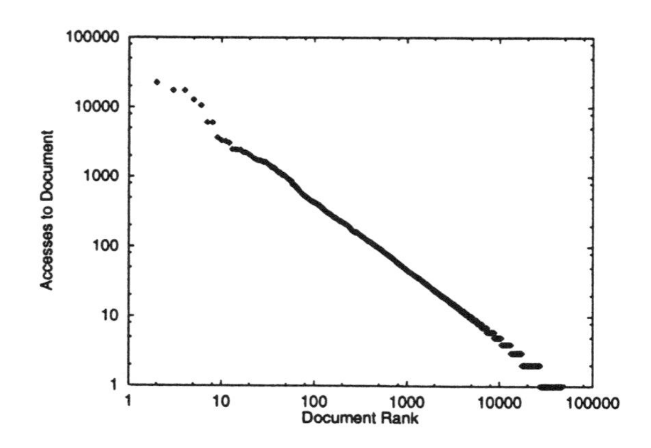

Figure 13: Zipf's Law Applied To Web Documents

3.5 Zipf's Law

Another instance of power-law distributions in our data occurs as an instance of Zipf's law [Zip49, discussed in [Man83]]. Zipf's law was originally applied to the relationship between the number of references made to a word in a given text, and its order in a ranking based on the same measurement. It states that if one ranks the words used in a given text by their popularity P (frequency of use) then P is related to the word's rank ρ by

$$P \sim 1/\rho.$$

Note that this relationship is parameterless, i.e., ρ is raised to exactly -1, so that the nth most popular word is exactly twice as popular as the $2n$th most popular word. Zipf's law has subsequently been applied to other examples of popularity in the social sciences.

Our data shows that Zipf's law applies quite strongly to documents on the Web. This is demonstrated in Figure 13 for all 46,830 unique files listed in Table 1. The figure shows a log-log plot of the number of references to each file as a function of the file's rank in reference count. The tightness of the fit to a straight line is remarkable ($R^2 = 1.00$), as is the slope of the line: -0.986. Thus the exponent relating popularity to rank for Web documents in our data is very nearly -1, as predicted by Zipf's law.

4. Implications for Traffic

One of the important implications of heavy-tailed file size distributions for network engineering lies in their connection to traffic self-similarity. Previous work has shown that network traffic, considered as a time series representing bytes or packets per unit time, typically shows self-similar characteristics with a scaling parameter $H > 1/2$ [LTWW94, WPT98]. Intuitively, this means that traffic shows noticeable "bursts" (sustained periods above or below the mean) at a wide range of time scales — perhaps at all scales of interests to network engineers.

Heavy-tailed distributions have been suggested as a cause of self-similarity in network trafic. The authors in [WTSW95] show that if traffic is constructed as the sum of many ON/OFF processes, in which individual ON or OFF periods are independently drawn from a heavy-tailed distribution, then the resulting traffic series will be asymptotically self-similar. If the distribution of ON or OFF times is heavy-tailed with parameter α, then the resulting series will be self-similar with $H = (3 - \alpha)/2$. If both ON and OFF times are heavy-tailed, the resulting H is determined by whichever distribution is heavier-tailed, i.e., has the lower α; see [WPT98] for further details.

In the context of the World Wide Web, we can consider individual ON/OFF processes to be analagous to Mosaic sessions. Each Mosaic session can be considered to be either silent, or receiving transmitted data at some regular rate. This is a simplification of a real Web environment, but it indicates that if transmission durations are heavy-tailed, then it is likely that the resulting traffic will be self-similar in nature.

In [PKC96] it is shown experimentally that heavy-tailed file size distributions are *sufficient* to produce self-similarity in network traffic. In that study a simple WAN was simulated in considerable detail, including the effects of the network's transmission and buffering characteristics, and the effects of the particular protocols used to transfer files. The results showed that if a network is used to repeatedly transfer files whose sizes are drawn from a heavy-tailed distribution, then the resulting traffic patterns exhibit self-similar characteristics, and the degree of self-similarity as measured by H is linearly related to the α of the file size distribution. In fact, the network traffic measured in this study is analyzed in [CB97] and is shown to exhibit characteristics consistent with self-similarity.

While transmission times correspond to ON times, the size distribution of OFF times (corresponding to times when the browser is not actively transferring a file) is also important. [CB95] contains further analyses of the data in this paper showing that silent times appear to exhibit heavy-tailed characteristics with α approximately in the range of 1.5. Thus, since the transmission time distribution appears to be heavier tailed than the silent time distribution, it seems more likely to be the primary determinant of Web traffic self-similarity.

Since Web traffic is currently responsible for more than half the traffic on the global Internet, the presence of strong self-similarity in Web traffic has implications for the performance of the Internet as a whole. In [ENW96, PKC97] it is shown that the presence of self-similarity with large values of H in network traffic can have severe performance effects when the network has significant buffering. Buffering refers to the use of storage in the network to temporarily hold packets while they wait for transmission. Buffer use is related to the burstiness of traffic, because when a burst occurs, transmission channels can become overloaded, and packets need to be buffered (queued) while waiting for transmission channels to become available.

When traffic is strongly self-similar in nature, bursts can occur at a wide range of timescales. When very long bursts occur, many packets may require buffering. There are two negative effects that can result. First, packets stored in a large buffer will wait for long periods before they can be transmitted. This is the problem of *packet delay.* Second, since buffers are finite, the demand placed on them by a large burst may exceed their capacity. In this case, networks discard these packets leading to the problem of *decreased throughput* (because network bandwidth must be used to retransmit packets). Both problems lead to delays in transmitting files; in the Web, this is perceived by the user as an unresponsive browser. That is, because of long bursts in network traffic, users experience long delays in transfers, and the network appears to perform in an unresponsive manner.

5. Conclusion

The explosive growth of the World Wide Web has made it essential that network engineers understand the Web's characteristics. Since the Web is a system for organizing, delivering, and displaying data in the form of files, some of the Web's most important characteristics relate to how files are distributed in terms of size.

In this paper we've described characteristics of files in the Web, concentrating on five datasets: 1) transmission times of Web files; 2) the set of file requests made by users; 3) the set of file transfers that resulted from cache misses; 4) the set of unique files contained in the request set; and 5) a sample of the set of available files in the Web. We've given evidence that each of these datasets exhibits a set of sizes that is consistent with a heavy-tailed distribution.

A number of questions are raised by this study. The influence of the local workload (an undergraduate and graduate academic computing environment) on our data is not known. Furthermore, the nature of the Web and its traffic is evolving. For example, caching in the Web is growing more sophisticated; in current caching schemes, even cache hits may generate some network traffic. In addition, an increasing amount of data transmitted via the Web is intended for only a single viewing, and therefore cannot be cached.

One of the important consequences of heavy-tailed distributions in network engineering lies in their relationship to traffic self-similarity. The presence of a heavy-tailed distribution of transfer times may represent a cause for the observed phenomenon of traffic self-similarity. We've indicated that traffic self-similarity has serious negative effects on network performance. Since the Web is currently the largest contributor of traffic to the Internet, the self-similarity of Web traffic and its possible causes is an important issue.

A thread running through our study is the attempt to trace the causal relationships of Web traffic self-similarity to heavy-tailed transmission durations, and from there to the characteristics of Web files. In analyzing these relationships, we have argued that the presence of caching in the Web has the effect of making the set of transmitted files relatively insensitive to the set of file requests, and distributionally similar to the set of available files.

References

[BHK+91] Mary G. Baker, John H. Hartman, Michael D. Kupfer, Ken W. Shirriff, and John K. Ousterhout. Measurements of a distributed file system. In *Proceedings of the Thirteenth ACM Symposium on Operating System Principles*, pages 198–212, Pacific Grove, CA, October 1991.

[Bra96] Tim Bray. Measuring the web. In *Proceedings of the Fifth International World Wide Web Conference*, Available from `http://www5conf.inria.fr`, May 1996.

[CB95] Mark E. Crovella and Azer Bestavros. Explaining World Wide Web traffic self-similarity. Technical Report TR-95-015 (Revised), Boston University Department of Computer Science, October 1995.

[CB97] Mark E. Crovella and Azer Bestavros. Self-similarity in World Wide Web traffic: Evidence and possible causes. In *IEEE/ACM Transactions on Networking*, 5(6), December 1997.

[CBC95] Carlos A. Cunha, Azer Bestavros, and Mark E. Crovella. Characteristics of WWW client-based traces. Technical Report TR-95-010, Boston University Department of Computer Science, April 1995.

[ENW96] A. Erramilli, O. Narayan, and W. Willinger. Experimental queueing analysis with long-range dependent packet traffic. *IEEE/ACM Transactions on Networking*, 4(2):209–223, April 1996.

[Flo86] Richard A. Floyd. Short-term file reference patterns in a UNIX environment. Technical Report 177, Computer Science Dept., University of Rochester, 1986.

[fSA] National Center for Supercomputing Applications. Mosaic software. Available at `ftp://ftp.ncsa.uiuc.edu/Mosaic`.

[Hal] Internet Town Hall. The internet traffic archives. Available at `http://town.hall.org/Archives/pub/ITA/`.

[Hil75] B. M. Hill. A simple general approach to inference about the tail of a distribution. *The Annals of Statistics*, 3:1163–1174, 1975.

[Irl94] Gordon Irlam. Unix file size survey — 1993. Available at `http://www.base.com/gordoni/ufs93.html`, September 1994.

[LTWW94] W.E. Leland, M.S. Taqqu, W. Willinger, and D.V. Wilson. On the self-similar nature of Ethernet traffic (extended version). *IEEE/ACM Transactions on Networking*, 2:1–15, 1994.

[Man83] Benoit B. Mandelbrot. *The Fractal Geometry of Nature*. W. H. Freedman and Co., New York, 1983.

[PF94] Vern Paxson and Sally Floyd. Wide-area traffic: The failure of poisson modeling. In *Proceedings of SIGCOMM '94*, 1994.

[PKC97] Kihong Park, Gi Tae Kim, and Mark E. Crovella. On the Effect of Traffic Self-Similarity on Network Performance. In *Proceedings of the SPIE International Conference on Performance and Control of Network Systems*, November, 1997.

[PKC96] Kihong Park, Gi Tae Kim, and Mark E. Crovella. On the relationship between file sizes, transport protocols, and self-similar network traffic. In *Proceedings of the Fourth International Conference on Network Protocols (ICNP'96)*, October 1996.

[Reg] Regents of the University of California. www-stat 1.0 software. Available from `http://www.ics.uci.edu/WebSoft/wwwstat/`.

[Res98] Sidney Resnick. Why non-linearities can ruin the heavy tailed modeler's day. In this volume, 1998.

[RS96] Sidney Resnick and Catalin Starica. Tail index estimation for dependent data. Technical Report 1174, School of OR&IE, Cornell University, 1996.

[Sat81] M. Satyanarayanan. A study of file sizes and functional lifetimes. In *Proceedings of the Eighth ACM Symposium on Operating System Principles*, December 1981.

[WPT98] Walter Willinger, Vern Paxson, and Murad S. Taqqu. Self-similarity and heavy tails: Structural modeling of network traffic. In this volume, 1998.

[WTSW95] Walter Willinger, Murad S. Taqqu, Robert Sherman, and Daniel V. Wilson. Self-similarity through high-variability: Statistical analysis of Ethernet LAN traffic at the source level. In *Proceedings of ACM SIGCOMM '95*, pages 100–113, 1995.

[Zip49] G. K. Zipf. *Human Behavior and the Principle of Least-Effort*. Addison-Wesley, Cambridge, MA, 1949.

Department of Computer Science and Department of Mathematics
Boston University
Boston, MA 02215

Email: crovella@cs.bu.edu, murad@math.bu.edu, best@cs.bu.edu

Self-Similarity and Heavy Tails: Structural Modeling of Network Traffic

Walter Willinger, Vern Paxson and Murad S. Taqqu [1 2 3]

Abstract

High-resolution traffic measurements from modern communications networks provide unique opportunities for developing and validating mathematical models for aggregate traffic. To exploit these opportunities, we emphasize the need for *structural* models that take into account specific physical features of the underlying communication network structure. This approach is in sharp contrast to the traditional *black box* modeling methodology from time series analysis that ignores, in general, specific physical structures. We demonstrate, in particular, how the proposed structural modeling approach provides a direct link between the observed self-similarity characteristic of measured aggregate network traffic, and the strong empirical evidence in favor of heavy-tailed, infinite variance phenomena at the level of individual network connections.

1. Introduction

Recent empirical studies of high-resolution traffic measurements from a variety of different working communication networks (e.g., see [LTWW1, LTWW2, PF1, PF2]) have provided ample evidence that actual network traffic is *self-similar* or *fractal* in nature, i.e., bursty over a wide range of time scales. This observation is in sharp contrast to commonly made modeling choices in today's traffic engineering theory and practice, where exponential assumptions still dominate and are only able to reproduce the bursty behavior of measured traffic either on a prespecified time scale or over a very limited range of time scales. That an observer can easily distinguish between traffic patterns predicted by currently used traffic models and actual, measured traffic traces (data sets) from today's networks challenges traditional approaches to traffic modeling. The purpose of this paper is to demonstrate how the self-similar nature of network traffic at the macroscopic level (i.e., the aggregation of traffic generated by all active hosts on

[1]W. Willinger and M. S. Taqqu were partially supported by the NSF grant NCR-9404931. M. S. Taqqu was also supported by the NSF grant DMS-9404093 at Boston University. Vern Paxson was supported by the Director, Office of Energy Research, Scientific Computing Staff, of the U.S. Department of Energy under Contract No. DE-AC03-76SF00098.

[2]*AMS Subject classification:* 60K30, 60G18, 60E07.

[3]*Keywords:* Long-range dependence, infinite variance, limit theorems, Pareto, stable distribution, computer network traffic models, local area network, wide area network.

the network) leads to new insights into the traffic dynamics at the microscopic level (i.e., the traffic patterns generated by the individual hosts).

To this end, we consider two of the most commonly encountered network environments, *local area networks (LANs)* and *wide area networks (WANs)*. Instead of applying the traditional *black box* approach often used in time series analysis, we focus on *structural* models. These take into account specific features of the underlying network structure and hence provide a physical explanation for observed phenomena such as self-similarity.

LANs were introduced in the mid-1970's to interconnect data processing equipment (host computers, file servers, printers etc.) in office and R&D environments, or within university departments. One of the most popular LAN technologies is Ethernet (e.g., see [MB]), and measured Ethernet traces were used in the original studies by Leland et al. [LTWW1, LTWW2] to demonstrate the self-similar nature of LAN traffic. Here we show how LAN traffic self-similarity gives rise to structural models that can be reduced to simple *ON/OFF* sources (also known as packet trains [JR]) with the distinctive feature that their *ON*- and/or *OFF*-periods are heavy-tailed with infinite variance. Convergence results for such *ON/OFF* processes (and superpositions thereof) provide a direct link between self-similar characteristics at the macroscopic level and the heavy-tailed phenomena observed at the microscopic level, that is, between the aggregate traffic stream and the traffic patterns displayed by individual source-destination pairs. We summarize in this paper the main features of the structural modeling approach to LAN traffic developed in [WTSW2, WTSW1], and validate the underlying assumptions against an additional set of Ethernet LAN source-destination pairs.

In contrast to LANs, WANs provide interconnectivity between users (e.g., host machines on different LANs) that reside, in general, in different geographic regions. The best-known WAN is the worldwide *Internet*, a global network connecting tens of millions of hosts and users. Evidence of WAN traffic self-similarity has been reported in recent studies by Paxson and Floyd [PF1, PF2], who, in an analysis of a number of different WAN traffic traces, showed the inadequacy of traditional exponential (Poisson) traffic models in describing many crucial aspects of WAN traffic behavior. Subsequently, a number of attempts have been made at providing structural models for WAN traffic, including the same reduction to simple *ON/OFF* models for individual source-destination pairs as mentioned earlier for LAN traffic (e.g., see [WTSW2, WTSW1]), and a description of WAN traffic at the level of individual applications, e.g., TELNET, FTP, and HTTP (see [PF2, CB]). Here we summarize the different attempts, and report on recently obtained limit theorems as well as on new statistical evidence that support a structural modeling approach for WAN traffic connecting the empirically observed self-similarity at the macroscopic level directly with infinite variance phenomena at the microscopic level. In this case, microscopic refers to the level of individual applications.

Thus, for LANs as well as for WANs, self-similarity of the aggregate network traffic results directly from structural models that mimic traffic dynamics at lower network levels and identify at those levels traffic characteristics that can be validated against actual high-resolution traffic measurements. The resulting models are simple and parsimonious, provide engineering insights and possess a number of desirable properties that are robust under constantly changing network conditions. They are therefore of considerable practical relevance for a wide range of network engineering tasks.

The paper is organized as follows. In Section 2, we focus on LAN traffic, introduce some commonly used terminology, and follow essentially [WTSW2] in deriving and validating a structural modeling approach for aggregate LAN traffic. Section 3 outlines the derivation of a structural model for WAN traffic, and includes a relevant limit theorem for self-similar processes as well as preliminary but persuasive empirical evidence in favor of the proposed approach. In Section 4, we discuss future work and illustrate the practical importance of structural models for the traffic engineering of today's high-speed networks.

2. Self-Similarity and Heavy Tails in LAN Traffic

We describe an *ON/OFF* model for the traffic transmitted between two typical LAN hosts and validate it by analyzing Ethernet LAN traces at the level of individual source-destination pairs.

2.1 Self-similarity

There are a number of different, not equivalent, definitions of self-similarity. The standard one states that a continuous-time process $Y = \{Y(t), t \geq 0\}$ is *self-similar* (with self-similarity parameter H) if it satisfies the condition:

$$Y(t) \stackrel{d}{=} a^{-H} Y(at), \quad \forall t \geq 0, \quad \forall a > 0, \ 0 < H < 1, \tag{2.1}$$

where the equality is in the sense of finite-dimensional distributions. The canonical example of such a process is Fractional Brownian Motion (see [ST, Be]), Brownian Motion if $H = 1/2$. While a process Y satisfying (2.1) can never be stationary (stationary requires $Y(t) \stackrel{d}{=} Y(at)$), Y is typically assumed to have stationary increments.

A second definition of self-similarity, more appropriate in the context of standard time series theory, involves a stationary sequence $X = \{X(i), i \geq 1\}$. Let

$$X^{(m)}(k) = 1/m \sum_{i=(k-1)m+1}^{km} X(i), \quad k = 1, 2, \ldots, \tag{2.2}$$

be the corresponding aggregated sequence with level of aggregation m, obtained by dividing the original series X into non-overlapping blocks of size m and averaging over each block. The index, k, labels the block. If X is the increment process of a self-similar process Y defined in (2.1), that is, $X(i) = Y(i+1) - Y(i)$, then for all integers m,

$$X \stackrel{d}{=} m^{1-H} X^{(m)}. \tag{2.3}$$

A stationary sequence $X = \{X(i), i \geq 1\}$ is called *exactly self-similar* if it satisfies (2.3) for all aggregation levels m. This second definition of self-similarity is closely related to the first, with $mX^{(m)}(\cdot)$ corresponding to $Y(a\cdot)$.

A stationary sequence $X(i), i \geq 1$ is said to be *asymptotically self-similar* if (2.3) holds as $m \to \infty$. Similarly, we call a covariance-stationary sequence $X(i), i \geq 1$ *exactly second-order self-similar* or *asymptotically second-order self-similar* if $m^{1-H} X^{(m)}$ has the same variance and autocorrelation as X, for all m, or as $m \to \infty$. Note that for a Gaussian process, the last two definitions (i.e., self-similarity and second-order self-similarity) are equivalent.

2.2 LANs, LAN Traffic and Ethernet LAN Traffic Measurements

The last ten years have seen a tremendous growth in the number of LANs, reflecting the need to link users together and to offer connectivity to common resources such as file servers and printers. The 10 megabits/sec (Mbps) Ethernet, a multi-access system for local computer networking with distributed control, has been and remains the workhorse LAN technology. By using properly instrumented monitoring hardware, it is possible to record the time stamp and header information of every single (complete) packet that is put on the monitored Ethernet cable by any of the host stations. Such high-resolution Ethernet LAN traffic measurements over week-long periods have been reported in [FL], where a "typical" day results in about 20-30 million Ethernet packets, or about 2 gigabytes worth of data.

By treating the recorded Ethernet packets as black boxes and only using the time stamp information, Leland et al. [LTWW1, LTWW2] showed that measured aggregate Ethernet LAN traffic (i.e., number of packets or bytes sent over the Ethernet by all active host stations per time unit), with its mean subtracted, is consistent with second-order statistical self-similarity; that is, Ethernet LAN traffic measured over milliseconds and seconds exhibits the same second-order statistics as Ethernet LAN traffic measured over minutes or over even larger time scales. Intuitively, this scale invariance of measured Ethernet LAN traffic manifests itself in the *absence* of a characteristic *burst length*; Ethernet traffic is bursty on all (or a wide range of) time scales, and plotting it over different time scales results in similar-looking pictures, all of which feature a distinctive "burst-within-burst" structure (for details, see [LTWW2]).

When trying to "explain" this empirically observed self-similarity in terms of simpler quantities, a structural modeling approach has been proposed in [LTWW2] and studied in greater detail in [WTSW2] (for related work, see also [LT, WTSW1, HRS]). In short, by using the time stamp information as well as the Ethernet source and destination addresses contained in the recorded header information of each packet seen on the Ethernet, it is possible to separate the aggregate Ethernet LAN traffic into the individual components representing traffic flows between each active pair of host computers, or *source-destination pairs*. At the level of individual source-destination pairs, simple traffic models such as *ON/OFF sources* or *packet train models* have been very popular. Informally, these models assume that a source alternates between an "active" state (or *ON*-period) and "idle" state (or *OFF*-period). During *ON*-periods, packets are sent at a constant rate, and during *OFF*-periods, no packets are transmitted. The group of packets sent during an *ON*-period are termed a "train," and the lull between two trains (i.e., the *OFF*-period) is termed the "intertrain gap." Traditionally, the successive *ON*-periods as well as the successive *OFF*-periods are assumed to be independent and identically distributed (i.i.d.) and independent from each other. Thus, the only stochastic elements in describing *ON/OFF* sources are the distributions that govern the lengths of the *ON*- and *OFF*-periods, respectively. In Section 2.3 below, we provide the details of a limit theorem that states that the superposition of many such *ON/OFF* sources will capture the empirically observed self-similar nature of measured aggregate Ethernet LAN traffic, provided that the distribution of either the *ON*- or *OFF*-periods of an individual source-destination pair has infinite variance. The assumptions under which Theorem 2.1 below holds will be validated in Section 2.3 against measured traffic flows of individual Ethernet source-destination pairs.

2.3 A Limit Theorem for Aggregate Traffic

We consider first a single *ON/OFF* source and focus on the stationary binary time series $\{W(t), t \geq 0\}$ it generates. $W(t) = 1$ means that there is a packet at time t and $W(t) = 0$ means that there is no packet. Viewing $W(t)$ as the reward at time t, we have a reward of 1 throughout an *ON*-period, then a reward of 0 throughout the following *OFF*-period, then 1 again, and so on. The length of the *ON*-periods are i.i.d., those of the *OFF*-periods are i.i.d., and the lengths of *ON*- and *OFF*-periods are independent. The *ON*- and *OFF*-period lengths may have different distributions. An *OFF*-period always follows an *ON*-period, and it is the pair of *ON*- and *OFF*-periods that defines an inter-renewal period. (Such a process is sometimes referred to as an *alternating renewal* process.)

Suppose now that there are M such i.i.d. *ON/OFF* sources. Since each source m sends its own sequence of packet trains, it has its own reward sequence $\{W^{(m)}(t),\ t \geq 0\}$. The superposition or cumulative packet count at time t is $\sum_{m=1}^{M} W^{(m)}(t)$. Rescaling time by a factor T, consider

$$W_M^*(Tt) = \int_0^{Tt} \left(\sum_{m=1}^{M} W^{(m)}(u) \right) du,$$

the aggregated cumulative packet counts in the interval $[0, Tt]$. We are interested in the statistical behavior of the stochastic process $\{W_M^*(Tt),\ t \geq 0\}$ for large M and T. This behavior depends on the distributions on the *ON*- and *OFF*-periods, the only elements we have not yet specified.

In considering the possible distributions, we are motivated by the empirically derived fractional Brownian motion model for measured aggregate cumulative packet traffic in Ethernet LANs in [LTWW2, No], or equivalently, by its increment process, the so-called fractional Gaussian noise model for aggregate traffic (i.e., number of packets per time unit). They model deviations from the mean value. Accordingly, we want to choose the *ON* and *OFF* distributions in such a way that, as $M \to \infty$ and $T \to \infty$, $\{W_M^*(Tt),\ t \geq 0\}$ adequately normalized is $\{\sigma_{\lim} B_H(t),\ t \geq 0\}$, where $\sigma_{\lim}$ is a finite positive constant and B_H is *fractional Brownian motion*, the only Gaussian process with stationary increments that is self-similar [ST]. The parameter $1/2 \leq H < 1$ is called the *Hurst parameter* or the *index of self-similarity*.

To specify the distributions of the *ON*- and *OFF*-periods, let

$$f_1(x),\ \ F_1(x) = \int_0^x f_1(u)du,\ \ \overline{F}_1(x) = 1 - F_1(x),\ \ \mu_1,\ \ \sigma_1^2$$

denote the probability density function, cumulative distribution function, complementary (or tail) distribution, mean length and variance of an *ON*-period, and let $f_2, F_2, \overline{F}_2, \mu_2, \sigma_2^2$ correspond to an *OFF*-period. Assume that as $x \to \infty$, both of the following statements hold:

$$\text{either } \overline{F}_1(x) \sim \ell_1 x^{-\alpha_1} L_1(x) \text{ with } 1 < \alpha_1 < 2 \text{ or } \sigma_1^2 < \infty,$$

and

$$\text{either } \overline{F}_2(x) \sim \ell_2 x^{-\alpha_2} L_2(x) \text{ with } 1 < \alpha_2 < 2 \text{ or } \sigma_2^2 < \infty,$$

where $\ell_j > 0$ is a constant and $L_j > 0$ is a slowly varying function at infinity, that is, $\lim_{x\to\infty} L_j(tx)/L_j(x) = 1$ for any $t > 0$. We also assume that either probability densities exist or that $F_j(0) = 0$ and F_j is non-arithmetic. Note that the mean μ_j is always finite but the variance σ_j^2 is infinite when $\alpha_j < 2$. For example, F_j could be Pareto, i.e. $\overline{F}_j(x) = K^{\alpha_j} x^{-\alpha_j}$ for $x \geq K > 0$, $1 < \alpha_j < 2$ and equal 0 for $x < K$, satisfying the first clause of the "either–or"; or it could be exponential, satisfying the second clause (finite variance). Observe that the distributions F_1 and F_2 of the *ON*- and *OFF*-periods are allowed to be different. One distribution, for example, can have a finite variance, the other an infinite variance.

Before stating the main result, we introduce the following normalization factors and limiting constants that prove convenient in formulating a single limit theorem that holds under the different assumptions on F_1 and F_2. When $1 < \alpha_j < 2$, set $a_j = \ell_j(\Gamma(2-\alpha_j))/(\alpha_j - 1)$. When $\sigma_j^2 < \infty$, set $\alpha_j = 2, L_j \equiv 1$ and $a_j = \sigma_j^2$. The normalization factors and the limiting constants in the theorem below depend on whether

$$\Lambda = \lim_{t\to\infty} t^{\alpha_2-\alpha_1} \frac{L_1(t)}{L_2(t)}$$

is finite, 0, or infinite. If $0 < \Lambda < \infty$, set $\alpha_{\min} = \alpha_1 = \alpha_2$,

$$\sigma_{\lim}^2 = \frac{2(\mu_2^2 a_1 \Lambda + \mu_1^2 a_2)}{(\mu_1+\mu_2)^3 \Gamma(4-\alpha_{\min})}, \quad \text{and} \quad L = L_2;$$

if, on the other hand, $\Lambda = 0$ or $\Lambda = \infty$, set

$$\sigma_{\lim}^2 = \frac{2\mu_{\max}^2 a_{\min}}{(\mu_1+\mu_2)^3 \Gamma(4-\alpha_{\min})}, \quad \text{and} \quad L = L_{\min},$$

where min is the index 1 if $\Lambda = \infty$ (e.g. if $\alpha_1 < \alpha_2$) and is the index 2 if $\Lambda = 0$, max denoting the other index. Under the conditions stated above, the following holds:

Theorem 2.1 *For large M and T, the aggregate cumulative packet process $\{W_M^*(Tt),\ t \geq 0\}$ behaves statistically like*

$$TM\frac{\mu_1}{\mu_1+\mu_2}t + T^H\sqrt{L(T)M}\sigma_{\lim} B_H(t)$$

where $H = (3-\alpha_{\min})/2$ and $\sigma_{\lim}$ is as above. More precisely,

$$\mathcal{L}\lim_{T\to\infty} \mathcal{L}\lim_{M\to\infty} \frac{1}{T^H L^{1/2}(T) M^{1/2}} \left(W_M^*(Tt) - \frac{\mu_1 M T t}{\mu_1+\mu_2} \right) = \sigma_{\lim} B_H(t), \tag{2.4}$$

where $\mathcal{L}\lim$ means convergence in the sense of the finite-dimensional distributions.

For a proof and for discussions of special cases and generalizations of Theorem 2.1, see [TWS]. Heuristically, Theorem 2.1 states that the mean level $TM(\mu_1/(\mu_1+\mu_2))t$ provides the main contribution for large M and T. Fluctuations from that level are given by the fractional Brownian motion $\sigma_{\lim} B_H(t)$ scaled by a lower order factor $T^H L(T)^{1/2} M^{1/2}$. As stated in [TWS], it is essential that the limits be performed in the order indicated. Also note that $1 < \alpha_{\min} < 2$ implies $1/2 < H < 1$, i.e., long-range dependence. Thus, the main ingredient that is needed to obtain an $H > 1/2$ is the heavy-tailed property

$$\overline{F}_j(x) \sim \ell_j x^{-\alpha_j} L_j(x), \quad \text{as } x \to \infty,\ 1 < \alpha_j < 2 \tag{2.5}$$

for the *ON*- or *OFF*-period; that is, a hyperbolic tail (or power law decay) for the distributions of the *ON*- or *OFF*-periods with an α between 1 and 2.

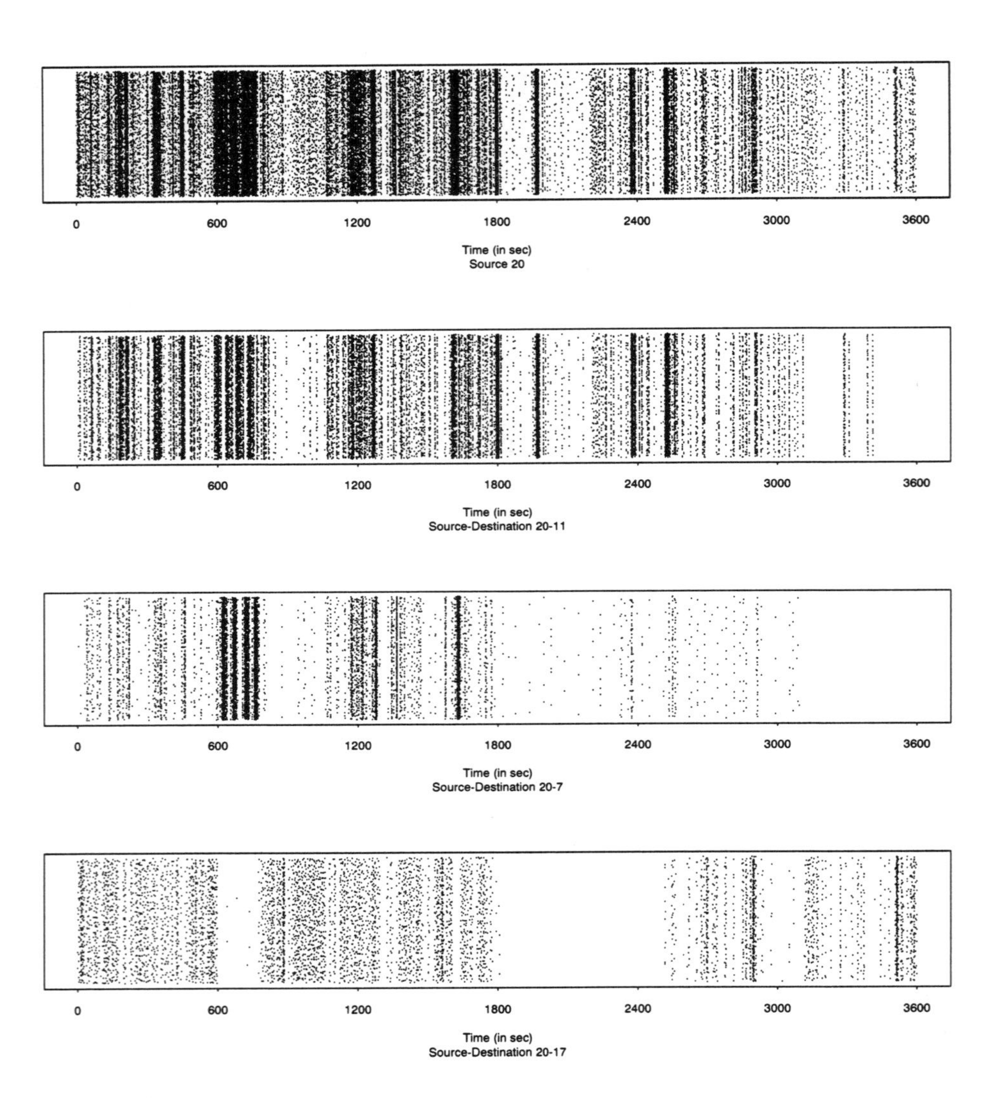

Figure 1: Textured plots of packet arrival times for (top to bottom) source 20 and source-destination pairs 20-11, 20-7, and 20-17.

2.4 Statistical Analysis of Ethernet LAN Traffic at the Level of Individual Source-Destination Pairs

In view of Theorem 2.1, in order to validate the proposed structural modeling approach for Ethernet LAN traffic in terms of the superpositions of many *ON/OFF* source-destination traffic flows, one must check whether measured traffic data are consistent with (i) the *ON/OFF* traffic model assumption for individual sources or source-destination pairs and (ii) the crucially important heavy-tailed property (2.5) for the corresponding *ON*- or *OFF*- periods. To this end, we make use of a data set known as the "busy hour" of a trace of August 1989 Ethernet LAN traffic. This data set was presented, analyzed and shown to be consistent with second-order self-similarity (with a Hurst parameter of $H \approx 0.90$ for the time series representing the packet counts per 10 milliseconds) in [LTWW1].

We first partition this hour-long sequence of Ethernet packets generated by all active Ethernet hosts into their distinct Ethernet source and destination addresses. We find that during the given hour 105 Ethernet hosts sent or received packets over the network. With regard to individual source-destination pairs, only 748 out of $104 \cdot 105 = 10,920$ possible pairs (about 6.8%) were actually sending or receiving packets.

In order to assess the *ON/OFF* nature of traffic generated by an individual source or source-destination pair, we use *textured dot strip plots* or simply *textured plots*, originally introduced in [TT]. The idea of textured plots is to display one-dimensional data points in a strip in an attempt to show all data points individually. Thus, if necessary, the points are displaced vertically by small amounts that are partly random, partly constrained. The resulting textured dot strip facilitates a visual assessment of changing patterns of data intensities in a way other better-known techniques such as histogram plots, one-dimensional scatterplots, or box-plots are unable to provide, especially in the presence of extreme values.

Figure 1 shows four textured plots associated with source 20 (other sources result in similar plots), each point in the plots representing the time of a packet arrival. Source 20 contributed 2.67% to the overall number of packets and sent data to 13 different destinations. The top plot in Figure 1 represents the textured plot corresponding to the arrival times of all packets originating from source 20 (there are 37,582 packets), and the subsequent 3 panels result from applying the textured plot technique to the arrival times of all packets originating from source 20 and destined for hosts 11, 7, and 17, respectively. These three source-destination pairs are responsible for 20,152, 7,497, and 5,511 packets, respectively, and make up about 88% of all the packets generated by source 20. Figure 1 supports the observations that the traffic generated by a reasonably active individual Ethernet host (e.g., source 20, top panel) is itself bursty in nature, and that a "typical" individual source-destination pair (e.g., source-destination pair 20-11, second panel) exhibits an apparent bursty or *ON/OFF* structure.

However, Figure 1 also shows that there is a clear ambiguity associated with identifying a "typical" burst length or *ON/OFF*-period in the traffic patterns generated by an individual Ethernet host or an individual Ethernet source-destination pair. We will show below that this ambiguity is a natural consequence of the heavy-tailed property (2.5) and can be exploited when inferring property (2.5) from a given set of data.

To determine whether or not a given data set is consistent with the heavy-tailed property (2.5) and, if in the affirmative, to estimate a range for the parameter α that characterizes the power law decay in (2.5), we make extensive use of *complementary distribution plots* (related to the *qq-plot* method discussed in [KR]) and *Hill plots* (derived from Hill's method, e.g., see [Hi, RS]). We assume that the reader is familiar with both methods, but would like to emphasize that although both methods are reasonably well understood in theory, they can perform quite erratically in practice. Since theoretical results for these procedures (e.g., confidence intervals for the Hill estimator) are known to hold only under conditions that often cannot be validated in practice (see for example [Re]), it is preferable to use data-intensive heuristics. As a result, we typically end up with strong empirical evidence on whether or not property (2.5) holds but without precise point estimates for α.

As an illustration, consider the traffic generated by the Ethernet source-destination pair 20-17 (Figure 1, bottom panel); the traffic streams from other source-destination pairs yield similar results (see below). Given its packet arrival process, it is natural to define an *OFF*-period to be *any interval of length longer than t seconds that does not contain any packet arrival*; this in turn, defines the *ON*-periods unambiguously. Of course, the ambiguity experienced earlier when trying to "eyeball" *ON/OFF*-periods based on the textured plots in Figure 1 has now been replaced by the equivalent problem of selecting the "right" threshold value t in the formal definition of an *OFF*-period. In the packet train terminology, the problem is to decide in a coherent manner on the "appropriate" intertrain gap, i.e., on deciding when the "departure" of the previous train took place and when the "arrival" of the next one occurs.

However, we show below that *as far as property (2.5) is concerned, it does not matter how the OFF-periods or intertrain distances (and subsequently, the ON-periods or packet train lengths) have been defined.* In other words, property (2.5) is robust under a wide range of choices for the threshold value t. In the case of the *OFF*-periods, the reason is the well-known scaling property of distributions that satisfy the power tail condition (2.5): if the distribution of the random variable U satisfies (2.5) and t denotes a threshold value, then for sufficiently large u, t with $u > t$,

$$P(U > u \mid U > t) \sim (\frac{u}{t})^{-\alpha}, \quad 1 < \alpha < 2. \tag{2.6}$$

Thus, the tail behavior of the (conditional) distributions of U given $U > t$, for dif-

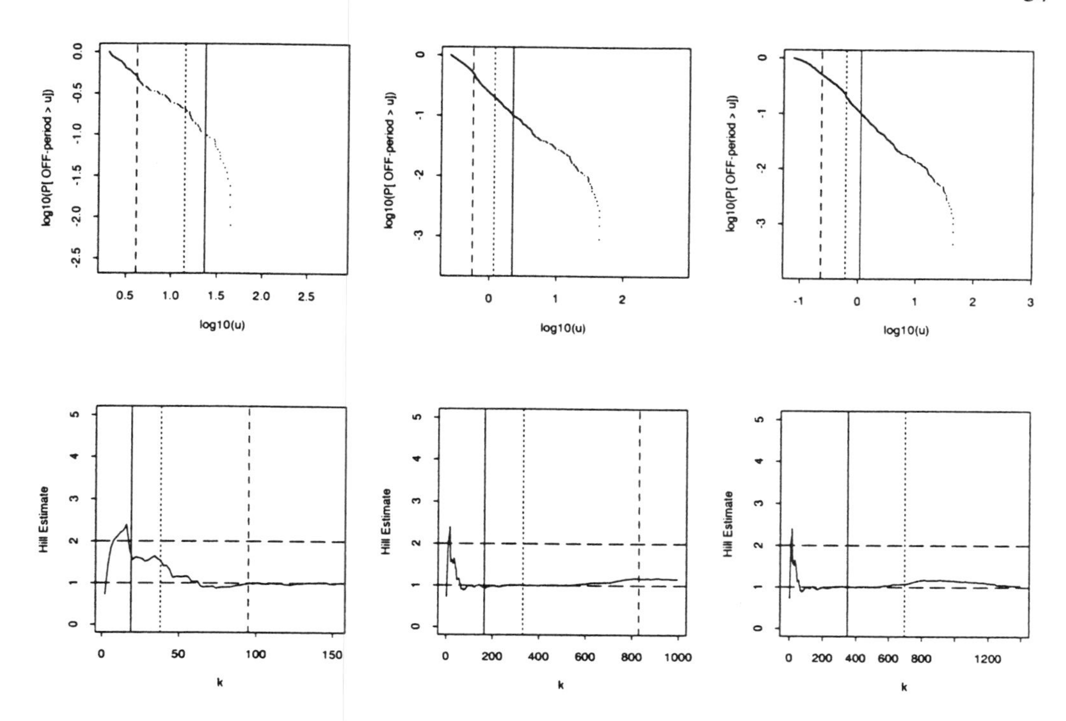

Figure 2: Robustness property of the *OFF*-periods (for source-destination pair 20-17), for threshold values (from left) $t = 2.0$ sec, .25 sec, .075 sec

ferent choices of the threshold t, differs only by a scaling factor and hence gives rise to complementary distribution plots (on log-log scale) with identical asymptotic slopes but different intercepts. In the case of the Ethernet host pair 20-17, this is illustrated in Figure 2, where we show the complementary distribution plots (top row) and corresponding Hill plots (bottom row) for three different ways of defining *OFF*-periods.

More specifically, we chose t-values that span 3 orders of magnitude, namely $t = 2.0$ sec (left column, 190 observations), $t = 0.25$ sec (middle column, 1,658 observations), and $t = 0.075$ sec (right column, 3,484 observations). The three vertical lines in the complementary distribution plots in Figure 2 indicate that 10%, 20% and 50% of all the data points are to the right of the solid, dotted and dashed line, respectively; in the Hill plots, the solid, dotted and dashed lines indicate that 10%, 20% and 50% of the largest order statistics have been included in the computation of the Hill estimator. Figure 2 confirms the robustness of property (2.5) under the different choices of t, with an estimated α-value between 1.0 to 1.3.

To explain the robustness property of the *ON*-periods with respect to a wide range of threshold values t, recall that a t-*ON* period (i.e., an *ON*-period that was obtained using the threshold value t) typically consists of a number of s-*ON* and s-*OFF* periods where $s < t$. However, if the s-*ON*/*OFF* periods satisfy relation (2.5), then so does

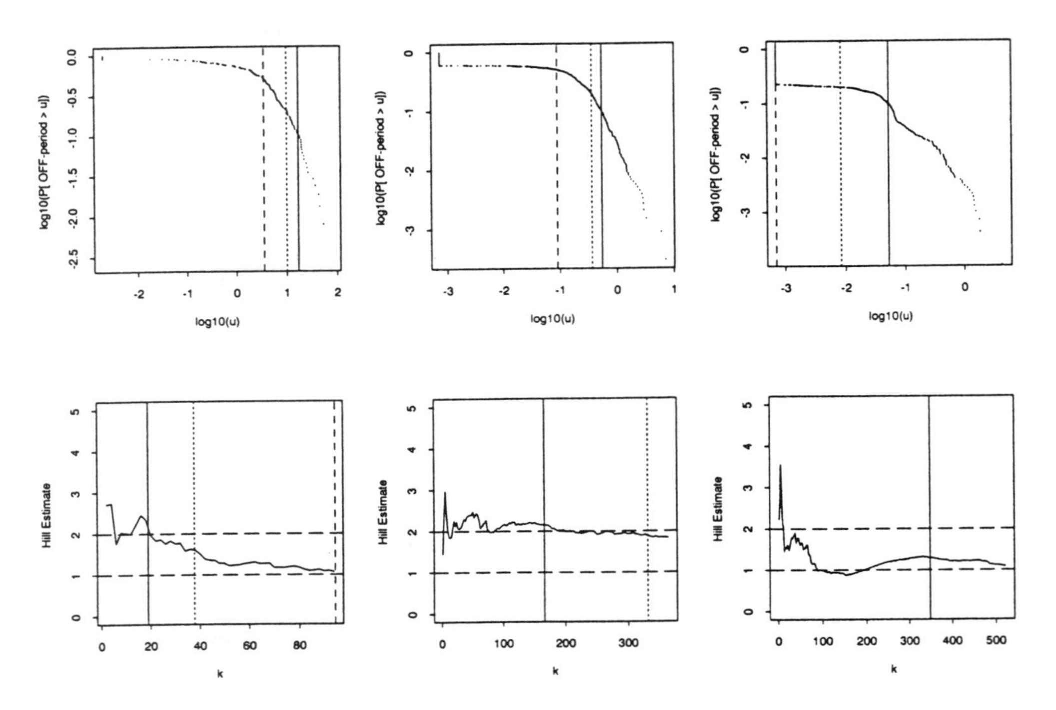

Figure 3: Robustness property of the *ON*-periods (for source-destination pair 20-17), for threshold values (from left) $t = 2.0$ sec, .25 sec, .075 sec

their sum. Subsequently, when interested in the power-law decay for the distribution of the *ON*-periods, fragmentation into smaller *ON*/*OFF*-periods as the threshold value t decreases should have minor impact and suggests that the *ON*-periods are generally robust under a wide range of t-values.

This property is illustrated in Figure 3, where we consider the *ON*-periods corresponding to the three different choices of *OFF*-periods that were used in Figure 2 for the pair 20-17 (i.e., $t = 2.0$ sec, 0.25 sec, 0.075 sec). As can be seen, the estimate of the α-parameter that characterizes the power law decay in (2.5) for the *ON*-periods remains within the interval $[1, 2]$, even though t varies from seconds to 100 milliseconds to milliseconds. Source-destination pairs other than 20-17, as well as individual Ethernet hosts, yield similar results and provide convincing evidence that property (2.5) for the *ON*/*OFF*-periods of the individual sources and source-destination pairs is robust under a wide range of choices of threshold values. The results of a full-fledged analysis of the 181 most active source-destination pairs (out of a total of 748 active pairs) that make up 93% of all the packets in this data set are shown in Figure 4.

Using the same thresholding procedure as in Figures 2 and 3, we checked each of the 181 source-destination pairs for the presence or absence of property (2.5) in their corresponding sequences of *ON*- and *OFF*-periods. We determined, for each of the 181 source-destination pairs, two ranges of α-values (one for the corresponding *ON*-periods, another for the *OFF*-periods) that are consistent with the data and insensitive to the particular definition of *ON/OFF*-periods. More precisely, we categorize the *ON/OFF*-nature of each source-destination pair, according to whether the resulting α-estimates fall within the intervals (0, .85), (.75, 1.35), (1.25, 1.75), (1.65, 2.25) or (2.25, 2.75). These represent the cases "definitely below 1.0", "around 1.0", "somewhere in the middle of the interval (1,2)", "around 2.0", and "definitely above 2.0 or inconclusive", respectively. For example, we classified the *ON* and *OFF* periods of source-destination pair 20-17, shown in Figures 2 and 3, as "somewhere in the middle of the interval (1,2)" and "around 1.0." For the vast majority of source-destination pairs, the categorization process worked well, especially for the *OFF*-periods. While not all *ON/OFF*-periods fitted this framework, the number of inconclusive cases was low, under 10%.

In Figure 4 we plot for each of the 181 source-destination pairs its load (in bytes, on log scale) against the range of α-values that is consistent with its traffic trace. As is seen, in the case of the *ON*-periods (top plot), the α-estimates consistent with the data cover pretty much the whole interval $(1, 2)$. In comparison, the lower plot in Figure 4 shows that in the case of the *OFF*-periods, α-estimates in the lower part of the interval $(1, 2)$ clearly dominate the picture. Thus, the evidence of infinite-variance *ON*-periods is not as consistently strong as that for the *OFF*-periods. Nevertheless, the mathematical results in Section 2.3 still apply since they allow the α-values for the *ON*- and *OFF*-periods to be different, or even one of the distributions to have finite variance.

To sum up, Figure 4 provides strong statistical evidence in favor of our physical explanation for the empirically observed self-similarity property of aggregate Ethernet LAN traffic. Our analysis shows that the data at the source-destination level are consistent with the *ON/OFF* model and that the distributions of the corresponding *ON/OFF*-periods tend to satisfy the heavy-tailed property (2.5). One may question the assumption made in Theorem 2.1 that W_M^* is the superposition of *independent ON/OFF* sources. Hosts that compete for available (finite) network resources and protocols that determine which host machine can send how many bytes at what times are bound to introduce dependencies among the individual source-destination pair; in fact, a careful examination of Figure 1 shows segments in the textured plots of source-destination pairs 20-11 and 20-7 with interchanging burst- and idle-period, indicating periods when there are obvious interactions between individual source-destination pairs. It is likely, however, that the conclusion of Theorem 2.1 will continue to hold if the sources are weakly dependent.

3. Self-Similarity and Heavy Tails in WAN Traffic

Research on structural models for wide-area network traffic is still at the preliminary stage. We focus on Internet-related traffic, consider a number of well-known applications

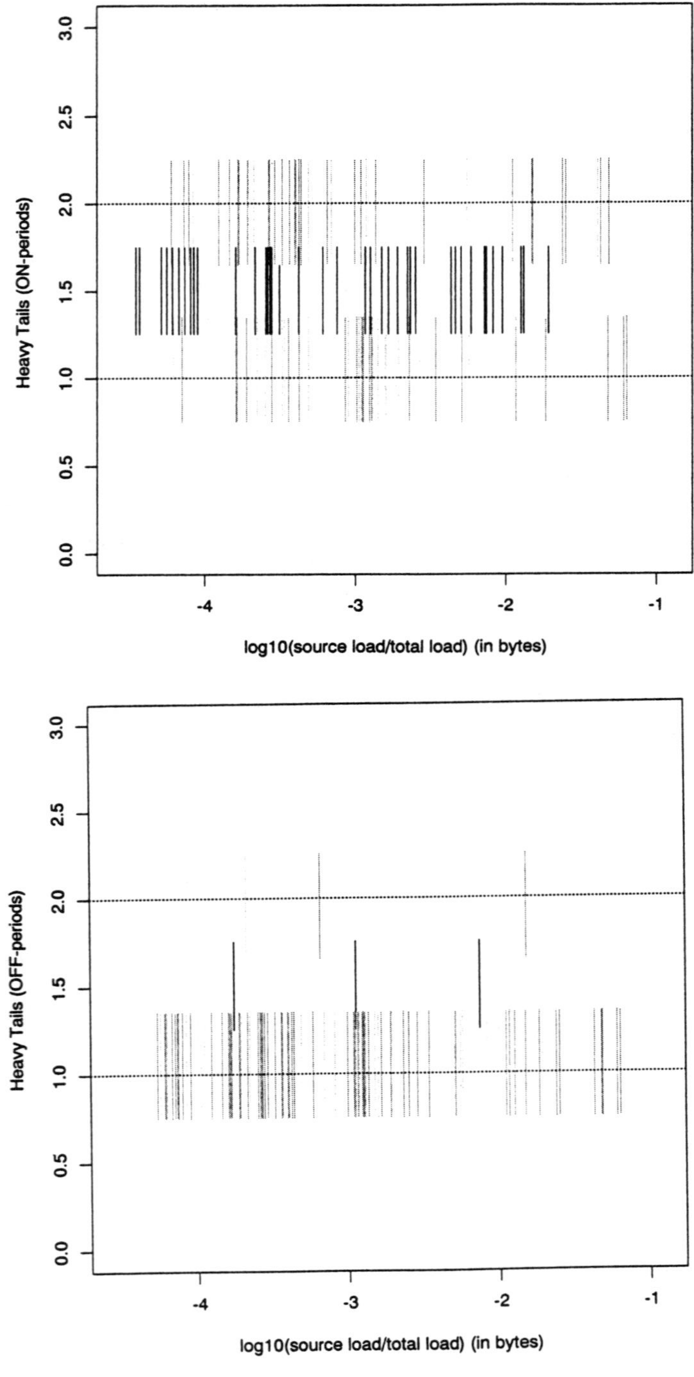

Figure 4: Summary plot of ranges for the α-estimates for the *ON*-periods (top) and *OFF*-periods (bottom) of the 181 most active source-destination pairs, as a function of their loads (in bytes, on log scale).

(e.g., FTP, HTTP, TELNET), discuss structural modeling approaches and comment on their validation.

3.1 WANs, WAN Traffic and WAN Traffic Measurements

Wide-area networks (or *WANs*) arose in the 1970's as a means for interconnecting computers located in geographically distant locations. WANs are often a collection of distinct (independently administered) networks using possibly different interconnection hardware. The best known example is the Internet, which has experienced sustained explosive growth, increasing in size by 80–100% per year for well over a decade. Presently, the Internet encompasses some 100,000 different networks and 10,000,000 hosts.

WANs differ from LANs in a number of fundamental ways. They are much more heterogeneous, making it difficult to predict what traffic conditions one might encounter in any particular situation. Another crucial difference is that speed-of-light limitations have an immense impact on WAN engineering, because the *time constants* associated with obtaining feedback on network conditions are measured from tens of milliseconds to seconds, instead of microseconds as in the case of LANs. This makes it much more difficult for WAN applications to adapt to current network conditions, and can create *congestion*, that is, a performance degradation due to the overloading of some of the components in the end-to-end chain of networks between two WAN hosts. If a component has no more buffer memory for temporarily storing the data packets it receives, it will drop them instead of forwarding them. Thus, the reliable transfer of data across a WAN requires a sophisticated "transport" protocol. This protocol must ensure that data packets are retransmitted when lost, and must avoid unnecessary retransmission which would further add to congestion.

For the Internet, the dominant transport protocol is the Transmission Control Protocol (TCP). The WAN traffic traces discussed below were obtained by recording the header of every TCP packet[4] at network links connecting various institutions to the Internet "backbone." TCP is not the only protocol in use on the Internet; indeed, a growing proportion of the traffic is "multicast" (often used for sending digitized audio and video), in which a single sender transmits to multiple receivers. These relatively new applications, however, have been less studied than the more well-established Internet applications that use TCP, so we confine ourselves here to the latter.

[4]Some of the analysis is based on data sets that recorded only the "connection-begun" and "connection-finished" packets that delimit each TCP connection.

3.2 WAN Traffic at the Application Level

The profile of dominant Internet applications keeps changing over time. Currently, the most significant applications on the Internet are: file transfer (FTP); structured information retrieval (HTTP; the "World Wide Web"); remote login (TELNET); electronic mail; and Network News. Just four years ago, HTTP traffic was virtually nonexistent. There also exists no "typical" WAN application "mix": the dominant WAN applications vary greatly from site to site. For example, a recent study [Pa] found that in January 1991, the proportion of traffic volume (total bytes) due to the Network News application at two different California sites varied from 15% at one of them to 60% at the other. Thus in general one must be wary of assuming that a particular WAN link trace reflects "typical" traffic. One way of dealing with this difficulty is to partition the traffic according to the different applications. This is easy to do when examining a trace of WAN traffic, because the TCP packet header includes a "port" number that identifies different applications. Here we will concentrate exclusively on TELNET, FTP and HTTP.

FTP and HTTP are "bulk transfer" applications, whose main task is to move a predetermined amount of data from one Internet host to another. While in a LAN environment bulk transfer is relatively straightforward, in a WAN setting it becomes significantly more complex, because of transient congestion and the dynamics of TCP. As a result, useful traffic models associated with WAN bulk transfers are rare and ad-hoc, at best. Based on the recent empirical evidence of self-similar features in measured aggregate WAN traffic (see [PF1, PF2]), we propose here a structural modeling approach for WAN bulk transfer that connects WAN traffic characteristics at the macroscopic (i.e., aggregate) level to the microscopic (i.e., application) level by focusing on typical bulk transfer features such as arrival patterns, the amount of data transferred, or duration (which depends not only on the amount of data transferred, but also on the network conditions encountered during the transfer). The resulting structural models suggest that for some problems of practical interest, capturing the fine details of bulk transfer (that is, the precise manner in which packets are transmitted during a session, which is mainly determined by transient congestion within the WAN and the dynamics of TCP) can be avoided; note however that these fine details are of great importance when studying, for example, how network controls affect traffic.

In contrast to FTP and HTTP, TELNET is an "interactive" application. The packets sent from the host initiating the connection to the receiving host are determined by the pattern of keystrokes made by the TELNET user. As such, these patterns can be expected to be fairly robust in the presence of widely varying networking conditions, and our structural modeling approach for TELNET traffic will focus on identifying "typical" features in these patterns.

3.3 Some Limit Results for Aggregate WAN Traffic

A natural modeling approach for aggregate WAN traffic is based on the idea of "separation of time scale." That is, there exist two distinct processes of interest: the start times of *sessions* (where a session consist of one or more related network connections), and the arrival process of *packets* within a session. One can observe empirically this separation when plotting Figure 1 for "typical" source-destination pairs in a WAN environment (not shown here): transmission starts at some random point in time ("start of a session"), packets are transmitted (in some bursty fashion) for some time, and then transmission stops ("end of session") until the next session begins. While session arrivals can, in general, be identified unambiguously, defining a "typical" burst of packets within a session is as ambiguous as for a host pair in a LAN environment (see Section 2). In the following, we present some known approaches for traffic modeling that explicitly mimic this two-stage procedure. We then comment on their relevance for capturing empirically observed WAN traffic characteristics.

We first recall a construction due to Cox [Co], also known as an *immigration death process* or *$M/G/\infty$ queueing model.* Cox's construction has been suggested for modeling modern communication network traffic in [LTWW2], and was considered explicitly for WAN traffic in [PF2]. In the context of WAN traffic, the Cox construction assumes that sessions (e.g., FTP, HTTP, TELNET) arrive according to a Poisson process, transmit packets deterministically at a constant rate (i.e., in a "fluid" fashion) during their "lifetime" or session length, and then cease transmitting packets. Note that once the Poisson nature of session arrivals is taken for granted (for further discussion of Poisson session arrivals, see Section 3.4), the only stochastic element left unspecified is the distribution of session lengths or durations. We shall choose it so as to capture in a parsimonious manner the empirically observed long-range dependence property of aggregate WAN traffic (see [PF2]), which in turn is related to statistical self-similarity. More precisely, working in discrete time, let X_n denote the the number of customers in the system at time n in the $M/G/\infty$ model, or equivalently, the total number of packets generated by all the sessions that are active at time n (assuming that packets are transmitted one per time unit during the lifetime of a session). Let $(f_n)_{n\geq 1}$, $F(n) = \sum_{k\leq n} f(k)$, $\overline{F} = 1 - F$ and μ be, respectively, the probability mass function, cumulative distribution function, tail distribution and mean of the session length and assume that as $n \to \infty$, F satisfies the heavy-tailed property (2.5), that is,

$$\overline{F}(n) \sim n^{-\alpha} L(n), \quad \text{as } n \to \infty, \ 1 < \alpha < 2 \,. \tag{3.7}$$

Under these conditions, Cox [Co] (see also [PF2]) obtained the following result:

Theorem 3.2 *The aggregate packet process $X = (X_n : n = 0, 1, 2, \ldots)$ exhibits long-range dependence. More precisely, denoting by $r(k)$ the autocorrelation function* *of X, we have*

$$r(k) = \mu^{-1} \sum_{n=k}^{\infty} \overline{F}(n) \sim Ck^{1-\alpha} L(k), \text{ as } k \to \infty, \tag{3.8}$$

for some constant $C > 0$. Moreover, the degree of long-range dependence (i.e., Hurst parameter) is given by $H = (3 - \alpha)/2$.

The main ingredient is the heavy-tailed property (3.7) of the session durations. Intuitively, this property implies that the length of a "typical" session is highly variable (*infinite variance*), i.e., exhibits fluctuations over a wide range of time scales. This basic characteristic at the application level manifests itself at the network level through property (3.8). This property implies that the aggregate traffic process X is *second-order asymptotically self-similar*, that is, when viewed over sufficiently large time scales, the second-order statistical properties of X remain essentially unchanged, and the traffic looks "similar" over a wide range of time scales.

While appealing in its simplicity, Cox's construction suffers from a number of shortcomings that limits its immediate applicability to WAN traffic modeling. First, the Poisson nature of session arrivals often turns out to be too restrictive in practice. Second, and more importantly, the applications that currently contribute major portions to WAN traffic, namely FTP and HTTP, are known (see [PF2, CB]) to transmit their packets *not* at a constant rate, but in a highly bursty manner, mainly as a result of transient network congestion and TCP dynamics. Concerning the first shortcoming, replacing the Poisson assumption in Cox's construction by a general renewal structure for the start time of sessions is straightforward and introduces greater flexibility, since it requires only that the session interarrival times are i.i.d. With regard to relaxing the assumption of a fixed and constant packet rate for the duration of an entire session, we refer to recent work by Kurtz [Ku], which provides new insights into this difficult problem. Briefly, Kurtz considers a large number of sessions, each of which has a random starting time and duration. Associated with each active session is a cumulative input process, that is, a nondecreasing stochastic process $Y = (Y(t), t \geq 0)$ such that $Y(t)$ represents the cumulative number of packets contributed during the first t time units during an active session; the total length of time τ that a session is active is often modeled separately from Y. For example, $(Y(t),\ 0 \leq t \leq \tau)$ with $Y(t) = t$ means constant rate, as in Cox's construction; a $(Y(t),\ 0 \leq t \leq \tau)$ that has partly slope 1 and partly slope 0, describes *ON/OFF* behavior as in Section 2, but restricted to within a session; and a $(Y(t),\ 0 \leq t \leq \tau)$ that is piecewise linear with different slopes (including slope 0) is a natural candidate for capturing actual TCP dynamics.

Kurtz's main results show that the same limiting regime holds for an appropriately normalized version of the aggregate packet process under very different assumptions on the fine structure of packet arrivals within sessions (assuming that the starting time of sessions is Poisson): in one case, constant rate is assumed; in the other case, Y is only required to have stationary, ergodic increments. In both cases, the limiting process is fractional Brownian motion, the only Gaussian process with stationary increments that is (exactly) self-similar. Moreover, as in Cox's construction, the essential ingredient for the self-similarity of the limiting process is the heavy-tailed property (3.7) of τ, i.e., the infinite variance assumption on the duration of the individual sessions. For further details and proofs, see [Ku].

3.4 Statistical Analysis of WAN Traffic at the Application Level

Validating the structural modeling approaches to WAN traffic, suggested by Cox's construction (see Theorem 3.2) or Kurtz's, requires checking whether or not measured WAN traffic at the application level is consistent with (i) Poisson (or more general, renewal) session arrival instants and (ii) session lifetimes that satisfy the fundamental heavy-tailed property (3.7). Detailed information concerning the arrival pattern of packets within a session is important for assessing whether a statistically self-similar limiting process is appropriate for accurately describing actual WAN traffic. To this end, we report on empirical studies and provide empirical evidence in favor of the proposed structural models in Section 3.3, especially of Kurtz's construction with its flexible intra-session packet arrival structure. In terms of WAN applications, we focus on measured FTP, HTTP and TELNET traffic traces; while FTP traffic still consumes a major part of available WAN capacity, HTTP traffic continues to increase in volume and has begun to replace FTP as the dominant WAN traffic contributor. TELNET, on the other hand, is a service qualitatively different from these two, with much less demand for bandwidth, but generating a high volume of packets, often one per user keystroke.

First, with regard to the stochastic properties of network session arrivals, measurement studies show that, not surprisingly, the arrival instants of network sessions exhibit a clear diurnal cycle. For example, TELNET sessions peak during afternoon hours and reach a low in early morning hours, almost identical in nature with the calling pattern observed in traditional telephony [DMRW]. While this observation rules out a homogeneous Poisson model for session arrivals, Paxson and Floyd [PF2] have shown that the arrivals of both TELNET and FTP sessions are well-modeled by nonhomogeneous Poisson processes, with rates that are constant over an hour, but are allowed to change from one hour to the next. The arrivals of HTTP sessions have not been studied in this regard. The difficulty here is determining, from a WAN link trace, when an HTTP session begins. Because of details in how the applications are structured, this determination is straightforward for FTP and TELNET.

Next we use measured FTP and HTTP traffic to illustrate the empirical evidence in

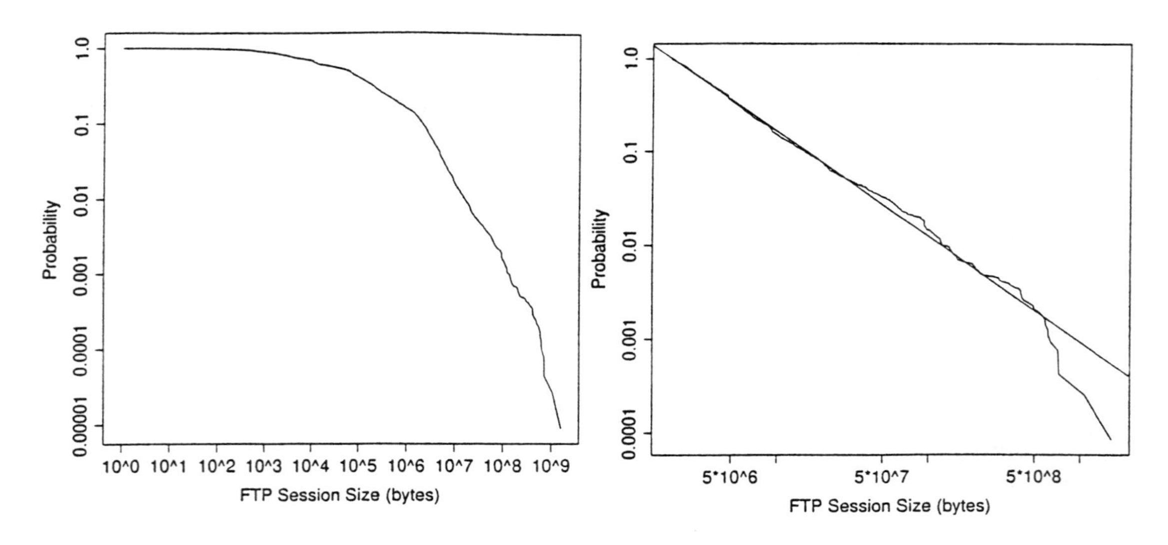

Figure 5: Log-log complementary plot of FTP session sizes: full data set (left), upper tail (right).

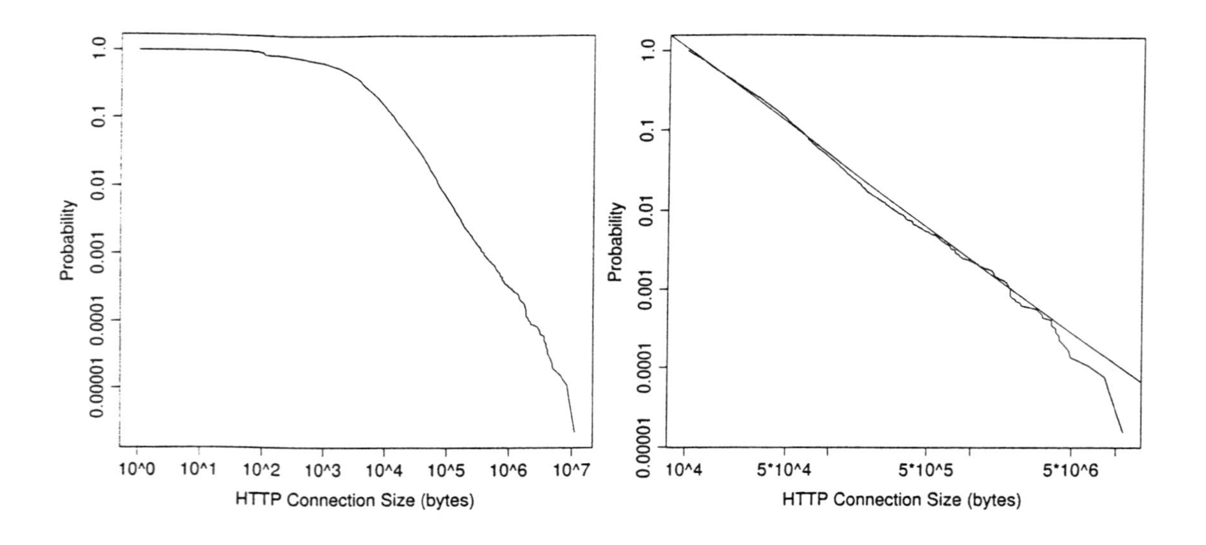

Figure 6: Log-log complementary plot of HTTP connection sizes: full data set (left), upper tail (right).

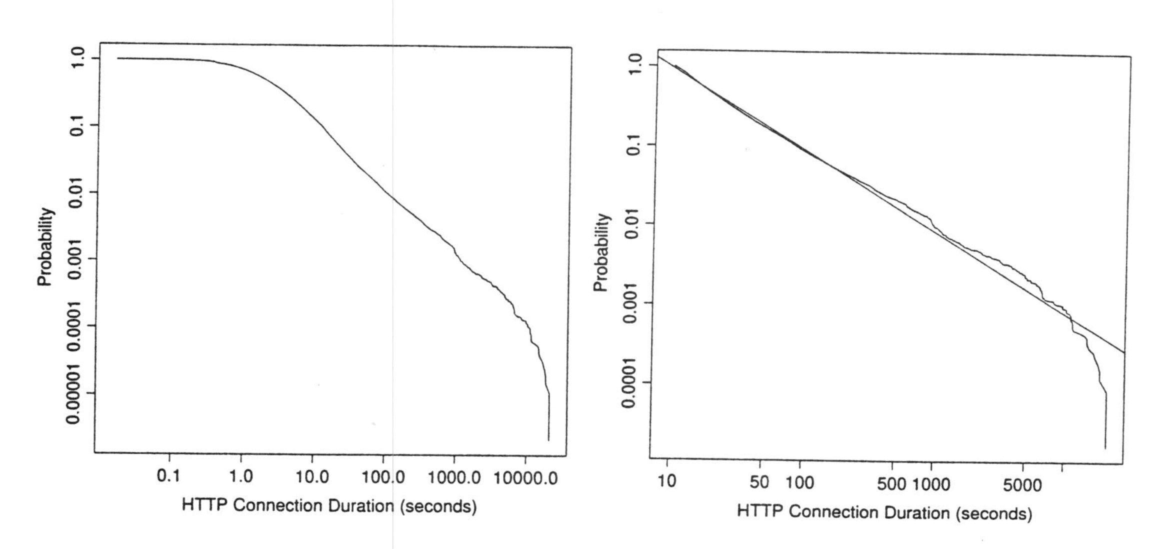

Figure 7: Log-log complementary plot of HTTP connection durations: full data set (left), upper tail (right).

favor of the heavy-tailed property (3.7) of session lifetimes extracted from WAN link traces. We start by postulating that session lifetimes are proportional to "session sizes," where a session size is characterized by the total number of bytes transmitted during the session. (Clearly, matching lifetime and size of a session is an over-simplification, and this aspect will be discussed in more detail below.) For FTP traffic extracted from a mid-1996 30-day WAN traffic trace, Figure 5 shows a log-log complementary distribution plot of the FTP session sizes, for all sessions (left, 56,421 observations) and for only those transferring more than two megabytes (right, 5,886 observations). This distribution closely matches those studied in [Pa], which found the distribution bodies to be well-modeled using the log-normal family of distributions. But the upper tail of the distribution (right plot), containing about 10% of all the data points, provides a very good fit with assumption (3.7), with an estimated α value around 1.1 (the straight line corresponds to a least-squares fit yielding $\alpha = 1.13$). Note that this is a case where the graphical technique (i.e., least-squares fit) by itself is compelling, mainly because of the large number of observations in the right tail; as can be expected from the appearance of the log-log complementary distribution plots in Figure 5, properly performed Hill estimation simply confirms the findings obtained via the graphical method.

The patterns of WAN HTTP *session* lifetimes and sizes have not been characterized because of the above-mentioned difficulty in extracting this data from a trace of traffic on a WAN link. However, using HTTP *connection* measurements (where a HTTP session is typically made up of many HTTP connections), we obtain strong empirical evidence in favor of the heavy-tailed property (3.7) for HTTP session size and – making the same assumption as above for FTP sessions – for HTTP session lifetimes. Indeed, for a 1996 24-hour measurement period, Figure 6 depicts log-log complementary distribution plots for all HTTP connection sizes (left, 226,386 observations) and for only those transmitting more than 10 kilobytes (right, 32,630 observations). In this case, the upper tail of the distribution contains 14% of the data points, is fully consistent with property (3.7), and yields an α estimate of about 1.35. Figure 7 shows the same information for HTTP connection lifetimes. Here, the upper tail consists of all connections that lasted for more than 10 seconds. The tail contains 28,469 data points (13%), and results in an α estimate of about 1.0. Both figures are clear indications of the heavy-tailed nature of HTTP session lifetimes, but they also illustrate that equating size with duration is not wholly accurate.

To accurately couple session size with session duration or lifetime requires a detailed understanding of the nature of the packet arrival process within a session; as mentioned earlier, such an understanding is also crucial for validating our proposed structural modeling approaches for WAN traffic. To illustrate, consider TELNET traffic, for which extensive packet level measurements exist. For example, a study of WAN traffic at three different sites by Danzig et al. [DJCME] found that the distribution of TELNET packet interarrival times within a single TELNET session was invariant across the three sites.

In a subsequent WAN study, Paxson and Floyd [PF2] confirmed this finding, observing the same distribution for data captured at a fourth site three years later [PF2]. What is striking about this distribution is that unlike the exponential models widely assumed for user keystrokes, it is unambiguously Pareto, with an α estimate around, or even slightly below, 1.0.

Figure 8 shows a log-log complementary plot of the distribution; the straight line corresponds to a least-squares fit to the upper 75% of the distribution, giving $\alpha = 0.90$. Adopting the conservative position of neglecting the evidence of infinite mean interarrival times (for a discussion about self-similarity and distributions with infinite mean, see [Ma1]), the results of these traffic studies show that for TELNET, the assumption – required for Cox's construction – of a constant packet arrival rate during a session is a large stretch, irrespective of the possible session lifetime characteristics. On the other hand, Kurtz's construction applies directly and suggests, for example, modeling the intra-session packet dynamics via a simple *ON/OFF* process, with the *OFF*-periods corresponding to the packet interarrival times and the *ON*-periods to the transmission times of the individual TELNET packets.

Based on extensive analyses of FTP traffic traces by Paxson and Floyd [PF2] and HTTP traffic measurements by Crovella and Bestavros [CB], similar arguments apply in the context of structural models for FTP and HTTP traffic: the packet arrival instants

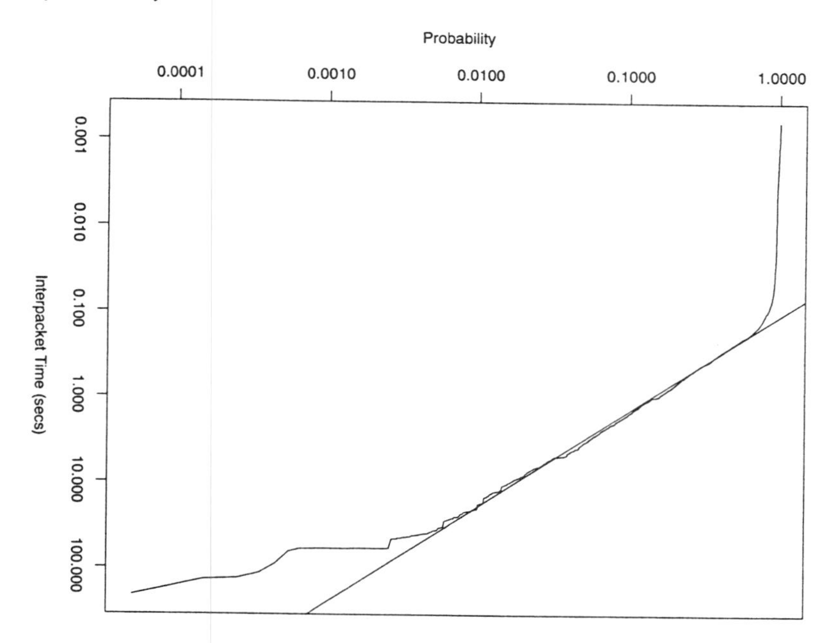

Figure 8: Log-log complementary plot of the full distribution of packet interarrival times for TELNET sessions.

within FTP and HTTP sessions show highly complex structures and seriously question the assumption of a constant packet rate. On the other hand, Kurtz's construction offers the ability to explicitly model the observed very general intra-session packet arrival behavior with, for example, versions of *ON/OFF* process that allow for different packet arrival rates during the different *ON*-periods, and thus mimic actual TCP dynamics in a real network setting. The two constructions let us choose between capturing only large-scale behavior, but retaining a simpler model (i.e., Cox's), or delving into finer-scale behavior, too, at the cost of a more complex model (Kurtz's). Both structural models are capable of capturing the empirically observed self-similar features in measured WAN traffic traces and explaining them in terms of infinite variance phenomena exhibited by the major applications contributing to aggregate WAN traffic.

4. Conclusions

Previous studies of modern communication network traffic (e.g., LAN and WAN traffic) have conclusively ruled out the possibility of modeling network *packet* arrivals using traditional Poisson processes [JR, Gu, FL, DJCME] – there is simply far too much correlation among packet arrivals for any hope of consistency with the Poisson assumption of independence. One major source of correlation is the prevalence of packets being sent in batches. Another is the correlations between batches introduced by the network itself, e.g., TCP's rate adaptation.

In this paper, we demonstrate how the recently observed self-similarity features of modern network packet traffic [LTWW1, PF1, CB] result in structural traffic models that (i) have a physical meaning in the network context, (ii) highlight the predominance of heavy tails in the packet arrival patterns generated by the individual source-destination pairs or by the major applications that make up the aggregate packet traffic, and (iii) provide fundamental insight into how individual network connections behave. For LAN traffic, the proposed structural models are based on a construction by Willinger, Taqqu, Sherman, and Wilson, which is a modification of one proposed by Mandelbrot [Ma2] in an economic setting. For WAN traffic, the proposed models are based on two related constructions by Cox and Kurtz; all three constructions relate (exactly or asymptotically second-order) self-similarity at the macroscopic or network level directly to heavy-tailed behavior with infinite variance at the microscopic or connection/packet level.

While Cox's construction focuses largely on global features at the level of individual network connections and offers little flexibility for modeling the fine structure of packet arrivals within a connection, the constructions of Willinger *et al.* and Kurtz offer the possibility of modelling packet level dynamics in greater detail. Despite this difference, all three constructions result in self-similar traffic dynamics when the aggregate packet traffic is viewed on large time scales and suggest that for a number of network engineering problems, the fine structure of the packet arrival instants within individual network connections is of minor importance, provided the global connection characteristics exhibit heavy-tails. Clearly, however, this fine structure is bound to play an important role when considering aggregate packet traffic over small time scales and when trying to understand the impact of network protocols on the nature of the traffic through, for example, TCP dynamics. In this context, an important unanswered question concerns the meaning of "large" and "small" time scales from a practical perspective. A closely related issue which requires further study is the adequacy of self-similar limiting processes for describing actual network traffic that necessarily consists of only finitely many connections which, in turn, transmit packets at different rates and which can only be observed over a finite range of time scales. Another important but difficult aspect of traffic modeling that is left for future work concerns the finite link capacities in a network. The models for aggregate network traffic on a particular link, described in this paper, do not include the fact that the link has finite capacity. Yet it is exactly this finite capacity that drives the TCP dynamics and couples the different simultaneous connections sharing the link.

Finally, the question remains: Where do these heavy tails come from? For possible answers, we refer to empirical studies reported, for example, in [PF2, CB, CTB, WTSW2], that relate the observed heavy tails at the connection level directly to heavy-tailed phenomena observed in the sizes of the documents that reside on a typical file or WWW server, and to heavy-tailed characteristics exhibited by human-computer interactions. We must leave the explanation of these latter phenomena to human factors experts and researchers in the cognitive sciences.

References

[Be] J. Beran, *Statistics for Long-Memory Processes* Chapman & Hall, New York, 1994.

[CB] M. E. Crovella and A. Bestavros, Self-Similarity in World Wide Web Traffic: Evidence and Possible Causes, in: *Proc. ACM/SIGMETRICS'96, May 1996.*

[CTB] M. E. Crovella and M. S. Taqqu and A. Bestavros, Heavy-tailed Probability Distributions in the World Wide Web, in: *This volume.*

[Co] D. R. Cox, Long-Range Dependence: A Review, in: *Statistics: An Appraisal*, ed. H. A. David and H. T. David, Iowa State University Press, 1984, 55–74.

[DJCME] P. Danzig, S. Jamin, R. Cáceres, D. Mitzel, and D. Estrin, An Empirical Workload Model for Driving Wide-area TCP/IP Network Simulations, *Internetworking: Research and Experience* **3** (1992), 1–26.

[DMRW] D. E. Duffy, A. A. Mcintosh, M. Rosenstein and W. Willinger, Statistical Analysis of CCSN/SS7 Traffic Data from Working CCS Subnetworks, *IEEE Journal on Selected Areas in Communications* **12** (1994), 544–551.

[FL] H. Fowler and W. E. Leland, Local Area Network Traffic Characteristics, with Implications for Broadband Network Congestion Management, *IEEE Journal on Selected Areas in Communications* **9** (1991), 1139–1149.

[Gu] R. Gusella, A Measurement Study of Diskless Workstation Traffic on an Ethernet, *IEEE Transactions on Communications* **38** (1990), 1557–1568.

[HRS] D. Heath, S. I. Resnick and G. Samorodnitsky, Heavy Tails and Long-Range Dependence in On/Off Processes and Associated Fluid Models, *Mathematics of Operations Research* (1997) (to appear).

[Hi] B. M. Hill, A Simple General Approach to Inference about the Tail of a Distribution, *The Annals of Statistics* **3** (1975), 1163–1174.

[JR] R. Jain and S. A. Routhier, Packet Trains – Measurements and a New Model for Computer Network Traffic, *IEEE Journal on Selected Areas in Communications* **4** (1986), 986–995.

[KR] M. Kratz and S. I. Resnick, The qq-Estimator and Heavy Tails, *Stochastic Models* **12** (1996), 699–724.

[Ku] T. G. Kurtz, Limit Theorems for Workload Input Models, in: *Stochastic Networks: Theory and Applications*, ed. F. P. Kelly, S. Zachary and I. Ziedins, Clarendon Press, Oxford, UK, 1996, 339–366.

[LTWW1] W. E. Leland, M. S. Taqqu, W. Willinger and D. V. Wilson, On the Self-Similar Nature of Ethernet Traffic, in: *Proc. ACM/SIGCOMM'93, San Francisco, 1993, Computer Communications Review* **23** (1993), 183–193.

[LTWW2] W. E. Leland, M. S. Taqqu, W. Willinger and D. V. Wilson, On the Self-Similar Nature of Ethernet Traffic (Extended Version), *IEEE/ACM Transactions on Networking* **2** (1994), 1–15.

[LT] S. B. Lowen and M. C. Teich, Fractal Renewal Processes Generate $1/f$ Noise, *Physical Review E* **47** (1993), 992–1001.

[Ma1] B. B. Mandelbrot, Self-Similar Error Clusters in Communications Systems and the Concept of Conditional Stationarity, *IEEE Transactions on Communications Technology* **COM-13** (1965), 71–90.

[Ma2] B. B. Mandelbrot, Long-Run Linearity, Locally Gaussian Processes, H-Spectra and Infinite Variances, *International Economic Review* **10** (1969), 82–113.

[MB] R. M. Metcalfe and D. R. Boggs, Ethernet: Distributed Packet Switching for Local Computer Networks, *Communications of the ACM* **19** (1976), 395–404.

[No] I. Norros, A Storage Model with Self-Similar Input, *Queueing Systems And Their Applications* **16** (1994), 387–396.

[Pa] V. Paxson, Empirically-Derived Analytic Models of Wide-Area TCP Connections, *IEEE/ACM Transactions on Networking* **2** (1994), 316–336.

[PF1] V. Paxson and S. Floyd, Wide-Area Traffic: The Failure of Poisson Modeling, in: *Proc. ACM/SIGCOMM'94, London, 1994, Computer Communications Review* **24** (1994), 257–268.

[PF2] V. Paxson and S. Floyd, Wide-Area Traffic: The Failure of Poisson Modeling, *IEEE/ACM Transactions on Networking* **3** (1995), 226–244.

[Re] S. I. Resnick, Heavy Tail Modeling and Teletraffic Data, *Annals of Statistics*, **25** (1997), 1805–1869. With discussions and rejoinder.

[RS] S. I. Resnick and C. Starica, Consistency of Hill's Estimator for Dependent Data, *Journal of Applied Probability* **32** (1995), 139–167.

[ST] G. Samorodnitsky and M. S. Taqqu, *Stable Non-Gaussian Processes: Stochastic Models with Infinite Variance*, Chapman and Hall, New York, 1994.

[TT] J. W. Tukey and P. A. Tukey, Strips Displaying Empirical Distributions: I. Textured Dot Strips, Bellcore Technical Memorandum, 1990.

[TWS] M. S. Taqqu, W. Willinger and R. Sherman, Proof of a Fundamental Result in Self-Similar Traffic Modeling, *Computer Communication Review* **27** (1997), 5–23.

[WTSW1] W. Willinger, M. S. Taqqu, R. Sherman and D. V. Wilson, Self-Similarity Through High-Variability: Statistical Analysis of Ethernet LAN Traffic at the Source Level, in: *Proc. ACM/SIGCOMM'95, Cambridge, MA, 1995, Computer Communications Review* **25** (1995), 100–113.

[WTSW2] W. Willinger, M. S. Taqqu, R. Sherman and D. V. Wilson, Self-Similarity Through High-Variability: Statistical Analysis of Ethernet LAN Traffic at the Source Level (Extended Version), *IEEE/ACM Transactions on Networking*, **5** (1997), 71–86.

AT&T Labs-Research, Florham Park, New Jersey 07932
Email: walter@research.att.com

Lawrence Berkeley National Laboratory, University of California, Berkeley, CA 94720
Email: vern@ee.lbl.gov

Department of Mathematics, Boston University, Boston, MA 02215-2411
Email: murad@math.bu.edu

Heavy Tails in High-Frequency Financial Data

Ulrich A. Müller, Michel M. Dacorogna and Olivier V. Pictet

Abstract

We perform a tail index estimation of financial asset returns in two markets: the foreign exchange market and the interbank market of cash interest rates. Thanks to the high-frequency of the data, we obtain good estimates of the tail indices and we are able to analyze their stability with time aggregation.

Our analysis confirms that the variance of the return is finite but points to the non-convergence of the kurtosis. Both financial markets present similar tail behavior of the returns. A study of the extreme risks reveals the need to depart from the Gaussian assumption by taking the fat tails fully into account. A study of tails under temporal aggregation, also investigating data from theoretical price formation processes, shows that ARCH-type processes represent the true behavior better than unconditional distribution models.

1. Introduction

"How heavy are the tails of financial asset returns?" The answer to this question is not only the key to evaluating risk in financial markets but also to accurately model the process of price formation. In a companion paper we present a bootstrap method that allows us to estimate with a good degree of accuracy the tail index of theoretical processes without a-priori knowledge of the theoretical value of this index. We present in this paper a study of extreme risk by empirically determining the tail index of returns applying this method to high-frequency financial data.

Evidence of the heavy tails presence in financial asset return distributions are plentiful since the seminal work of Mandelbrot on cotton prices (Mandelbrot, 1963). Mandelbrot advanced the hypothesis of an underlying stable distribution on the basis of an observed invariance of the return distribution across different frequencies and the apparent heavy tails of the distribution. Nevertheless, a controversy is still going on in the financial research community as to whether the second moment of the distribution of returns converges. This question is central to many models in finance since they heavily rely on the existence of the variance of returns (σ^2). The risk in financial markets has often been associated with the variance of returns since portfolio theory was developed. From the Sharpe ratio for measuring portfolio performance to option pricing models, the variable σ is always present.

Another important motivation of this study is the need to evaluate extreme risks in financial markets. Recently, the problem of risk in these markets has become topical following a few unexpected big losses, as in the case of Barings or Daiwa. The Bank of International Settlements has edicted rules to be followed by banks to control their risks but most of the current models for assessing risks are based on the assumption that financial assets are distributed according to a normal distribution. In the Gaussian model the evaluation of extreme risks is directly related to the variance, but in the case of fat-tailed distributions this is no longer the case.

In a recent paper (Dacorogna et al., 1994), results for the tail estimation of foreign exchange (FX) data were presented using the same method as in this paper. These results confirmed that the second moment of the distribution exists for foreign exchange rate returns, but a question still remains concerning the behavior of the tail index under aggregation. We discuss here further results about the time aggregation problem, using an operational time scale to treat the seasonal patterns of market activity (Dacorogna et al., 1993). Results on the tail index of cash interest rates (IR) are also presented, together with an evaluation of how current models of asset returns reproduce extreme risks.

In section 2, our particular data set is described with the method of constructing homogeneous time series from irregular quote arrivals. In section 3, empirical distributions are presented as a function of the time interval at which the returns are computed and properties related to the importance of the tails are discussed. The tail estimation is described in section 4 for a set of FX rates and for the 3 and 6 month cash interest rates of four major currencies. The expected extreme returns are computed for particular probabilities and presented against some model processes in section 5. Conclusions are drawn in section 6.

2. High frequency data for financial assets

The data used in this study come from two financial markets: the foreign exchange (FX) market and the interbank money market for cash interest rates. For both of these markets, the bid and ask offers of major financial institutions are conveyed to customers' screens by large data suppliers such as Reuters, Telerate or Knight-Ridder and the deals are negotiated over the telephone or through automatic dealing systems such as Reuters-2000. The markets have no business hour limitations. Any market maker can submit new bid/ask prices; many large institutions have branches worldwide so that trading is continuous. Nevertheless, the bid/ask prices are seen to emanate from particular banks in particular locations and the deals are entered into dealers' books in particular institutions. These prices are termed *ticks*. They cover the market worldwide and 24 hours a day,

FX rate	ticks	IR rate	ticks
USD-DEM	11,059,640	USD 3m	44,905
USD-JPY	5,299,018	USD 6m	45,530
GBP-USD	4,256,965	DEM 3m	20,468
USD-CHF	4,311,767	DEM 6m	20,365
USD-FRF	2,555,526	JPY 3m	17,479
USD-ITL	1,312,563	JPY 6m	17,898
USD-NLG	1,857,179	GBP 3m	12,718
XAU-USD	1,526,783	GBP 6m	13,074
XAG-USD	442,025	CHF 3m	15,181
		CHF 6m	13,494

Table 1: The different time series and their corresponding number of valid raw data records for the period from January 1, 1987 to June 30, 1996.

including weekends, with more than 300 financial institutions as contributors.

We have been collecting tick-by-tick data since 1986. The recorded prices are composed of three quantities:

1. the time t_j at which the price has been recorded (in Greenwich Mean Time) [1],
2. the bid price $p_{bid,j}$,
3. and the ask price $p_{ask,j}$,

plus information concerning the origin of the published price. The sequence of the tick recording times t_j is unequally spaced. We have analyzed the main empirical features of these data in a series of papers (Müller et al., 1990; Dacorogna et al., 1993; Guillaume et al., 1994a). This paper will only discuss stylized facts related to the problem of fat tails. In the case of the cash interest rates, we do not collect a price *per se* but rather a percentage rate r_{bid} or r_{ask}. To convert this into a price we compute $p = 1 + r/100$ where r is given in percents.

Our database now contains more than *50 million ticks* of more than 600 time series. It covers almost every day of the year except for rare failures of our system or of the data supplier which lead to data holes. The database is completed with daily prices from other data suppliers since June 1, 1973 for FX rates and since January 2, 1979 for

[1] so this time stamp is not affected by any daylight saving change.

the cash interest rates. The sample we choose to study is composed of nine and a half years of tick-by-tick data from January 1, 1987 to June 30, 1996 for the FX rates. In Table 1, we give the number of raw data records present in the different databases for these periods. For the interest rates, the number of raw data is two to three orders of magnitude smaller than for FX rates and the data is concentrated in the morning of the European markets. This prevents us from fully using high-frequency data in the case of interest rates; we choose rather to concentrate on daily returns. This choice avoids interpolation problems. Besides, such a frequency allows us to also use our daily data since 1979 to complete the sample.

Our study focuses on the distribution of returns which are computed from the bid and ask prices through the logarithmic *middle price*, x_j,

$$x(t_j) \equiv x_j \equiv \frac{\log p_{ask,j} + \log p_{bid,j}}{2}. \tag{1}$$

Here we take the average of the logarithms instead of the logarithm of the average. The numerical difference between these two definitions is insignificant. Our definition has the advantage of behaving symmetrically when the price is inverted (e.g., 1 DEM expressed in USD instead of 1 USD expressed in DEM). Differences in logarithmic prices, x_j, are unitless and thus allow direct comparisons between currencies. Consistent with definition (1), we define the return r_i measured over a fixed time interval Δt as,

$$r(t_i) \equiv r_i \equiv x(t_i) - x(t_i - \Delta t) \tag{2}$$

where $x(t_i)$ is the sequence of logarithmic middle prices spaced equally in time.

One important issue in the case of intra-daily data is the use of the right *time-scale*. Contrary to daily and weekly data, tick-by-tick data are irregularly spaced in time, t_j. However, most statistical analyses rely upon the use of data regularly spaced in time, t_i. For obtaining price values at a time t_i within a data hole or in any interval between ticks, we use linear interpolation between the previous price at t_{j-1} and the next one at t_j, with $t_{j-1} < t_i < t_j$. As advocated in (Müller et al., 1990), linear interpolation is the appropriate method for interpolating in a series with independent random increments for most types of analyses. An alternative method, taking the last valid price before the hole as representative for the hole interval, must be avoided in a study of extreme values, as it would lead to a spurious large price change from the last valid price within the hole to the first real price after the hole.

This exceptionally large database, in the order of many million prices, contains some rare but aberrant values caused by technical and human errors. Some spurious large price movements can arise due to these aberrant outliers. An analysis of extreme price movements is more sensitive against such errors than other statistical studies. A full discussion of this problem can be found in (Dacorogna et al., 1994). We note here that two types of filtering errors can be made. The first type of error arises due to including false prices in the analysis; that is, the case of underfiltering. The second type is because of rejecting valid prices in the analysis; that is, the case of overfiltering. Both error types influence our analysis. It is not always possible to determine whether an apparent valid price is realistic or not. The database quotes are market maker quotes and are not always trading prices. Some extreme prices might be valid in the sense of being serious market maker quotes although nobody used them in a real transaction. If the validity is unclear, no filter can be perfect. Therefore, several different filters have been alternatively used in the computation of the results. For all reasonable filters tested, we obtained similar results. We conclude that the final results are not essentially biased by the choice of our filter.

We record t_j on the *physical time* scale, but this is not self-evident. Many other authors take samples of daily data as their equally spaced time series but ignore the weekends and holidays, so the spaces are actually unequal. In fact, these authors implicitly use another scale that reflects the time of active markets and can be called *business time*. Their series can be called equally spaced only on that new time scale. Intra-day studies face an identical but aggravated problem. For instance in (Feinstone, 1987), the author uses a more complex business time which includes only the business hours of the IMM in Chicago (8:48 to 13:18, Chicago Time) but does not discuss why this choice was made. Since then, high-frequency data has become more popular in financial studies and the need for using an appropriate time scale to reflect the changing market activity for statistical analysis has become more apparent. In this study, we present results computed on the *business time* scale (ϑ-time scale) developed in (Dacorogna et al., 1993). The advantage of the ϑ-time is to remove the seasonal heteroskedasticity due to changing market activity (Müller et al., 1990). In Figure 1, we show an example of the use of the ϑ-time scale by plotting three months of hourly returns both in physical time scale and in this new time scale. The returns in physical time present clear seasonalities due to the weekends, while the returns plotted in ϑ-time look more homogeneous because each time interval has the same expectation of activity. The extreme returns also look different because by changing the time scale, the phase has also changed and some extreme movements which were fully in one interval are now split over two time intervals.

To obtain aggregated data, a time series with a larger time interval is generated by aggregating the returns measured on the basic time interval. We choose these larger intervals as multiples of the basic time interval in any time scale (in this case, it is the ϑ-time). The basic time interval is chosen to be 30 minutes (in ϑ-time) for this study. The overlap factors are 2, 6 and 6 for 2 hours, 6 hours and 1 day respectively as in the Monte-Carlo study of the companion paper. So, the results for this time intervals are actually averages over 2, 6 and 6 sets of data differing only in the phases of their time grids. By including different phases in the estimation, we increase its precision (Müller, 1993).

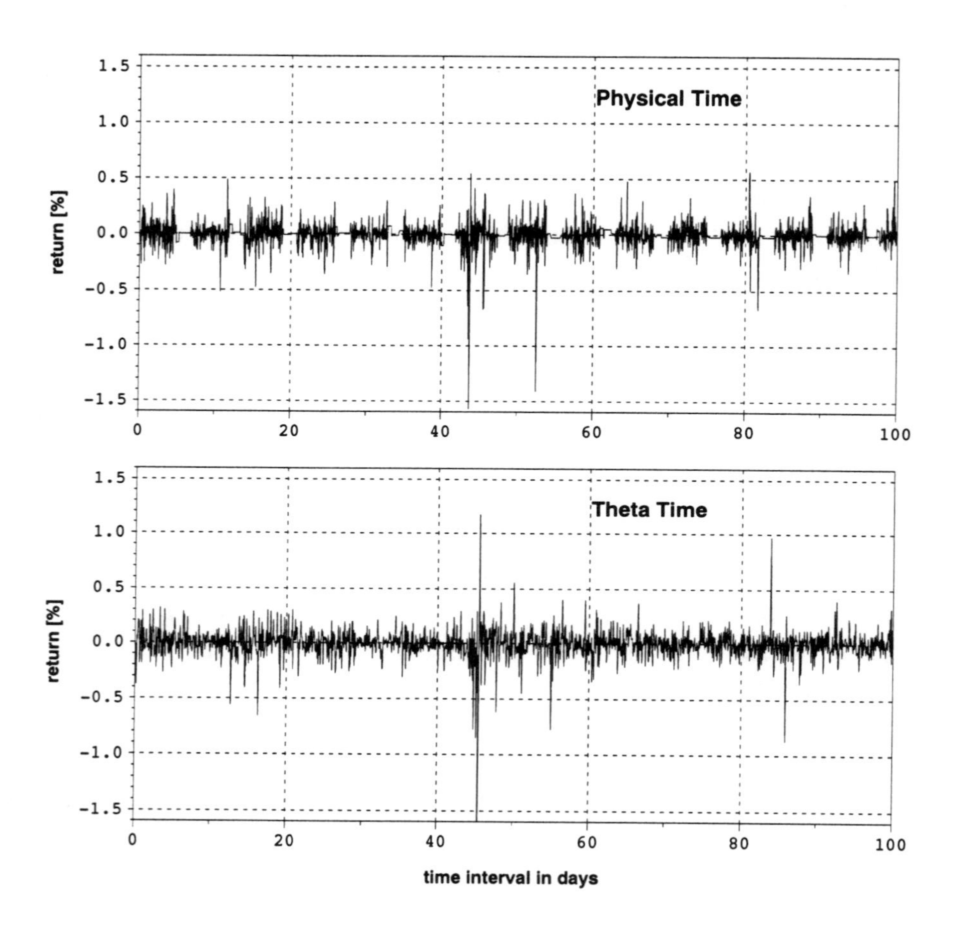

Figure 1: The hourly returns for USD/DEM from 03.06.96 00:00:00 to 11.09.96 00:00:00 are plotted using the physical time scale and the ϑ-time scale. Note also the extreme events that are clearly visible on both graphs.

3. Empirical distribution and autocorrelation of returns

The variety of opinions about the distributions of FX price changes and their generating process is broad. Some authors claim the distributions to be close to Paretian stable (McFarland et al., 1982), some to Student-t distributions (Boothe and Glassman, 1987) and some reject any single distribution (Calderon-Rossel and Ben-Horim, 1982). There is, however, one point on which everybody agrees: FX returns are fat-tailed. Similar observations have been made on interest rates and other financial assets.

In Table 2, we present the moments of the distributions for the five major exchange rates. The means are about two orders of magnitude smaller than the standard deviations and the absolute values of the skewness are, except in very few cases, significantly smaller than one. We can conclude from those facts that the empirical distribution is almost symmetric. For the shortest time intervals, the estimates of the kurtosis are clearly higher than 3, which is the expected value for a Gaussian distribution. Another interesting feature is that all the rates show the same general behavior, a decreasing kurtosis with increasing time intervals. At intervals of around 1 week, the kurtosis is rather close to the Gaussian value. The variance values for the short-time intervals are somewhat higher than the values we obtain with physical time (Müller et al., 1990; Guillaume et al., 1994a) since with ϑ-time, we have equalized the volatility and removed the periods where the market is inactive. For 1 day and 1 week intervals, the variance values are very similar in both time scales. The mean values are slightly negative (except for GBP/USD where the currencies are inverted) since during this period we have experienced an overall decline of the USD.

The quote frequency of cash interest rates is about two orders of magnitude lower than that of the FX rates (see Table 1). That is why we present the moments only for 24-hour and 1-week intervals in Table 4. The general behavior is similar that of FX rates with few exceptions: the variance tends to be much lower (one order of magnitude) and the kurtosis is higher than in Table 2 and diminishes more slowly with time aggregation. Even for 1 week intervals, we still see kurtosis values that significantly deviate from the Gaussian value. Although small[2], the skewness tends to depart more from zero than those of FX rates. This might be related, as we shall see later in the paper, to a low tail index α that increases the error on the estimation of the skewness. The market for cash interest rates is much less liquid than the foreign exchange market; this reflects in its low quote frequency and variance. Another difference that plays a role in this computation is the fact that the rates are reported with a lower precision (2 decimal digits) than for the FX rate (4 decimal digits) and the changes are only made in steps of 1/32 percent.

On the same samples and ϑ-scales as in the tables, we compute the empirical

[2]The case of JPY is a special one because the two first years present much higher interest rates than the rest of the sample.

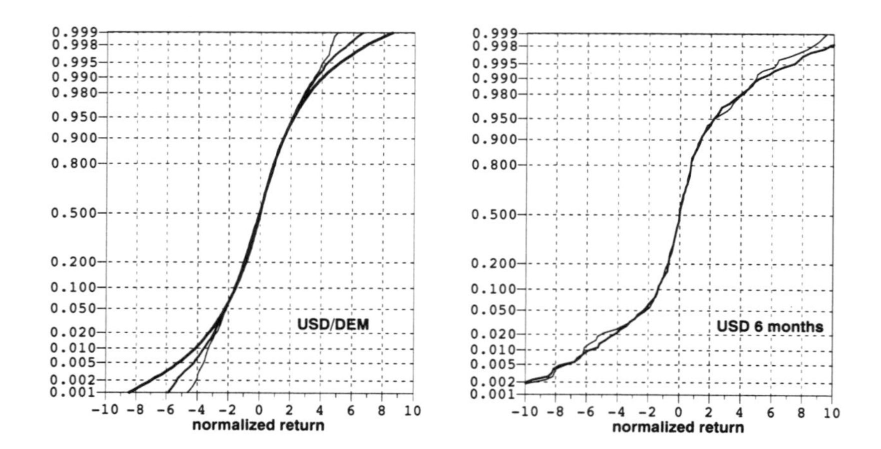

Figure 2: The empirical cumulated distributions for USD/DEM and USD 6 months are shown for different time intervals: 30 minutes, 1 day and 1 week for USD/DEM and 1 day and 1 week for USD 6 months. The fat lines are for the shortest time intervals. The distributions are computed on the same samples as in Table 2 and 3.

distributions. Three empirical distributions for USD/DEM and two for the USD 6-month cash interest rate are plotted in Figure 2.

The cumulative frequency is drawn on the scale of the cumulative Gaussian probability function. Normal distributions have the form of a straight line in this representation. This is approximately the case for the cumulative distribution of weekly returns whose kurtosis is slightly more than 3 for the USD/DEM (see Table 4). The distributions of 30-minute and 24-hour returns, however, are distinctly fat-tailed and the kurtosis values in Table 4 are high (46.10 for 30 minutes and 6.04 for 24 hours). The remarkable feature here is that the shape of the distribution is not preserved under time aggregation as it was the case for the cotton prices in (Mandelbrot, 1963). These distributions are not stable with time aggregation.

In the case of interest rates, both distributions look remarkably alike besides the statistical fluctuations. The change in shape that we observe on the USD/DEM is not visible here and it is not possible to reject the hypothesis of the distribution being stable under time aggregation. This fact is related to the kurtosis remaining high for weekly returns (14.48 in Table 4). Moreover, the cumulative distributions look slightly more fat-tailed than for the FX rates which is also confirmed by the values for the kurtosis in Tables 4 and 4. One should note that the empirical distributions of the cash interest rate are more noisy than those of the FX rates. This is due to both the precision of the quotes and their low frequency. There are not many differences between the currencies

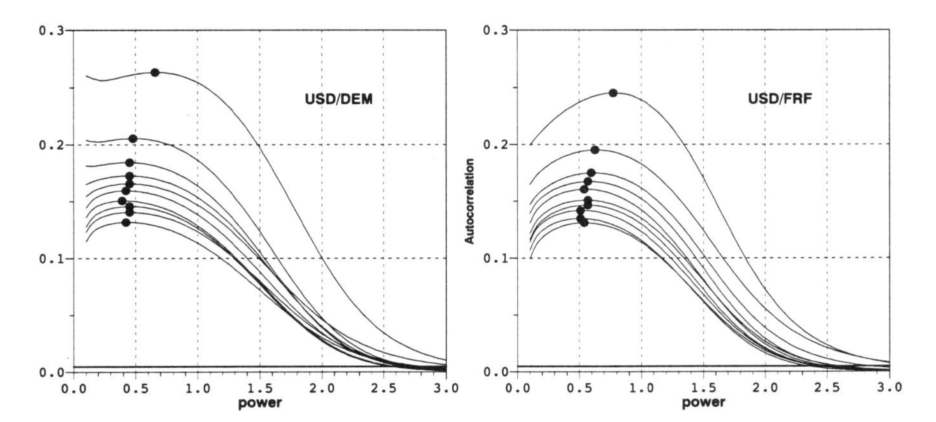

Figure 3: The first 10 lags of the autocorrelation function of $|r|^d$ as a function of the power d for USD/DEM and USD/FRF (first lag on top, 10th at the bottom). The maxima are shown by the bullet sign (•). The returns are measured over 30m (in ϑ-time). The horizontal lines represent the 95% significance level of a random walk.

except a slightly higher variance for JPY where the cash interest rates have experienced large changes in the early years of our sample (1979 and 1980). In Figure 2 we present only two examples for the most liquid assets in these two financial markets, but similar distributions are found for other FX rates and interest rates as well.

From these empirical facts it becomes clear that the tails of the distribution play an important role in the statistical properties of financial asset returns. To illustrate this, we study the autocorrelation function of powers of absolute returns as a function of the exponent of this power. Although financial asset returns do not present significant a autocorrelation, their absolute values are known to exhibit a strong one, stronger than the squared returns (Taylor, 1986; Müller et al., 1990). Some studies on the influence of the power of absolute returns on the autocorrelation have been published (Granger and Ding, 1995). The authors of this study conclude that a power 1 gives the highest autocorrelation.

The example of Figure 3 shows how the tails influence one of the most important properties found in the statistics of financial data. Varying the power of the absolute returns comes back to varying the relative importance of extreme events in the statistics. Increasing the power increases the influence of extreme events in the statistics.

In Figure 3, we see that the autocorrelation, for the 10 lags considered, decreases when the influence of the large returns are increased. In other words, extreme events are less correlated with each other than the average or small absolute returns. From this study, it seems that the heteroskedasticity is mainly due to the average behavior, not the extreme events. This is better represented by a low power d smaller than 1; the maximum autocorrelation is for values of d close to 1/2.

Studying the tails themselves is thus important for understanding how autocorrelation behaves and what can be learned from them. In particular, we would like to examine whether the hypothesis of $\alpha > 2$ is compatible with our data. Fortunately, we have seen in the companion paper that it is possible to determine the tail index of empirical time series to a reasonable accuracy with the bootstrap method. We want to apply it in this paper to the cases of FX rates and cash interest rates.

4. Estimation of the tail indices of financial assets

The tail estimates for seven USD rates, five cross rates (not involving the USD) and two commodity prices are reported in Table 4 as a function of the time intervals on which the returns are measured. These time intervals are computed in ϑ-time which varies compared to physical time according to the market activity. However, a 1-week interval in ϑ-time is roughly equal to 1 week in physical time[3].

All the point estimates are above 3 and below 5 for the shortest horizons. Hence we can safely assume that the second moment of the distribution is finite; but the existence of the fourth moment is rejected at the 95% level in a number of cases. These results are in line with the evidence from data sets that are orders of magnitude smaller, such as in (Hols and De Vries, 1991) and (Loretan and Phillips, 1994). We also see no noticeable difference in behavior between the USD rates, the cross rates and the two commodity prices. Overall the data display moderately heavy tails. If we except the daily returns which we shall discuss later in the section, the tail indices are remarkably stable under aggregation. They are much more so in ϑ-time than in physical time (Dacorogna et al., 1994). The case of the longest horizons (6 hours and especially 1 day) illustrates the problems discussed in the companion paper: an accurate estimate of the tail index can only be achieved with a sufficiently large number of data. Gold and silver and most of the cross rates are exceptions. Their tail index is also stable up to a 1 day horizon but with increasing errors.

The results presented in Table 4 can well explain the feature shown in Figure 3: the positive autocorrelation of absolute returns is stronger than that of squared returns. The tail index α being almost always below 4, the fourth moment of the distribution is unlikely to converge[4]. In the computation of the autocorrelation of squared returns, the denominator requires the convergence of the fourth moment while the denominator of the autocorrelation of absolute returns only requires the convergence of the second moment, which exists if α is larger than 2. Indeed, besides the empirical evidence of

[3]For a full description and discussion of ϑ-time the reader is referred to (Dacorogna et al., 1993).

[4]The authors in (Loretan and Phillips, 1994) come to a similar conclusion when examining the tail behavior of daily closing prices for FX rates.

Figure 3, we find that the difference between the autocorrelations of absolute returns and squared returns grows with increasing sample size. This difference computed on twenty years of daily data is much larger than that computed on only eight years. For a lag of 9 days, we obtain autocorrelations of 0.11 and 0.125 for the absolute returns over 8 and 20 years respectively, while we obtain 0.072 and 0.038 for the squared returns, showing a strong decrease when going to a larger sample. The same effect as for the daily returns is found for the autocorrelation of twenty-minute returns, where we compared a nine-year sample to half-yearly samples.

In Table 5, we present the results of α estimations for the interbank money market cash interest rate for 5 different currencies and 2 maturities. Although presenting overall lower α's, the results are close to those of the FX rates. The message seems to be the same: fat tails, finite second moment[5] and non-converging fourth moment. We find also a relative stability of the tail index with time aggregation. The estimations for daily returns give more consistent values than in the case of the FX rates. Yet, the estimations are somehow more noisy, as one would expect from data of lower frequency. This is reflected in the high errors displayed in Table 5. The tail index estimation is quite consistent but can significantly jump from one time interval to the next as it is the case for GBP and CHF 6 months. This market is much less liquid than the FX market. Interest rate markets with higher liquidity can be studied in terms of interest rate derivatives which are traded in futures markets such as LIFFE in London or SIMEX in Singapore.

To study the stability of the tail index under aggregation, we use similar methods as in the companion paper: we observe how the estimates change with varying sample size. We do not know any theoretical tail index for empirical data, but we compare the estimates with the estimation done on the "best" sample we have, 30-minute returns from January 1987 to June 1996.

To study the tail index of daily returns, we use an extended sample of daily data from June 1, 1978 to June 30, 1996. The small-sample bias can be studied by comparing the results to the averages of the results from two smaller samples: one from June 1, 1978 to June 30, 1987 and another from June 1, 1987 to June 30, 1996. These two short samples together cover the same period as the large sample. The results[6] are given in Table 6. When going from the short samples to the long 18-year sample, we see a general decrease of the estimated α towards the values reached for 30-minute returns. The small-sample bias is thus reduced, but probably not completely eliminated.

[5]In most of the cases except perhaps the 6 months interest rate for JPY α is significantly bigger than 2. In the JPY case, if the first two years are removed, we get back to values for α around 3.

[6]For the short samples we do not give the errors since we present only the average of both samples. The errors are larger for the short samples than for the long sample.

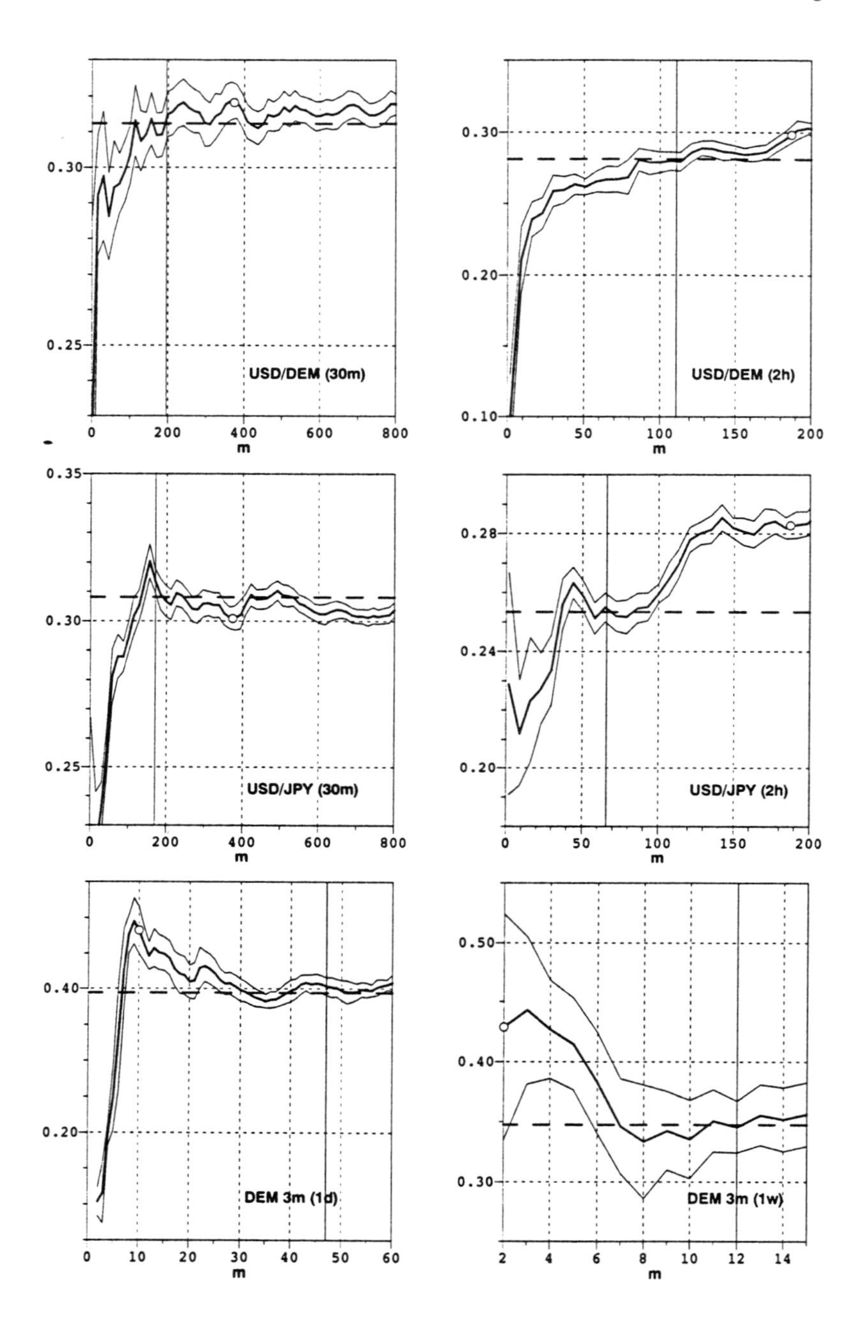

Figure 4: The Hill estimator is plotted as a function of the maximum order statistics. The thick curves represent the average over 10 samples computed using the jackknife method and the two thin curves the standard deviation around the average. The white circles indicate the starting value before the bootstrap and the vertical and horizontal lines the final average values after bootstrapping the data.

We conclude that, at least for the FX rates against the USD, the α estimates from daily data are not accurate enough even if the sample covers up to 18 years. The case of gold and silver is somehow different since the market has experienced huge fluctuations in the early 80s which have disappeared since then. The picture in this case is blurred by the changing market conditions.

A similar exercise on 30-minute returns reinforces the obtained conclusions. In this case, the shorter samples are from June 1, 1988 to June 30, 1992 and from June 1, 1992 to June 30, 1996. The large sample is again the union of the two shorter samples. A certain small-sample bias is found also for the 30-minute returns of most rates comparing the two last columns of Table 4, but this bias is rather small. This is an expected result since the number of 30-minute observations is much larger than that of the daily observations.

In a special study, we examine the sensitivity to the variable m, the maximum order statistics, used in the tail index estimations shown in Tables 4 and 5. This study relies heavily on the theoretical developments and discussions of the companion paper. In particular, the notation and definitions are taken from it.

We show in Figure 4 the variation of the Hill estimator as a function of m, for USD/DEM, USD/JPY and DEM 3 months and for two time intervals. This figure illustrates the procedure that is used to estimate the tail index. The thick curve represents the average values of the Hill estimator as a function of m for the different samples used by the jackknife method. The thin lines around it represent the standard error over these samples.

The starting values for the maximum order statistics, the m_0 value of the companion paper (section 5), are marked with a white circle and the resulting average value for $\hat{m}$ and $\hat{\gamma}$ (eqs. (53) and (55) of the companion paper) are marked by a vertical and a horizontal dashed line respectively. Note that the intersections between these two straight lines are not on the curve of the average $\gamma(m)$'s because they are averages of values with different maximum order statistics m (see the companion paper for the formal derivation of the estimator).

We find that for the FX rates the bootstrap procedure usually pushes the choice of m further into the tails than the initial choice. In rare cases such as the 3-month interest rate for DEM, however, m is pushed away from the extreme tail. For very small ms, the Hill estimator becomes noisier than for large values of m, as expected. Therefore, it is difficult to assess by only looking at the graphs what m should be chosen. Yet, the resulting $\hat{\gamma}$'s are close to values that we would choose by human eye. This confirms the result of the companion paper. The bootstrap procedure manages to pick reasonable values for the estimator although all the results strongly depend on the (in practice) unknown second order behavior of the tail expansion.

In some cases though, the bootstrap does not find the intuitively perceived optimal value. For example in the USD/JPY case, we see that the Hill estimator for the 2 hour interval has $\hat{\gamma}$ values close to the 30 minutes up to m values around 120 but lower values for smaller m. The initial γ_0-value would be very close to the 30-minute result but the bootstrap procedure pushes m to the noisy part of the curve.

5. Extreme risks in financial markets

From the practitioners' point of view, one of the most interesting questions is that tail studies can answer is: what are the extreme movements that can be expected in financial markets? Have we already seen the largest ones or are we going to experience even larger movements? Are there theoretical processes that can model the type of fat tails that comes out of our empirical analysis? The answers to such questions are essential for good risk management of financial exposures. It turns out that we can partially answer them here. Once we know the tail index, we can apply an extreme value theory *outside* our sample to consider possible extreme movements *which have not yet been observed historically*. This can be achieved by a computation of the quantiles with exceedance probabilities[7].

Let us consider, as in the companion paper, the expansion of the asymptotic cumulative distribution function from which the X_i observations are drawn as

$$F(x) \;=\; 1 - a\,x^{-\alpha}[1 + b\,x^{-\beta}]\,. \tag{3}$$

We denote by x_p and x_t quantiles with respective exceedance probabilities p and t. Let n be the sample size and choose $p < 1/n < t$; that is, x_t is inside the sample, while x_p is not observed. By definition (we concentrate on the positive tail),

$$p \;=\; ax_p^{-\alpha}[1 + bx_p^{-\beta}]\,, \quad t \;=\; ax_t^{-\alpha}[1 + bx_t^{-\beta}]\,. \tag{4}$$

Division of the two exceedance probabilities and rearrangement yields

$$x_p \;=\; x_t \left(\frac{t}{p}\right)^{1/\alpha} \left(\frac{1 + bx_p^{-\beta}}{1 + bx_t^{-\beta}}\right)^{1/\alpha}\,. \tag{5}$$

Given that t is inside the sample, we can replace t by its empirical counterpart m/n, say; that is, m equals the number of order statistics X_i which are greater than X_t. An estimator for x_p is then as follows

$$\hat{x}_p \;=\; X_m \left(\frac{m}{np}\right)^{\hat{\gamma}}\,. \tag{6}$$

where m equals the $\hat{m}$ obtained from the tail estimation corresponding to $\hat{\gamma}$.

To write this estimator we ignore the last factor on the right hand side of eq. (5). This would be entirely justified in case of the Pareto law when $b = 0$. Thus $\hat{x}_p$ is based

[7] We follow here the approach developed in (Dacorogna et al., 1994).

on the same philosophy as the Hill estimator: for a m sufficiently small relative to n, the tails of (3) are well approximated by those of the Pareto Law, and hence (6) is expected to do well. In fact, it is possible to prove (de Haan et al., 1994) that, for the law (3)

$$\frac{\sqrt{m}}{\ln \frac{m/n}{p}} \left(\frac{\hat{x}_p}{x_p} - 1 \right) \tag{7}$$

has the same limiting normal distribution as the Hill estimator. Expression (7) gives us a way to estimate the error of our quantile computation.

In Table 7, we show the result of a study of extreme risk using eq. (6) to estimate the quantiles for returns over 6 hours[8]. This time interval is somehow shorter than an overnight position (in ϑ-time) but is somehow a compromise between the accuracy of the tail estimation and the length of the interval needed by risk managers. The first part of the table is produced by Monte-Carlo simulations of synthetic data where the process was first fitted to the 30-minute returns of the USD/DEM time series. The second part is the quantile estimation of the FX-rates as a function of the probability of the event happening once every year, once every 5 years and so on. Because we use here a sample of 9 years, the first two columns represent values that have been actually seen in the data set. Although the probabilities we use here seem very small, some of the extreme risk shown in Table 7 may be experienced by a trader during his active life.

An interesting piece of information displayed in Table 7 is the comparison of empirical results and results obtained from theoretical models. The models are built by fitting them to the USD/DEM 30-minute returns. For the normal distribution, we define the variance as that of our sample of 30-minute returns for USD/DEM; the same is true for the Student-t distribution. For the GARCH(1,1) model (Bollerslev, 1986), we use the standard maximum likelihood fitting procedure to determine the parameters (Guillaume et al., 1994b) and use the GARCH equation to generate synthetic time series. The same procedure is used for the HARCH model (Müller et al., 1995).

The model results are computed using the average $\hat{m}$ and $\hat{\gamma}$ obtained by estimating the tail index of 10 sets of synthetic data for each of the models for the aggregated time series over 6 hours. As expected, the normal distribution model fares poorly as far as the extreme risks are concerned. Surprisingly, this is also the case for the Student distribution with 3 degrees of freedom. The GARCH(1,1) model gives results closer to the ones of USD/DEM but still underestimates by a significant amount the risks. The HARCH model, however, is overestimating the extreme risk. This is probably due to its long memory which does not allow the process to modify sufficiently its tail behavior under aggregation. Obviously, further studies need to be pursued to assess how well the HARCH model can *predict* extreme risks but, in general, ARCH-type processes seem

[8]To keep the table easy to read, we do not report the errors on this estimations but they are available upon request.

to capture better the tail behavior of FX rates than the simple unconditional distribution models.

The advantage of having a model for the process equation is that it allows to use a dynamic definition of the movements and it can hopefully provide an early warning in case of turbulent situations.

Paradoxically, in situations more represented by the center of the distribution (non-extreme quantiles), the usual Gaussian-based models would overestimate the risk. Our study is valid for the tails of the distribution, but it is known that far from the tails the normal distribution produces higher quantiles than are actually seen in the data.

6. Conclusion

The combined use of high-frequency data and business time scales allows us to explore the tail behavior of financial asset returns with a good degree of accuracy. The bootstrap method described in the companion paper gives a consistent picture of the tails in financial asset return distributions. We report tail index estimations for two different markets: the foreign exchange market and the interbank money market. We present results of tail estimations for different time intervals and see that the tail indices vary only slightly when changing the data frequency. The limitations arise when the degree of aggregation becomes too high and does not leave enough observations to use the subsample bootstrap method meaningfully.

This study confirms that the second moments of the empirical distributions of FX rates are finite and that this is also true for interbank cash interest rates. In fact, the tail properties of the two sets of financial assets are remarkably similar. In both cases, the finiteness of the fourth moment is either clearly rejected or not demonstrated. This sheds doubts on the usefulness of autocorrelation studies of squared returns and other methods requiring finite fourth moments.

We have seen in the last part of this paper that price movements as high as 3% may occur even at time intervals of 6 hours in ϑ-time. Such intervals are not even long enough to cover an overnight period. The tail index seems to be quite constant under aggregation. Thus, current models for estimating extreme risks based on the Gaussian assumption are likely to underestimate risks considerably.

References

Bollerslev T., 1986, *Generalized autoregressive conditional heteroskedasticity*, Journal of Econometrics, **31**, 307–327.

Boothe P. and Glassman D., 1987, *The statistical distribution of exchange rates, empirical evidence and economic implications*, Journal of International Economics, **22**, 297–319.

Calderon-Rossel J. R. and Ben-Horim M., 1982, *The behavior of the foreign exchange rates, empirical evidence and economic implications*, Journal of International Business Studies, **13**, 99–111.

Dacorogna M. M., Müller U. A., Nagler R. J., Olsen R. B., and Pictet O. V., 1993, *A geographical model for the daily and weekly seasonal volatility in the FX market*, Journal of International Money and Finance, **12**(4), 413–438.

Dacorogna M. M., Pictet O. V., Müller U. A., and de Vries C. G., 1994, *The distribution of extremal foreign exchange rate returns in extremely large data sets*, Internal document UAM.1992-10-22, Olsen & Associates, Seefeldstrasse 233, 8008 Zürich, Switzerland.

de Haan L., Jansen D. W., Koedijk K. G., and de Vries C. G., 1994, *Safety first portfolio selection, extreme value theory and long run asset risks*, In J. Galambos, J. Lechner and E. Simiu, eds., Extreme value Theory and Applications, Kluwer, Dordrecht, 471–488.

Feinstone L. J., 1987, *Minute by minute: efficiency, normality, and randomness in intra-daily asset prices*, Journal of applied econometrics, **2**, 193–214.

Granger C. and Ding Z., 1995, *Some properties of absolute return: An alternative measure of risk.*, Annales d'Economie et de Statistique, **40**, 67–91.

Guillaume D. M., Dacorogna M. M., Davé R. D., Müller U. A., Olsen R. B., and Pictet O. V., 1997, *From the Bird's Eye to the Microscope: A Survey of New Stylized facts of the intra-daily Foreign Exchange Markets*, by Dominique M. Guillaume, Michel M. Dacorogna, Rakhal D. Davé, Ulrich A. Müller, Richard B. Olsen and Olivier V. Pictet. Finance and Stochastics, **Vol. 1**, p.95–129.

Guillaume D. M., Pictet O. V., and Dacorogna M. M., 1994b, *On the intra-day performance of GARCH processes*, Internal document DMG.1994-07-31, Olsen & Associates, Seefeldstrasse 233, 8008 Zürich, Switzerland.

Hols M. C. and De Vries C. G., 1991, *The limiting distribution of extremal exchange rate returns*, Journal of Applied Econometrics, **6**, 287–302.

Loretan M. and Phillips P. C. B., 1994, *Testing the covariance stationarity of heavy-tailed time series*, Journal of Empirical Finance, **1**(2), 211–248.

Mandelbrot B. B., 1963, *The variation of certain speculative prices*, Journal of Business, **36**, 394–419.

McFarland J. W., Petit R. R., and Sung S. K., 1982, *The distribution of foreign exchange price changes: trading day effects and risk measurement*, The Journal of Finance, **37**(3), 693–715.

Müller U. A., 1993, *Statistics of variables observed over overlapping intervals*, Internal document UAM.1993-06-18, Olsen & Associates, Seefeldstrasse 233, 8008 Zürich, Switzerland.

Müller U. A., Dacorogna M. M., Davé R. D., Olsen R. B., Pictet O. V., and von Weizsäcker J. E., 1995, *Volatilities of Different Time Resolutions – Analyzing the Dynamics of Market Components*, Journal of Empirical Finance, **4**, No. 2–3, 213–240.

Müller U. A., Dacorogna M. M., Olsen R. B., Pictet O. V., Schwarz M., and Morgenegg C., 1990, *Statistical study of foreign exchange rates, empirical evidence of a price change scaling law, and intraday analysis*, Journal of Banking and Finance, **14**, 1189–1208.

Taylor S. J., 1986, *Modelling Financial Time Series*, J. Wiley & Sons, Chichester.

Authors' address: Olsen & Associates, Seefeldstrasse 233, CH-8008 Zurich, Switzerland

fx rate	time interval	mean	variance	skewness	kurtosis
USD/DEM	30 minutes	$-1.40 \cdot 10^{-6}$	$7.53 \cdot 10^{-7}$	0.60	46.10
	6 hours	$-1.68 \cdot 10^{-5}$	$8.42 \cdot 10^{-6}$	0.27	11.75
	24 hours	$-6.62 \cdot 10^{-5}$	$3.48 \cdot 10^{-5}$	0.12	6.04
	1 week	$-4.65 \cdot 10^{-4}$	$2.41 \cdot 10^{-4}$	0.17	4.24
USD/JPY	30 minutes	$-2.20 \cdot 10^{-6}$	$7.16 \cdot 10^{-7}$	−0.05	24.02
	6 hours	$-2.65 \cdot 10^{-5}$	$7.98 \cdot 10^{-6}$	−0.17	11.64
	24 hours	$-1.06 \cdot 10^{-4}$	$3.13 \cdot 10^{-5}$	−0.16	7.06
	1 week	$-7.57 \cdot 10^{-4}$	$2.22 \cdot 10^{-4}$	−0.23	4.29
GBP/USD	30 minutes	$2.33 \cdot 10^{-7}$	$6.56 \cdot 10^{-7}$	−0.14	21.49
	6 hours	$2.79 \cdot 10^{-6}$	$7.77 \cdot 10^{-6}$	−0.42	10.89
	24 hours	$9.86 \cdot 10^{-6}$	$3.13 \cdot 10^{-5}$	−0.34	7.50
	1 week	$8.53 \cdot 10^{-5}$	$2.38 \cdot 10^{-4}$	−0.84	7.32
USD/CHF	30 minutes	$-1.53 \cdot 10^{-6}$	$9.20 \cdot 10^{-7}$	0.52	66.76
	6 hours	$-1.82 \cdot 10^{-5}$	$1.04 \cdot 10^{-5}$	0.12	10.09
	24 hours	$-7.17 \cdot 10^{-5}$	$4.24 \cdot 10^{-5}$	−0.04	5.68
	1 week	$-5.06 \cdot 10^{-4}$	$2.85 \cdot 10^{-4}$	0.07	3.90
USD/FRF	30 minutes	$-1.28 \cdot 10^{-6}$	$6.13 \cdot 10^{-7}$	0.27	28.01
	6 hours	$-1.54 \cdot 10^{-5}$	$7.59 \cdot 10^{-6}$	0.21	9.89
	24 hours	$-6.04 \cdot 10^{-5}$	$3.10 \cdot 10^{-5}$	0.09	5.92
	1 week	$-4.25 \cdot 10^{-4}$	$2.19 \cdot 10^{-4}$	0.19	4.32

Table 2: The four first moments of the return distribution at different time intervals for the major currencies against the USD. The estimations are perfomed on samples from January 1, 1987 to June 30, 1996.

ir rate	time interval	mean	variance	skewness	kurtosis
USD 3m	24 hours	$-1.27 \cdot 10^{-5}$	$2.41 \cdot 10^{-6}$	−0.16	24.72
	1 week	$-8.88 \cdot 10^{-5}$	$2.20 \cdot 10^{-5}$	−0.53	14.98
USD 6m	24 hours	$-1.04 \cdot 10^{-5}$	$1.98 \cdot 10^{-6}$	−0.20	20.49
	1 week	$-7.33 \cdot 10^{-5}$	$1.71 \cdot 10^{-5}$	−0.82	14.48
DEM 3m	24 hours	$-1.72 \cdot 10^{-7}$	$7.93 \cdot 10^{-7}$	0.39	28.68
	1 week	$-7.35 \cdot 10^{-7}$	$5.62 \cdot 10^{-6}$	0.22	18.80
DEM 6m	24 hours	$-4.76 \cdot 10^{-7}$	$7.80 \cdot 10^{-7}$	0.22	33.52
	1 week	$-3.99 \cdot 10^{-6}$	$5.35 \cdot 10^{-6}$	0.10	11.90
JPY 3m	24 hours	$6.06 \cdot 10^{-7}$	$1.28 \cdot 10^{-6}$	1.23	43.74
	1 week	$4.24 \cdot 10^{-6}$	$7.65 \cdot 10^{-6}$	2.80	36.97
JPY 6m	24 hours	$-2.47 \cdot 10^{-6}$	$9.94 \cdot 10^{-7}$	0.50	46.29
	1 week	$-1.73 \cdot 10^{-5}$	$6.22 \cdot 10^{-6}$	2.42	28.04
GBP 3m	24 hours	$-1.41 \cdot 10^{-5}$	$1.69 \cdot 10^{-6}$	0.44	47.91
	1 week	$-1.00 \cdot 10^{-4}$	$1.10 \cdot 10^{-5}$	0.77	9.31
GBP 6m	24 hours	$-1.52 \cdot 10^{-5}$	$1.74 \cdot 10^{-6}$	−0.09	17.13
	1 week	$-1.07 \cdot 10^{-4}$	$1.22 \cdot 10^{-5}$	0.02	6.70
CHF 3m	24 hours	$4.98 \cdot 10^{-6}$	$1.58 \cdot 10^{-6}$	−0.38	16.40
	1 week	$3.68 \cdot 10^{-5}$	$1.10 \cdot 10^{-5}$	−0.30	7.58
CHF 6m	24 hours	$5.34 \cdot 10^{-6}$	$1.29 \cdot 10^{-6}$	−0.29	17.07
	1 week	$3.84 \cdot 10^{-5}$	$8.27 \cdot 10^{-6}$	−0.45	8.78

Table 3: The four first moments of the return distribution at different time intervals for the short term cash interest rates for major currencies. The estimations are performed on samples going from January 2, 1979 to June 30, 1996.

rates	30m	1h	2h	6h	1d
USD-DEM	3.18 ±0.42	3.24 ±0.57	3.57 ±0.90	4.19 ±1.82	5.70 ±4.39
USD-JPY	3.19 ±0.48	3.65 ±0.79	3.80 ±1.08	4.40 ±2.13	4.42 ±2.98
GBP-USD	3.58 ±0.53	3.55 ±0.65	3.72 ±1.00	4.58 ±2.34	5.23 ±3.77
USD-CHF	3.46 ±0.49	3.67 ±0.77	3.70 ±1.09	4.13 ±1.77	5.65 ±4.21
USD-FRF	3.43 ±0.52	3.67 ±0.84	3.54 ±0.97	4.27 ±1.94	5.60 ±4.25
USD-ITL	3.36 ±0.45	3.08 ±0.49	3.27 ±0.79	3.57 ±1.35	4.18 ±2.44
USD-NLG	3.55 ±0.57	3.43 ±0.62	3.36 ±0.92	4.34 ±1.95	6.29 ±4.96
DEM-JPY	3.84 ±0.59	3.69 ±0.87	4.28 ±1.49	4.15 ±2.20	5.33 ±3.74
GBP-DEM	3.33 ±0.46	3.67 ±0.70	3.76 ±1.17	3.73 ±1.59	3.66 ±1.70
GBP-JPY	3.59 ±0.63	3.44 ±0.70	4.15 ±1.32	4.35 ±2.27	5.44 ±4.12
DEM-CHF	3.54 ±0.54	3.28 ±0.54	3.44 ±0.82	4.29 ±1.84	4.21 ±2.43
GBP-FRF	3.19 ±0.46	3.33 ±0.62	3.37 ±0.90	3.41 ±1.27	3.34 ±1.65
XAU-USD	4.47 ±1.15	3.96 ±1.13	4.36 ±1.82	4.13 ±2.22	4.40 ±2.98
XAG-USD	5.37 ±1.55	4.73 ±1.93	3.70 ±1.52	3.45 ±1.35	3.46 ±1.97

Table 4: Estimated tail exponent for the main FX rates, gold (XAU), and silver (XAG), using the bootstrap and jackknife methods. Here β is chosen to be 1 and $c = 0$. The time intervals are measured in ϑ-time.

curr.	maturity	1d	1w	maturity	1d	1w
USD	3m	4.03 ±2.99	3.53 ±3.46	6m	4.10 ±2.84	3.50 ±3.07
DEM	3m	2.54 ±0.73	2.88 ±1.63	6m	2.39 ±0.76	2.62 ±1.82
JPY	3m	3.16 ±2.07	3.43 ±3.01	6m	2.03 ±0.85	3.60 ±3.53
GBP	3m	2.61 ±0.84	3.86 ±3.78	6m	4.04 ±2.64	6.65 ±7.53
CHF	3m	3.69 ±2.41	5.24 ±5.13	6m	3.02 ±1.26	7.46 ±7.31

Table 5: Estimated tail exponent for 5 different currencies and two maturities, using the bootstrap and jackknife methods. Here β is chosen to be 1 and $c = 0$. The time intervals are measured in ϑ-time.

FX rates	daily returns		30m returns	
	short samples	6/78 – 6/96	6/88 – 6/96	short samples
USD-DEM	4.84	4.34 ±2.46	3.27 ±0.50	3.29
USD-JPY	7.81	5.69 ±3.94	3.86 ±0.71	3.94
GBP-USD	4.79	4.35 ±3.02	3.37 ±0.53	3.57
USD-CHF	5.24	4.15 ±2.71	3.63 ±0.55	3.61
USD-FRF	4.48	4.37 ±2.85	3.52 ±0.54	3.59
USD-ITL	3.82	3.97 ±1.94	3.38 ±0.44	3.56
USD-NLG	4.17	4.05 ±1.98	3.56 ±0.66	3.57
XAU-USD	3.65	3.88 ±2.53	4.24 ±0.99	4.00
XAG-USD	3.94	3.40 ±1.92	4.12 ±0.75	3.54

Table 6: Estimated tail exponent for the main FX rates, gold (XAU), and silver (XAG) on different samples for both daily and 30-minute returns. The time intervals are measured in ϑ-time.

	Probabilities p					
	1/1y	1/5y	1/10y	1/15y	1/20y	1/25y
Models:						
Normal	0.4%	0.5%	0.6%	0.6%	0.7%	0.7%
Student 3	0.5%	0.8%	1.0%	1.1%	1.2%	1.2%
GARCH(1,1)	1.5%	2.1%	2.4%	2.6%	2.7%	2.9%
HARCH	1.8%	2.9%	3.5%	4.0%	4.3%	4.6%
USD Rates:						
USD-DEM	1.7%	2.5%	3.0%	3.3%	3.5%	3.7%
USD-JPY	1.7%	2.4%	2.9%	3.2%	3.4%	3.6%
GBP-USD	1.6%	2.3%	2.6%	2.9%	3.1%	3.2%
USD-CHF	1.8%	2.7%	3.1%	3.5%	3.7%	4.0%
USD-FRF	1.6%	2.3%	2.8%	3.0%	3.3%	3.4%
USD-ITL	1.8%	2.8%	3.4%	3.8%	4.1%	4.4%
USD-NLG	1.7%	2.5%	2.9%	3.2%	3.4%	3.6%
Cross rates:						
DEM-JPY	1.3%	1.9%	2.2%	2.5%	2.6%	2.8%
GBP-DEM	1.1%	1.7%	2.1%	2.3%	2.5%	2.6%
GBP-JPY	1.6%	2.3%	2.7%	3.0%	3.2%	3.4%
DEM-CHF	0.7%	1.0%	1.2%	1.3%	1.4%	1.5%
GBP-FRF	1.1%	1.8%	2.2%	2.5%	2.7%	2.9%

Table 7: Extreme risks over 6 hours for model distributions produced by Monte-Carlo simulations of *synthetic* time series fitted to USD/DEM compared to empirical data studied through a tail estimation for FX rates.

Stable Paretian Modeling in Finance: Some Empirical and Theoretical Aspects[1]

Stefan Mittnik, Svetlozar T. Rachev and Marc S. Paolella

Abstract

The paper discusses consequences of the α–stable assumption in financial modeling. Aspects relevant for financial theory and empirical finance are considered. We also present empirical evidence favoring the α–stable hypothesis over the normal assumption as a model for unconditional, conditional homoskedastic and conditional heteroskedastic asset return distributions.

1. Introduction

The distributional form of returns on financial assets has important implications for theoretical and empirical analyses in economics and finance. For example, asset, portfolio and option pricing theories are typically based on distributional assumptions. In empirical tests, statistical inference concerning, for example, the efficient market hypothesis, the excess–volatility question, or the validity of option pricing models may be sensitive to the distributional assumptions for the returns of the underlying assets. The work of Mandelbrot [Man1], [Man2], [Man3], [Man4], and Fama [F1] has sparked considerable interest in studying the empirical distribution of returns on common stocks as well as other financial assets. While earlier theories had been based on the normal distribution, the excess kurtosis found in Mandelbrot's and Fama's investigations led them to reject the normal assumption and propose the *α–stable* distribution (also referred to as *stable Paretian* distribution) as a distributional model for asset returns. In subsequent years, a number of empirical investigations have lend support to this conjecture (see, for example, [MR2], [CMMR], [MRC], and the references therein).

In addition to providing a better empirical fit, the stable Paretian assumption has several desirable theoretical properties. Detailed accounts of properties of stably distributed random variables can be found in [ST] and [JW]. An important desirable property is the fact that stable Paretian distributions have *domains of attraction*. In general, any decision (inference) based on observed data is a continuous functional on the space of

[1]The research of the first author was supported by the Deutsche Forschungsgemeinschaft. It was conducted while the second author visited the Christian Albrechts University at Kiel with support from the Alexander–von–Humboldt Foundation for U.S. Scientists.

distributions that govern the data. In practice one cannot expect that observed data follow *exactly* the "ideal" distribution specified by the modeler. The distributional model represents only an approximation of the distribution underlying the observed data. This problem gives rise to the crucial question of what is the domain of applicability of the specified model. Could it happen that slight perturbations in the data give rise to substantially different distributional outcomes and, thus, different conclusions? Loosely speaking, any distribution in the domain of attraction of a specified stable distribution will have properties which are close to the ones of the stable distribution. Consequently, decisions will, in principle, not be affected by adopting an "idealizing" stable distribution as the distributional model instead of the true distribution. Moreover, it is possible to check whether or not a distribution is in the domain of attraction of a stable distribution by examining only the tails of the distribution, because only these parts specify the domain of attraction properties of the distribution. Thus, the continuity (stability) of the adopted model is valid for any distribution with the appropriate tail.

A second attractive aspect of the stable Paretian assumption is the *stability property*. This is desirable because it implies that each stable distribution has an "index of stability" (shape parameter), which remains the same regardless of the scale (sampling interval) adopted. The index of stability plays the role of the "compound" parameter that governs the main properties of the underlying distribution. When using stable distributions, we need not be concerned with the question of which of the parameters is most relevant. The index of stability can be regarded as an overall parameter, which can be employed for inference and decision making.

In [MR2] and [MR3] we provide an overview of the *alternative stable* probabilistic schemes considered in modeling returns of individual assets (see Table 1 for a summary). In Table 1, X_i stands for the return on an asset in period t_0+i. We assume that $X_1, X_2, \dots$ are i.i.d. real-valued r.v.'s. We write

$$X_1 \stackrel{d}{=} a_n(X_1 \circ X_2 \circ \cdots \circ X_n) + b_n, \tag{1.1}$$

where $\stackrel{d}{=}$ denotes equality in distribution, $\circ$ stands for summation, min, max, or multiplication, n is a deterministic or random integer, $a_n > 0$, and $b_n \in \mathbf{R}$. The standard nonrandom summation scheme, i.e., $\circ$ stands for $+$ and n is a deterministic integer, produces the stable Paretian distribution. The maximum and minimum schemes lead to extreme-value distributions. For the minimum scheme, the bull distribution is one of these extreme-value distributions. The multiplication scheme yields the multiplication-stable distribution of which the log-normal is the basic example.

In the geometric schemes, where n is random, we set $n = T(p)$, with $T(p)$ representing a geometrically distributed r.v. with parameter $p \in (0, 1)$ independent of X_i, i.e.,

$$P(T(p) = k) = (1-p)^{k-1}p, \qquad k = 1, 2, \dots . \tag{1.2}$$

Table 1: Stable Probabilistic Schemes

Scheme	Stability Property[a]
Summation	$X_1 \stackrel{d}{=} a_n(X_1 + \cdots + X_n) + b_n$
Maximum	$X_1 \stackrel{d}{=} a_n \max_{1 \le i \le n} X_i + b_n$
Minimum	$X_1 \stackrel{d}{=} a_n \min_{1 \le i \le n} X_i + b_n$
Multiplication	$X_1 \stackrel{d}{=} A_n(X_1 X_2 \cdots X_n)^{C_n}$
Geometric Summation	$X_1 \stackrel{d}{=} a(p)(X_1 + \cdots + X_{T(p)}) + b(p)$
Geometric Maximum	$X_1 \stackrel{d}{=} a(p) \max_{1 \le i \le T(p)} X_i + b(p)$
Geometric Minimum	$X_1 \stackrel{d}{=} a(p) \min_{1 \le i \le T(p)} X_i + b(p)$
Geometric Multiplication	$X_1 \stackrel{d}{=} A(p)(X_1 X_2 \cdots X_{T(p)})^{C(p)}$

[a]Notation "$X_1 \stackrel{d}{=}$" stands for "equality in distribution."

In (1.2), $T(p)$ could be interpreted as the moment at which the probabilistic structure governing the asset returns breaks down. The *breakdown* could, for example, be due to new information affecting future fundamentals of the underlying asset in a presently unknown manner. Thus, the stability properties of r.v.'s X_i are only preserved up to period $T(p)$, the moment of the breakdown.

Here, we focus on the non-Gaussian stable Paretian case. We discuss possible consequences of the stable Paretian assumption for financial modeling. We consider implications of both financial theory and empirical finance. Before doing so, we provide in Section 2 additional empirical evidence favoring the stable Paretian hypothesis over the normal assumption. In our empirical analyses we go beyond the usual empirical analyses in the literature, which focus on the unconditional distribution of asset returns, and examine conditional homoskedastic (i.e., constant–conditional–variance) and heteroskedastic (i.e., varying–conditional–variance) distributions. The *conditional* distributions are of interest, because asset returns typically exhibit temporal dependence. They are obtained by conditioning on past returns and, thus, allowing for certain dynamic structures. Section 3 discusses theoretical aspects of stable models in finance. Here, we briefly allude to the problems of risk assessment, option pricing, portfolio management, and asset pricing. Some implications for empirical modeling are addressed in Section 4. We restrict ourselves to the questions of sampling distributions based on stable variates, adaptations of some test statistics, and testing for random–walk–like behavior of univariate and multivariate time series.

2. Some Additional Empirical Evidence

Financial modeling frequently involves information on past market movements. Examples include investors employing chart analysis or technical analysis to derive investment decisions, or researchers assessing the efficiency of financial markets. In such cases, it is not the *unconditional* return distribution which is of interest, but the *conditional* distribution, which is conditioned on information contained in past return data, or a more general information set. The class of autoregressive moving average (ARMA) models is a natural candidate for conditioning on the past of a return series. These models have the property that the conditional distribution is homoskedastic. In view of the fact that financial markets frequently exhibit volatility clusters, the homoskedasticity assumption may be too restrictive. As a consequence, conditional heteroskedastic models, such as Engle's [En] autoregressive conditional heteroskedastic (ARCH) models and the generalization (GARCH) of Bollerslev [Bo], possibly in combination with an ARMA model, referred to as an ARMA–GARCH model, are now common in empirical finance. It turns out that ARCH-type models driven by normally distributed innovations imply unconditional distributions which themselves possess heavier tails. Thus, in this respect, ARCH models and α–stable distributions can be viewed as competing hypotheses.

Despite the desirable theoretical and empirical attractiveness of α–stable distributions, they have almost exclusively been used to model unconditional asset return distributions, and not their more relevant conditional counterparts. This has been mainly due to the practical difficulties in implementing the α–stable model. One of the foremost problems has been parameter estimation. Quantile estimators are commonly employed in empirical work. They are based on a small number of order statistics and, thus, discard a substantial amount of information contained in a sample.

In this section, we examine unconditional, homoskedastic conditional, and heteroskedastic conditional distributions of several asset return series and we compare how compatible the normal and non-normal α–stable hypotheses are with the data at hand. It should be noted that conditional stable distributions are not necessarily α–stable. For a discussion of conditional laws of multivariate stable distributions with dependence structure we refer to [ST], Ch. 4.

To obtain conditional models, we estimate both ARMA and ARMA–GARCH models under normal and α–stable assumptions, employing (approximate) conditional maximum-likelihood (ML) estimation. The ML estimation is *conditional*, in the sense that, when estimating, for example, an ARMA(p, q) model, we condition on the first p realizations of the sample, $r_p, r_{p-1}, \ldots, r_1$, and, since $\alpha > 1$ holds in our empirical applications, set innovations $\varepsilon_p, \varepsilon_{p-1}, \ldots, \varepsilon_{p-q+1}$ to their unconditional mean $\mathbf{E}(\varepsilon_t) = 0$.

The estimation of all α–stable models is *approximate* in the sense that the α–stable density function, $S_{\alpha,\beta}(\delta, c)$, is approximated via fast Fourier transformation (FFT) of the α–stable characteristic function (ch.f.),

$$\int_{-\infty}^{\infty} e^{itx} dH(x) = \begin{cases} \exp\{-c^{\alpha}|t|^{\alpha}[1 - i\beta \text{sign}(t) \tan \frac{\pi\alpha}{2}] + i\delta t\}, & \text{if } \alpha \neq 1, \\ \exp\{-c|t|[1 + i\beta \frac{2}{\pi} \text{sign}(t) \ln |t|] + i\delta t\}, & \text{if } \alpha = 1, \end{cases} \tag{2.3}$$

where H is the distribution function corresponding to $S_{\alpha,\beta}(\delta, c)$; α $(0 < \alpha \leq 2)$ is the *index of stability*; and β $(-1 \leq \beta \leq 1)$, δ $(\delta \in \mathbf{R})$ and c $(c > 0)$ are the *skewness*, *location* and *scale* parameters, respectively. Our ML estimation essentially follows that of [Du], but differs in that we numerically approximate the α–stable density by an FFT of the ch.f. rather than some series expansion. Considering, for example, the unconditional case, the ML estimate of $\theta = (\alpha, \beta, c, \delta)$ is obtained by maximizing the logarithm of the likelihood function

$$L(\theta) = \prod_{t=1}^{T} S_{\alpha,\beta}\left(\frac{r_t - \delta}{c}\right) c^{-1}.$$

As [Du] shows, the resulting estimates are consistent and asymptotically normal with the asymptotic covariance matrix of $T^{1/2}(\hat{\theta} - \theta_0)$ being given by the inverse of the Fisher information matrix. The standard errors of the estimates reported below are obtained by evaluating the Fisher information matrix at the ML point estimates. For details on our α–stable ML estimation we refer to [MRDC] and [MR4].

Here, we investigate three financial time series whose behavior is rather typical for financial return data:

(i) The daily AMEX Composite index from September 1, 1988 to July 28, 1994, with sample size $n = 1810$.

(ii) The daily AMEX OIL index from September 1, 1988 to July 28, 1994, with sample size $n = 1810$.

(iii) The daily DM/US$ exchange rate from January 2, 1973 to July 28, 1994, with $n = 5401$.

The levels and returns[2] of each series are shown in Figure 1.

Table 2 summarizes the basic statistical properties of the three return series. A negative skewness statistic indicates that the distribution is skewed to the left, i.e., compared to the right tail, the left tail is elongated. The kurtosis statistic reflects the peakedness of the center compared to that of the normal distribution, so that a value near three would be indicative of normality. Although formal tests could in principle be

[2]We use the standard convention and define the return r_t in period t by $r_t = (\ln P_t - \ln P_{t-1}) \times 100$, where P_t is the price of the asset at time t.

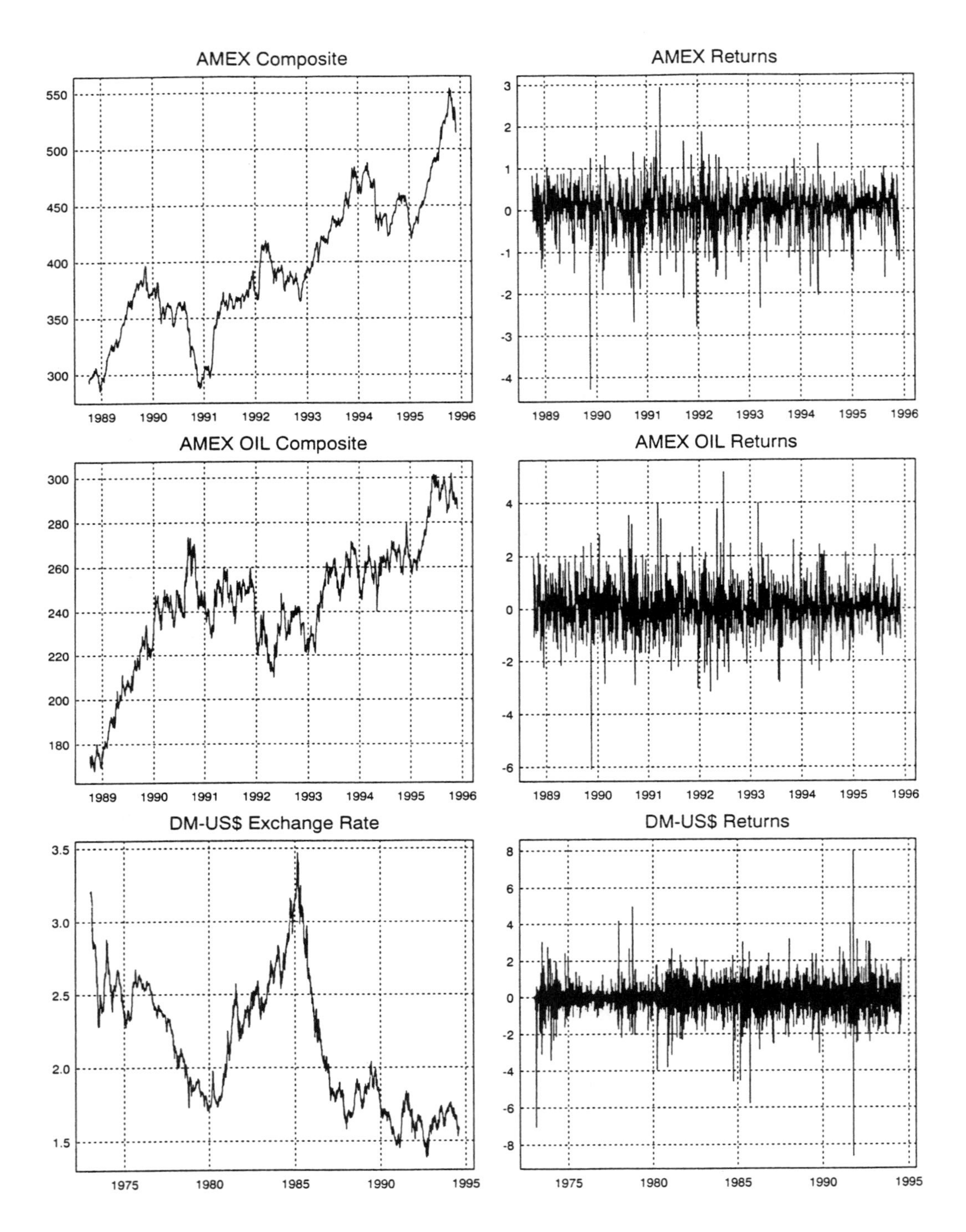

Figure 1: Levels and Returns of Time Series

Table 2: Statistical Properties of Returns

	Mean	Std. Dev.	Skewness	Kurtosis
AMEX	0.0311	0.5173	-0.8916	8.120
AMEX OIL	0.0283	0.8488	0.1369	6.637
DM/US$	-0.0132	0.7353	-0.3210	13.63

conducted, it should be kept in mind that under the Paretian stable hypothesis, second and higher moments do not exist, rendering such tests useless. The numbers in Table 2 are fairly typical for daily asset return data and indicate considerable deviation from normality. Formal tests confirm this as well.

We employ two criteria for comparing candidate distributions. The first is the maximum likelihood value obtained from the (conditional) ML estimation. The ML value may be viewed as an overall measure of goodness of fit and allows us to judge which candidate is more likely to have generated the data. From a Bayesian viewpoint, given large samples and assuming equal prior probabilities for two candidate distributions, the ratio of the maximum log likelihood values of two competing models represents the asymptotic posterior odds ratio of one candidate relative to the other (see [Ze] and [BG]). The second criterion is the Kolmogorov distance (KD)

$$\rho = \sup_{x \in \mathbf{R}} |F_s(x) - \hat{F}(x)|,$$

where $F_s(x)$ denotes the empirical sample distribution and $\hat{F}(x)$ is the estimated distribution function. It is a robust measure in the sense that it focuses only on the maximum deviation between the sample and fitted distributions. Note that the ML value of nested models will necessarily increase as they become more general. This may not be the case with the KD value.

2.1 Unconditional Distribution

ML estimation of the three unconditional return distributions led to the estimates given in Table 3, where the parameter labels correspond to ch.f. (2.3). Note that the scale for the normal case is given by the standard deviation and not by parameter c in (2.3), so that the entries cannot be compared directly. For each asset, the estimated shape parameter of the $\alpha-$ stable distribution is well below $\alpha = 2$, the value for the normal. The standard deviations of $\hat{\alpha}$ suggest that the normal hypothesis is inappropriate for all three assets. This is supported by the goodness-of-fit measures reported in Table 4. Both the maximum log–likelihood values and the Kolmogorov distances clearly indicate that, in all three cases, the α–stable distribution dominates that of the normal.

Table 3: Estimates of Unconditional Distributions[a]

	Parameters			
	Index	Location	Scale	Skewness
AMEX:				
Normal	2	0.0311	0.5173	0
	—	(0.0121)	(0.0081)	—
α-stable	1.728	0.0171	0.2996	-0.5860
	(0.0386)	(0.0142)	(0.0059)	(0.0898)
AMEX OIL:				
Normal	2	0.0283	0.8488	0
	—	(0.0200)	(0.0122)	—
α-stable	1.776	0.0345	0.5132	0.1873
	(0.0358)	(0.0214)	(0.0123)	(0.1241)
DM/US$:				
Normal	2	-0.0132	0.7353	0
	—	(0.0136)	(0.0070)	—
α-stable	1.727	-0.0100	0.4189	-0.0127
	(0.0232)	(0.0087)	0.0061)	(0.0149)

[a] Standard deviations are given in parentheses.

To provide a visual impression, Figure 2 shows the empirical densities obtained via kernel density estimation (dashed lines), along with fitted normal distributions (left panels) and fitted α–stable distributions (right panels). The graphs are consistent with the goodness-of-fit measures reported in Table 4. For all three cases, the α–stable model provides a much closer approximation to the empirical densities.

2.2 Conditional Homoskedastic Distribution

The unconditional results in the previous section ignore possible temporal dependencies in the return series. Traditionally, serial dependence in time series has been modeled with an ARMA structure. Such models allow conditioning of the process mean on past realizations, and have been proven successful for the short-term prediction of time series.

A peculiar feature of ARMA models is that the conditional (prediction error) variances are, in fact, independent of past realizations; i.e., they are conditionally ho-

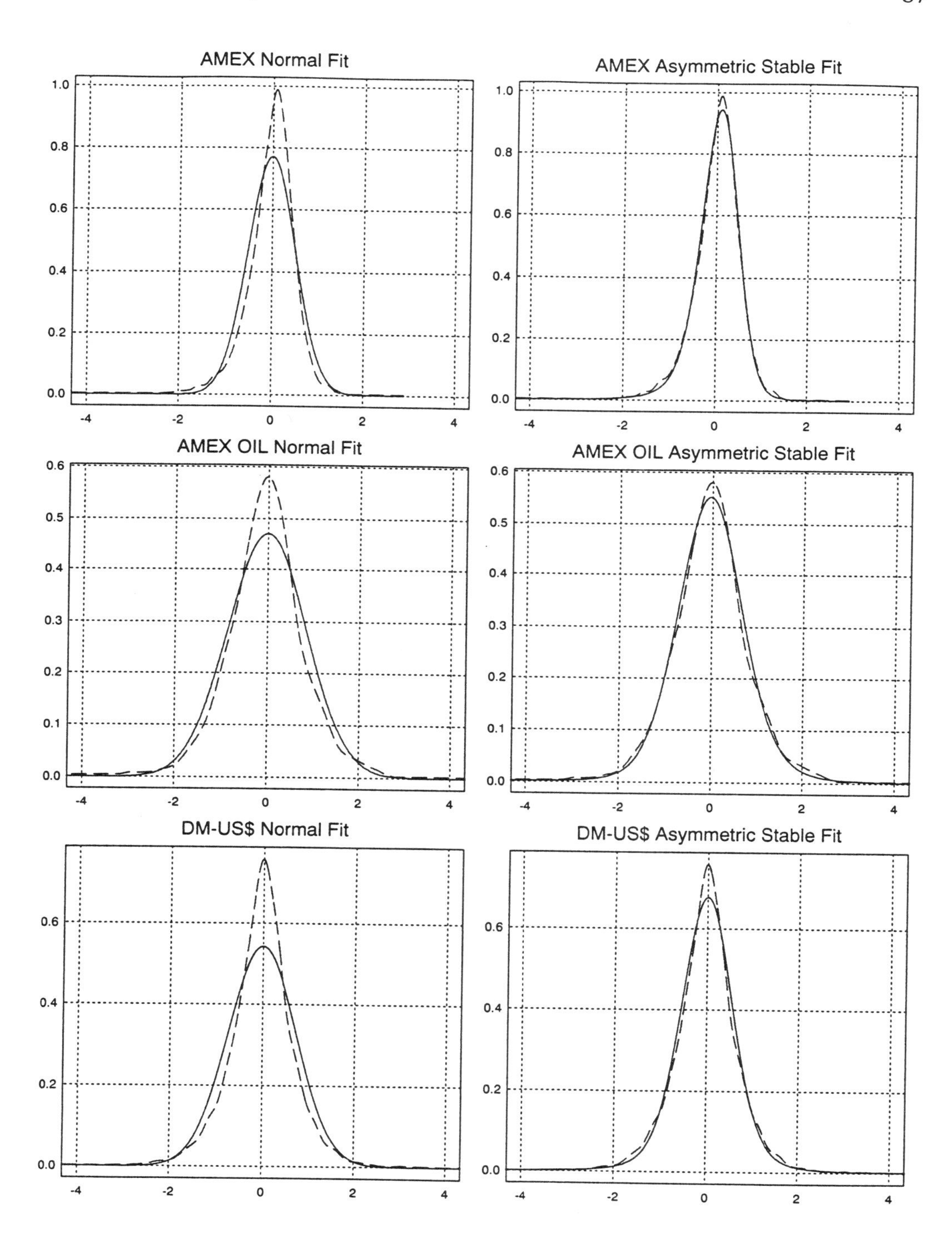

Figure 2: Fitted Unconditional Normal and Stable Densities
— fitted – – empirical

Table 4: Goodness of Fit of Unconditional Distributions

	log-lik		KD	
	Normal	Stable	Normal	Stable
AMEX	-1374	-1255	7.07	1.93
AMEX OIL	-2270	-2194	5.57	1.85
DM/US$	-6001	-5597	5.86	2.07

moskedastic. The assumption of conditional homoskedasticity is commonly violated in financial data, where we typically observe volatility clusters, implying that a large absolute return is often followed by more large absolute returns, which more or less slowly decay. Such behavior can be captured by ARCH or GARCH models (see [En] and [Bo]). They express the conditional variance as an explicit function of past information and permit conditional heteroskedasticity.

In this subsection we consider the homoskedastic case. The conditional heteroskedastic distribution is investigated in the next subsection.

An ARMA model of autoregressive order p and moving average order q is of the form

$$r_t = \mu + \sum_{i=1}^{p} a_i r_{t-i} + \varepsilon_t + \sum_{j=1}^{q} b_j \varepsilon_{t-j}, \tag{2.4}$$

where $\{\varepsilon_t\}$ is a white noise process.

To specify the orders p and q in (2.4), we followed standard Box–Jenkins identification techniques (see, for example, [BJ], [Wei], [BD], and [Jo]) and inspected sample autocorrelation functions (SACFs) and sample partial autocorrelation functions (SPACFs) of the return series, as shown in Figure 3. The exponentially decaying SACF and the single large spike in the SPACF corresponding to the AMEX series strongly suggest the appropriateness of an AR(1) structure. The correlograms for the other two series are not as easy to interpret. Both the SACF and SPACF for the AMEX Oil returns exhibit two relatively large spikes at lags 1 and 4, possibly suggesting either a subset AR(4) or MA(4), with the second and third coefficients restricted to zero. We opted for the subset AR(4), as both under the normal and stable Paretian assumption on the innovation process, it resulted in a higher likelihood value. The correlograms for the DM-US$ return series exhibit a less clear pattern, with a relatively large negative first order component, along with small, though persistant, positive correlation occuring for the next few lag times. Within the linear ARMA class of models, it appears that a parsimonious parameterization is restricted to either an AR(1) or an MA(1) model. As with the AMEX Oil series, a better fit was achieved using an AR(1) model.

It should be noted that we used Gaussian asymptotic confidence intervals to determine the significance of SACF and SPACF spikes. Recall that, although the SACF provides a consistent estimate of the theoretical ACF in the presence of α-stable data, it is not asymptotically normally distributed (see [DR]). Brockwell and Davis ([BD], pp. 540-2) provide examples, however, showing that with Cauchy data (i.e., symmetric 1-stable), the usual variance bounds obtained from Bartlett's formula are close to the true ones, the latter being somewhat smaller, so that standard identification rules might still be approximately applicable.

Arguably more disturbing is the fact that the presence of ARCH invalidates the standard asymptotic theory of the SACF, so that inference from Figure 3 is potentially misleading (see [Di] and [BHL]). For two of the series however, namely AMEX and AMEX OIL, the SACFs and PACFs depicted in Figure 3 were qualitatively the same as those obtained by using the GARCH filtered residuals, as detailed in the next subsection. Those for the DM-US$ did in fact differ somewhat, although not decisively so.[3]

These findings, together with the general guidelines of [Jo] (see p. 381), cautioning the modeler not to be overly "mechanical" when applying Box–Jenkins–type identification techniques, led us to the following specifications for the conditional homoskedastic models:

(i) AMEX: AR(1);

(ii) AMEX OIL: Subset AR(4) with $a_2 \equiv a_3 \equiv 0$;

(iii) DM/US$: AR(1).

Conditional ML estimation was used to estimate both the normal and stable ARMA(p, q) models, whereby we took the first p values of the return series to be fixed. The parameter estimates of the fitted conditional distributions are reported in Table 5. Comparing the results to those of the fitted unconditional distributions, we see little differences in the parameters common to both, with two exceptions, both of which pertain to the AMEX data. Firstly, the location parameter μ for the conditional model has changed considerably, which is not surprising in light of the relatively strong AR(1) component. Secondly, the skewness parameter for the α-stable parameterization has also changed quite a bit. One should keep in mind, however, that as α increases towards 2, the effect of the skewness parameter diminishes. For α near 1.8, even a skewness component of -0.8 is very mild, so that the large change is somewhat illusory.

Table 6 reports the goodness of fit of the estimated ARMA–based conditional distributions. For the AMEX and AMEX OIL series, we observe a significant increase in

[3]For the DM-US$ return series with the GARCH component removed, there was less evidence of an AR(1) component. This is also reflected in the estimates in Table 7, showing that a_1 is insignificant. It was decided to keep this term, because the likelihood ratio test is, in fact, significant, and the remaining parameters change insignificantly when it is omitted.

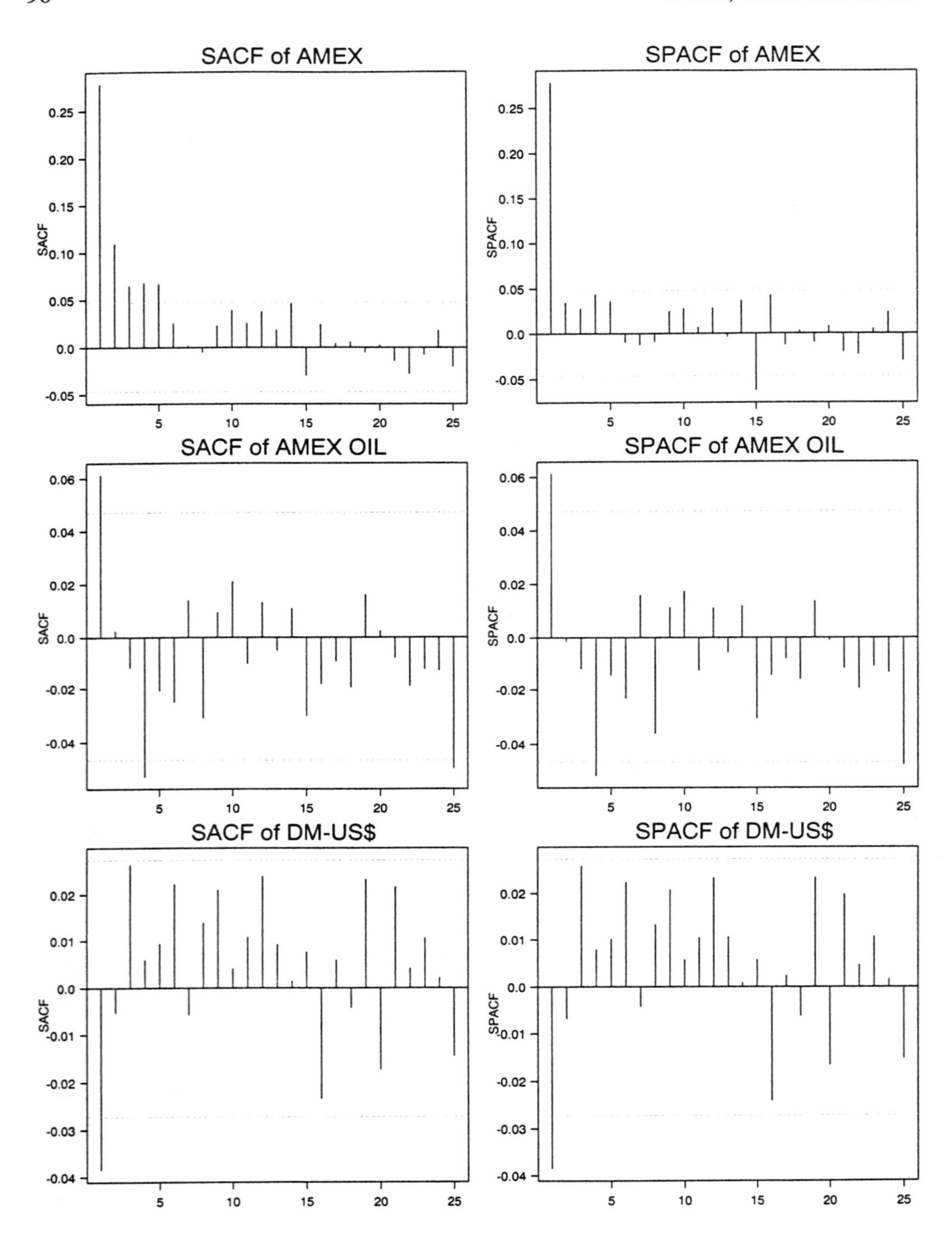

Figure 3: Sample Correlation Functions of Returns

Table 5: Estimates of Conditional Homoskedastic Distributions[a]

	ARMA Parameters[b]		Distribution Parameters		
	μ	a_1 a_4	Index	Scale	Skewness
AMEX:					
Normal	0.0219 (0.0111)	0.2781 (0.0224)	2 —	0.4966 (0.0078)	0 —
α-stable	0.0045 (0.0121)	0.2490 (0.0210)	1.796 (0.0287)	0.2970 (0.0058)	-0.8146 (0.0018)
AMEX OIL:					
Normal	0.0283 (0.0198)	0.0607 (0.0234) -0.0527 (0.0230)	2 —	0.8448 (0.0130)	0 —
α-stable	0.0346 (0.0211)	0.0543 (0.0232) -0.0495 (0.0363)	1.778 (0.0363)	0.5110 (0.0119)	0.1837 (0.1300)
DM-US$:					
Normal	-0.0137 (0.0100)	-0.0385 (0.0134)	2 —	0.7347 (0.0070)	0 —
α-stable	-0.0103 (0.0087)	-0.0157 (0.0128)	1.727 (0.0229)	0.4188 (0.0060)	-0.0138 (0.0160)

[a]Standard deviations are given in parentheses.

[b]The fourth–order AR parameter applies only to AMEX-OIL returns.

the likelihood, for both the normal and Paretian stable models. For the DM-US$ series, the additional AR(1) component adds very little, particularly in the stable case. With the exception of the normal case for the Amex series, the KD values remain virtually unchanged. Analogous to the results for the unconditional distributions, both the maximum log–likelihood values and the KDs again indicate the dominance of the α–stable distributions over the normal.

2.3 Conditional Heteroskedastic Distribution

An ARCH or GARCH model extends the mean equation, here (2.4), by assuming that

$$\varepsilon_t = c_t u_t, \tag{2.5}$$

Table 6: Goodness of Fit of Conditional Homoskedastic Distributions

	log-lik		KD	
	Normal	Stable	Normal	Stable
AMEX	-1300	-1187	5.99	1.79
AMEX OIL	-2257	-2181	5.12	1.91
DM-US$	-5997	-5596	5.84	2.15

where, in the normal case, $u_t \sim N(0, 1)$ and

$$c_t^2 = \omega + \sum_{i=1}^{r} \alpha_i \varepsilon_{t-i}^2 + \sum_{i=1}^{s} \beta_i c_{t-i}^2, \qquad c_t > 0. \tag{2.6}$$

[PMR] and [MRP] proposed the stable GARCH model and derived necessary and sufficient conditions for stationarity. The model takes the form

$$c_t = \omega + \sum_{i=1}^{r} \alpha_i |\varepsilon_{t-i}| + \sum_{i=1}^{s} \beta_i c_{t-i}, \qquad c_t > 0, \tag{2.7}$$

where u_t is α–stable with $\alpha > 1$.

A standard approach to detecting GARCH-dependencies in a time series, y_t, is to compute the SACF of the squared series, y_t^2. Figure 4 shows the SACF and PACF of the squared returns. For the AMEX Composite and the DM-US$ exchange rate, standard Box-Jenkins methodology would suggest the need for a mixed model, i.e., one with r and s both greater than zero. As is common in financial GARCH modeling (see, for example, [BCK]), it was found that $r = s = 1$ was adequate in capturing the correlation structure for these two squared series. The AMEX OIL squared returns exhibit an ARCH(1) structure.

For each series, the conditional mean was modeled with the same ARMA parameterization as used in the previous section. To summarize, we specified the following conditional heteroskedastic models for the three return series:

(i) AMEX: AR(1)-GARCH(1,1);

(ii) AMEX OIL: Subset AR(4)-GARCH(1,0)

(iii) DM/US$: AR(1)-GARCH(1,1).

The parameters in the ARMA and GARCH equations of a model are jointly estimated via conditional ML, where we assumed that the scaled innovations, $u_t = \varepsilon_t / c_t$, are either i.i.d. normal or i.i.d. α–stable and c_t satisfies GARCH recursions (2.6) or (2.7), respectively. The parameter estimates are reported in Table 7. We observe that the

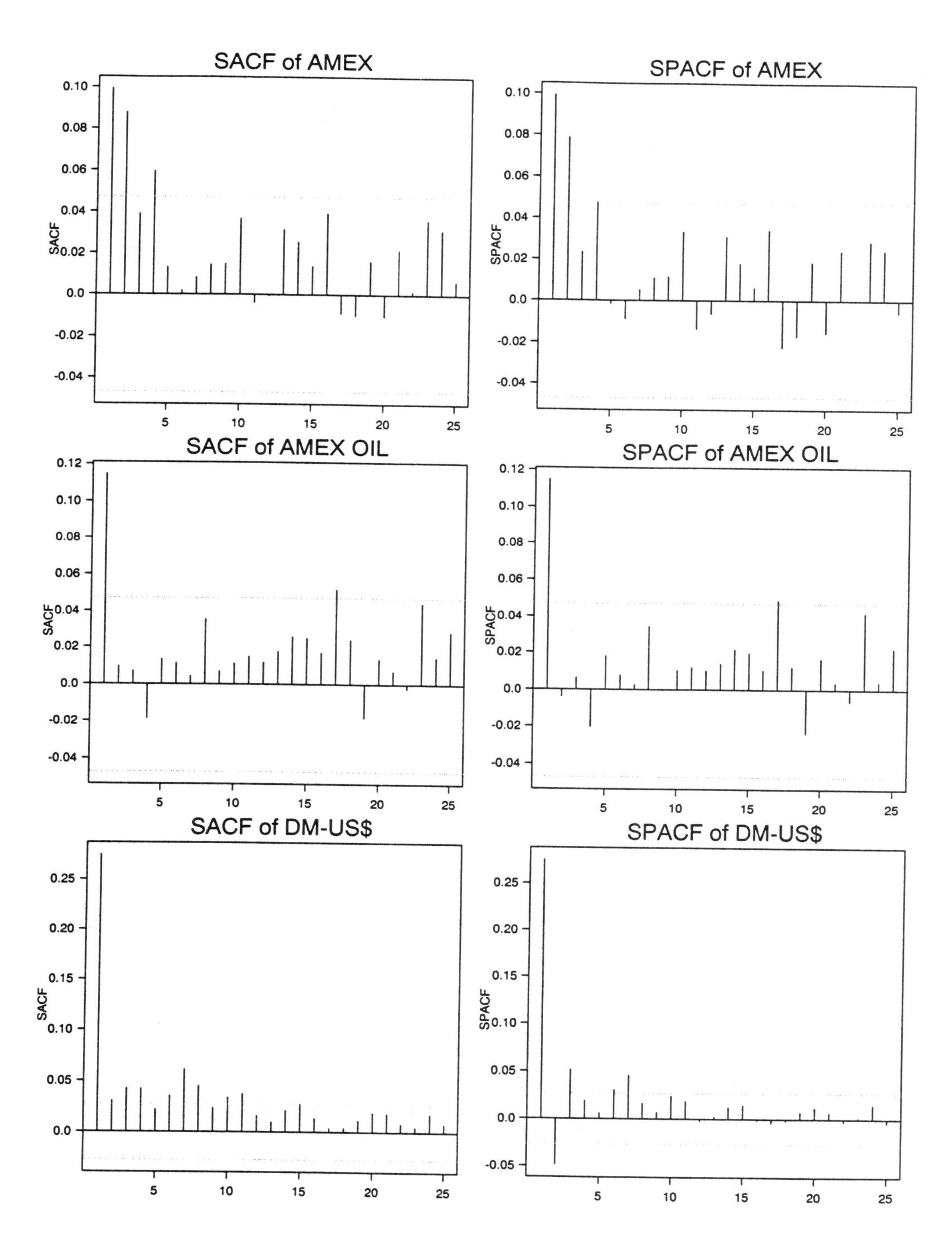

Figure 4: Sample Correlation Functions of Squared Returns

Table 7: Estimates of Conditional Heteroskedastic Distributions[a]

	ARMA Parameters[b]		GARCH Parameters			Distribution Parameters	
	μ	a_1 a_4	ω	α_1	β_1	Index	Skewness
AMEX:							
Normal	0.0326	0.2904	0.0856	0.1484	0.5079	2	0
	(0.0115)	(0.0261)	(0.0202)	(0.0312)	(0.0992)	—	—
α-stable	0.0279	0.2545	0.0403	0.0844	0.7664	1.852	-0.815
	(0.0122)	(0.0170)	(0.0055)	(0.0100)	(0.0225)	(0.028)	(0.145)
AMEX OIL:							
Normal	0.0249	0.0729	0.6602	0.0702	0	2	0
	(0.0217)	(0.0255)	(0.0252)	(0.0237)	—	—	—
		-0.0496					
		(0.0230)					
α-stable	0.0305	0.0532	0.4825	0.0484	0	1.782	0.1432
	(0.0231)	(0.0237)	(0.0120)	(0.0102)	—	(0.015)	(0.0200)
		-0.0475					
		(0.0243)					
DM/US$:							
Normal	-0.0044	0	0.0078	0.0944	0.9052	2	0
	(0.0004)	—	(0.0003)	(0.0023)	(0.0011)	—	—
α-stable	-0.0139	0	0.0073	0.0860	0.8816	1.827	-0.0510
	(0.0069)	—	(0.0017)	(0.0062)	(0.0093)	(0.019)	(0.0138)

[a]Standard deviations are given in parentheses.
[b]The fourth–order AR parameter applies only to AMEX-OIL returns.

estimates of stable index α, which correspond now to the scaled innovations, u_t, are larger than those for the unconditional distributions in Tables 3 and 5. This is what one expects, as ARCH/GARCH components absorb a portion of the excess kurtosis of the unconditional distribution.

Table 8 reports the goodness of fit of the estimated ARMA–GARCH–based conditional heteroskedastic distributions. Allowing for GARCH components clearly improves the fits of all models in terms of the log–likelihood values. The KD values decrease for all three series using the normal assumption, but increase slightly for two of the three under the stable assumption.

Figure 5 shows the SACFs corresponding to both the ARMA–GARCH residuals themselves (left panels) and their squares (right panels).[4] One sees that the parsimoniously parameterized ARMA–GARCH models are capable of extracting the majority of the outstanding serial correlation exhibited by both the mean and variance of the asset

[4]The associated SPACFs were qualitatively similar and are not shown.

Table 8: Goodness of Fit of Conditional Heteroskedastic Distributions

	log-lik		KD	
	Normal	Stable	Normal	Stable
AMEX	-1257	-1149	4.53	2.77
AMEX OIL	-2245	-2178	4.86	2.03
DM/US$	-5476	-5183	4.81	1.29

returns.

Similar to Figure 2, Figure 6 shows the kernel and fitted densities of the residuals corresponding to both the normal and α-stable conditional heteroskedastic models. As with the previous fits, the results demonstrate again the clear dominance of the α–stable distribution over the normal, even after having removed the GARCH component. Particularly for the DM-US$ exchange rate series, however, we see that for both the normal and α–stable cases, the conditional heteroskedastic model significantly outperforms its unconditional counterpart. Thus, a combination of fat-tailed innovations and a GARCH structure appears necessary to successfully account for the excess kurtosis in some series. This agrees with the findings in [Bo], who proposed the use of the Student's t distribution in conjunction with GARCH models.

3. Stable Models in Finance

We now address theoretical aspects of stable models in finance. Here, we can only allude to a small selection of topics, namely risk and portfolio management, option valuation, and asset pricing. More detailed expositions of these and further topics on stable finance can be found in [MR4] and references therein.

3.1 Risk Assessment, Option Pricing and Portfolio Management

Univariate models are used to capture the marginal distribution of an *individual asset* or index. Univariate stable fits of asset or index returns are reported in [MR2], [MRC] and [CMMR] and references therein. These fits show that the normal assumption is inappropriate for returns computed on a weekly or daily basis or for even higher frequencies.

From these estimated stable distributions, one can obtain more realistic *risk assessments*. Especially, the question of how frequently one can expect the occurrence of large price movements or stock market crashes can be addressed more realistically with the stable Paretian than with the normal assumption (see, for example, [GR4],[HV], [JV]).

A second issue which typically involves univariate models is the problem of *option valuation*. The question of pricing options under stable assumptions is rather complex

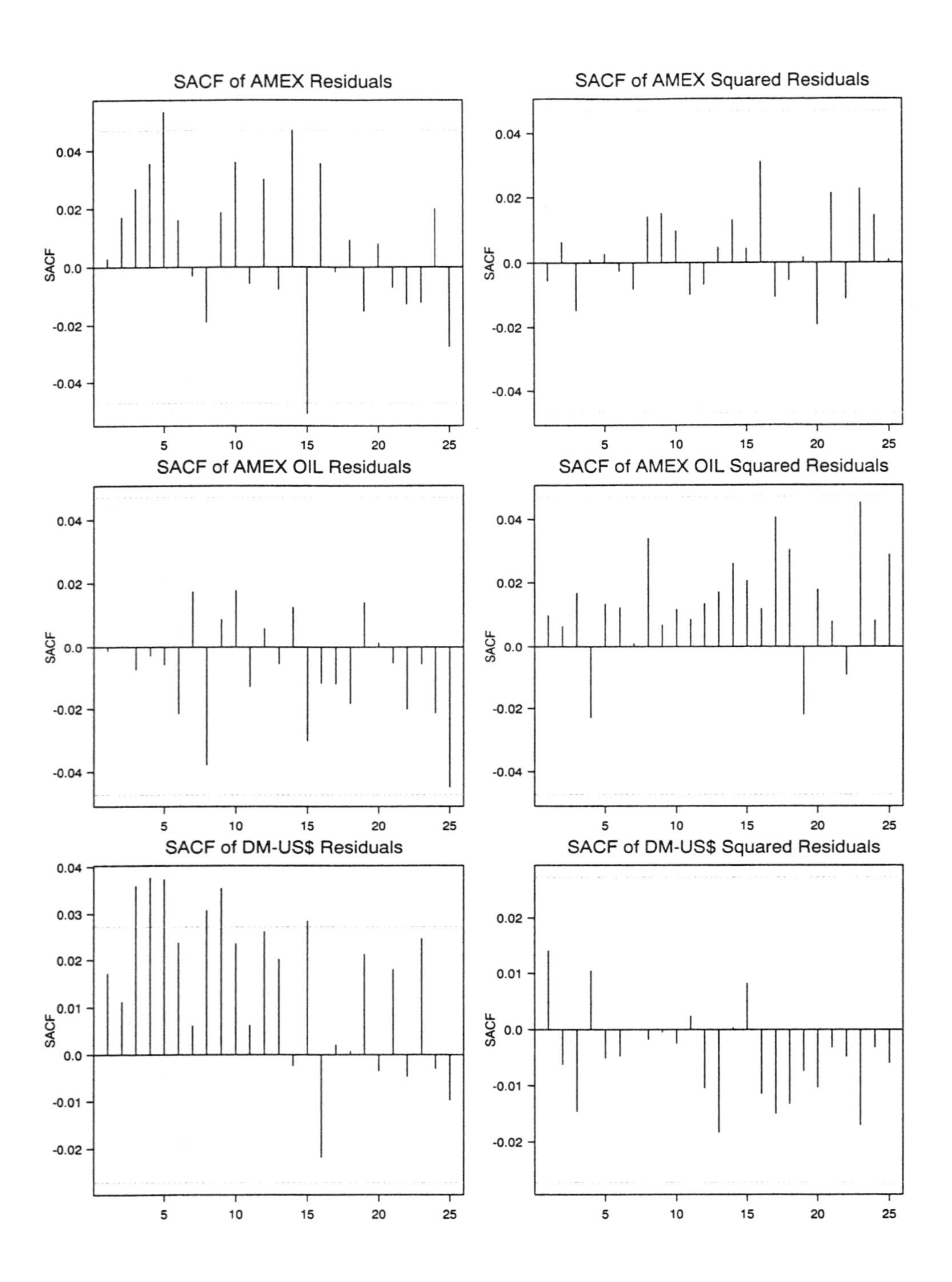

Figure 5: SACFs of Conditional Heteroskedastic Residuals

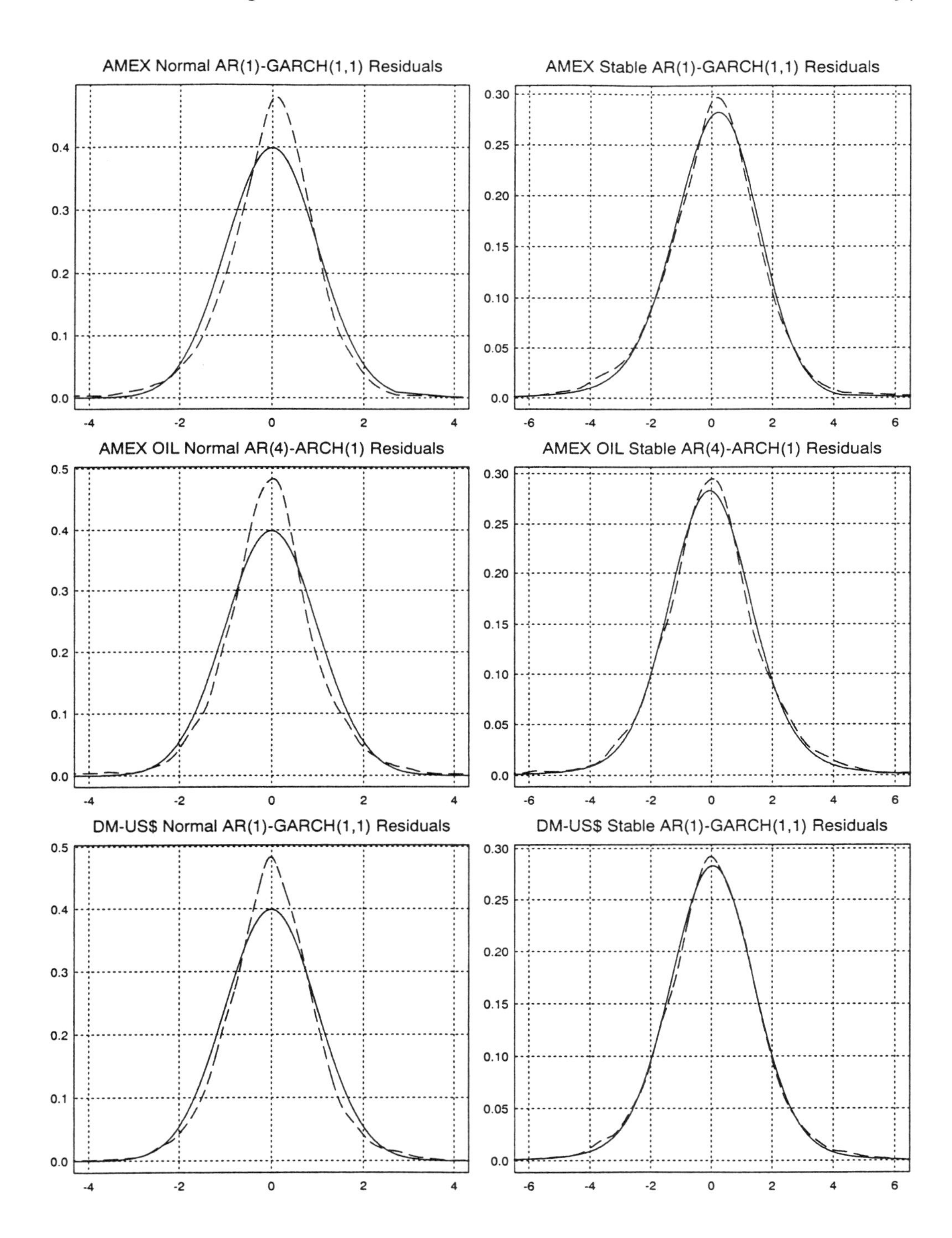

Figure 6: Fitted Conditional Heteroskedastic Normal and Stable Densities
— fitted – – empirical

and has been addressed in [KaR], [JW] and [HPR2]. For results on the *generalized binomial model* and the *generalized Mandelbrot–Taylor model* (cf. [MT]) we refer to [RS] and [KaR]. Option pricing for infinitely divisible returns, which encompasses the stable hypothesis as a particular case, is treated in [RR]; for details we refer to [MR4].

The crucial feature of the Mandelbrot–Taylor model is that the return process $(W(T))_{T\geq 0}$, is measured in relation to the transaction volume and not physical or calendar time. Transaction volume is assumed to follow Brownian motion with zero drift and variance v^2. The cumulative volume $(T(t))_{t\geq 0}$, i.e., the number of transactions up to calendar time t, is assumed to follow a positive, increasing in time $\frac{\alpha}{2}$-stable process, $1 < \alpha < 2$, with scale parameter $\nu > 0$. The subordinated process $Z(t) = W(T(t))$, representing the return process with respect to calendar time, is then an α-stable Lévy motion with ch.f.

$$\mathbf{E}e^{i\theta Z(t)} = e^{-t|\sigma\theta|^{\alpha}}, \tag{3.8}$$

where $\sigma^{\alpha} = \nu(v^2/2)^{\alpha/2}/\cos(\frac{\pi\alpha}{4})$.

A discrete version of the Mandelbrot–Taylor model and various alternative approaches to derivative pricing are considered in [RS], [KaR], [HPR1], [HPR2], and [MR4].

While much work has been devoted to investigating the univariate distribution of individual asset return series, *multivariate distributions* of sets of assets have rarely been studied. Some results in this area can be found in [Pr1], [Pr2] (Chapter 12), [Zi], [MR1]; see also the discussion in [MR3]). The lack of multivariate modeling efforts is somewhat surprising, given that modern portfolio management and asset–pricing theories involve distributional properties of sets of investment opportunities. In portfolio theory, for example, the Markowitz model assumes that returns of alternative investments have a joint multivariate distribution whose relevant properties are described by its mean vector and covariance matrix. In fact, the optimal composition of an investor's portfolio depends crucially on the covariances between the individual returns. The evaluation of the risk of a *portfolio of assets* with α-stable distributed returns involves multivariate models. This issue has been addressed in [RX], [ChR] and [GR3]. The problem of how to determine the dependence structure in a portfolio of stable distributed assets corresponds to that of deriving the covariance structure in a multivariate normal portfolio. This issue has been considered in [LRS1] and [LRS2].

3.2 Asset–Pricing Under Stable Paretian Laws

The implications of having a portfolio with multivariate stable Paretian returns have been investigated in [GR1], [GR2] and [GR3]. Here, we focus on the Capital Asset Pricing Model (CAPM)—in particular the issue of computing the so–called beta coefficient—and the Arbitrage Pricing Theory (APT). We also address the question of testing the

CAPM and APT.

The *Capital Asset Pricing Model (CAPM)* was introduced by [Sh] and [Li]. It states that, given certain market assumptions, the mean return of asset i is given by

$$\mathbf{E}(R_i) = \rho_0 + \beta_{im}(\mathbf{E}(R_m) - \rho_0), \tag{3.9}$$

where ρ_0 represents the return of the riskless asset; R_m is the random return of the market portfolio (i.e., the portfolio of all marketed assets) and β_{im}—known as the "beta" of asset i—is $\mathrm{cov}(R_i; R_m)/\mathrm{var}(R_m)$.

The CAPM was the first attempt to explain the asset return behavior (with one factor) and has experienced considerable theoretical developments in the last thirty years. [Mer] added a temporal dimension to CAPM by modeling asset returns by a diffusion process, and [Ch] showed that the hypothesis of normality could be replaced with the weaker one of finite variance. But neither the CAPM nor its extensions seemed satisfactory when tested empirically (see, for example, [AM] and [BF]). All these papers assume square integrability or, more strongly, normality of asset returns. If this is not the case, the statistical test may suffer from inconsistency.

The stable Paretian law was initially a widely accepted alternative to normality. [F2] established a CAPM for symmetric stable Paretian[5] returns. It is of the form

$$R_i = \rho_i + b_i\delta + \epsilon_i \tag{3.10}$$

where δ and ϵ_i are independent and symmetric α-stable Paretian. Fama showed that in this case, the "beta" coefficient in (3.9) is given by

$$\beta_{im} = \frac{1}{\sigma(R_m)} \frac{\partial\sigma(R_m)}{\partial(\lambda_{im})}, \tag{3.11}$$

where $R_m = \Sigma_i \lambda_{im} R_i$, with $\Sigma_i \lambda_{im} = 1$, represents the return of the market portfolio, and $\sigma(\cdot)$ is the scale parameter of the return under consideration. [Ros2] also claimed that a CAPM-like formula would still hold for stable-distributed returns when restriction (3.10) for independence of δ and ϵ_i is not satisfied. However, he does not provide an expression for beta analogous to (3.11) for this more general case. In fact, with stable Paretian returns, it is not possible to compute beta in this fashion or to conduct straightforward tests of the CAPM.

In [GR1] and [GR2] a testable version for β_{im} is provided in terms of the "covariation" $[R_i, R_m]_\alpha$ between R_i and R_m and the "variation" of R_m, $[R_m, R_m]_\alpha$.[6] Then, analogous to the normal case, one obtains

$$\beta_{im} = \frac{[R_i, R_m]_\alpha}{[R_m, R_m]_\alpha}.$$

[5] As in [F2], we assume $1 < \alpha < 2$ from now on. As we have pointed out, $\alpha \leq 1$ does not seem to be encountered in financial return data.

[6] For a systematic treatment of variation and covariation we refer to [ST].

For details, see [GR3] and [MR4].

In response to the CAPM's empirical failures, [Ros1] suggested a linear multi-factor pricing model, the so-called *Arbitrage Pricing Theory (APT)*. The APT implies that if the return of asset i, R_i, is of the form $\mathbf{E}(R_i) + \beta_{i1}\delta_1 + \cdots + \beta_{ik}\delta_k + \epsilon_i$ (here the δ_j's are the *factors* and the ϵ_i's are the *idiosyncratic risks*), then, under usual assumptions, the mean return of the i-th asset is

$$\mathbf{E}(R_i) = \rho_0 + \beta_{il}\rho_1 + \cdots + \beta_{ik}\rho_k, \tag{3.12}$$

where ρ_j is the *risk premium* linked to factor δ_j. The idea of the APT is that the mean return is not tied to its total variance, but only to that portion of the variance that is due to the market, because the idiosyncratic part can be diversified.

The Arbitrage Pricing Theory (APT) can be viewed as a multi-index generalization of the CAPM. [Co] treats the APT for situations where asset returns are defined in normed vector spaces. Without going into details, it should be noted that there are two versions of the APT for α-stable distributed returns, a so-called equilibrium and an asymptotic version. [Co] provides the proof for the former, while [GR1], [GR2] and [GR3] for the latter.

We briefly summarize the APT. It assumes that the returns of all assets are due to a finite number of common sources or risk factors, represented by random variables $\delta_1, \cdots, \delta_k$. The return of asset i, R_i, is assumed to be linearly related to the factors, namely

$$R_i = \mathbf{E}(R_i) + \beta_{i1}\delta_1 + \beta_{i2}\delta_2 + \cdots + \beta_{ik}\delta_k + \epsilon_i, \tag{3.13}$$

where ϵ_i represents asset-specific risk. The ϵ_i's are assumed to be mutually independent and also independent of factors $\delta_1, \cdots, \delta_k$. Given some additional assumptions, the APT shows that each risk, δ_j, has premium ρ_j associated with it, determining the expected return of asset i by

$$\mathbf{E}(R_i) = \rho_0 + \beta_{i1}\rho_1 + \beta_{i2}\rho_2 + \cdots + \beta_{ik}\rho_k. \tag{3.14}$$

The difficulty in testing the APT empirically is the determination of the risk factors. In the Gaussian case, a maximum likelihood based approach to multi–factor analysis can be adopted. Unfortunately, this approach cannot be generalized to stable Paretian laws. [GR1] and [GR3] developed a method for this case.

4. Empirical Modeling Under Stable Paretian Laws

We now turn to practical aspects of modeling under the stable Paretian assumption—in particular, implications for inference when disturbances are stable Paretian. Specifically, we address the problems of testing for structural breaks, unit roots and cointegration in time series. For a more detailed overview, see [RMK1].

4.1 Sampling Distributions Based on Paretian Random Variates

The distribution of sums of squared r.v.'s with heavy-tailed distributions is studies in [MKR1]. Considering r.v.'s in the domain of attraction of a stable Paretian law, the limiting distribution, as the degrees of freedom approach infinity, is derived. It turns out that the limiting distribution of sums of squared standard α-stable r.v.'s, after a suitable normalization, is an $\alpha/2$-stable *subordinator*, i.e., a positive r.v. (Z) with Laplace transform

$$\mathbf{E}e^{-\theta Z} = \exp\left\{-\frac{4-\alpha}{\alpha}\Gamma\left(1-\frac{\alpha}{2}\right)\theta^{\frac{\alpha}{2}}\right\}, \qquad \theta > 0,$$

(see also [DR]). The finite-degrees-of-freedom behavior for stable Paretian variates is simulated. Response surface techniques are employed to compactly summarize the simulation results for quantiles relevant for hypothesis testing.

A straightforward application of the arguments in [MKR1] leads to limiting theorems for the student's t-distribution for stable variates. In a similar line of work, [Ru1] derived the asymptotic distribution of the F-statistic for α-stable r.v.'s.

4.2 Modifications of Some Test Statistics

If we adopt the general class of Paretian distributions for cross section or time series models, a number of asymptotic results for estimators and diagnostic statistics must be either appropriately modified or newly developed. For example, [KMR] studied the properties of the *cumulative sum of squares* test procedure (CUSUM) for time-constancy of the unconditional variance of time series under the stable Paretian assumption. [Ru2] investigated the asymptotic distribution of the Box-Pierce statistic for random variables with infinite variance.

[RMK2] proposes modified tests for outliers in order to identify influential observations in the presence of α-stable distributions and provided the rate-of-convergence and limiting distributions of different statistics for outliers under the Gaussian and Paretian stable assumptions. As expected, it turns out that under the Paretian assumption, the generalized test procedures for outliers are much more robust than those based on the Gaussian assumption.

[MR2] discusses alternative stable distributions, which capture ARCH effects in time series processes. [MKR2] generalizes the Lagrange-multiplier-type test for linearity under the α-stable assumption and illustrates in an empirical example that test results depend crucially on the distributional assumption of the innovation process.

Table 9: Data generating processes and regressions for unit–root tests

Case	DGP	Estimated Regression	t-statistic
I	$y_t = y_{t-1} + u_t$	$\Delta y_t = \beta y_{t-1} + u_t$	t_β
II	$y_t = y_{t-1} + u_t$	$\Delta y_t = \mu + \beta y_{t-1} + u_t$	$t_{\beta,\mu}$
III	$y_t = \mu + y_{t-1} + u_t$	$\Delta y_t = \mu + \delta\tau + \beta y_{t-1} + u_t$	$t_{\beta,\mu,\tau}$

4.3 Inference in Time Series with Unit-Root and Regressions with Integrated Variables

Integrated economic time series processes have been investigated in [Gr] and [NP], where strong evidence for nonstationarity in U.S. macroeconomic time series is reported. Studying this issue under the α-stable assumption [MRK] provide finite–sample critical values and the asymptotic distributions of t-statistics arising in tests for unit roots (cf. [CT] and [Ph]). Table 9 summarizes test statistics commonly used in empirical work. [RMK3] consider Cases I and II of Table 9 under stable assumptions.

[EG] introduced the multivariate concept of cointegration. Two time series are called cointegrated when they both have a unit root, but a linear combination of them is stationary. [RMK3] studies the problem of statistical inference in cointegrating regressions, considering various characteristic-exponent combinations for the distributions of the cointegration residual process and the innovation process of the exogenous variable. The cointegrated regression model is of the form

$$y_t = \beta x_t + u_t, \qquad t = 1, \ldots, T, \tag{4.15}$$

where the x_t's are generated by random-walk process

$$x_t = x_{t-1} + e_t. \tag{4.16}$$

The assumptions on the cointegration process generating pair $\{u_t, e_t\}$ are:

(i) $\{u_t\}_1^\infty$ is in the domain of attraction (DA) of an α-stable law, $0 < \alpha \leq 2$;

(ii) $\{e_t\}_1^\infty$ is in the DA of α'-stable law, $0 < \alpha' \leq 2$;

(iii) Treating $\{u_t, e_t\}_1^\infty$ as a sequence of i.i.d. random pairs, we suppose that there exist vectors

$$\mathbf{a}_T = (a_T, a'_T) \in \mathbf{R}^2_+ \quad \text{and} \quad \mathbf{b}_T = (b_T, b'_T) \in \mathbf{R}^2$$

such that

$$\left(\frac{1}{a_T}\sum_{t=1}^T u_t, \frac{1}{a'_T}\sum_{t=1}^T e_t\right) - \mathbf{b}_T \overset{w}{\Longrightarrow} \mathbf{L}(1), \tag{4.17}$$

where $\mathbf{L}(1) = (U(1), U'(1))$ is a bivariate vector with stable components; $U(1)$ is an α-stable r.v., while $U'(1)$ is an α'-stable r.v.

In view of the tail behavior of the generating sequence $\{u_t, e_t\}$, [RMK3] considers the cointegrated regression model 4.15, 4.16 in the following five cases:

Case I: Both $\{u_t\}$ and $\{e_t\}$ are in the DA of the normal law. In this case we have $\alpha = \alpha' = 2$ and $\mathbf{L}(s) = (U(s), U'(s))$, $s > 0$, determined by the functional limit theorem derived from (4.17) is a two-dimensional Brownian motion.

Case II: $\{u_t\}$ has infinite–variance and is in the DA of an α-stable Paretian (non-Gaussian) law and $\{e_t\}$ are in the DA of a normal law. In this case, $0 < \alpha < \alpha' = 2$. Then, the two components of $\{u_t, e_t\}$ will be independent;[7] $U(s)$ is an α-stable Lévy motion and $U'(s)$ is a Brownian motion.

Case III: $\{u_t\}$ are in the DA of the normal law, while $\{u_t\}$ are in the DA of a Paretian α'-stable law. Here, $0 < \alpha' < \alpha = 2$, and $U(s)$ and $U'(s)$ are independent Brownian and α'-stable motions, respectively.

Case IV: Both $\{u_t\}$ and $\{e_t\}$ are in the DA of Paretian α-stable r.v.'s, i.e., $0 < \alpha = \alpha' < 2$. Then, $\mathbf{L} = (U, U')$ becomes a bivariate α-stable Lévy motion.

Case V: Both $\{u_t\}$ and $\{e_t\}$ are in the DA of the (α, α')-stable Paretian law. However, here $0 < \alpha_1, \alpha_1 < 2$, $\alpha_1 \neq \alpha_2$. Then, $\mathbf{L}(s)$, $s > 0$, is a process with stationary infintely divisible increments. The probability law of $\mathbf{L}$ is completely determined by the characteristic function of (u_t, e_t) is given by $\mathbf{E}e^{i(\mathbf{L}(s),\tau)} = e^{ih(\tau)}$, where

$$\begin{aligned} h(\tau) &= i(\mathbf{a}, \tau) + \frac{1}{2}(\mathrm{B}(\tau), \tau) + \int_{\|\vartheta\|>1} (e^{i(\tau,\vartheta)} - 1)\nu(d\vartheta) \\ &+ \int_{0<\|\vartheta\|\leq 1} \left(e^{i(\tau,\vartheta)} - 1 - i(\tau, \vartheta)\right) \nu(d\vartheta), \end{aligned}$$

with "Gaussian part" $B = 0$.

5. Conclusions

We have seen that the stable Paretian law is a more realistic assumption for the asset return data under consideration. This holds for the unconditional, conditional homoskedastic and the conditional heteroskedastic distributions. It allows us to specify more realistic models than permitted with the normal assumption, without having to

[7]For relevant facts about (α, α')-stable laws see [RG],[RX] and [MRR].

resort to alternatives, such as the t-distribution, which does not possess the appealing stability properties of the α-stable distribution. The empirical results show that simply assuming a fat–tailed unconditional distribution is not sufficient to capture the features of the return data. In addition to the fat-tailed innovations, GARCH or ARMA–GARCH structures are needed to successfully capture temporal dependencies.

To allow for these empirical facts in practical work in finance, we have provided a brief overview of various aspects of theoretical and empirical financial modeling under the Paretian hypothesis. These results, together with all other presented in this volume, provide a good base for the field of stable Paretian finance.

References

[AM] J. Affleck-Graves and B. Mac Donald, Multivariate Tests of Asset Pricing: The Comparative Power of Alternative Statistics, *Journal of Financial and Quantitative Analysis* 25 (1990), 163–185.

[BHL] A.K. Bera, M.L. Higgins and S. Lee, Interaction Between Autocorrelation and Conditional Heteroskedasticity: A Random Coefficient Approach, *Journal of Business and Economic Statistics* 10 (1992), 133–142.

[BG] R.C. Blattberg and N.J. Gonedes, Stable and Student Distributions for Stock Prices, *Journal of Business* 47 (1974), 244–280.

[BF] M. Blume and I. Friend, A New Look at the Capital Asset Pricing Model, *Journal of Finance* 28 (1973), 19–33.

[BJ] G.E.P. Box and G.M. Jenkings, *Time Series Analysis: Forecasting and Control, 2nd ed.*, San Francisco:: Holden-Day, 1976.

[Bo] T. Bollerslev, Generalized Autoregressive Conditional Heteroskedasticity, *Journal of Econometrics* 31 (1986), 307–327.

[BCK] T. Bollerslev, R. Chou, and K. Kroner, ARCH Modeling in Finance: A Review of the Theory and Empirical Evidence, *Journal of Econometrics* 52 (1992), 5–59.

[BD] P.J. Brockwell and R.A. Davis, *Time Series: Theory and Methods, 2nd. ed.*, New York: Springer-Verlag (1991).

[Ch] G. Chamberlain, A Characterization of the Distributions that Imply Mean Variance Utility Functions, *Journal of Economic Theory* 29 (1983), 985–988.

[CT] N.H. Chan and L.T. Tran, On the First Order Autoregressive Process With Infinite Variance, *Econometric Theory* 5 (1989), 354–362.

[ChR] B. N. Cheng and S. T. Rachev, Multivariate Stable Future Prices, *Mathematical Finance* 5 (1995), 133–153.

[CMMR] G. Chobanov, P. Mateev, S. Mittnik and S. T. Rachev, Modeling the Distribution of Highly Volatile Exchange-Rate Time Series, in: *Time Series*, ed. P. Robinson et al., Springer, 1996, 130–144.

[Co] G. Connor, A Unified Beta Pricing Theory, *Journal of Economic Theory* 34 (1984), 13–31.

[DR] R.A. Davis and S.I. Resnick, Limit theory for the Sample Covariance and Correlation Functions of Moving Averages, *Annals of Statistics* 14 (1986), 533–558.

[Di] F. Diebold, Testing for Serial Correlation in the Presence of ARCH, *Proceedings of the Business and Economic Statistics Section, American Statistical Association* (1986), 323–328.

[Du] W.H. DuMouchel, On the Asymptotic Normality of the Maximum–Likelihood Estimate when Sampling from a Stable Distribution, *Annals of Statistics* 1 (1973), 948–957.

[En] R. Engle, Autoregressive Conditional Heteroskedasticity with Estimates of the Variance of United Kingdom Inflation, *Econometrica* 50 (1982), 987–1007.

[EG] R.F. Engle and C.W.J. Granger, Cointegration and Error Correction Representation, Estimation and Testing, *Econometrica* 55 (1987), 251–276.

[F1] E. Fama, The Behavior of Stock Market Prices, *Journal of Business* 38 (1965), 34–105.

[F2] E. Fama, Risk, Return and Equilibrium, *Journal of Political Economy* 78 (1970), 30–55.

[GR1] B. Gamrowski and S. T. Rachev, Stable Models in Testable Asset Pricing, in: *Approximation, Probability and Related fields*, Plenum Press, New York, 1994, 315–320.

[GR2] B. Gamrowski and S. T. Rachev, The Implementation of Stable Laws in Financial Models: A Practical Approach, *Technical Report*, University of California at Santa Barbara, Dept. of Statistics and Applied Probability, Santa Barbara, USA, 1994.

[GR3] B. Gamrowski and S. T. Rachev, Financial Models Using Stable Laws, in: *Applied and Industrial Mathematics*, ed. Yu. V. Prohorov, 1995, 2, 556–604.

[GR4] B. Gamrowski and S. T. Rachev, Testing the Validity of Value-at-Risk Measures, in: *Applied Probability*, ed. C.C. Heyde et al., Springer, 1996, 307–320.

[Gr] C.W.J. Granger Some Properties of Time Series Data and Their Use in Econometric Model Specification, *Journal of Econometrics* 16 (1981), 121–130.

[HP] J. M. Harrison and S. R. Pliska, Martingales and Stochastic Integrals in the Theory of Continuous Trading, *Stochastic Processes Appl. II* (1981), 215–260.

[HV] M.C.A.B. Hols and C.G. de Vries, The Limiting Distribution of Extremal Exchange Rate Returns, *Journal of Applied Econometrics* 6 (1991), 287–302.

[HPR1] S. R. Hurst, E. Platen and S. T. Rachev, A Comparison of Subordinated Asset Pricing Models, *Research Report*, Centre for Mathematics and its Applications, School of Mathematical Sciences, Australian National University, Canberra ACT0200, Australia, 1995.

[HPR2] S. R. Hurst, E. Platen and S. T. Rachev, Option Pricing for Asset Returns Driven by Subordinated Process, *Technical Report* 81, Department of Statistics and Applied Probability, UCSB, Santa Barbara, CA 93106, USA, 1995.

[JW] A. Janicki and A. Weron, *Simulation and Chaotic Behaviour of α Stable Stochastic Processes*, Marcel Dekker, New York, 1994.

[JV] D.W. Jansen and C.G. de Vries, On the Frequency of Large Stock Returns: Putting Booms and Busts into Perspective, *Review of Economics and Statistics* 73 (1983), 18–24.

[Jo] J. Johnston, *Econometric Methods*, 3rd. ed., Mc–Graw Hill Book Company, Singapore, 1984.

[KaR] R. L. Karandikar and S. T. Rachev, A Generalized Binomial Model and Option Formulae for Subordinated Stock-Price Processes, *Probability and Mathematical Statistics* 15 (1995), 427–446.

[KMR] J.-R. Kim, S. Mittnik and S.T. Rachev, The CUSUM Test Based on OLS-residuals When Disturbances are Heavy-tailed, *Unpublished Manuscript*, Institute of Statistics and Econometrics, Christian Albrechts University at Kiel, 1996.

[LRS1] M.-L. T. Lee, S. T. Rachev and G. Samorodnistky, Association of Stable Random Variables, *Ann. Probability* 18 (1990), 1759–1764.

[LRS2] M.-L. T. Lee, S. T. Rachev and G. Samorodnistky, Dependence of Stable Random Variables, in: *Stochastic Inequalities*, IMS Lecture Notes-Monograph Series, 22 (1993), 219–234.

[Li] J. Lintner, The Valuation of Risk Assets, and the Selection of Risky Investments in Stock-Portfolios and Capital Budgets, *Review of Economics and Statistics* 47 (1965), 13–37.

[Man1] B. B. Mandelbrot, Sur Certain Prix Spéculatifs: Faits Empiriques et Modèle Basé sur les Processes Stables Additifs de Paul Lévy, *Comptes Rendus* 254 (1962), 3968–3970.

[Man2] B. B. Mandelbrot, New Methods in Statistical Economics, *Journal of Political Economy* 71 (1963), 421–440.

[Man3] B. B. Mandelbrot, The Variation of Certain Speculative Prices, *Journal of Business* 26 (1963), 394–419.

[Man4] B. B. Mandelbrot, The Variation of Some Other Speculative Prices, *Journal of Business* 40 (1967), 393–413.

[MT] B. B. Mandelbrot and M. Taylor, On the Distribution of Stock Price Differences, *Operations Research* 15 (1967), 1057–1062.

[Mer] R. Merton, An Intertemporal Capital Asset Pricing Model, *Econometrica* 41 (1973), 967–886.

[MKR1] S. Mittnik, J.-R. Kim and S.T. Rachev, Chi-square-type Distributions for Heavy-tailed Variates, *Econometric Theory* 14 (1998), 339–354.

[MKR2] S. Mittnik, J.-R. Kim and S.T. Rachev, Testing Linearity when Disturbances are Heavy-tailed, *Unpublished Manuscript*, Institute of Statistics and Econometrics, Christian Albrechts University at Kiel, 1996

[MR1] S. Mittnik and S. T. Rachev, Alternative Multivariate Stable Distributions and Their Applications to Financial Modeling, in: *Stable Processes and Related Topics*, ed. S. Cambanis et al., Birkhäuser, Boston, 1991, 107–119.

[MR2] S. Mittnik and S. T. Rachev, Modeling Asset Returns with Alternative Stable Distributions, *Econometric Reviews* 12 (1993), 261–330.

[MR3] S. Mittnik and S. T. Rachev, Reply to Comments on "Modeling Asset Returns with Alternative Stable Distributions" and some Extensions, *Econometric Reviews* 12 (1993), 347–389.

[MR4] S. Mittnik and S. T. Rachev, Financial Modeling and Option Pricing with Alternative Stable Models, to appear in: *Wiley Series in Financial Economics and Quantitative Analysis*, 1997.

[MRC] S. Mittnik, S. T. Rachev and D. Chenyao, Distribution of Exchange Rates: A Geometric Summation-stable Model, to appear in: *Proceeding of the 1996 International Conference on Data Analysis*, Sofia: Bulgarian Academy of Science (1996).

[MRDC] S. Mittnik, S. T. Rachev, T: Doganoglu and D. Chenyao, Maximum Likelihood Estimation of Stable Paretian Models, Working Paper, Institute of Statistics and Econometrics, Christian Albrechts University at Kiel, 1996.

[MRK] S. Mittnik, S. T. Rachev and J.–L. Kim, Testing for Unit-roots in the Presence of Finite-variance Disturbances, Working Paper, Institute of Statistics and Econometrics, Christian Albrechts University at Kiel, (1996).

[MRP] S. Mittnik, S. T. Rachev and M. Paolella, Integrated Stable GARCH Processes, Working Paper, Institute of Statistics and Econometrics, Christian Albrechts University at Kiel, (1996).

[MRR] S. Mittnik, S. T. Rachev and L. Rüschendorf, Test of Association Between Multivariate Stable Vectors, to appear in *Mathematical and Computer Modelling*, 1998.

[NP] Nelson, C.R. and C.I. Plosser, Trends and Random Walks in Macroeconomic Time Series, *Journal of Monetary Economics* 10 (1982), 139–162.

[PMR] A.K. Panorska, S. Mittnik and S.T. Rachev, Stable GARCH Models for Financial Time Series, *Applied Mathematics Letters* 8 (1995), 33–37.

[Ph] P.C.B. Phillips, Time Series Regression with a Unit Root and Infinte–Variance Errors, *Econometric Theory* 6 (1990), 44–62.

[Pr1] S. J. Press, Estimation of Univariate and Multivariate Stable Distributions, *Journal of the American Statistical Association* 67 (1972), 842–846.

[Pr2] S. J. Press, *Applied Multivariate Analysis*, 2nd edn., Robert E. Krieger, Malabar, 1982.

[RMK1] S. T. Rachev, S. Mittnik and J.–L. Kim, Modeling Economics Relationships Driven by Heavy–tailed Innovation Processes: A Survey of Some Recent Results, *Communications in Statistics: Stochastic Models*, 13 (1997), 841–866.

[RMK2] Rachev, S.T., S. Mittnik and J.-R. Kim (1996), Testing Outliers when Observations are Heavy-tailed, Unpublished Manuscript, Institute of Statistics and Econometrics, Christian Albrechts University at Kiel.

[RMK3] S.T. Rachev, S. Mittnik and J.–R. Kim, Statistical Inference in Time Series with Unit Roots in the Presence of Heavy-tailed Error Processes, Unpublished Manuscript, Institute of Statistics and Econometrics, Christian Albrechts University at Kiel, (1996).

[RR] S. T. Rachev and L. Rüschendorf, Models for Option Prices, *Teor. Veroyatnost. i. Primenen.* 39 (1994), 150–199.

[RS] S. T. Rachev and G. Samorodnitsky, Option Pricing Formulae for Speculative Prices Modelled by Subordinated Stochastic Processes, *Pliska* 19 (1993), 175–190.

[RX] S.T. Rachev and H. Xin, Test on Association of Random Variables in the Domain of Attraction of Multivariate Stable Law, *Probability and Math. Statist.* 14 (1993), 125–141.

[RG] S. I. Resnick and P. Greenwood, A Bivariate Stable Characterization and Domains of Attraction, *Journal Mult. Anal.* 10 (1979), 206–221.

[Ros1] S. A. Ross, The Arbitrage Theory of Capital Asset Pricing, *Journal of Economic Theory* 13 (1976), 341–360.

[Ros2] S. A. Ross, Mutual Fund Separation in Financial Theory – The Separating Distributions, *Journal of Economic Theory* 17 (1978), 254–286.

[Ru1] R. Runde, A Note on the Asymptotic Distribution of the F-statistic for Random Variables with Infinite Variance, *Statistics & Probability Letters* 18 , 1993, 9–12.

[Ru2] R. Runde, The Asymptotic Null Distribution of the Box-Pierce Q-statistic for Random Variables with Infinite Variance – With an Application to German Stock Returns, Unpublished Manuscript, Department of Statistics, University of Dortmund, (1996).

[ST] G. Samorodnitsky and M. Taqqu, *Stable Non-Gaussian Random Processes: Stochastic Models with Infinite Variance*, New York: Chapman and Hall, 1994.

[Sh] W. F. Sharpe, Capital Asset Prices: A Theory of Market Equilibrium under Conditions of Risk, *Journal of Finance* 19 (1964), 425–442.

[Wei] William W.S. Wei, *Time Series Analysis: Univariate and Multivariate Methods*, Addison-Wesley, 1990.

[Ze] A. Zellner, An Introduction to Bayesian Inference in Econometrics, John Wiley & Sons, New York, (1971).

[Zi] W. T. Ziemba, Choosing Investment Portfolios When the Returns Have Stable Distributions, in: *Mathematical Programming in Theory and Practice*, ed. P. L. Hammer and G. Zoulendijl, North-Holland, 1974, 443–482.

Institute of Statistics and Econometrics
Christian Albrechts University at Kiel
Olshausenstr. 40
D–24098 Kiel, Germany

Department of Statistics and Applied Probability
University of California at Santa Barbara
Santa Barbara, CA 93106, USA

Institute of Statistics and Econometrics
Christian Albrechts University at Kiel
Olshausenstr. 40
D–24098 Kiel, Germany

Risk Management and Quantile Estimation

Franco Bassi, Paul Embrechts[1] *and Maria Kafetzaki*

Abstract

Questions concerning risk management in finance and premium calculation in non-life insurance often involve quantile estimation. We give an introduction to the basic extreme value theory which yields a methodological basis for the analysis of such questions.

1. Introduction

The following two dates are important with respect to risk management:

- February 1, 1953,
- January 28, 1986.

Upon asking "why?" to audiences of bankers and/or (re)insurers, we hardly ever got the right answer. First of all, during the night of February 1, 1953, at various locations the sea dykes in the Netherlands collapsed during a severe storm, causing major flooding in large parts of coastal Holland and killing over 1800 people. An extremal event (a record surge) caused a protective system (the sea dykes) to break down.

In the wake of this disaster, the Dutch government established the Delta-committee. Under van Dantzig, statistical problems discussed included the task of answering the following question: "What would the dyke heights have to be (at various locations) so as to prevent future catastrophic flooding with a sufficiently high probability?" Though the above formulation is a gross simplification of the questions discussed, it pays to look a bit more in detail at the issues underlying a possible answer. To (over)simplify matters even further, suppose we are given data $X_1, \ldots, X_n$ interpreted as iid observations on maximum annual wave heights with common distribution function F. The iid assumption is clearly only a first approximation especially now where, due to changing climatological factors, more general models will have to be looked at more carefully. The Dutch government required the Delta-committee to estimate the necessary dyke heights which would imply a so-called t-year event for t sufficiently large. In engineering language, a t-year event related to a random variable (rv) X with distribution function (df) F corresponds to a value

$$x_t = F^{\leftarrow}(1 - t^{-1})$$

[1] This paper is based on a talk given by the second author at Olsen and Associates, Zurich.

where $F^{\leftarrow}$ denotes the generalised inverse of F, i.e.,

$$F^{\leftarrow}(t) = \inf\{x \in \mathbb{R} : F(x) \geq t\}.$$

Its obvious interpretation is that, with probability $1 - t^{-1}$, the level x_t will not be superceded. Formulated differently, x_t is the $(1 - t^{-1})$ upper quantile of the df F. In the case of the dyke problem, what were the available historical data? The basic data consisted of observations on maximum annual heights. Moreover, the previous record surge occurred in 1570 with a recorded value of (NAP+4)m, where NAP stands for normal Amsterdam level. The 1953 flooding corresponded to a (NAP+3.85)m surge. The statistical analysis in the Delta-report led to an estimated $(1 - 10^{-4})$-quantile of (NAP+5.14)m for the yearly maximum, based on a total of 166 historical observations. We most strongly recommend the paper by de Haan [Ha] for further reading on this topic. Clearly, the task of estimating such a 10 000-year event leads to

estimating well beyond the range of the data.

The only way in which such a problem can be solved is by making extra assumptions on the underlying model, i.e., on the df F of the annual maxima. Below some such models will be introduced.

The previous example already stressed the importance of estimating quantiles well beyond the range of the data. The second date, January 28, 1986 corresponds to the tragic explosion of the Space Shuttle Challenger. Our account concerning the search for possible causes is taken from the splendid book by Feynman [Fe] and the statistical analysis in Dalal, Fowlkes and Hoadley [DFH]. In his contribution to the Presidential Committee investigating the Challenger disaster, Richard Feynman found out that a potential cause may have been the insufficient functioning of so-called O-rings used for sealing the various segments of the external rocket boosters. The latter was mainly due to the exceptionally low temperature ($31^o F$) the night before launching. The data-set plotted in Figure 1 plays an important role in the discussion in [DFH].

Each plotted point represents a shuttle flight that experienced thermal distress on the O-rings; the x-axis shows the ground temperature at launch and the y-axis shows the number of O-rings that experienced some thermal distress. From an analysis of the effect of (low) temperature on O-ring performance the night before the accident, using the data in Figure 1a, one decided that "The temperature data are not conclusive for predicting primary O-ring blowby". After the accident a presidential commission [PC], p.145, noted a mistake in the analysis of the thermal-distress data (Figure 1a), namely, that the flights with zero incidents were left off the plot because it was felt that these flights did not contribute any information about the temperature effect. Including these data

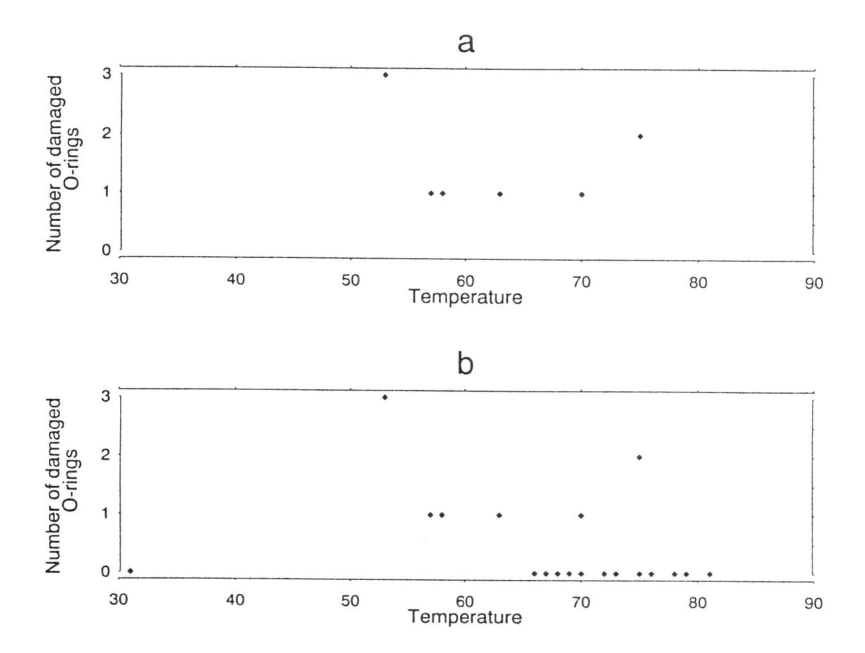

Figure 1: *Temperature versus number of O-rings having some thermal distress. Panel b includes flights with no incidents.*

(see Figure 1b), a significant dependence between O-ring damages and temperature was concluded. In [DFH] a detailed statistical analysis, based on a logistic regression model, is performed for the data in Figures 1a and 1b. The outcome of this showed a very significant increase in the failure probability in the case of Figure 1b as compared to Figure 1a. Richard Feynman summarised his findings as follows:

> It appears that there are enormous differences of opinion as to the probability of the failure with loss of vehicle and human life. The estimates range from roughly 1 in 100 to 1 in 100 000. The higher figures come from management. As far as I can tell "engineering judgement" means that they are just going to make up numbers!... It was clear that the numbers were chosen so that when you add up everything you get 1 in 100,000.

The resulting message from the above for risk management should be clear!

Always look very critically at the available data.

The basic aim of the present paper is to show that relevant methodology from extreme value theory offers a modelling tool which helps answer some of the technical questions in modern risk management. Our paper should be viewed as a key opening the door to

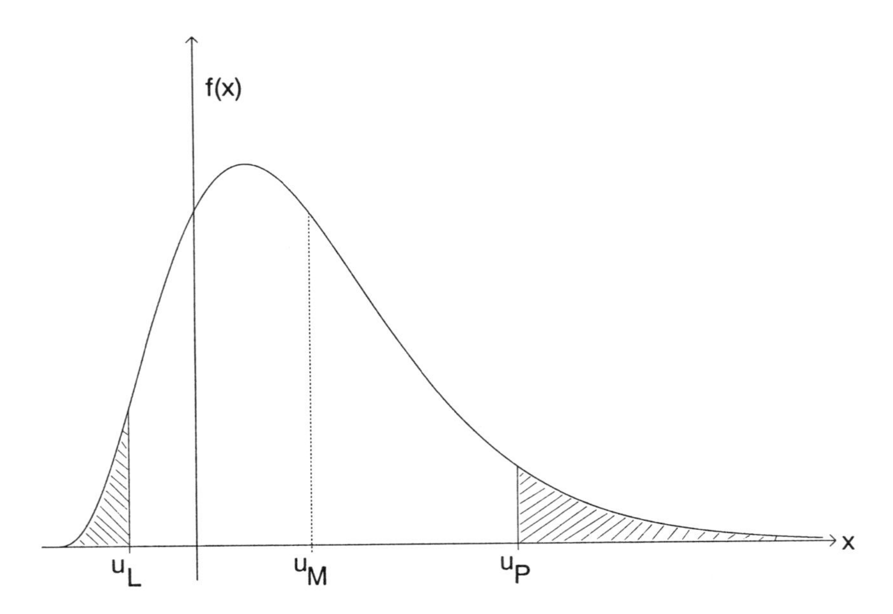

Figure 2: Possible P&L-density with α_L-quantile u_L, $(1 - \alpha_P)$-quantile u_P, for $0 < \alpha_P, \alpha_L < 1$, and mean u_M.

this essential toolkit. The reader is referred to the other papers in this volume for further details on some of the techniques mentioned. A general overview for insurance and finance relevant extreme value theory, together with many examples and references, is to be found in Embrechts, Klüppelberg and Mikosch [EKM].

2. On Value-at-Risk

One of the catchwords now frequently heard at all levels of financial institutions is "VaR" standing for Value-at-Risk. Though not uniquely describing one well-defined mathematical concept, in general terms VaR comes about as follows. Suppose a rv X with df F and density f describes the profit and losses of a certain financial instrument. The latter can be a particular asset, a portfolio or even the global financial exposure of a bank. Typically, the so-called Profit-Loss (P&L-) distribution may look like the example given in Figure 2.

A typical VaR calculation (for a portfolio say) has the following ingredients: "Calculate a possible extreme loss resulting from holding the portfolio for a fixed period (say10 days) using as a measure of underlying riskiness the volatility over the last 100 days". The notion of *extreme loss* is translated as the left 1 or 5%-quantile u_L for the P&L-distribution. Hence, in its easiest form, VaR estimates u_L for sufficiently small values of α_L. As in the dyke example, this quantile estimation will typically be outside (or at the edge of) the range of available data. Moreover, management may also be interested in setting targets at the profit side of the P&L-curve, i.e., it wants estimates for u_P. One step further would involve questions concerning the estimation of the size of under- (or over-) shoot of the levels u_L, respectively u_P, given that these events have occurred. Consequently, we want to estimate the following functions:

- $P(X - u_L \leq -x \mid X < u_L) = \theta(x \mid u_L), \quad x \geq 0,$
- $P(X - u_P > x \mid X > u_P) = \theta(x \mid u_P), \quad x \geq 0.$

In the finance context, $\theta(x \mid u_L)$ is often referred to as the *shortfall distribution*.

One of the basic messages of the present paper is that modern extreme value theory offers the necessary tools and techniques for answering the following questions:

- What are potential parametric models for the P&L-distribution function?
- Given such a model how can we estimate the quantiles u_L (VaR) and u_P?
- What are the resulting models for the conditional probabilities $\theta(\cdot \mid u_L)$ and $\theta(\cdot \mid u_P)$? How can these be estimated efficiently?
- Can these procedures be extended to more complex situations including non-stationarity (e.g. trends, seasonality), the use of exogenous economic variables, a multivariate set-up,...?

The good news is that all of the above questions can be answered. A detailed discussion however would be beyond the scope of this paper. Below we have listed some of the terminology which every risk management group should be aware of.

A survival kit on quantile estimation

- Quantiles
- QQ-plots
- The mean excess function
- The generalised extreme value distribution
- The generalised Pareto distribution
- The point process of exceedances
- The POT-method: peaks over thresholds
- The Hill and Pickands estimators, together with their ramifications and extensions

Though this list is not complete, it contains some of the relevant items from extreme value theory. Omitted, for instance, are methods for continuous-time stochastic

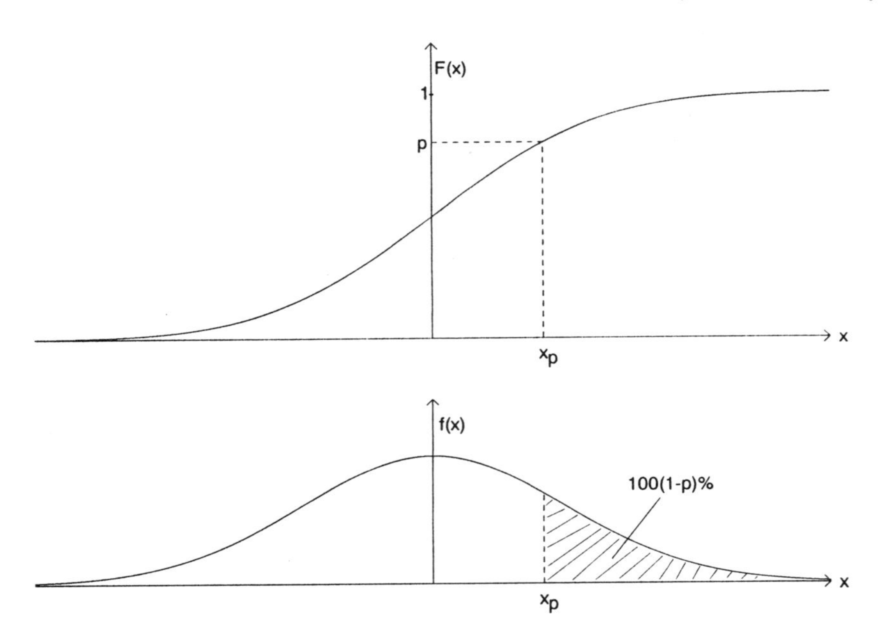

Figure 3: Graphical representation of the p-quantile; once by means of the distribution function and once by means of the density.

processes and, more importantly, multivariate techniques. The reader is referred to Embrechts et al. [EKM] for an extensive list of references and indications for further reading. In the following sections, we explain some of the items above more in detail.

3. Basic definitions

Suppose F is the distribution function of a real-valued random variable X. If F has a density, we denote it by f. The (right) tail of F is denoted by $\bar{F} = 1 - F$. In the sequel we only concentrate on $\bar{F}(x)$ for $x \geq 0$; similar results may be obtained for left tails, i.e., $F(-x)$ for $x \geq 0$ or for the two-sided tail $T(x) = 1 - F(x) + F(-x),\ x \geq 0$. Given a threshold probability $p \in (0, 1)$, the p-quantile x_p of F is defined as (see Figure 3):

$$x_p = F^{\leftarrow}(p) = \inf\{x \in \mathbb{R} : F(x) \geq p\}.$$

A key question now concerns the statistical inference of x_p, typically for p close to 1. Suppose the data $X_1, \ldots, X_n$ are iid with df F. An obvious estimator uses the so-called empirical distribution function

$$F_n(x) = \frac{\#\{i : 1 \leq i \leq n \text{ and } X_i \leq x\}}{n}, \quad x \in \mathbb{R}.$$

From the Glivenko-Cantelli theorem we know that with probability 1,

$$\sup_{x\in\mathbb{R}} |F_n(x) - F(x)| \to 0, \qquad n \to \infty.$$

Consequently, for given $p \in (0, 1)$, we may estimate $x_p = F^{\leftarrow}(p)$ by

$$\hat{x}_{p,n} = F_n^{\leftarrow}(p).$$

The properties of this estimator can most easily be expressed in terms of the ordered data. Denote by $X_{k,n}$ the kth largest observation. For the sake of simplicity, we assume that F is continuous so that with probability 1, there are no ties in $X_1, \ldots, X_n$. The elements of the ordered sample

$$X_{1,n} = \max(X_1, \ldots, X_n) \geq X_{2,n} \geq \cdots \geq X_{n,n} = \min(X_1, \ldots, X_n)$$

are called the *order statistics*. One readily verifies that, for $k = 1, \ldots, n$,

$$\hat{x}_{p,n} = F_n^{\leftarrow}(p) = X_{k,n}, \quad 1 - \frac{k}{n} < p \leq 1 - \frac{k-1}{n}. \tag{3.1}$$

Using the Central Limit Theorem (CLT), it is not difficult to show that for a df F with continuous density f with $f(x_p) \neq 0$ and $k = k(n)$ so that $n - k = np + o(n^{\frac{1}{2}})$,

$$\hat{x}_{p,n} \sim \mathrm{AN}\left(x_p, \frac{p(1-p)}{nf^2(x_p)}\right). \tag{3.2}$$

Using a slight, though obvious abuse of notation, $\mathrm{AN}(\mu, \sigma^2)$ stands for asymptotically normal with mean μ and variance σ^2. Also, using the iid assumption on $X_1, \ldots, X_n$, one immediately obtains, for $i < j$:

$$P\left(X_{j,n} \leq x_p < X_{i,n}\right) = \sum_{r=i}^{j-1} \binom{n}{r} p^{n-r}(1-p)^r. \tag{3.3}$$

Either from (3.2) or (3.3), one can obtain confidence intervals (approximate in the case of (3.2)) for the required quantiles. However notice the main problems.

- In the case of (3.2), asymptotics for $n \to \infty$ are used. This allows for the estimation of p close to 1 if sufficient data are available. A crucial choice to be made is that of $k = k(n)$.

- In the case of (3.3), exact confidence intervals can be obtained for n not too large. One very much relies on the fact that p is such that x_p lies *within* the range of the data.

Although the above approaches can be refined, one clearly faces basic problems for p-values such that $1 - p \ll n^{-1}$. As stated in the Section 1, in order to estimate quantiles for p close to 1 we must assume extra (parametric) conditions on F. Standard methods from statistical data fitting can (and should) be used:

- likelihood techniques,
- goodness-of-fit procedures,
- Bayesian methods.

Two graphical techniques deserve special mention. First of all the so-called *QQ- (or quantile-) plot* as a graphical goodness-of-fit procedure. Suppose $F(.;\theta)$ is a parametric model to be fitted to $X_1, \ldots, X_n$, resulting in an estimate $\hat{\theta}$ and hence a fitted model $\hat{F} = F(.;\hat{\theta})$. The plot of the points

$$\left\{\left(X_{k,n}, \hat{F}^{\leftarrow}(p_{k,n})\right) : k = 1, \cdots, n\right\}$$

for an appropriate plotting sequence $(p_{k,n})$ is the QQ-plot of the data $X_1, \ldots, X_n$ when fitting the parametric family $F(.;\theta)$. One often takes $p_{k,n} = (n-k+1)/(n+1)$, or some variant of this. A good fit results in a straight line of the plot, whereas model deviations (like heavier tails, skewness, outliers, ...) can easily be diagnosed. An example is shown in Figure 4.

A second graphical technique, used mainly in biostatistics, insurance and reliability, is the so-called *mean excess function (mef)*

$$e(u) = E\left(X - u | X > u\right), \quad u \geq 0, \tag{3.4}$$

and its empirical counterpart

$$e_n(u) = \frac{\sum_{i=1}^{n} (X_i - u)^+}{\sum_{i=1}^{n} I_{\{X_i > u\}}}. \tag{3.5}$$

Here $y^+ = \max(y, 0)$ and I_A denotes the indicator function of A, i.e., $I_A(x) = 0$ or 1 depending on whether $x \notin A$ or $x \in A$. By plotting $\{(u, e_n(u)) : u \geq 0\}$, as in Figure 5, one easily distinguishes between short- and long-tailed distributions. Figure 6 shows an application of these graphical techniques to a real data-set. For applications in insurance, see for instance Hogg and Klugman [HK] and references therein.

We would like to stress that both the QQ-plot and the mef belong to the realm of exploratory data analysis. The QQ-plot gives a quick assessment of the plausibility of certain distributional model assumptions. More importantly, it quickly shows how the

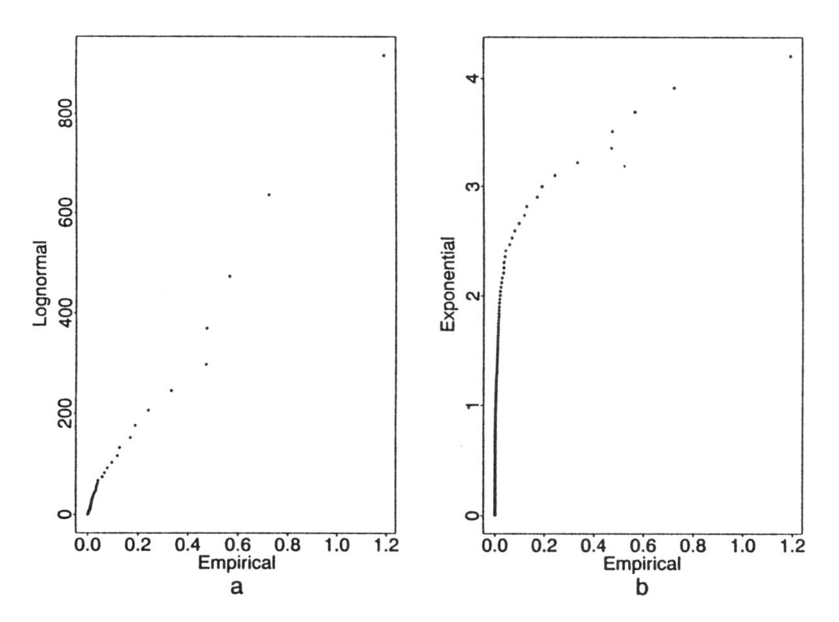

Figure 4: QQ-plot for lognormal versus lognormal(a) and versus the exponential distribution (b). The sample size is 200. One sees in the right picture that curvature appears, the empirical quantiles grow faster than the theoretical ones: the simulated lognormal data is heavy-tailed in comparison to the exponential model.

data deviates from a model proposed. For instance concavity hints at a heavier-tailed model as is clearly shown in Figure 4 (b). Though mefs could serve a similar purpose, we tend to use them only to distinguish between heavy-tailed and short-tailed distributions. The former typically exhibit an increasing pattern, whereas the latter are either constant or decreasing; see Figure 5 (a). In the Figures 5 (b), (c) and (d) we have simulated 4 realisations of the mef for a given model. Especially in the heavy-tailed Pareto case, there is great instability towards the higher u-values. The latter is due to the sparseness of the data in that range. It clearly shows that mefs have to be treated with care.

4. Extremes versus sums

We find it pedagogically useful to compare and contrast the Central Limit Theorem with the main results from Extreme Value Theory. One reason for this is that the mathematical technology used in both cases is fairly similar. A second reason is that a comparison of both theories often leads to important new insight in extremal event problems at hand. For a detailed discussion on this, together with many examples, see Embrechts, Klüppelberg and Mikosch [EKM].

The general formulation of the CLT goes as follows. Suppose $X_1, \ldots, X_n$ are iid random variables with $df\ F$. Define the partial sum $S_n = X_1 + \cdots + X_n$. The CLT

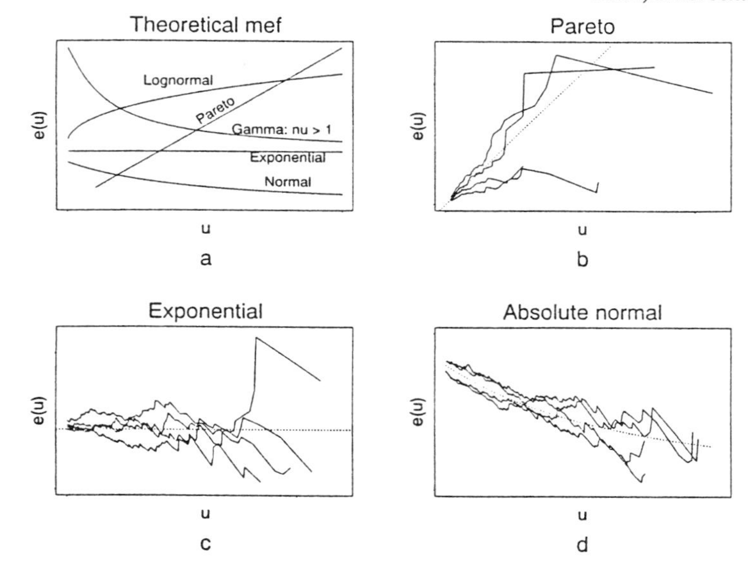

Figure 5: Theoretical mefs (a) and some examples of empirical mefs for simulated data (n=*200*). Normal in Figure a stands for the absolute value of a normally distributed rv. In Figure b the empirical mef of 4 simulated Pareto (1.5) distributed data-sets are plotted, Figure c corresponds to Exp(*1*), and d to the absolute value of a normally distributed rv. The dotted line shows the theoretical mef of the underlying distribution.

now solves the following problems.

(CLT 1) Given F, find constants $a_n > 0$ and $b_n \in \mathbb{R}$ so that

$$\frac{S_n - b_n}{a_n} \xrightarrow{d} Y, \quad n \to \infty, \tag{4.6}$$

where Y is a non-degenerate rv with df G say.

(CLT 2) Characterise the df G of Y in (4.6) above .

(CLT 3) Given a possible df G in (4.6), find all dfs F which satisfy (4.6) (the domain of attraction problem), and characterise the sequences (a_n) and (b_n).

All the above points can be solved and yield results typically belonging to the *sum world*. We all know the special case when $\sigma^2 < \infty$; indeed then $a_n = \sqrt{n}\sigma$, $b_n = n\mu$, $G = \mathrm{N}(0, 1)$, the standard normal, and all df fs with finite second moment are attracted to N(0, 1). An essential ingredient in the analysis of the above problems are the so-called functions of *regular variation*. First of all, a positive, Lebesgue measurable function $L : \mathbb{R}^+ \to \mathbb{R}^+ \setminus \{0\}$ is *slowly varying* if

$$\text{for all } \lambda > 0, \quad \lim_{x \to \infty} \frac{L(\lambda x)}{L(x)} = 1. \tag{4.7}$$

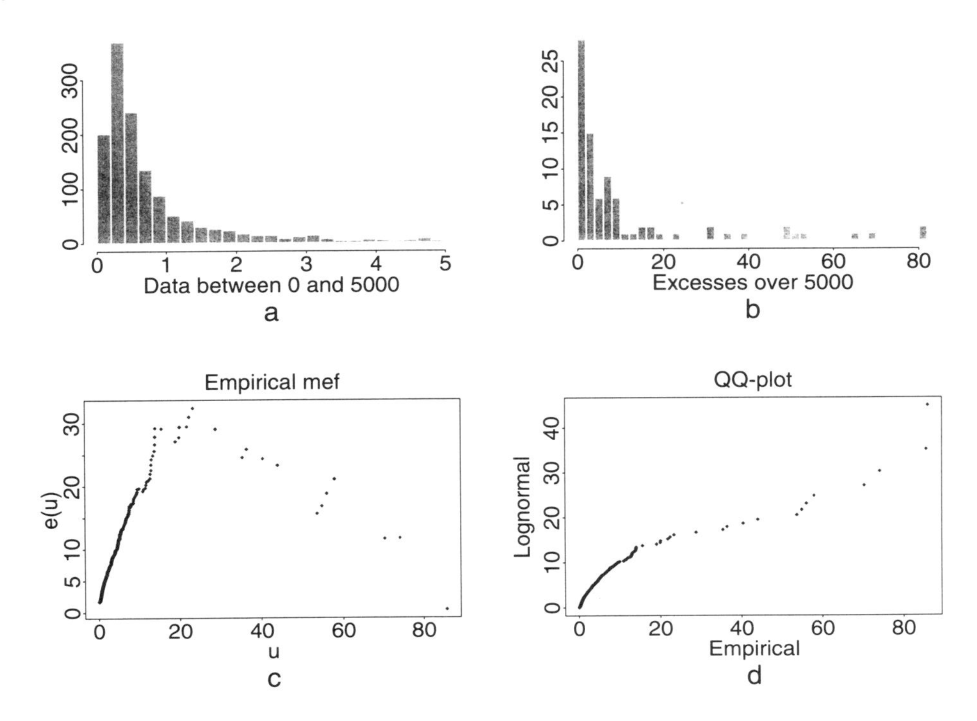

Figure 6: Graphical representation of fire insurance data (*in units of 1 000 Sfr*). The sample size is 1 438. First two histogramms are plotted: one with the "smaller" data (a) and one with the higher data (b). A QQ-plot against the lognormal (see (d)) shows that the data are more heavy-tailed. Indeed (c) hints at a Pareto like model.

Notation: $L \in \mathcal{R}_0$. A positive, Lebesgue measurable function h is *regularly varying with index* $\alpha \in \mathbb{R}$ if $h(x) = x^\alpha L(x)$. Notation: $h \in \mathcal{R}_\alpha$. For a detailed analysis of $\mathcal{R}_\alpha$, together with a discussion of many applications to probability theory, including the CLT, see Bingham, Goldie and Teugels [BGT].

The full solution of (4.6) leads to the class of *stable distributions*, elements of which are mainly characterised by a parameter $\alpha \in (0, 2]$. The case $\alpha = 2$ corresponds to the normal distribution, the case $\alpha = 1$ leads to the Cauchy distribution with density $f(x) = (\pi(1 + x^2))^{-1}$, $x \in \mathbb{R}$. Whenever $\alpha < 2$, $\sigma^2 = \infty$ and $E|X| = \infty$ for $\alpha < 1$, so that stable distributions with $\alpha < 2$ are sometimes used as models for heavy-tailed log-returns, a fact used by Mandelbrot [Ma] to promote these models in the context of finance.

Parallel to the above sum-world, similar questions can now be asked upon replacing S_n by $M_n = X_{1,n} = \max(X_1, \ldots, X_n)$. The following theorem (going back to Fisher and Tippett [FT]) summarises the for us most important results.

Theorem 4.1 *Suppose* $X_1, \ldots, X_n$ *are iid with* df F. *If there exist constants* $a_n >$

0 and $b_n \in \mathbb{R}$ so that

$$\frac{M_n - b_n}{a_n} \xrightarrow{d} Y, \quad n \to \infty, \tag{4.8}$$

where Y is non-degenerate with df G, then G is of one of the following types :

1. *(Gumbel)*

$$\Lambda(x) = \exp\left\{-e^{-x}\right\}, \quad x \in \mathbb{R},$$

2. *(Fréchet)*

$$\Phi_\alpha(x) = \begin{cases} 0 & \text{if } x \le 0 \\ \exp\left\{-x^{-\alpha}\right\} & \text{if } x > 0 \end{cases} \quad \text{for } \alpha > 0,$$

3. *(Weibull)*

$$\Psi_\alpha(x) = \begin{cases} \exp\left\{-(-x)^{\alpha}\right\} & \text{if } x < 0 \\ 1 & \text{if } x \ge 0 \end{cases} \quad \text{for } \alpha > 0.$$

□

The limit distribution in (4.8) is determined up to type: two dfs F_1 and F_2 are of the same type if there exist constants $a_1 \in \mathbb{R}$, $a_2 > 0$ so that $F_2(x) = F_1((x - a_1)/a_2)$, i.e., F_1 and F_2 belong to the same location and scale family.

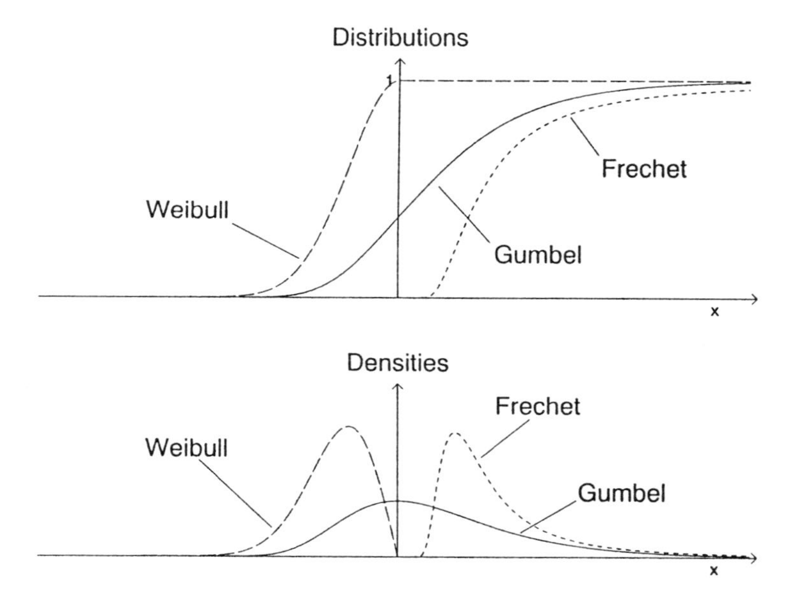

Figure 7: Distribution functions and densities of the Gumbel, Fréchet (with $\alpha = 2$) and Weibull (with $\alpha = 2$).

*df*s of the type of Λ, Φ_α and Ψ_α are called the (*generalised*) *extreme value distributions*. A useful parametrisation of these three classes is $H_{\xi,\mu,\beta}$ where

$$H_{\xi,\mu,\beta}(x) = \exp\left\{-\left(1+\xi\frac{x-\mu}{\beta}\right)^{-\frac{1}{\xi}}\right\}, \quad 1+\xi\frac{x-\mu}{\beta} \geq 0. \tag{4.9}$$

The adjective "generalised" has no special meaning here; $H_{\xi,\mu,\beta}$ is merely a useful reparametrisation of the Gumbel-, Fréchet- and Weibull-types. This terminology has become widely accepted. The case $\xi = 0$ in (4.9) has to be interpreted as $\xi \to 0$, resulting in the Gumbel type. The df $H_{\xi,\mu,\beta}$ is referred to as the *generalised extreme value* (GEV) *distribution with shape parameter* $\xi \in \mathbb{R}$, *location parameter* $\mu \in \mathbb{R}$ and *scale parameter* $\beta > 0$. The standard case $\mu = 0$, $\beta = 1$ will be denoted by $H_\xi = H_{\xi,0,1}$. The case $\xi < 0$ corresponds to the Weibull, $\xi = 0$ to the Gumbel and $\xi > 0$ to the Fréchet distribution.

Similar to the CLT, one may try to find a solution to the *maximum domain of attraction problem*, i.e., given H_ξ, find conditions on F so that (4.8) holds for appropriate sequences (a_n) and (b_n). In this case we say that F belongs to the *maximum domain of attraction of* H_ξ *with norming constants* a_n, b_n. The latter is denoted by $F \in MDA(H_\xi)$. For extremes the solution turns out to be much more dependent on the fine behaviour of $1 - F$. Two examples may illustrate this in the case of the Gumbel distribution.

1. Suppose $F = \mathrm{Exp}(1)$, then as $n \to \infty$,

$$M_n - \ln n \xrightarrow{d} \Lambda.$$

2. Suppose $F = \mathrm{N}(0, 1)$, then as $n \to \infty$,

$$(2\ln n)^{\frac{1}{2}}\left(M_n - (2\ln n)^{\frac{1}{2}} + \frac{1}{2}\frac{\ln\ln n + \ln 4\pi}{(2\ln n)^{\frac{1}{2}}}\right) \xrightarrow{d} \Lambda.$$

For applications in insurance and finance, the Gumbel and the Fréchet family turn out to be the most important models for extremal events. The domain of attraction condition for the Fréchet takes on a particularly easy form. The following theorem is due to Gnedenko.

Theorem 4.2 *Suppose* $X_1, \ldots, X_n$ *are iid with* df F. *Then there exist sequences* (a_n) *and* (b_n) *so that* (4.8) *holds for* Y *Fréchet distributed with parameter* $\alpha > 0$ *(i.e.,* $F \in MDA(H_{1/\alpha}) = MDA(\Phi_\alpha)$ *) if and only if*

$$1 - F(x) = x^{-\alpha}L(x), \quad x > 0, \tag{4.10}$$

with L *slowly varying as defined in* (4.7). □

We have seen that from a methodological point of view, the class of distribution functions with regularly varying tail ((4.10)) enter very naturally as standard models in the area of extreme value theory. From a statistical point of view, the specification (4.10) is a semi-parametric one: parametric in α and non-parametric in L. It should be clear from the defining property (4.7) that a conclusive test for slow variation of $x^{\alpha}\bar{F}(x)$ in any practical situation is impossible. However, based on the general asymptotic theory, (4.10) is not just a valiant assumption.

A useful property which trivially holds for a positive rv with df satisfying (4.10) is that for $\beta > \alpha, EX^{\beta} = \infty$ whereas for $\beta < \alpha, EX^{\beta} < \infty$. It now turns out that parameter estimates for α in (4.10) or alternatively for the range of divergent versus convergent moments rather consistently points at values for $\alpha \in (1,2)$ for claim size distributions in non-life insurance, whereas in finance, often log-return distributions yield $\alpha \in (3,4)$. The former implies the divergence of second moments, whereas the latter corresponds to infinite fourth moment. References on this topic are for instance Hogg and Klugman [HK] for loss data in insurance, and Taylor [Tay] for data in finance. Embrechts, Klüppelberg and Mikosch [EKM] contains numerous examples and references for further reading.

The above suggests which parametric models to use in practice when estimating quantiles beyond the range of the data: a possible answer is the GEV distribution $H_{\xi},\ \xi > 0$. This observation forms the cornerstone to the use of extreme value theory in finance and insurance applications.

Suppose now that $X_1, \ldots, X_n$ are iid with $df\ F \in MDA(H_{\xi}),\ \xi \in \mathbb{R}$. Recall the definition of p-quantile, $F(x_p) = p$. Then

$$\begin{aligned} P\left(a_n^{-1}(M_n - b_n) \le x\right) &= P\left(M_n \le a_n x + b_n\right) \\ &= F^n\left(a_n x + b_n\right) \longrightarrow H_{\xi}(x), \quad n \to \infty, \end{aligned}$$

whence, by using Taylor expansion,

$$n\bar{F}(u) \approx \left(1 + \xi \frac{u - b_n}{a_n}\right)^{-\frac{1}{\xi}},$$

where $u = a_n x + b_n$. A resulting tail estimator for $\bar{F}(u)$ could take on the form

$$\widehat{\bar{F}(u)} = \frac{1}{n}\left(1 + \hat{\xi}\frac{u - \hat{b}_n}{\hat{a}_n}\right)^{-\frac{1}{\hat{\xi}}},$$

yielding as an estimator for the quantile x_p

$$\hat{x}_p = \hat{b}_n + \frac{\hat{a}_n}{\hat{\xi}}\left((n(1-p))^{-\hat{\xi}} - 1\right).$$

Here $\hat{a}_n$, $\hat{b}_n$ and $\hat{\xi}$ are suitable estimators of a_n, b_n and ξ. As a candidate estimator for the crucial shape parameter ξ one often encounters the so-called *Hill estimator*

$$\hat{\xi}^{(H)}_{n,k} = \frac{1}{k}\sum_{j=1}^{k} \ln X_{j,n} - \ln X_{k+1,n},$$

where $k = k(n)$ has to be chosen in an appropriate way. Another classical estimator for ξ is the *Pickands estimator*

$$\hat{\xi}^{(P)}_{n,k} = \frac{1}{\ln 2} \ln \frac{X_{k,n} - X_{2k,n}}{X_{2k,n} - X_{4k,n}},$$

where $k = k(n)$, as in the Hill case. The choice of k turns out to be the Achilles heel of these estimators. In order to achieve weak consistency of these estimators, one needs $k \to \infty$ and $k/n \to 0$. More complicated growth conditions on k are required in order to obtain strong consistency and asymptotic normality. We prefer to plot the estimator accross a wide range of k-values and look (hope?) for stability. For a detailed discussion of the properties of $\hat{\xi}^{(H)}$, $\hat{\xi}^{(P)}$ and related estimators we refer to other papers in this volume and to Embrechts et al. [EKM] and the references therein. An illustration of these estimators is to be found in Figure 8.

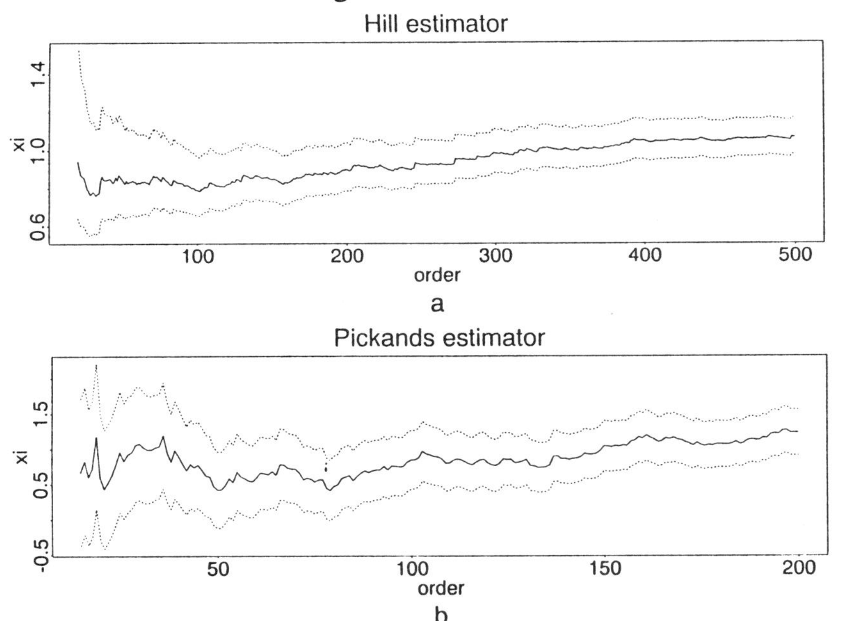

Figure 8: The Hill and Pickands estimator for the fire data in Figure 6. Both plots suggest a value for ξ in the range (0.5,1.5) corresponding to an estimate for α in the range (0.6,2).

The basic message of the above is that extreme value theory, through results like

Theorems 4.1 and 4.2, yields a possible methodological basis on which to base quantile estimation.

In Section 2 on Value-at-Risk we discussed the importance of estimating the conditional dfs $\theta(\cdot|u_L)$ and $\theta(\cdot|u_P)$, also referred to as *shortfall distributions* in the finance literature. The key result needed to treat this issue is given in Theorem 4.3 known under the name of *Gnedenko-Pickands-Balkema-de Haan Theorem.* In order to state it in its most general form, we first define the right-endpoint x_F of a df F by

$$x_F = \sup\{x \in \mathbb{R} : F(x) < 1\}.$$

Clearly $x_\Lambda = x_{\Phi_\alpha} = \infty$, $x_{\Psi_\alpha} = 0$. One can also show that for $F \in MDA(\Phi_\alpha)$, $x_F = \infty$, whereas for $F \in MDA(\Psi_\alpha)$, $x_F < \infty$. The former result is one of the reasons why in insurance and finance, the Fréchet distribution and its MDA provide the relevant models: they allow for unbounded upper values. Skeptics may eliminate the Fréchet distribution precisely because of the fact that $x_F = \infty$, on the basis of the somewhat tired argument that everything in the universe is bounded. Besides throwing away the normal distribution with the same argument, the main point however is that one can often clearly distinguish between data for which the larger values seem to cluster toward some well-defined upper limit, whereas in genuinely heavy-tailed data there does not seem to be such a clustering effect. Especially in applications in (re)insurance it may be much more prudent to model without an a priori fixed, finite x_F. Therefore all standard models used in non-life insurance have $x_F = \infty$. In what follows, we assume without loss of generality that $x_F \geq 0$. The following notion of *excess distribution* $F_u(x)$ is also important:

$$F_u(x) = P\left(X - u \leq x \mid X > u\right), \quad x \geq 0.$$

In the VaR example of Section 2,

$$\theta(x|u_P) = \bar{F}_{u_P}(x), \quad x \geq 0.$$

In the sequel we will use a new class of distributions. The *generalised Pareto distribution* with parameters $\xi \in \mathbb{R}$, $\beta > 0$ is defined as follows:

$$G_{\xi,\beta}(x) = \begin{cases} 1 - \left(1 + \xi \frac{x}{\beta}\right)^{-\frac{1}{\xi}}, & \xi \neq 0, \\ 1 - \exp\left\{-\frac{x}{\beta}\right\}, & \xi = 0. \end{cases}$$

where $x \geq 0$ for $\xi \geq 0$ and $0 \leq x \leq -\beta/\xi$ for $\xi < 0$. Figure 9 plots the densities of the GPD for some values of the parameter ξ.

We are now ready to state the main result concerning the estimation of the overshoot probabilities $\theta(\cdot|\cdot)$ in Section 2.

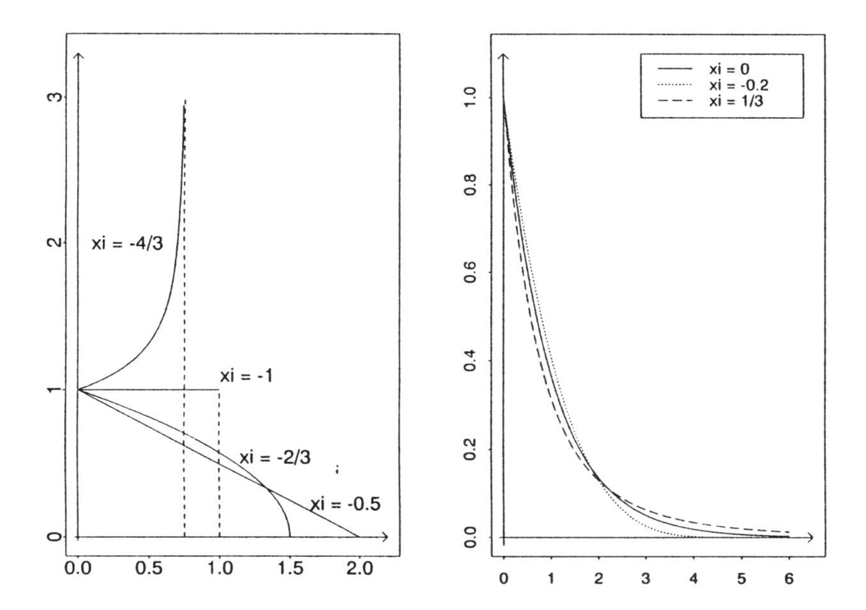

Figure 9: Densities of some generalised Pareto distributions. The parameter ξ is given in the pictures, β is always 1.

Theorem 4.3 *Suppose F is a df with excess distributions F_u, $u \geq 0$. Then, for $\xi \in \mathbb{R}$, $F \in MDA(H_\xi)$ if and only if there exists a positive measurable function $\beta(u)$ so that*

$$\lim_{u \uparrow x_F} \sup_{0 \leq x \leq x_F - u} |\bar{F}_u(x) - \bar{G}_{\xi,\beta(u)}(x)| = 0. \tag{4.11}$$

□

Remark that in most cases interesting to us, $x_F = \infty$ so that (4.11) yields an approximation to $\bar{F}_u$ by a generalised Pareto distribution; *the so-called threshold u has to be taken large* ($u \uparrow \infty$). From this approximation we immediately obtain an upper-tail estimator. Indeed, for $x, u \geq 0$,

$$\bar{F}(u + x) = \bar{F}_u(x)\bar{F}(u) \approx \bar{G}_{\xi,\beta(u)}(x)\bar{F}(u).$$

Based on $X_1, \ldots, X_n$ iid with df F, we obtain as a natural estimator (writing $\beta = \beta(u)$):

$$\bar{F}\widehat{(u + x)} = \bar{G}_{\hat{\xi},\hat{\beta}}(x)\frac{N_u}{n}, \tag{4.12}$$

where $N_u = \#\{i : 1 \leq i \leq n,\ X_i > u\}$. The estimators $\hat{\xi}$ and $\hat{\beta}$ can for instance be obtained through maximum likelihood estimation.

For further details see Embrechts et al. [EKM] and references therein. We would like to stress that $\hat{\xi}$ and $\hat{\beta}$ are functions of u! From the estimate for $\bar{F}$ above one can immediately obtain the necessary quantile estimator. For instance, an estimator for the p-quantile $x_p = F^{\leftarrow}(p)$ becomes

$$\hat{x}_p = u + \frac{\hat{\beta}}{\hat{\xi}}\left(\left(\frac{n}{N_u}(1-p)\right)^{-\hat{\xi}} - 1\right). \tag{4.13}$$

In practice, its construction involves the following steps:

1. The data are $X_1, \ldots, X_n$, and for given threshold u, denote N_u the number of exceedances of the level u.

2. Determine u large enough so that the approximation (4.11) is valid. One graphical tool is based on the mef, i.e., look for u-value above which the mef "looks linear". In the case of the fire insurance data (Figure 6(c)) relatively small u-values could be considered. A first choice of $u = 10$ may be good.

3. Use maximum likelihood theory to fit the GPD to the excesses over u (= 10, say). From this, a tail fit (4.12) and an estimator for the quantile x_p ((4.13)) are obtained. In Figure 10 two of such plots are shown.

4. Repeat step 3. for various u-values. This implies refitting a model for each new u. A summary plot of the type

$$\{(u, \hat{x}_p) : u \geq 0\}$$

gives a good idea on the (non-)stability of an estimation procedure as a function of the excesses used.

The statistical theory underlying the above methodology runs under the acronym POT: Peaks Over Threshold method. A nice introduction is given in Davison and Smith [DS]. The software used in the above analysis was written in S-plus by Alexander McNeil and is available via `http://www.math.ethz.ch/~mcneil/`.

5. Conclusion

Going back to the "Survival kit on quantile estimation" given in Section 2, we have touched upon all the items listed. Much more can, and indeed should be said. This volume and similar ones will hopefully contribute to making extreme value methodology more widely known in fields like insurance and finance. Natural or man-made disasters, crashes on the stockmarket or other extremal events form part of society. The methodology given above may be useful in contributing towards a sound scientific discussion concerning their causes and effects.

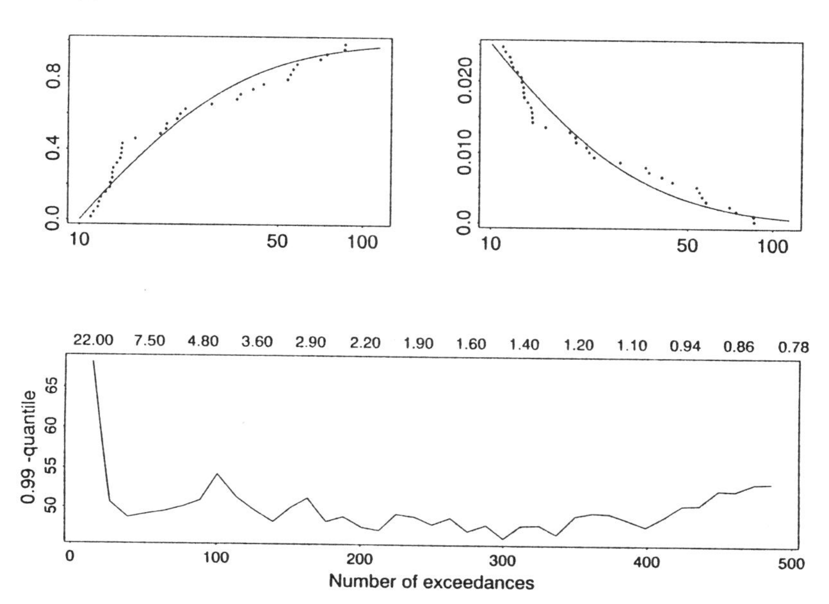

Figure 10: Analysis of the fire data: the top left figure contains the fit for the excess distribution $F_u(x - u)$ where $u = 10$; the top right figure gives the fit for $\bar{F}(u + x)$, again for $u = 10$. The estimated shape parameter equals 0.43. Both figures have logarithmic horizontal scale. Finally, the bottom figure yields a quantile plot $\{(u, \hat{x}_p) : u \geq 0\}$ for $p = 0.99$. Note that in the bottom x-axes the number of exceedances is given, on the other hand on the top x-axes the corresponding threshold is reported.

Acknowledgment. We would like to thank an anonymous referee for the very careful reading of a first version of the paper. His/her comments have lead to numerous improvements.

References

[BGT] N. H. Bingham, C. M. Goldie, J. L. Teugels, *Regular Variation.* Cambridge University Press, Cambridge (1987).

[DFH] S. R. Dalal, E. B. Fowlkes and B. Hoadley, Risk analysis of the space shuttle: pre-Challenger prediction of failure. *Journal of the American Statistical Association* **408** (1989), 945-957.

[DS] A. Davison and R. L. Smith, Models for exeedances over high thresholds (with discussion). *J. Roy. Statist. Soc.* **Ser. B 52** (1990), 393-442.

[EKM] P. Embrechts, C. Klüppelberg, and T. Mikosch, *Modelling Extremal Events for Insurance and Finance.* Springer, Berlin (1997).

[Fe] R. Feynman, *What do you care what other people think?* Bantam books (1989).

[FT] R. A. Fisher and L. H. C. Tippett, Limiting forms of the frequency distribution of the largest or smallest number of the sample. *Proc. Cambridge Philos. Soc.* **24**, (1928), 180-190.

[Ha] L. de Haan, Fighting the arch-enemy with mathematics. *Statistica Neerlandica* **44** (1990), 45-68.

[HK] R. V. Hogg and S. A. Klugman, *Loss Distributions.* Wiley, New York (1984).

[Ma] B. Mandelbrot, The variation of certain speculative prices. *Journal of Business*, Univ. Chicago **36**, (1963), 394-419.

[PC] *Report of the Presidential Commission on the Space Challenger Accident*, **Vols. 1&2**, Washington DC (1986).

[Tay] S. J. Taylor, *Modelling Financial Time Series.* Wiley, Chichester (1986).

Department of Mathematics, ETHZ, CH-8092 Zürich, Switzerland

II. Time Series

Analysing Stable Time Series

Robert J. Adler, Raisa E. Feldman and Colin Gallagher [1]

Abstract

We describe how to take a stable, ARMA, time series through the various stages of model identification, parameter estimation, and diagnostic checking, and accompany the discussion with a goodly number of large scale simulations that show which methods do and do not work, and where some of the pitfalls and problems associated with stable time series modelling lie.

1. Introduction

There are three major stages in the now standard "Box-Jenkins" time series modelling techniques for Gaussian time series: Model identification, parameter estimation, and diagnostic checking.

In many ways, the techniques behind these three stages really only involve two bags of tricks, since most diagnostic checks rely on testing whether or not the fitted residuals, after parameter estimation, behave like a white noise sequence. This, of course, is tantamount to identifying a model for the residuals, and so takes us, more or less, back to stage one.

In this paper we will concentrate on a variety of issues related to ARMA model identification in the stable setting. The paper by Calder and Davis in this volume [CD] describes the parameter estimation problem, so that the two papers, together, should give a good overview of the overall ARMA problem and be of some assistance to a practitioner who wishes to analyse a particular series. We shall also have something to say about parameter estimation, for one specific technique.

There are no new theorems in this paper or even new ways of thinking about things. Rather we have tried to collect in one place a number of results that are rather widely scattered, and to investigate their practical efficiency on replicates of synthetic data. Some of the results are somewhat surprising, and some more than a little worrying. Many beg further and deeper theoretical investigation.

The bottom line will be that while, in principle, the standard Gaussian Box-Jenkins techniques [BJ], [BD] do carry over to the stable setting, in practice a great deal of care needs to be exercised.

[1] Research supported in part by Office of Naval Research, grants N00014-94-1-0191, N00014-93-1-0800 and N00014-96-1-0739

Results in a similar vein can also be found in the paper [R2] in this volume, as well as [R1] and [FR]. These papers treat real as well as synthetic data, and general heavy-tailed rather than purely stable series.

Finally, before we start, we should determine precisely what we mean by the various stable parameters by defining a stable distribution with parameters $(\alpha, \beta, \sigma, \mu)$ as usual via its characteristic functions, as follows:

$$E\left(e^{itZ}\right) = \begin{cases} \exp\left\{-\sigma^\alpha |t|^\alpha \left(1 - i\beta(\operatorname{sign} t)\tan\frac{\pi\alpha}{2}\right) + i\mu t\right\} & \text{if } \alpha \neq 1, \\ \exp\left\{-\sigma|t|\left(1 + \frac{2i\beta}{\pi}(\operatorname{sign} t)\ln|t|\right) + i\mu t\right\} & \text{if } \alpha = 1, \end{cases} \tag{1.1}$$

where $0 < \alpha \leq 2$, $\sigma \geq 0$, $-1 \leq \beta \leq 1$, and $\mu \in \Re$. We shall denote such a distribution by writing $Z \sim S_\alpha(\sigma, \beta, \mu)$.

2. Preliminary data analysis – Is it stable?

The first question is whether or not our data is "heavy-tailed", in some general sense, and if so, whether or not it is stable. We shall not be interested in the possibility of heavy-tailed, but non-stable data, for a number of reasons:

1. If the data is in the domain of attraction (cf. [FE]) of a stable distribution, then, in general, large sample techniques are identical to those for the purely stable situation.

2. In the domain of attraction case, the difference between a stable and non-stable model lies in the central region of the distribution. If one is using stable or other heavy-tailed techniques, this is generally not the region of interest.

3. In the case of heavy tails, *not* in a stable domain of attraction, there are comparatively few reliable techniques around (see [R1, FR] for further details and discussion).

We shall also make one significant simplification throughout this paper: We shall always work with examples in which, in terms of (1.1), $\beta = \mu = 0$; i.e., with centered and symmetric variables. This simplification is common in most of the theory that we quote, although unfortunately it is not always justified in practice. However, we doubt that it has much *qualitative* effect on the phenomena we shall look at. That this is definitely the case in some situations is born out by [KN2].

2.1 Graphing the series

The simplest, most obvious, and often most powerful, techniques for detecting stable data are also unfortunately those with very little theory behind them. They start with a

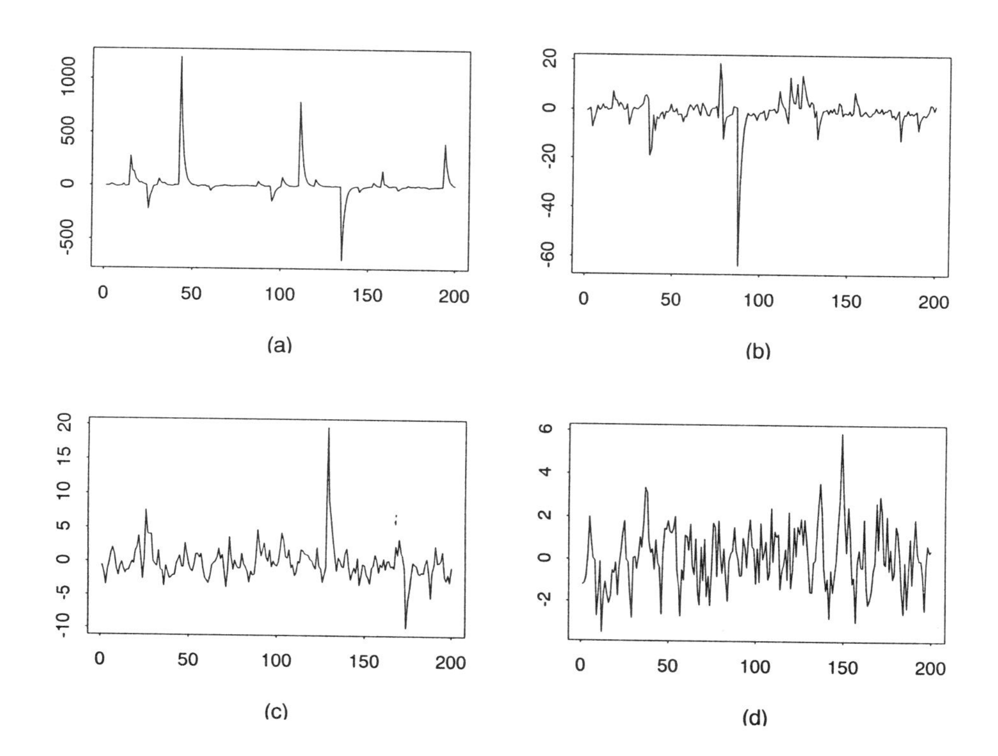

Figure 1: Four AR(1), stable, time series, with increasing values of α. (a) $\alpha = 0.6$, (b) $\alpha = 1.2$, (c) $\alpha = 1.8$, (d) $\alpha = 2.0$

visual inspection of the data in a search for highly "inhomogeneous" data, in the (non-technical) sense that one or several observations dominate the rest. This is generally so notable that on graphing the time series, most of the data is "squeezed" onto the horizontal axis by the automatic scaling of the plotting routine.

An example of this is given if Figure 1, where plots of four stable time series are given. Each follows the AR(1) model

$$X_t - .5X_{t-1} = Z_t, \tag{2.1}$$

where the $\{Z_t\}$ are symmetric i.i.d. stables with scaling parameter $\sigma = 1$. It is obvious that all three of the stable cases are qualitatively different to the final, Gaussian, case.

Of course, in each of these cases, it would be hard to distinguish on a graphical basis between a purely stable series and a Gaussian series with the occasional outlier (cf. the examples in [MI]). This requires looking more carefully at further distributional information.

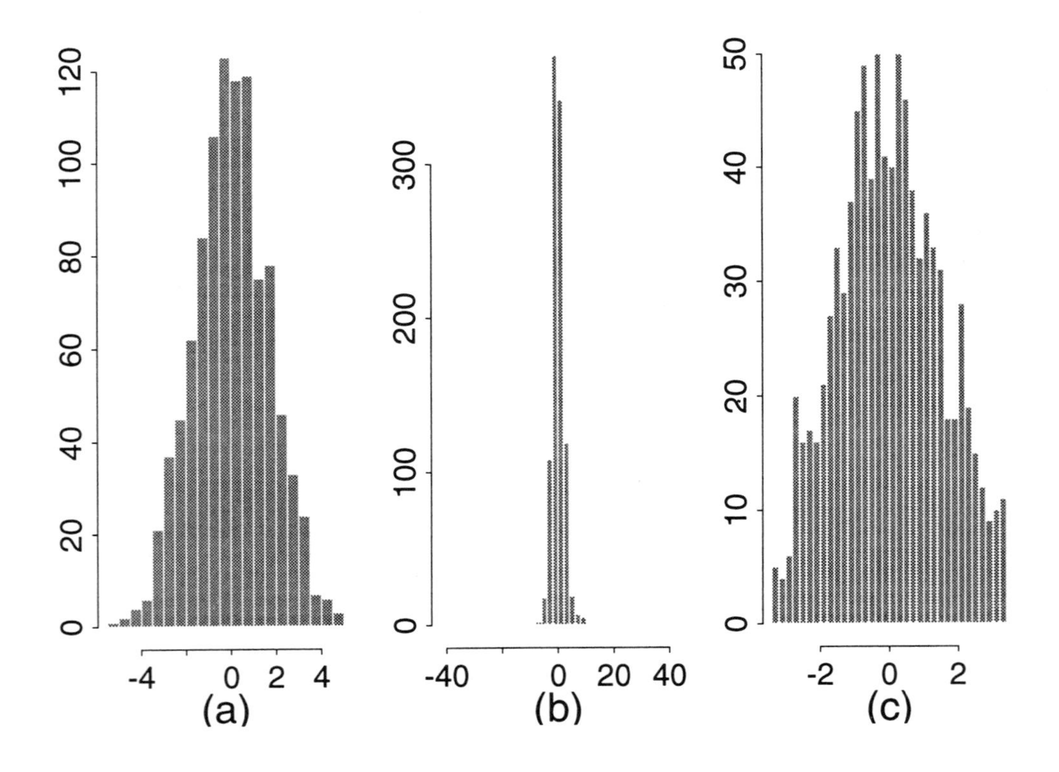

Figure 2: Histograms of (a) Gaussian, (b) Stable, $\alpha = 1.6$, and (c) Truncated stable time series.

2.2 The histogram

Essentially the same information as is obtained by graphing the series can be gleaned from the histogram of the data. What is lost in the histogram is, of course, the temporal structure of the data, but what is more obvious is the presence or absence of symmetry.

Figure 2 shows a histogram from data generated by the same model as in (2.1), but now only for two cases, $\alpha = 2$ (Gaussian) and $\alpha = 1.6$, and for series of length 1,000.

Two factors should be noted here. The first is that, despite the fact that the sample size is quite large, the Gaussian case is much further from the traditional bell curve than one would expect with i.i.d. data. But this is correlated data so that the laws of large numbers take longer to come into play.

The second is a repetition of the phenomenon mentioned above, about automatic scaling "spoiling" the graph. In (b) a few outliers are so large that the entire histogram is squeezed into a few bars in the center. When the largest and the smallest 5% of the data is truncated, as in (c), the shape of the graph changes dramatically.

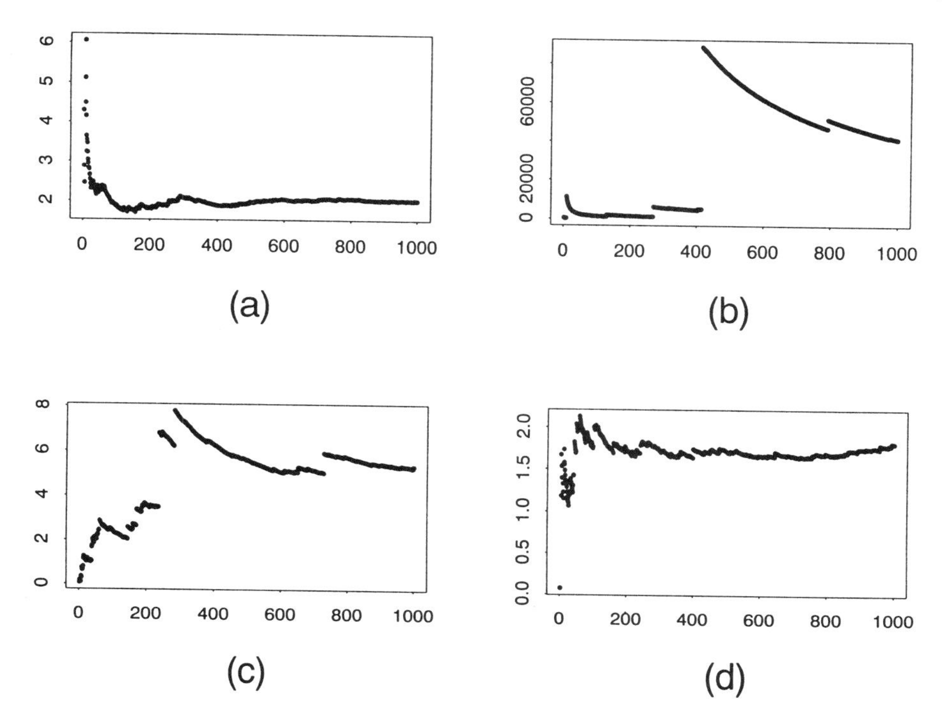

Figure 3: Cumulative variance plots for (a) Gaussian, (b) $\alpha = 0.8$, (c) $\alpha = 1.6$ and (d) Student t ARMA(1,1) processes.

2.3 The "converging variance" test

One of the oldest tests for determining whether data has infinite variance is the trick of plotting the sample variance S_n^2, based on the first n observations, as a function of n. If the data comes from population with finite variance, S_n^2 should converge to a finite value. Otherwise, it should diverge as n grows, and the graph typically shows large jumps.

Although this test was originally designed for i.i.d. data, it also works well for correlated data, as long as the order of the observations is first randomised, so as to destroy dependencies that might lead to trends and jumps with other explanations.

Figure 3 contains graphs for this test for two stable ($\alpha = 0.8, 1.6$), Gaussian, and "t" (with 4 degrees of freedom) processes. (By the last we mean an ARMA process in which the innovations Z_t have a Student t distribution. This is an interesting case, since it gives a distribution with much heavier than Gaussian tails, but in the Gaussian domain of attraction.) For variety, we took the ARMA(1,1) model $X_t - 0.3X_{t-1} = Z_t + .5Z_{t-1}$, with series of length 1,000.

The divergence of S_n^2, as n grows, and the irregularity of the graphs in the two stable cases are very marked.

2.4 Preliminary estimation of the stable parameters

One last and natural step before entering the time series arena proper is the estimation of the various stable parameters.

There are a number of techniques available for estimating tail decay, most of which are built around the so-called "Hill estimator", and many of which, while based on sound theory, turn out to be far from satisfactory in practice. (cf. [R1, R2] or [PDM] in this volume, for details.)

This is one of the reasons for being prepared to assume a specifically stable model rather than one generically heavy-tailed. In the stable case there exists an excellent estimator of the parameters due to Hu McCulloch [MC]. This estimator, which is based in essence on fitting tabulated quantiles of stable distributions, works for $\alpha \in [0.6, 2.0]$ and $\beta \in [-1, 1]$ (which covers most of the cases met in practice) and all values of the other parameters. The estimator of the location parameter, however, can be inaccurate near $\alpha = 1$.

The McCulloch estimator was originally designed for, and indeed works best on, i.i.d. data. Nevertheless, some initial information on the parameters, especially α, is generally required for model identification so that one has no choice but to work with the time series data.

A typical example of the precision of the McCulloch estimator of α is given in Table 1, in which the results for estimating α from 1,000 observations from the MA(2) model $X_t = Z_t + .5Z_{t-1} - .3Z_{t-2}$ are presented for various α. The accuracy of the estimator is, we believe, truly impressive.

For each iteration we estimate α from the simulated innovations, and then from the time series. For the final column, we estimated the MA parameters using Whittle's estimator, described in Section 4, and computed the residuals. We then estimated α a third time using these residuals, expecting (incorrectly) that this estimation would be better than from the raw time series. We repeated this process 10,000 times. Clearly the estimates obtained from the innovations are the best, but, perhaps rather surprisingly, McCulloch's technique seems to work better when applied to the original time series rather than to the residuals.

Interestingly, however, McCulloch's estimator does not seem to work as well for AR processes as it does for pure moving averages, at least for small values of α, the main problem being in the substantially increased sample variance, rather than in the bias.

There are a number of possible explanations for this, although we are not certain which, if any, is real. Two candidates for consideration are:

(i) It may simply be due to divergences, numerical and other, as α becomes close to the region where the estimator is not supposed to work.

(ii) In the Gaussian case, the correlation structure is much stronger in the AR(2) model than in the MA(2), and this should affect estimator variance. Translating "correlation" to "dependence" in the stable case may create a similar problem. However, when estimating α from the residuals obtained from the Whittle parameter estimates, the simulations indicate that the estimates are superior in the AR case (at least for $\alpha \geq 1$).

Estimation of alpha, MA(2)			
α	*Innovations*	*Time series*	*Residuals*
.6	0.633 (0.0397)	0.665 (0.0796)	0.679 (0.1010)
.8	0.804 (0.0363)	0.811 (0.0524)	0.822 (0.0617)
1	1.002 (0.0439)	1.007 (0.0593)	1.014 (0.0666)
1.2	1.203 (0.0506)	1.206 (0.0648)	1.209 (0.0724)
1.4	1.402 (0.0582)	1.405 (0.0690)	1.407 (0.0769)
1.6	1.606 (0.0711)	1.608 (0.0791)	1.610 (0.0888)
1.8	1.808 (0.0935)	1.810 (0.0977)	1.812 (0.1023)
2	1.953 (0.0653)	1.954 (0.0659)	1.952 (0.0683)

Table 1: Mean and standard deviation (in parentheses) of 10,000 estimates of alpha using simulated innovation sequence, corresponding time series and estimated residuals.

Table 2 gives the result of a similar study for the model $X_t - .8X_{t-1} + .7X_{t-2} = Z_t$.

Note that these positive results become less than satisfactory when one leaves the permissible parameter region for the estimator, and in the region of $\alpha = 1$. We simulated a sample of length 1000 from a symmetric Cauchy distribution ($\alpha = 1$, $\mu = 0$) and used McCulloch's estimator to estimate the location parameter μ. The mean of 50 of such independent estimates of location was -6.17. Similarly, estimates of α for values of $\alpha < 0.6$ also give poor and misleading results. For example, the average of 100 estimates of α based on independent samples of 1000 was 5.23 for $\alpha = 0.3$, and 0.95 for $\alpha = 0.5$.

Estimation of alpha, AR(2)		
α	*Time series*	*Residuals*
.6	0.836 (0.7122)	0.790 (0.6434)
.7	0.758 (0.2976)	0.740 (0.2822)
.8	0.843 (0.1040)	0.818 (0.0702)
1	1.033 (0.0946)	1.009 (0.0561)
1.2	1.213 (0.0975)	1.207 (0.0544)
1.4	1.408 (0.1007)	1.405 (0.0588)
1.6	1.612 (0.1044	1.608 (0.0719)
1.8	1.813 (0.1079)	1.811 (0.0941)
2	1.954 (0.0670)	1.954 (0.0655)

Table 2: Mean and standard deviation (in parentheses) of 10,000 estimates of alpha using simulated time series and estimated residuals.

3. Model Identification

The first step when fitting data $\{X_1, X_2, \dots, X_n\}$ into a linear ARMA(p, q) time series model

$$X_t - \phi_1 X_{t-1} - \cdots - \phi_p X_{t-p} = Z_t + \theta_1 Z_{t-1} + \cdots + \theta_q Z_{t-q}, \tag{3.1}$$

with i.i.d. innovations $\{Z_t\}$, normal or stable, is the identification of the lag parameters p and q.

In the Gaussian case model identification techniques are based on analysis of the sample autocorrelation function (ACF)

$$\hat{\rho}(h) = \sum_{t=1}^{n-h} X_t X_{t+h} \Big/ \sum_{t=1}^{n} X_t^2, \quad h = 1, 2, \dots, \tag{3.2}$$

or its mean-corrected version,

$$\tilde{\rho}(h) = \sum_{t=1}^{n-h} (X_t - \bar{X})(X_{t+h} - \bar{X}) \Big/ \sum_{t=1}^{n} (X_t - \bar{X})^2, \tag{3.3}$$

where $\bar{X} = n^{-1}(X_1 + \ldots + X_n)$, and of the equally familiar sample partial autocorrelation function (PACF). To make some details below easier to follow, we define this in detail:

Consider the AR(p) model

$$X_t - \phi_1 X_{t-1} - \cdots - \phi_p X_{t-p} = Z_t, \tag{3.4}$$

$1-\phi_1 z-\ldots-\phi_p z^p \neq 0$, $|z| \leq 1$. Let $R_p = [R_{i,j}]_{i,j=1}^p = [\rho(i-j)]_{i,j=1}^p$ be the $p \times p$ matrix of ACFs computed under a finite variance assumption, $\boldsymbol{\rho} = (\rho(1), \ldots, \rho(p))'$, $\boldsymbol{\phi} = (\phi_1, \ldots, \phi_p)'$. Then the Yule-Walker matrix equation is

$$R_p \boldsymbol{\phi} = \boldsymbol{\rho}, \tag{3.5}$$

and the Yule-Walker estimate of $\boldsymbol{\phi}$ is then defined as the solution of

$$\hat{R}_p \hat{\boldsymbol{\phi}} = \hat{\boldsymbol{\rho}}, \tag{3.6}$$

where $\hat{R}_p = [\hat{\rho}(i-j)]_{i,j=1}^p$ and $\hat{\boldsymbol{\rho}} = (\hat{\rho}(1), \ldots, \hat{\rho}(p))'$.

To define the PACF in the AR case, we consider vectors $\boldsymbol{\phi}_m^* = (\phi_1^*, \ldots, \phi_m^*)$, where $\phi_i^* = \phi_i$ for $i \leq p$ and $\phi_i^* = 0$ when $i > p$. For this vector we write Yule-Walker equation (3.5) as $R_m \boldsymbol{\phi}^* = \boldsymbol{\rho}_m$, where $\boldsymbol{\rho}_m = (\rho(1), \ldots, \rho(m))$. The PACF at lag m, ϕ_{mm}, is then defined as the m-th component of the vector $\boldsymbol{\phi}_m^* = R_m^{-1} \boldsymbol{\rho}$. Similarly, the sample PACF function at lag m, $\hat{\phi}_{mm}$, is defined as the m-th component of the vector

$$\hat{\boldsymbol{\phi}}_m^* = \hat{R}_m^{-1} \hat{\boldsymbol{\rho}}. \tag{3.7}$$

In the heavy-tailed case the ACF and PACF do not exist, but we can still use equations (3.2) – (3.7) to define their sample equivalents.

In the general ARMA case, the PACF at lag h is defined as the sample correlation between the residuals of X_{t+h} and X_t after linear regression (under a finite variance assumption) on $X_{t+1}, X_{t+2}, \ldots, X_{t+h-1}$.

We shall use the notation $\tilde{\phi}_{kk}$ for the PACF's when the centered ACFs $\tilde{\rho}$'s of (3.3) are used in the Yule-Walker equation (3.6) instead of the non-centered $\hat{\rho}$'s of (3.2). For most of the simulations we shall prefer to use the centered variables, and so will need to assume that $\alpha > 1$.

We shall see, basically via simulation, that both the ACF and PACF provide excellent tools for studying stable time series. It is perhaps rather surprising that although second moments are infinite in the stable case, the tools that we are used to from the Gaussian case are still available, albeit with some modifications.

The following subsection sets up the theory underlying the use of the ACF and PACF. We then provide some tables for hypothesis testing, followed by two sections on numerics. The final subsection looks at the use of the Akaike Information Criterion in the stable setting.

3.1 The basic theory

The theoretical basis for the usage of the sample ACF for the identification of the order q of a MA(q) time series is the following fundamental result of [DR2] (we follow here [BD] Theorem 13.3.1):

Theorem 3.1 *Let $\{Z_t\}$ be an iid symmetric sequence of α-stable random variables and let $\{X_t\}$ be the strictly stationary process,*

$$X_t = \sum_{j=-\infty}^{\infty} \psi_j Z_{t-j}, \tag{3.8}$$

where

$$\sum_{j=-\infty}^{\infty} |j||\psi_j|^\delta < \infty \quad \text{for some } \delta \in (0, \alpha) \cap [0, 1].$$

Define for such a process an analogue of the ACF function, namely

$$\rho(h) = \sum_j \psi_j \psi_{j+h} / \sum_j \psi_j^2, \; h = 1, 2, \ldots . \tag{3.9}$$

Then, for any positive integer h,

$$(n/\ln n)^{1/\alpha}(\hat{\rho}(1) - \rho(1), \ldots, \hat{\rho}(h) - \rho(h))' \Rightarrow (Y_1, \ldots, Y_h)', \tag{3.10}$$

where

$$Y_k = \sum_{j=1}^{\infty} (\rho(k+j) + \rho(k-j) - 2\rho(j)\rho(k)) \, S_j/S_0. \tag{3.11}$$

Here $S_0, S_1, \ldots$ are independent stable variables; S_0 is positive with $S_0 \sim S_{\alpha/2}(C_{\alpha/2}^{-2/\alpha}, 1, 0)$, and the $S_j \sim S_\alpha(C_\alpha^{-1/\alpha}, 0, 0)$ where

$$C_\alpha = \begin{cases} \frac{1-\alpha}{\Gamma(2-\alpha)\cos(\frac{\pi\alpha}{2})} & \text{if } \alpha \neq 1 \\ \frac{2}{\pi} & \text{if } \alpha = 1. \end{cases} \tag{3.12}$$

The marginal *distribution of each Y_k is somewhat simpler, and we have*

$$Y_k = \left(\sum_{j=1}^{\infty} |\rho(k+j) + \rho(k-j) - 2\rho(j)\rho(k)|^\alpha \right)^{1/\alpha} U/V, \tag{3.13}$$

where V and U are independent stable random variables with the same distributions as S_0 and S_1 in (3.11). When $\alpha > 1$ then (3.10) is also true when $\hat{\rho}(h)$ is replaced by its mean-corrected version, $\tilde{\rho}(h)$.

We consider now an example of this theorem in practice. Let X_t be the symmetric stable MA(q) process,

$$X_t = Z_t + \theta_1 Z_{t-1} + \cdots + \theta_q Z_{t-q}. \tag{3.14}$$

Then the above theorem implies that

$$(n/\ln n)^{1/\alpha}(\tilde{\rho}(h) - \rho(h)) \Rightarrow \left(1 + 2\sum_{j=1}^{q} |\rho(j)|^\alpha\right)^{1/\alpha} U/V, \quad h > q, \tag{3.15}$$

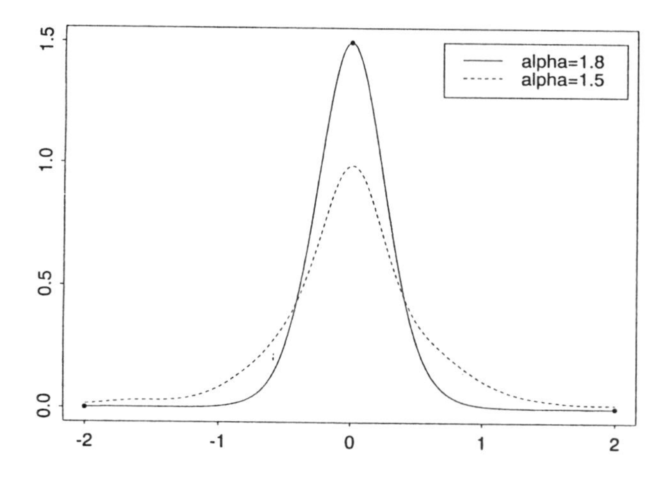

Figure 4: Kernel density estimates of U/V for $x \in [-2, 2]$ and $\alpha = 1.5$ and 1.8.

where the right hand side reduces to U/V if $q = 0$. Thus one can use the above results to plot confidence intervals for the ACF function and identify the parameter q, once the distribution of U/V is known.

A similar result also holds for the distribution of the PACF. Since $\hat{\rho} \xrightarrow{P} \rho$ and $\hat{R}_p \xrightarrow{P} R_p$, the consistency of the Yule-Walker estimates follows. In fact, the mean value theorem gives

$$\hat{\phi} - \phi = D(\hat{\rho} - \rho) + o_p(\hat{\rho} - \rho),$$

where D is the $p \times p$ matrix of partial derivatives of vector function $\psi(\mathbf{z}) := R_p(\mathbf{z})^{-1}\mathbf{z}$. Here $R_p(\mathbf{z}) = [z_{|i-j|}]_{i,j=1}^{p}$, $z_0 \equiv 1$ and $\phi = \psi(\rho)$. Theorem 3.1 then yields that

$$(n/\ln n)^{1/\alpha}(\hat{\phi} - \phi) \Rightarrow D(Y_1, \ldots Y_p)' \tag{3.16}$$

where the vector $(Y_1, \ldots Y_p)$ has distribution described by (3.11) - (3.13).

The limiting distribution of the PACF's is now given by (3.16) and (3.11) - (3.13), and so is, in general, rather complicated. However, when $p = 0$, the right hand side of (3.16) reduces to U/V, which is the same limit as for the ACF of white noise. Since in the null hypothesis case this is what is required to test which of the ϕ_{mm} are zero, we are in the fortunate situation of being able to use the same distribution twice.

3.2 Some important quantiles

In practice, the distribution of U/V cannot be computed theoretically, but only via simulation or numerical integration of the joint density of the vector (U, V) over an appropriate region.

Figure 4 gives the density of U/V for two values of α. Note the high tails of the distribution, which are high even in relation to stable distributions.

Table 3 contains the 97.5% quantiles of U/V (the distribution of U/V is symmetric) which for $\alpha < 2$ were found via simulation of 500,000 values of U/V using a corrected version of the S-plus routine for generating stable random variables. (cf. [ST] p. 43. The S-plus routine does not quite deliver what you might expect in the asymetric case!) Our value for $\alpha = 1$ coincides with the value quoted in [BD], p. 541, and is found via numerical integration.

Table 3: 97.5% quantiles of U/V

α	97.5% quantile	α	97.5% quantile
0.3	9.338e+04	1.3	3.865e+00
0.4	4.625e+03	1.4	2.814e+00
0.5	7.375e+02	1.5	2.059e+00
0.6	1.986e+02	1.6	1.516e+00
0.7	7.745e+01	1.7	1.096e+00
0.8	3.710e+01	1.75	9.280e-01
0.9	2.072e+01	1.8	7.637e-01
1.0	1.240e+01	1.9	4.765e-01
1.1	8.088e+00	2.0	1.960e+00
1.2	5.532e+00		

3.3 Estimating the lag parameters via the ACF and PACF

To see how useful the above results are, and to compare them to an attempt to estimate the parameters p and q assuming normality, we conducted the following, rather illuminating, double blind study.

We simulated 100 time series of length $n = 1000$ with symmetric stable innovations with $\alpha = 1.2$. The models were selected at random from the following five models:

$$\begin{aligned} X_t &= Z_t, \\ X_t &= Z_t - .7Z_{t-1}, \\ X_t &= Z_t - .7Z_{t-1} + .8Z_{t-2}, \\ X_t - .7X_{t-1} &= Z_t, \\ X_t - .7X_{t-1} + .8X_{t-2} &= Z_t. \end{aligned}$$

We plotted the ACF and PACF for each of these series, and then used these plots to try to identify the true model using the standard time series technique of looking at the plots and seeing how they behave relative to the 95% confidence intervals under a white noise null hypothesis. Since we did not want to assume α to be known, we used confidence intervals corresponding to the Gaussian ($\alpha = 2$) and Cauchy ($\alpha = 1$) cases. (In fact, the confidence intervals for the true $\alpha = 1.2$ and Cauchy cases were indistinguishable to the human eye. The Gaussian intervals, however, were about 28% shorter.) The conclusions were then compared against the true models.

The procedure showed 31% error when Gaussian limits were used and 17% error using Cauchy limits. Although the error rate clearly depends on experience of a person doing identification, it is clear from this study that using stable limits for heavy-tailed data reduces the error rate significantly.

3.4 On asymptotics or "a funny thing happenned on the way to ∞"

A fact often touted by stable time series theorists as a compensation for the difficulties generally associated with stable rather than Gaussian analysis is that the rate of convergence of $\tilde{\rho}(h) - \rho(h)$ to zero is of the order $O([n/\ln n]^{-1/\alpha}) = o(n^{-1/\beta})$ for all $\beta > \alpha$, which is considerably faster than in Gaussian case when the rate is on the order of $O(n^{-1/2})$. However, despite this comforting fact, there are some other, rather problematic, phenomena associated with this convergence, since the rate of convergence of the *distribution* of $\tilde{\rho}(h) - \rho(h)$ to the limiting distribution is actually very slow.

Before we mention some theory, consider Table 4 , which indicates how fast (or slow) the distribution of $\tilde{\rho}(1)$ converges to the theoretical one for the white noise model.

For values of $\alpha = 1.4, 1.7, 1.75$, and 1.8 and sample size n, we computed 10,000 coefficients $\tilde{\rho}(1)$ from independent white noise sequences with corresponding values of α and n and checked the percentage of times the coefficient was *not* within the nominal 95% confidence interval. (Of course, this should be 5%.) For determining confidence intervals we used three different distributions: a stable distribution with the correct α (as described above), a Cauchy distribution and Gaussian distribution. (i.e. we used either the correct value of α, or behaved as if $\alpha = 1$ or 2). The results show that when the correct stable distribution is used, the convergence to theoretical 5% error is very slow, and that for "small" sample sizes of the order of 1000 the Cauchy based limits actually give the best results.

The main reason behind this phenomenon seems to be the slow convergence of the distributions of stable averages to their limiting distributions, a fact that has a well documented history. (cf. [CH, HA, HW, JM] although none of these quite treats our setting.)

From the practical point of view, this phenomenon shows up in an interesting way in the numerics. Figure 5 shows the values of the upper limits of the confidence intervals for $(\tilde{\rho}(1) - \rho(1))$ for various α and n, based on the limiting distribution of Theorem 1 and a white noise time series. Note how, for "small" n, the confidence interval shrinks to a point as $\alpha \to 2$.

Table 4: Percent error for 95% confidence interval of $\tilde{\rho}(1)$, white noise				
α	n = 1000	n = 10,000	n = 500,000	n = 1,000,000
(i) α-stable limits				
1.4	1.43	2.28	3.08	3.27
1.7	4.37	4.58	4.68	3.67
1.75	6.53	5.35	5.40	5.60
1.8	10.55	8.65	7.44	6.29
(ii) Cauchy limits				
1.4	1.24	6.62	32.27	40.10
1.7	0.80	15.87	69.87	73.95
1.75	0.70	17.19	73.07	79.65
1.8	0.74	19.79	76.04	80.69
(iii) Gaussian limits				
1.4	2.91	2.17	0.85	0.89
1.7	3.46	3.36	2.14	1.35
1.75	3.56	3.23	2.72	2.52
1.8	4.22	3.68	2.90	2.41

The reason for this is not clear. One possibility is that for small n, the limiting distribution of U/V is not appropriate. What seems more reasonable, however, is that the norming constants used to obtain the limiting distribution, while asymptotically correct, are too large (in an α dependent fashion) for small and intermediate values of n.

It is not totally clear how to get around this problem without using something perhaps like a bootstrap. However, there is growing evidence that in the stable situation bootstrapping is also problematic (cf. [LPPR] in this volume.)

One possible approach comes out of Table 5, which explores the $n = 1000$ case for different α's. Our model is again white noise, and we record the percentage of times when the sample coefficient $\tilde{\rho}(1)$ is outside the 95% confidence interval; i.e., the percentage of wrong identifications of the model when the identification procedure is done by computer and is based on the value of $\tilde{\rho}(1)$ only. The confidence levels were computed based on the true α-stable distribution, as well as Cauchy and Gaussian distributions. The number of simulations for each case was $m = 10,000$.

It seems that the Gaussian limits perform poorest, at least for $\alpha < 1.7$, and that for all α, Cauchy limits not only perform better than the others but do quite well on an absolute scale. (The error here is never larger than 1.28%). Furthermore, for small α, there is no significant benefit in using the confidence limits based on the true stable distribution. But, for large α, the rate of convergence of $\tilde{\rho}(1)$ to its theoretical distribution is so slow that Cauchy or Gaussian distribution should be used. Comparison with Table 4

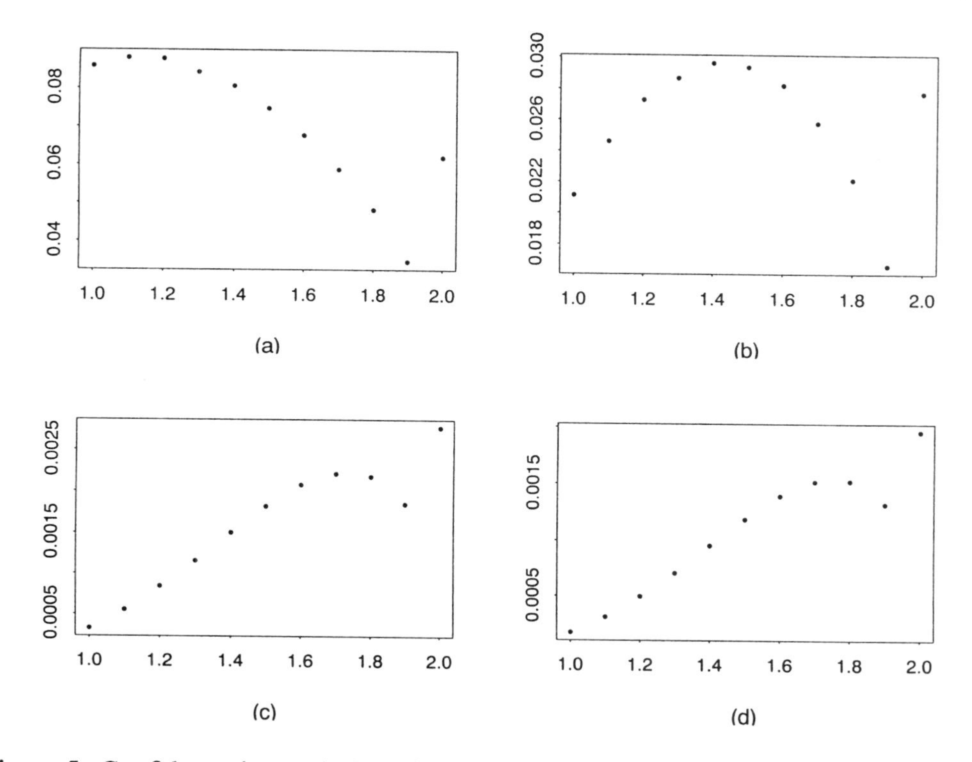

Figure 5: Confidence intervals for $(\tilde{\rho}(1) - \rho(1))$ for $1 \leq \alpha \leq 2$ and various values of n. (a) $n = 1,000$, (b) $n = 5,000$, (c) $n = 500,000$, (d) $n = 1,000,000$

strengthens this point.

In Table 6 we continue the theme of using Cauchy based bounds regardless of the true value of α. While this clearly gives a conservative test, it turns out that in practice it is not overly so, and the results here illustrate how amazingly well the Cauchy bounds perform in the identification of the MA(1) model $X_t = Z_t - .8Z_{t-1}$. Again, each specific case was run $m = 10,000$ times. In each run for sample size $n = 1000$ and given α we computed $\tilde{\rho}(1), \ldots, \tilde{\rho}(10)$ and recorded an "error" when $\tilde{\rho}(1)$ was within the confidence limits for white noise or when $\tilde{\rho}(1)$ was outside these limits but one of the coefficients $\tilde{\rho}(2), \ldots, \tilde{\rho}(10)$ was also outside the limits, so identification as MA(1) would not be called for. The confidence limits were taken to be (i) $(\ln n/n)^{1/\alpha}(U/V)$ with U and V coming from the true distribution, (ii) as in (i) but with U and V corresponding to Cauchy distribution and with $\alpha = 1$, (iii) $1.96/\sqrt{n}$, (iv) according to Bartlett's formula for MA(1) model with true value of $\rho(1)$. The whole procedure was performed by computer without human intervention.

Table 5: Percent error for 95% confidence interval of $\tilde{\rho}(1)$, $n = 1000$, WN			
α	*true α-stable distribution*	*Cauchy limits*	*Gaussian limits*
1.0	1.12	1.12	2.03
1.1	1.20	1.28	2.32
1.2	1.13	1.22	2.46
1.3	1.27	1.23	2.61
1.4	1.43	1.24	2.91
1.5	1.85	1.18	3.26
1.6	2.55	1.19	3.33
1.7	4.37	0.80	3.46
1.8	10.55	0.74	4.22
1.9	25.63	0.42	4.55
2.0	4.70	0.77	4.70

It is clear, although perhaps somewhat surprising, that Cauchy limits work the best and that the identification procedure works better for heavy-tailed series than for their finite variance counterparts.

Table 6: Percent of wrong identifications of MA(1) model				
α	*true α-stable*	*Cauchy*	*Gaussian*	*Bartlett*
1.0	10.31	10.31	17.36	12.97
1.1	11.61	12.13	20.36	14.81
1.2	11.48	12.08	22.15	15.38
1.3	12.96	12.44	25.10	16.64
1.4	14.89	12.53	27.75	17.80
1.5	20.33	13.80	31.99	19.71
1.6	28.19	13.60	35.52	20.80
1.7	45.59	13.50	40.33	22.32
1.8	68.55	13.85	43.49	23.95
1.9	91.59	14.86	48.39	26.17
2.0	52.59	17.21	52.59	29.73

We close this subsection with a some brief information on the asymptotic behaviour of the PACF, which, not surprisingly, is similar to that of the ACF. Table 7 is a PACF version of Table 5, generated in the same fashion, however with data on $\tilde{\phi}_{11}$ rather than $\tilde{\rho}(1)$ being tabulated.

Table 7: Percent error for 95% confidence interval of $\tilde{\phi}_{11}$, $n = 1000$, WN			
α	*true α-stable distribution*	*Cauchy limits*	*Gaussian limits*
1.0	1.07	1.07	1.93
1.1	1.27	1.30	2.39
1.2	1.13	1.25	2.44
1.3	1.43	1.36	2.62
1.4	1.44	1.20	3.11
1.5	1.87	1.30	3.23
1.6	2.45	1.08	3.33
1.7	4.64	0.84	3.86
1.8	10.52	0.70	4.35
1.9	25.93	0.47	4.13
2.0	4.84	0.83	4.84

3.5 The AIC criterion

In the finite variance case the prime criterion for automated model selection recommended by [BD] is the AICC, a modified version of Akaike's AIC criterion. An investigation of the AIC criterion in the infinite variance situation was carried out by [BH] and [KN1].

For an autoregressive model the AIC statistics are defined by

$$AIC(k) = n \ln \hat{\sigma}^2(k) + 2k,$$

where n is the sample size, and $\hat{\sigma}^2(k)$ is the estimate of the innovation variance obtained from Yule-Walker estimates for k-th order autoregressive sequence (cf. (3.6)). Then

$$\hat{p} = \text{argmin}_{k \leq K} AIC(k),$$

where K is an acceptable upper bound for p, is the corresponding estimate of the order p. [KN1] showed that this procedure is consistent for heavy-tailed situations.

To see how the AIC criterion works in practice for stable AR series, we performed 1000 simulations of the AR(1) model

$$X_t = 0.4X_{t-1} + Z_t, \tag{3.17}$$

for $\alpha = 0.5,\ 1.0,\ 1.75,\ 2.0$. The sample size was fixed at $n = 200$, which is rather small by stable standards, where larger sample sizes are required than in the Gaussian case.

In each run we assumed that we were looking for the best AR model, i.e., we looked for order p which minimised the AIC criterion. We then recorded the number of times that the correct order ($p = 1$) was correctly identified.

To check how the AIC criterion works for a MA model, we used the S-plus *arima.mle* routine which defines AIC as (-2) times the log Gaussian likelihood plus two times the number of parameters fit (see the definition in [BD], p. 304 or consult the S-plus manual). We performed 1000 runs for model:

$$X_t = Z_t + 0.8Z_{t-1} \tag{3.18}$$

with $\alpha = 0.5,\ 1.0,\ 1.75,\ 2.0,\ n = 200$. Again, in each run we assumed that we were looking for the best MA model, and recorded the number of correct identifications. The results of simulations for both models are given in Table 8 and the conclusion is that in both cases the AIC criterion works better the heavier the tails!

The MA case is particularly interesting, since we have no theoretical justification for applying the AIC based on the Gaussian likelihood to heavy-tailed data.

4. Parameter estimation for ARMA models: The Whittle estimator

Paper [CD] of this volume gives an extensive review of estimation techniques for linear processes with stable innovations. They present convincing evidence to the effect that LAD and MLE estimators are superior, in the heavy-tailed setting, to the estimators traditionally used in finite variance time series, such as least square or Whittle (periodogram) estimators. However, for both of these cases, the sampling distribution of the parameter estimates is, at least at the moment, numerically as well as theoretically intractable.

Table 8: Percent of correct model identifications made by AIC.

α	*AR(1) – (3.17)*	*MA(1) – (3.18)*
0.5	96.1	92.5
1.0	89.6	89.0
1.75	78.7	79.7
2.0	73.2	72.5

However, for the Whittle estimator, we have the following result of [MGKA]:

Theorem 4.1 *Let $\{X_t\}$ be a causal, invertible, α-stable, ARMA(p, q) process, with parameters $\beta = (\phi_1, \ldots, \phi_p, \theta_1, \ldots, \theta_q)'$. Let C denote the space of permissible parameter values: i.e., $C = \{\beta \in \Re^{p+q} : \phi(z) \cdot \theta(z) \neq 0$ for $|z| \leq 1$ and $\phi(\cdot),\ \theta(\cdot)$ have* *no common zeroes}. Define the polynomials*

$$\phi(z) = 1 - \phi_1 z - \ldots - \phi_p z^p, \qquad \theta(z) = 1 + \theta_1 z + \ldots + \theta_q z^q \tag{4.1}$$

and, for all $-\pi < \lambda \leq \pi$, *the "power transfer function"*

$$g(\lambda; \beta) = \frac{|\theta(e^{-i\lambda})|^2}{|\phi(e^{-i\lambda})|^2}. \tag{4.2}$$

Denote the self-normalized periodogram by

$$I_{n,X}(\lambda) = \frac{|\sum_{t=1}^{n} X_t e^{-i\lambda t}|^2}{\sum_{t=1}^{n} X_t^2}, \qquad -\pi < \lambda \leq \pi. \tag{4.3}$$

The periodogram, or "Whittle" estimator $\bar{\beta}_n$ *of the true, but unknown, parameter* β_0, *is found by minimizing*

$$\sigma_n^2(\beta) = \frac{2\pi}{n} \sum_j \frac{I_{n,X}(\lambda_j)}{g(\lambda_j; \beta)} \tag{4.4}$$

with respect to $\beta \in C$, *where the sum is taken over all frequencies* $\lambda_j = 2\pi j/n \in (-\pi, \pi]$. *Then,*

$$\left(\frac{n}{\ln n}\right)^{-1/\alpha} \left(\bar{\beta}_n - \beta_0\right) \Rightarrow 4\pi W^{-1}(\beta_0) \frac{1}{S_0} \sum_{k=1}^{\infty} S_k b_k, \tag{4.5}$$

where the $S_0, S_1, \ldots$ *are as in Theorem 1,* $W^{-1}(\beta_0)$ *is the inverse of the matrix*

$$W(\beta_0) = \int_{-\pi}^{\pi} \left[\frac{\partial ln\, g(\lambda; \beta_0)}{\partial \beta}\right] \left[\frac{\partial ln\, g(\lambda; \beta_0)}{\partial \beta}\right]^T d\lambda,$$

and

$$b_k = \frac{1}{2\pi} \int_{-\pi}^{\pi} e^{-ik\lambda} g(\lambda; \beta_0) \frac{\partial g^{-1}(\lambda; \beta_0)}{\partial \beta} d\lambda,$$

where g^{-1} *denotes the reciprocal of* g.

Self-normalization of the periodogram, as in (4.3), is essential for the proof of the above Theorem. However, in practice, in order to find the estimator $\bar{\beta}_n$ we minimize expression (4.4) when self-normalized periodogram $I_{n,X}(\lambda)$ is replaced by

$$\hat{I}_{n,X}(\lambda) = \left| n^{-1/\alpha} \sum_{t=1}^{n} X_t e^{-i\lambda t} \right|^2. \tag{4.6}$$

One of the useful aspects of this result is that the asymptotic sample distribution involved here is closely related to that which arises in the study of the ACF and PACF, so that simple extensions of the numerics required to find confidence intervals there also work here.

To see how close the theoretical asymptotic distribution of the Whittle estimator is to the true sample distribution when n is fixed, we simulated data using the AR(1) model (3.17). We ran three different sample sizes $n = 200,\ 1000$ and $10,000$ and two values of $\alpha = 1.5,\ 1.75$. For comparison we also simulated Gaussian data and used confidence intervals for the MLE, the asymptotic distribution of which coincides with the distribution of Whittle estimates in the finite variance case. We ran $m = 10,000$ simulations for each particular case.

The results, described in Table 9 and on Figure 6, show that the confidence intervals based on the limiting distribution (4.5) work very well even for the relatively small sample size of $n = 200$. The entries in Table 9 are the percent of estimates $\bar{\beta}_n \equiv \{\bar{\phi_1}\}_n$ which fall into the 95% confidence interval centered around the true parameter value. Although the numbers in the Gaussian row are the closest to theoretical 95%, the results for $\alpha = 1.75$ are impressive and unexpected, given the slow convergence rate of the distributions of the sample ACF and PACF. Figure 6 gives histograms for the Whittle estimate $\{\bar{\phi_1}\}_n$ for $\alpha = 1.75$ and the three different sample sizes used. The vertical bars represent 95% confidence levels.

Table 9: Percent of $\{\bar{\phi_1}\}_n$'s within 95% confidence interval; model (3.17)

α	*sample size n=200*	*sample size n=1000*	*sample size n=10000*
1.5	99.35	98.48	96.41
1.75	92.71	93.76	95.08
2	94.77	94.55	94.89

5. Diagnostic checking

After identifying and estimating the parameters of a times series model, it is always nice (although sometimes disconcerting) to see if the estimated model is really a good fit to the data. This is where diagnostic checking procedures come in. This usually involves identifying the residuals and seeing how well they match the distribution originally assumed for the innovations. In our case, this involves checking whether the residuals are i.i.d. under the assumption of an SαS distribution.

For infinite variance series we recommend the four following steps:

(i) Graph the residuals; the pattern should follow a white noise model.

(ii) Check that the ACF and PACF of the residuals are those of white noise.

The results and recommendations of Section 3 all apply here.

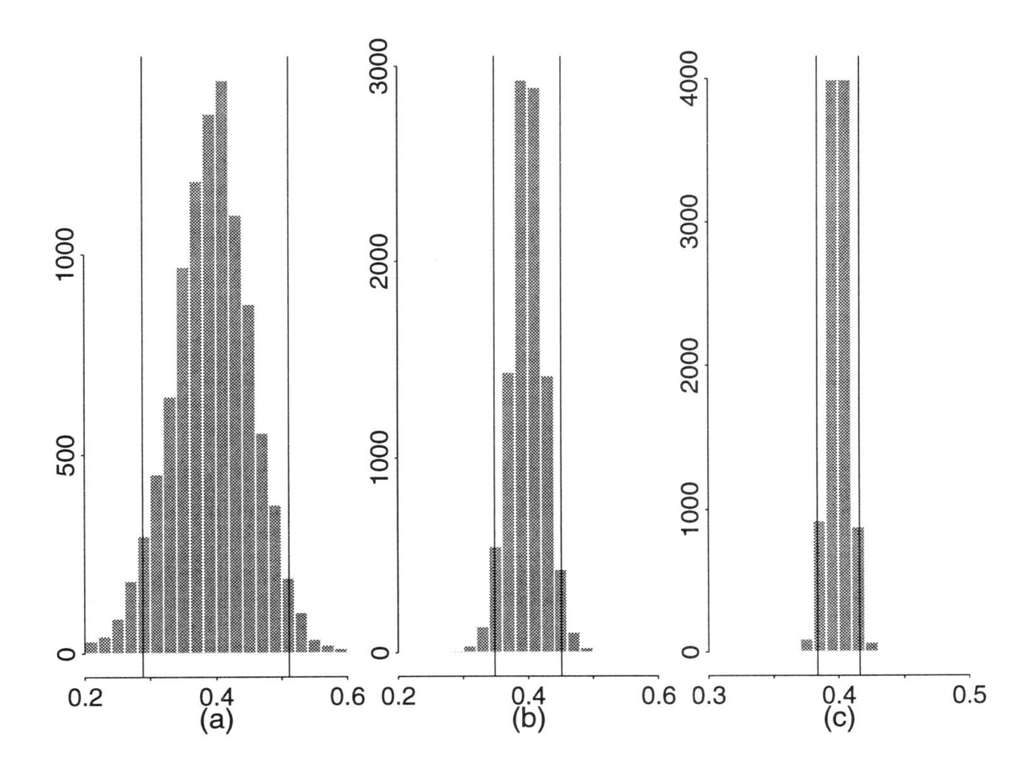

Figure 6: Histograms and theoretical 95% confidence intervals for Whittle estimate $\{\bar{\phi}_1\}_n$ of ϕ_1 in model (3.17), (a) $n = 200$, (b) $n = 1000$, (c) $n = 10,000$.

(iii) The Durbin-Watson test of [PL].

(iv) Various non-parametric tests, see [BD], p.312-313.

We ran a small simulation to see how well the second of these techniques actually worked. For models (3.17) and (3.18), stable parameters $\alpha = 1.5, 1.75, 2$, and sample sizes $n = 200, 1000, 10,000$ we simulated data, identified a model using the techniques of Section 3, estimated parameters (using the Whittle estimator) and then checked whether the residuals were consistent with stable white noise model.

Model identification was based on ACF/PACF analysis, unless more than one model seemed acceptable, in which case we differentiated between the models via the AIC. For the "small" sample sizes of $n = 200, 1000$ we used Cauchy based confidence limits for the ACF/PACF analysis, while for $n = 10,000$ we used Gaussian limits. In fitting the residuals, with sample sizes $n = 200, 1000$ we used both Cauchy and Gaussian limits.

Table 10: Diagnostic checking based on ACF/PACF; models (3.17)–(3.18)

Model	n	*Ident.*	$\hat{\beta}_n$	*Diag. Check (C)*	*Diag. Check(G)*
$\alpha = 1.5$					
(3.17)	200	AR(1) (C)	.355	WN	NOT WN
(3.18)	200	MA(1) (C)[1]	.867	WN	NOT WN
(3.17)	1000	AR(1)(C)[2]	.388	WN	WN
(3.18)	1000	MA(1) (C)	.785	WN	WN
(3.17)	10000	AR(1) (G)	.402	N/A	WN
(3.18)	10000	MA(1) (G)	.793	N/A	WN
$\alpha = 1.75$					
(3.17)	200	AR(1) (C)	.388	WN	WN
(3.18)	200	MA(1) (C)[3]	.835	WN	NOT WN
(3.17)	1000	AR(1) (C)	.400	WN	WN
(3.18)	1000	MA(1) (C)[4]	.820	WN	NOT WN
(3.17)	10000	AR(1) (G)	.404	N/A	WN
(3.18)	10000	MA(1) (G)	.797	N/A	NOT WN
$\alpha = 2.00$					
(3.17)	200	AR(1) (G)	.443	WN	WN
(3.18)	200	MA(1) (C)[5]	.767	WN	WN
(3.17)	1000	AR(1) (G)	.398	WN	WN
(3.18)	1000	MA(1) (C)	.822	WN	WN
(3.17)	10000	AR(1) (G)	.425	N/A	NOT WN
(3.18)	10000	MA(1) (G)	.801	N/A	WN

[1] *Gaussian and true* 1.5*-stable limits indicate MA(11). AIC indicates MA(1).*

[2] *Cauchy limits indicate AR(1) or ARMA(1,1). AIC indicates AR(1).*

[3] *Gaussian and true* 1.75*-stable limits indicate MA(5). AIC indicates MA(1).*

[4] *Gaussian limits indicate MA(5). True* 1.75*-stable limits indicate possibly MA(8). AIC indicates MA(1) or MA(5)(the value is almost the same). For reasons of parsimony we chose MA(1).*

[5] *True Gaussian limits indicate MA(13), while Cauchy limits indicate correct model MA(1). AIC indicates MA(1).*

The results summarized in Table 10 show that diagnostic checking based on ACF/PACF works as well in the stable case and in the Gaussian. The letters (C) or (G) in the table indicate that, respectively, Cauchy or Gaussian based confidence limits were used. The coefficient estimate $\hat{\beta}_n$ is $\bar{\phi}_1$ in the AR(1) (3.17) case, and $\bar{\theta}_1$ in the MA(1) (3.18) case.

As a guide to reading the table, consider the first line, which indicates that an

AR(1) series of length $n = 200$ and stable innovations with $\alpha = 1.5$ was generated according to the model (3.17). It was correctly identified as an AR(1) model, using Cauchy confidence intervals for the ACF and PACF. The estimate of the AR coefficient was .355. (The actual value was 0.4; cf. (3.17).) The residuals were then investigated. Using Cauchy confidence intervals they were judged to be white, which was not the case with Gaussian confidence intervals.

Although we have not studied it, we note that a stable analogue of the Durbin-Watson statistic has been developed in [PL]. Since this statistic essentially checks that the ACF of the residual sequence at lag 1 is that of white noise, its asymptotic distribution is given by Theorem 3.1.

Unfortunately no other, stronger tools (such as the χ^2 test in the finite variance case) are currently available for diagnostic checking in the stable situation. This would seem to be a promising and important direction for further research.

6. Acknowledgement

We are grateful to Dr. Aaron Gross who wrote the initial S-Plus programs on which this study was based.

Appendix: McCulloch's quantile estimator of stable parameters

In this section we describe the estimator developed by McCulloch ([MC]) for the indices α (of stability) and β (of skewness) of a stable distribution, which have been referred to in the body of the paper.

Let $X \sim S_\alpha(\sigma, \beta, \mu)$ and denote the p-th quantile of this distribution by X_p. McCulloch's estimator uses five quantiles to estimate $\alpha \in [0.6, 2.0]$ and $\beta \in [-1, 1]$, and is structured as follows:

Set

$$\Phi_1(\alpha, \beta) = \frac{X_{.95} - X_{.05}}{X_{.75} - X_{.25}},$$

$$\Phi_2(\alpha, \beta) = \frac{X_{.95} + X_{.05} - 2X_{.50}}{X_{.95} - X_{.05}}.$$

Since Φ_1 is monotonic in α and Φ_2 is monotonic in β (for fixed α) we can invert these functions to obtain

$$\alpha = \Psi_1(\Phi_1, \Phi_2),$$

$$\beta = \Psi_2(\Phi_1, \Phi_2).$$

McCulloch tabulated Ψ_1 and Ψ_2 for various values of Φ_1 and Φ_2.

To form an estimator, take a random sample from a stable distribution and define $\hat{\Phi}_1$ and $\hat{\Phi}_2$ by replacing the quantiles X_k by the corresponding sample quantiles $\hat{X}_k$. Since the sample quantiles are consistent for the population quantiles, $\hat{\Phi}_1$ and $\hat{\Phi}_2$ are consistent estimators of Φ_1 and Φ_2. Define

$$\hat{\alpha} = \Psi_1(\hat{\Phi}_1, \hat{\Phi}_2),$$

$$\hat{\beta} = \Psi_2(\hat{\Phi}_1, \hat{\Phi}_2).$$

Given a random sample from a stable distribution, we can now use the tables to find $\hat{\alpha}$ and $\hat{\beta}$.

To obtain estimates of the scale and location parameters, McCulloch defined similar functions using these same 5 quantiles, which were also tabulated for various scale and location values. These tables can be used in a similar fashion so as to obtain estimates for μ and for σ.

References

[BD] P. J. Brockwell and R. A. Davis, *Time Series: Theory and Methods*, Springer, New York, 1991.

[BH] R. Bhansali, Consistent order determination for processes with infinite variance, *J. R. Statist. Soc. B*, Vol. 50 (1988), 46–60.

[BJ] G. E. P. Box and G. M. Jenkins *Time Series Analysis: Forecasting and Control* Wiley, New York, 1976.

[CD] M. Calder and R. A. Davis, Inference for linear processes with stable noise, *this volume.*

[CH] G. Christoph, On some differences in limit theorems with a normal or a non-normal stable limit law, *Math. Nachr.*, Vol. 153 (1991), 247– 256.

[CMS] J. M. Chambers, C. L. Mallows and B. W. Stuck, A method for simulating stable random variables, *Journal of the American Statistical Association*, Vol. 71 (1976), 340–344.

[DR1] R. A. Davis and S. Resnick, Limit theory for moving averages of random variables with regularly varying tail probabilities, *Ann. Prob.*, Vol. 13 (1985), 179–195.

[DR2] R. A. Davis and S. Resnick, Limit theory for the sample covariance and correlation functions of moving averages, *Ann. Statist.*, Vol. 14 (1986), 533–558.

[FE] W. Feller, *An Introduction to Probability Theory and Its Applications*, Vol. II, 4th ed., Wiley, New York, 1983.

[FR] P. D. Feigin and S. R. Resnick, Pitfalls of fitting autoregressive models for heavy-tailed time series, *in preparation*.

[HA] P. Hall, Two-sided bounds on the rate of convergence to a stable law, *Z. Wahrsch. verw. Gebiete*, Vol. 57 (1981), 349–364.

[HW] L. Heinrich and Wr. Wolf, On the convergence of U-statistics with stable limit distribution *J. Mult. An.*, Vol. 44 (1993), 266–278.

[JM] A. Janssen and D. M. Mason, On the rate of convergence of sums of extremes to a stable law, *Probab. Theory and Rel. Fields*, Vol. 86 (1990), 253–264.

[KN1] K. Knight, Consistency of Akaike's information criterion for infinite variance autoregressive processes, *Ann. Statist.*, Vol. 17 (1989), 824–840.

[KN2] K. Knight, Rate of convergence of centred estimates of autoregressive parameters for infinite variance autoregressions, *J. Time Ser. Anal.*, Vol. 8 (1987), 51–60.

[LPPR] R. LePage, K. Podgórski, M. Ryznar, Bootstrapping signs and permutations for regression with heavy-tailed errors: a seamless resampling, *this volume*.

[MC] J.H. McCulloch, Simple consistent estimators of stable distribution parameters, *Communications in Statistics – Computation and Simulation*, Vol. 15 (1986), 1109–1136.

[MGKA] T. Mikosch, T. Gadrich, C. Klüppelberg and R.J. Adler, Parameter estimation for ARMA models with infinite variance innovations, *Annals of Statistics* Vol. 23 (1995), 305-326.

[MI] T. Mikosch, Periodogram estimates from heavy-tailed data, this volume.

[NS] C. L. Nikias and M. Shao, *Signal Processing with Alpha-Stable Distributions and Applications*, Wiley, New York, 1995.

[PDM] O. V. Pictet, M. M. Dacorogna, U. A. Müller, Behaviour of tail estimates under various distributions, *this volume*.

[PL] P. C. B. Phillips, and M. Loretan, The Durbin-Watson ratio under infinite-variance errors, *J. of Econometrics*, Vol. 47 (1991), 85–114.

[R1] S. Resnick, Heavy tail modelling and teletraffic data, *Ann. Statistics*, 1997.

[R2] S. Resnick, Why non-linearities can ruin the heavy-tailed modeller's day, *this volume*.

[ST] G. Samorodnitsky, and M. Taqqu, *Stable Non-Gaussian Random Processes*, Chapman-Hall, New York, 1994.

Robert J. Adler
Department of Statistics
UNC Chapel Hill
Chapel Hill, NC, 27599-3260
and
Faculty of Industrial Engineering
Technion, Haifa, 32000, Israel.

adler@stat.unc.edu
radler@ie.technion.ac.il

Raisa E. Feldman and Colin Gallagher
Statistics and Applied Probability
UC Santa Barbara
Santa Barbara, CA, 93106-2050

epstein@pstat.ucsb.edu
colin@pstat.ucsb.edu

Inference for Linear Processes with Stable Noise [1]

M. Calder and R. A. Davis

Abstract

In this paper M-estimators of the parameters of ARMA models with heavy-tailed or stable distributed noise are discussed. A general method for deriving the limit behavior of M-estimators is outlined, and three specific estimators: the least absolute deviation estimator, the least squares estimator, and the maximum likelihood estimator, are examined in detail. Simulations are then used to compare the estimators under a variety of conditions.

1. Introduction

Estimation of the parameters for a linear process typically relies only on second-order properties of the process described by either the autocovariance or spectral density functions. However, when the innovations driving the process are heavy-tailed, such estimators are no longer optimal. M-estimation provides a convenient and flexible method of estimation, which includes a number of important estimation procedures. In this paper three different M-estimators are compared: least-absolute deviations, least-squares, and maximum likelihood.

The models under consideration are all causal-invertible ARMA models of the form,

$$X_t = \phi_1 X_{t-1} + \cdots + \phi_p X_{t-p} + Z_t + \theta_1 Z_{t-1} + \cdots + \theta_q Z_{t-q}, \tag{1}$$

where $\{Z_t\}$ is an i.i.d. sequence of random variables with a common distribution function F. For the most part it will be assumed that F is a symmetric α-stable distribution although this property can be relaxed to large extent. The focus of this paper is to study estimators of the parameter vector $\boldsymbol{\beta} = (\phi_1, \ldots, \phi_p, \theta_1, \ldots, \theta_q)'$.

As shown in a number of papers, second-order based estimation methods for $\boldsymbol{\beta}$ perform surprisingly well when the innovations are heavy-tailed—much better than they do in the finite variance case. There are, however, very few papers in the literature that describe estimation procedures which take genuine advantage of the heavy-tailedness of the data. The results that follow all show that there are alternatives to second-order-based estimation methods that will give superior estimates when the innovations have heavy tails.

---

[1]This research was supported in part by NFS grant DMS-9504596

Two alternatives are examined through simulation, least absolute deviation, and maximum likelihood. The least-absolute deviations method is found to be both computationally efficient and nearly optimal in estimating the parameters of the ARMA model. Moreover, this method does not require knowledge (or estimation) of the parameters of the innovations distribution. Maximum likelihood is a standard and often preferred estimation procedure and serves as a barometer of performance for other estimation methods. There can be, however, severe limitations in using maximum likelihood estimation. For example, in our application, calculation of the likelihood is notoriously difficult due to the lack of a closed form expression of the stable density function. In addition, maximum likelihood implicitly assumes a specific innovations distribution, and unless this assumption is valid, maximum likelihood has no special allure.

The remaining sections of the paper are arranged as follows. Section 2 gives a brief description of M-estimates and describes a technique used to derive the asymptotic properties of M-estimators. Section 3 focuses on three particular M-estimators and presents previous results about their limiting behavior. Section 4 contains the results of several simulations meant to demonstrate the ideas of the previous sections. Concluding remarks are made in Section 5.

2. Review of M-estimation for ARMA models

M-estimation provides a general framework for many common estimation procedures. This section presents a review of M-estimation for the parameter vector $\boldsymbol{\beta}$ of the ARMA model given by (1).

Let $X_1, \ldots, X_n$ be observations generated from the ARMA model in (1) with true parameter vector $\boldsymbol{\beta}_0 = (\phi_{01}, \ldots, \phi_{0p}, \theta_{01}, \ldots, \theta_{0q})'$. An M-estimate $\hat{\boldsymbol{\beta}}_M$ of $\boldsymbol{\beta}_0$ is any parameter vector which minimizes the objective function,

$$\sum_{t=1}^{n} \rho(Z_t(\boldsymbol{\beta})), \tag{2}$$

where $\rho(x)$ is some suitably chosen loss function and $\{Z_t(\boldsymbol{\beta})\}$ is an estimate of the innovation sequence $\{Z_t\}$. Estimates of Z_t can be calculated for any particular vector of parameters $\boldsymbol{\beta} = (\phi_1, \ldots, \phi_p, \theta_1, \ldots, \theta_q)'$ via,

$$\begin{aligned}
Z_1(\boldsymbol{\beta}) &= X_1, \\
Z_2(\boldsymbol{\beta}) &= X_2 - \phi_1 X_1 - \theta_1 Z_1(\boldsymbol{\beta}), \\
&\vdots \\
Z_n(\boldsymbol{\beta}) &= X_n - \phi_1 X_{n-1} - \cdots - \phi_p X_{n-p} - \theta_1 Z_{n-1}(\boldsymbol{\beta}) - \cdots - \theta_q Z_{n-q}(\boldsymbol{\beta}).
\end{aligned}$$

The objective function is thus a function of the parameters through the estimated innovations $\{Z_t(\boldsymbol{\beta})\}$.

Many familiar estimation procedures fit into this scheme. Least-squares (LS) estimation has $\rho(x) = x^2$, while least absolute deviations (LAD) estimation has $\rho(x) = |x|$. Maximum likelihood estimation (MLE) is also a form of M-estimator with $\rho(x) = -\log p(x)$ where $p(\cdot)$ is the probability density function of the underlying innovations. Note that the objective function in (2) is not necessarily the exact log likelihood even if $p(\cdot)$ is the actual density of the innovations due to edge effects in $Z_t(\boldsymbol{\beta})$. (For more on this point see Section 3.)

The performance of these estimators can be evaluated through their limiting behavior. This limiting behavior is studied in detail for AR processes in Davis, Knight, and Liu (1992) and for ARMA processes in Davis (1996). The results are derived by first defining a centered and rescaled parameter vector,

$$\mathbf{u} = a_n(\boldsymbol{\beta} - \boldsymbol{\beta}_0), \tag{3}$$

where the scaling constants $\{a_n\}$ are given by,

$$a_n = \inf\{x : P(|Z_1| > x) \leq n^{-1}\}.$$

Then by substituting the vector $\mathbf{u}$ into (2) and centering, one obtains the process

$$W_n(\mathbf{u}) = \sum_{t=1}^{n} [\rho(Z_t(\boldsymbol{\beta}_0 + a_n^{-1}\mathbf{u})) - \rho(Z_t(\boldsymbol{\beta}_0))].$$

Minimizing $W_n(\mathbf{u})$ with respect to $\mathbf{u}$ is equivalent to minimizing the original objective function (2) with respect to $\boldsymbol{\beta}$.

Under certain conditions on the loss function $\rho(\cdot)$ (which exclude the least squares case) and the innovation sequence $\{Z_t\}$, the process $W_n(\mathbf{u})$ converges to a process $W(\mathbf{u})$ on $C(\mathbb{R}^{p+q})$ with a unique minimum $\hat{\mathbf{u}}$. Moreover, it can then be shown that a sequence of local minima, $\hat{\mathbf{u}}_n$, of $W_n(\mathbf{u})$ will converge in distribution to $\hat{\mathbf{u}}$. This is all that is needed to obtain the limiting distribution of $\hat{\boldsymbol{\beta}}_M$ since

$$a_n(\hat{\boldsymbol{\beta}}_M - \boldsymbol{\beta}_0) = \hat{\mathbf{u}}_n \xrightarrow{d} \hat{\mathbf{u}}.$$

The distribution of $\hat{\mathbf{u}}$, which depends on the loss function $\rho(\cdot)$ and the distribution of the innovations, is typically intractable. Davis and Wu (1994) discuss the use of resampling techniques to approximate the distribution of $(\hat{\boldsymbol{\beta}}_M - \boldsymbol{\beta}_0)$. The least squares estimate also converges in distribution when scaled and centered, however the scaling factor is typically of a smaller order than a_n above.

3. M-Estimate examples

The conditions under which convergence of the M-estimate occurs depends on the loss function $\rho(\cdot)$ and the distribution of the innovations sequence. This section examines the convergence properties of the LAD, LS, and MLE estimators and compares these estimates through their respective asymptotic scaling factors.

For simplicity, assume throughout this discussion that the innovations have a symmetric α-stable distribution with $\alpha \in (0, 2)$, even though the results presented have broader applicability. The more general case is considered in Davis (1996).

The LS estimator $\hat{\beta}_{LS}$ minimizes the sum of the squares of the estimated innovations. Least squares estimation is widely used in statistical problems since it is easy to implement and is optimal in some situations. However, when heavy-tailed errors are present, least squares methods generally perform poorly. The minimization of the squared errors can focus on a few large errors and ignore the rest of the data, resulting in inferior estimates.

The scaling factors for $\hat{\beta}_{LS}$ are of interest in their own right and also because the LS estimate is asymptotically equivalent to the Whittle estimate (see Mikosch, Gadrich, Klüppelberg, and Adler, 1995), and to the Yule-Walker estimate (see Davis, and Resnick, 1986). For $\hat{\beta}_{LS}$,

$$\frac{a_n^2}{b_n}(\hat{\beta}_{LS} - \beta_0) \xrightarrow{d} \eta,$$

for some random vector η, where

$$\begin{aligned} a_n &= \inf\{x : P(|Z_1| > x) \leq n^{-1}\} \\ &= (Cn)^{\frac{1}{\alpha}}, \\ b_n &= \inf\{x : P(|Z_1 Z_2| > x) \leq n^{-1}\} \\ &= (C^2 n \ln n)^{\frac{1}{\alpha}}, \end{aligned}$$

and C is a known constant. Thus the convergence rate for $\hat{\beta}_{LS}$ is

$$\frac{a_n^2}{b_n} = \left(\frac{n}{\ln n}\right)^{\frac{1}{\alpha}}.$$

The LAD estimator is obtained by minimizing the sum of the absolute values of the estimated innovations. Minimizing the absolute value of errors is a common alternative to minimizing squared errors when the errors have large outliers (i.e., are heavy-tailed). The outliers do not dominate the minimization procedure as they do in the least squares case. With the use of standard optimization algorithms, LAD estimation can be done quickly and efficiently, while using specialized algorithms allows LAD estimation to be performed at speeds that rival LS estimation.

The convergence of the LAD estimator is derived in Davis (1996) where it is shown that

$$a_n(\hat{\boldsymbol{\beta}}_{LAD} - \boldsymbol{\beta}_0) \xrightarrow{d} \boldsymbol{\tau}$$

for some random vector $\boldsymbol{\tau}$. The scaling constant is $a_n = (Cn)^{\frac{1}{\alpha}}$ as before. Comparing the convergence of $\hat{\boldsymbol{\beta}}_{LAD}$ to $\hat{\boldsymbol{\beta}}_{LS}$ it follows that

$$\frac{||\hat{\boldsymbol{\beta}}_{LAD} - \hat{\boldsymbol{\beta}}_0||}{||\hat{\boldsymbol{\beta}}_{LS} - \hat{\boldsymbol{\beta}}_0||} \xrightarrow{P} 0.$$

Therefore, asymptotically, $\hat{\boldsymbol{\beta}}_{LAD}$ dominates $\hat{\boldsymbol{\beta}}_{LS}$.

The exact maximum likelihood estimator is obtained by maximizing the joint density function of an observed sample, $(X_1, X_2, \dots, X_n)$, as a function of $\boldsymbol{\beta}$, say, $g(X_1, \dots, X_n; \boldsymbol{\beta})$. In general, direct computation of this joint density is intractable due to the interdependencies of the observations and the innovations sequence. However, by conditioning on the initial innovations and some unobserved data points a manageable approximation can be made. To this end, let

$$\begin{aligned} \mathbf{X}_0 &= (X_{-p+1}, \dots, X_{-1}, X_0), \\ \mathbf{Z}_0 &= (Z_{-q+1}, \dots, X_{-1}, Z_0), \end{aligned}$$

and suppose $(\mathbf{X}_0, \mathbf{Z}_0)$ has density $h(\cdot; \boldsymbol{\beta})$. Then the likelihood can be written as

$$g(X_1, \dots, X_n; \boldsymbol{\beta}) = \int\int g(X_1, \dots, X_n | \mathbf{x}_0, \mathbf{z}_0; \boldsymbol{\beta}) h(\mathbf{x}_0, \mathbf{z}_0; \boldsymbol{\beta}) d\mathbf{x}_0 d\mathbf{z}_0.$$

Now if it is assumed that n is large relative to both p and q, then

$$g(X_1, \dots, X_n | \mathbf{X}_0, \mathbf{Z}_0; \boldsymbol{\beta}) \approx g(X_1, \dots, X_n | \mathbf{X}_0 = \mathbf{0}_p, \mathbf{Z}_0 = \mathbf{0}_q; \boldsymbol{\beta}),$$

where $\mathbf{0}_p$ and $\mathbf{0}_q$ are zero vectors of length p and q, respectively. Substituting this into the expression for the likelihood gives,

$$\begin{aligned} g(X_1, \dots, X_n; \boldsymbol{\beta}) &= \int\int g(X_1, \dots, X_n | \mathbf{x}_0, \mathbf{z}_0; \boldsymbol{\beta}) h(\mathbf{x}_0, \mathbf{z}_0; \boldsymbol{\beta}) d\mathbf{x}_0 d\mathbf{z}_0 \\ &\approx \int\int g(X_1, \dots, X_n | \mathbf{x}_0 = \mathbf{0}_p, \mathbf{z}_0 = \mathbf{0}_q; \boldsymbol{\beta}) h(\mathbf{x}_0, \mathbf{z}_0; \boldsymbol{\beta}) d\mathbf{x}_0 d\mathbf{z}_0 \\ &= g(X_1, \dots, X_n | \mathbf{X}_0 = \mathbf{0}_p, \mathbf{Z}_0 = \mathbf{0}_q; \boldsymbol{\beta}) \\ &= \prod_{i=1}^{n} f(Z_i(\boldsymbol{\beta})), \end{aligned}$$

where $f(\cdot)$ is the density of the innovations. This simplification occurs because conditional on the event $\{\mathbf{X}_0 = \mathbf{0}_p, \mathbf{Z}_0 = \mathbf{0}_q\}$, the sequence $Z_i(\boldsymbol{\beta})$ represents the exact innovations sequence of the ARMA model. By taking log's the negative log-likelihood can be approximated by the objective function in (2), with $\rho(x) = -\log f(x)$. An estimator that minimizes such a loss function will be referred to as maximum likelihood even though it does not maximize the exact likelihood.

Unfortunately, it is unlikely that the distribution of the innovations will be known exactly. However, it is often possible to use a particular distribution in the loss function $\rho(\cdot)$ such that good estimates are obtained for a wide class of true innovations distributions. This practice has a long history in time series analysis. For instance, a Gaussian likelihood is often used to provide estimators, while no Gaussian assumption is made about the underlying process. (See Brockwell and Davis (1991) p. 255–256.) Also, the LAD estimator can be viewed as the maximum likelihood estimate when the innovations have a Laplace distribution. The density function used in the maximum likelihood may belong to a parametric family, in which case parameters of the family also require estimation. For our situation we assume that $f(\cdot)$ belongs to the family of symmetric α-stable densities, with scale parameter σ. Maximizing the likelihood relative to these parameters is discussed more fully in Section 4.

Convergence for the maximum likelihood estimator $\hat{\beta}_{MLE}$ will depend upon the density function chosen to form the loss function, and the (possibly different) density chosen to model the distribution of the innovations. Specifically, suppose the loss function $\rho(x)$ has derivative $\psi(x)$ that satisfies:

(a) The influence function $\psi(\cdot)$ satisfies a Lipschitz condition of order β; $|\psi(x) - \psi(y)| \leq K|x-y|^{\beta}$ where $\beta > \max(\alpha - 1, 0)$ and K is a constant.

(b) $E(|\psi(Z_1)|) < \infty$ if $\alpha < 1$.

(c) $E(\psi(Z_1)) = 0$ and $\text{Var}(\psi(Z_1)) < \infty$ if $\alpha \geq 1$.

Then (see Davis, 1996),

$$a_n(\hat{\beta}_M - \beta_0) \xrightarrow{d} \nu,$$

where ν is some random vector. (The convexity assumption in Davis (1996) is not required.)

Assume that the loss function is of the form $-\log\left(\sigma^{-1} f_{\lambda}(\frac{\cdot}{\sigma})\right)$ where $f_{\lambda}(\cdot)$ is a symmetric stable distribution, with exponent $\lambda \in (0, 2)$ fixed, and scale parameter $\sigma > 0$. For the present discussion, we assume, without loss of generality, that $\sigma = 1$. Now let the innovations sequence have a symmetric stable distribution, $f_{\alpha}(\cdot)$ (α not necessarily equal to λ). The conditions for convergence can be easily verified by using the well known properties for symmetric α-stable densities,

$$\begin{aligned} f_{\alpha}(x) &\sim C_0 x^{-(\alpha+1)} \\ f'_{\alpha}(x) &\sim C_1 x^{-(\alpha+2)} \\ f''_{\alpha}(x) &\sim C_2 x^{-(\alpha+3)} \end{aligned}$$

as $x \to \infty$, where the C_i's are known constants.

Condition (a) can be verified by bounding the derivative of $\psi(\cdot)$,

$$\psi'(x) = \frac{(f_\lambda'(x))^2 - f_\lambda(x) f_\lambda''(x)}{(f_\lambda(x))^2}.$$

Since $f_\lambda(x)$, $f_\lambda'(x)$, and $f_\lambda''(x)$ are bounded and $f_\lambda(x)$ is strictly positive, this derivative is bounded except possibly as $x \to \infty$. However using the polynomial form of $f_\lambda(x)$ and its derivatives gives, for large x,

$$\psi'(x) \sim K x^{-2},$$

and so $\psi(\cdot)$ satisfies a Lipschitz condition of order $1 > \max(\alpha - 1, 0)$.

The expectation in (b) can be handled similarly. Writing

$$\begin{aligned} E(|\psi(Z_1)|) &= \int_{-\infty}^{\infty} \left| \frac{f_\lambda'(x)}{f_\lambda(x)} \right| f_\alpha(x) dx \\ &= I_M + 2 \int_M^{\infty} \left| \frac{f_\lambda'(x)}{f_\lambda(x)} \right| f_\alpha(x) dx, \end{aligned}$$

where M is large and I_M is the (finite) center portion of the integral. Now, for M large enough, the integrand of the right hand term behaves as, $Cx^{-(\alpha+2)}$ and so the integral is finite, and hence the whole expectation is finite also.

The pair of conditions in (c) can be handled separately. Since $\psi(\cdot)$ is an odd function and Z_1 is a symmetric random variable, $E(\psi(Z_1)) = 0$. The variance of $\psi(Z_1)$ can be computed in the same way as the expectation in (b),

$$\begin{aligned} \text{Var}(\psi(Z_1)) &= \int_{-\infty}^{\infty} \left(\frac{f_\lambda'(x)}{f_\lambda(x)}\right)^2 f_\alpha(x) dx \\ &= I_M + 2 \int_M^{\infty} \left(\frac{f_\lambda'(x)}{f_\lambda(x)}\right)^2 f_\alpha(x) dx. \end{aligned}$$

Here the integrand of the right hand term behaves as $x^{-(\alpha+3)}$ and so the integral is finite, giving a finite variance. Applying the results of Davis (1996), we obtain

$$n^{\frac{1}{\alpha}}(\hat{\beta}_{MLE} - \beta_0) \xrightarrow{d} \nu,$$

where $\hat{\beta}_{MLE}$ is the M-estimate corresponding to the loss function $-\log f_\lambda(x)$.

The scaling factor for $\hat{\beta}_{MLE}$ is the same no matter what symmetric α-stable distribution is generating the innovations. Therefore any reasonable estimate of α should be adequate for determining the density to use in the loss function. This suggests one could simply use the Cauchy density ($\alpha = 1$, which has a closed form) in the loss function and still be assured that the estimator has good asymptotic properties. Alternatively, the maximum likelihood estimator could be obtained by simultaneously optimizing over α, σ, and $\boldsymbol{\beta}$ — this is the approach that we have taken.

The scaling constants for the maximum likelihood estimator are the same as for other M-estimators with loss functions satisfying (a)–(c) and as for the LAD estimator. Estimators with such scaling constants greatly outperform second-order based methods such as LS estimation. Further, in the heavy-tailed cases, MLE and LAD estimators achieve such high precision (see Figures 2 and 3) that relative efficiencies are no longer of great concern in choosing one of these estimators over another. Other considerations, such as computability and robustness come into play. We recommend the LAD over the MLE for the models considered here for these very reasons. While the performance of both estimators is excellent, the LAD estimator is much simpler and faster to compute. Moreover, the convergence of the LAD estimator is derived under weaker assumptions on the distribution of the innovations and would presumably find wider applicability. This will be borne out in the simulations that follow.

4. Simulations

This section presents the results of simulations involving the estimators of the previous section. The simulations will confirm the assertions made about the relative performance of the different estimators. First, however, some pragmatic issues that arise in the estimation of parameters of ARMA models with α-stable noise will be considered.

It has already been noted that the asymptotic scaling constants of $\hat{\boldsymbol{\beta}}_{LAD}$ and $\hat{\boldsymbol{\beta}}_{MLE}$ are of order $n^{\frac{1}{\alpha}}$ and those of $\hat{\boldsymbol{\beta}}_{LS}$ are of order $\left(\frac{n}{\log n}\right)^{\frac{1}{\alpha}}$. Both of these rates compare favorably to the typical finite variance scaling constants of $n^{\frac{1}{2}}$. The following idealized example provides some intuition into why estimates can be very good in the infinite variance case and also when they can be quite bad.

Consider an i.i.d. sequence of α-stable random variables. For such a sequence it is not unusual to see one extremely large value amongst several relatively small values. As an idealization then,

$$\{Z_t\} \approx \{\ldots, 0, 0, c, 0, 0, \ldots\}$$

for some c with $|c| \gg 0$.

Now consider estimating the parameter ϕ_0 in an AR(1) model, $X_t = \phi_0 X_{t-1} + Z_t$, driven by such innovations. The following sequences are produced:

t	Z_t	X_t	$Z(\phi)$
$\vdots$	$\vdots$	$\vdots$	$\vdots$
$m-2$	0	0	0
$m-1$	0	0	0
m	c	c	c
$m+1$	0	$\phi_0 c$	$(\phi_0 - \phi)c$
$m+2$	0	$\phi_0^2 c$	$(\phi_0^2 - \phi_0\phi)c$
$\vdots$	$\vdots$	$\vdots$	$\vdots$

The resulting objective function to be minimized is then

$$\sum_{t=1}^{n} \rho(Z_t(\phi)) \approx \sum_{k=1}^{n-m} \rho((\phi_0^k - \phi_0^{k-1}\phi)c),$$

which will clearly be minimized by having $\hat{\phi} = \phi_0$. Thus large isolated impulses will lead to very good estimates, a situation that will not occur in the finite variance case because all values tend to be on relatively the same scale.

It can happen, however, that extremely large values come in a pair, say at times $m, m+1$. In this case,

$$\{Z_t\} \approx \{\ldots, 0, 0, c_1, c_2, 0, 0, \ldots\}$$

and the objective function to be minimized is found to be

$$\sum_t \rho(Z_t(\phi)) \approx \rho(c_2 + (\phi_0 - \phi)c_1) + \sum_{k=1}^{n-m} \rho(\phi^{k-1}(\phi_0 - \phi)(c_2 + (\phi_0 - \phi)c_1)).$$

This loss function will not be minimized at $\hat{\phi} = \phi_0$, but instead at $\hat{\phi} \approx \frac{(c_2 + c_1\phi_0)}{c_1}$, which is an incorrect estimate.

Similar results hold for MA processes. Isolated large innovations yield very good estimates while certain combinations of large innovations lead to very poor estimates. Many methods exist in regression analysis for testing whether certain observations are having an undue influence on the parameter estimates. When estimating ARMA parameters with heavy-tailed innovations, similar cautions should be taken. Of course, in the following simulations we were interested in quantity and no special attention was paid to the quirks of specific realizations.

All of the simulations done here involve one of the following three models, which were also used by Mikosch et al. (1995):

$$\begin{aligned}
(M.1) \quad X_t &= 0.4X_{t-1} + Z_t, \\
(M.2) \quad X_t &= Z_t + 0.8Z_{t-1}, \\
(M.3) \quad X_t &= 0.4X_{t-1} + Z_t + 0.8Z_{t-1},
\end{aligned}$$

where $\{Z_t\}$ is an i.i.d. sequence of symmetric α-stable random variables with $\sigma = 1$. Three values of α were used for each of the models, $\alpha = 0.75$, $\alpha = 1.0$, and $\alpha = 1.75$. A scale parameter was introduced by multiplying the sequence $\{Z_t\}$ by $\sigma = 25$. Each estimate in the simulation is based upon a sequence $\{X_t\}$ of length 100. The same observation sequence was used to produce each estimator, $\hat{\beta}_{LAD}$, $\hat{\beta}_{LS}$, and $\hat{\beta}_{MLE}$, and 10,000 such sequences were simulated. For the models containing an AR component, the observations were generated by first setting $X_{-N} = 0$ and then the X_t, $t = -N+1, \ldots, 100$ were computed from the ARMA recursions. The observation sequence is then $\{X_t\}$ for $t = 1 \ldots 100$. By allowing a presample period of $N(= 100)$ observations, a stationary sequence is produced. For example, in our AR(1) model

$$X_t = \sum_{j=0}^{\infty} (0.4)^j Z_{t-j}$$

is approximated by the finite sum $\sum_{j=0}^{t+N} (0.4)^j Z_{t-j}$, which yields an error of, $\sum_{j=t+N+1}^{\infty} (0.4)^j Z_{t-j}$. The error is equal in distribution to $(0.4)^{t+N+1} X_1$, well below machine precision.

Estimates are then obtained by minimizing the objective function (2), repeated here,

$$\sum_{t=1}^{n} \rho(Z_t(\beta)),$$

where $\rho(\cdot)$ is the loss function: $|\cdot|$ for LAD estimation, $(\cdot)^2$ for LS estimation, and $-\log f(\cdot)$ for MLE estimation, and $\{Z_t(\beta)\}$ are the estimated innovations obtained recursively from the assumed parameter vector β.

The maximum likelihood estimator requires computation of the stable density, which in turn requires estimation of the index α and the scale parameter σ. This is accomplished by minimizing

$$\sum_{t=1}^{n} \log f_\alpha \left(\frac{Z_t(\beta)}{\sigma} \right) - n \log \sigma$$

as a function of β, α, and σ. Here $f_\alpha(\cdot)$ is a symmetric stable density with scale parameter 1. The minimization is done simultaneously for all of the parameters.

The first set of simulations are meant to compare the relative performances of the three different estimators, $\hat{\beta}_{LAD}$, $\hat{\beta}_{LS}$, and $\hat{\beta}_{MLE}$. Evaluation of the stable density presents serious numerical difficulties. Fortunately, a program for quickly and accurately evaluating the symmetric stable density (and distribution) is available in McCulloch (1994). The program is accurate for $\alpha \in [0.7, 2.0]$. It was somewhat abusive to use the program in the simulations with $\alpha = 0.75$, since during the optimization, calculations of the density at values of α less than 0.7 occur. It is a credit to the program that results were quite good anyway.

The results are tabulated in Tables 1–3 and graphed in the box plots of Figures 1–6. The box in each box plot is drawn from the 25th percentile to the 75th percentile of the estimates with the median dividing the box. The whiskers are drawn at the 5th and 95th percentiles. Notice that the scales of the plots are the same, that is, comparisons can be made between the box-plots for different models and different α's.

Model	True Values	$\hat{\beta}_{LAD}$	$\hat{\beta}_{MLE}$	$\hat{\beta}_{LS}$
M.1	$\phi = 0.4$	0.39358 (0.06873)	0.39463 (0.05987)	0.39176 (0.06814)
M.2	$\theta = 0.8$	0.79151 (0.05226)	0.79336 (0.04554)	0.79355 (0.05323)
M.3	$\phi = 0.4$	0.39239 (0.07861)	0.39853 (0.06746)	0.39349 (0.06746)
	$\theta = 0.8$	0.78357 (0.06152)	0.77854 (0.05612)	0.77854 (0.05612)

Table 1:
Mean and mean-absolute deviations (in parentheses) for $\hat{\beta}_{LAD}$, $\hat{\beta}_{MLE}$, and $\hat{\beta}_{LS}$, for models $M.1$, $M.2$, and $M.3$ with $\alpha = 1.75$.

Model	True Values	$\hat{\beta}_{LAD}$	$\hat{\beta}_{MLE}$	$\hat{\beta}_{LS}$
M.1	$\phi = 0.4$	0.39845 (0.01489)	0.39943 (0.01287)	0.39587 (0.04113)
M.2	$\theta = 0.8$	0.79231 (0.01539)	0.79096 (0.01610)	0.79318 (0.04102)
M.3	$\phi = 0.4$	0.39752 (0.02069)	0.40305 (0.01787)	0.39463 (0.04995)
	$\theta = 0.8$	0.78396 (0.02509)	0.78383 (0.02339)	0.78084 (0.05532)

Table 2:
Mean and mean-absolute deviations (in parentheses) for $\hat{\beta}_{LAD}$, $\hat{\beta}_{MLE}$, and $\hat{\beta}_{LS}$, for models $M.1$, $M.2$, and $M.3$ with $\alpha = 1.00$.

Model	True Values	$\hat{\beta}_{LAD}$	$\hat{\beta}_{MLE}$	$\hat{\beta}_{LS}$
M.1	$\phi = 0.4$	0.39907 (0.00543)	0.39984 (0.00370)	0.39604 (0.03042)
M.2	$\theta = 0.8$	0.79087 (0.01150)	0.77145 (0.03084)	0.79313 (0.03371)
M.3	$\phi = 0.4$	0.39736 (0.01130)	0.40090 (0.00628)	0.39462 (0.03780)
	$\theta = 0.8$	0.78185 (0.02161)	0.76566 (0.03639)	0.78297 (0.04878)

Table 3:
Mean and mean-absolute deviations (in parentheses) for $\hat{\beta}_{LAD}$, $\hat{\beta}_{MLE}$, and $\hat{\beta}_{LS}$, for models $M.1$, $M.2$, and $M.3$ with $\alpha = 0.75$.

The simulations show quite conclusively that the LAD and MLE estimation procedures are superior to LS estimation and that LAD estimation is nearly as good as MLE estimation. Although the difference is slight for $\alpha = 1.75$, as α decreases the difference becomes larger. For $\alpha = 0.75$, the spans of the box plots cannot even be drawn properly on the same scale. Based upon the asymptotic properties and the results of the simulations, it should be clear that when possible, LAD estimation is preferable to LS estimation when dealing with α-stable innovations.

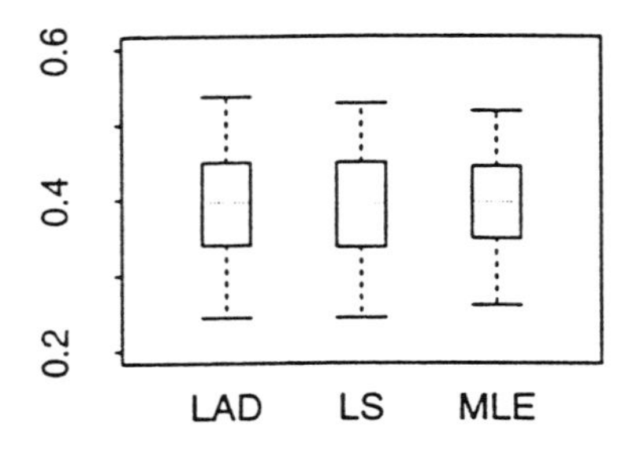

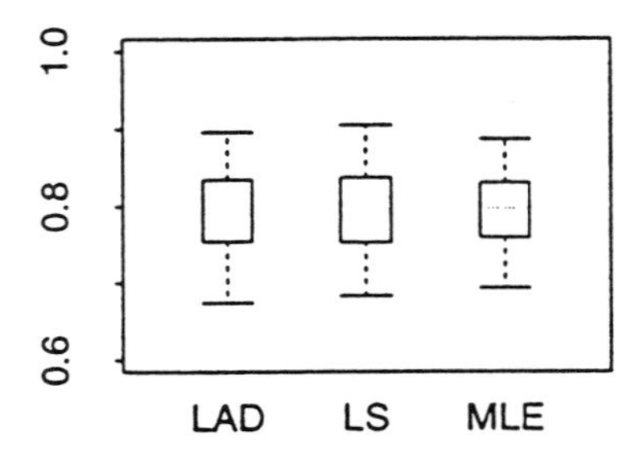

Figure 1: Distribution of estimates $\hat{\beta}_{LAD}$, $\hat{\beta}_{LS}$, and $\hat{\beta}_{MLE}$ for model (M.1) $X_t = 0.4X_{t-1} + Z_t$ (left), and (M.2) $X_t = Z_t + 0.8Z_{t-1}$ (right) with $\alpha = 1.75$.

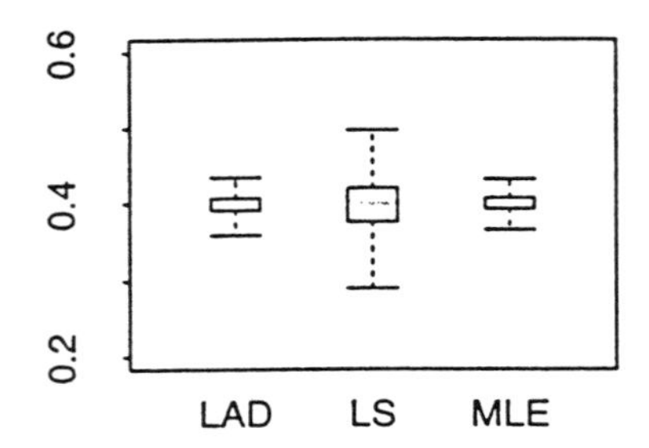

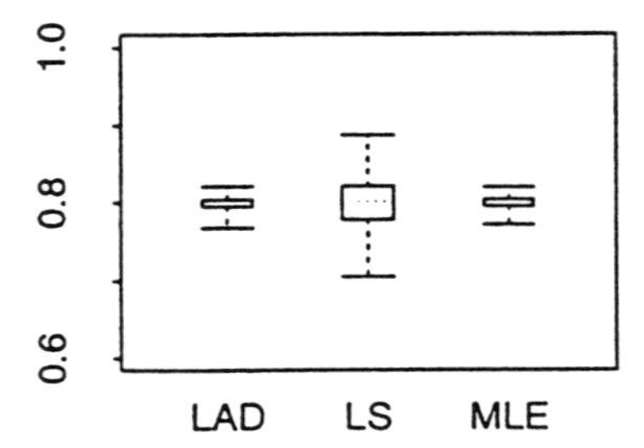

Figure 2: Distribution of estimates $\hat{\beta}_{LAD}$, $\hat{\beta}_{LS}$, and $\hat{\beta}_{MLE}$ for model (M.1) $X_t = 0.4X_{t-1} + Z_t$ (left), and (M.2) $X_t = Z_t + 0.8Z_{t-1}$ (right), with $\alpha = 1.0$.

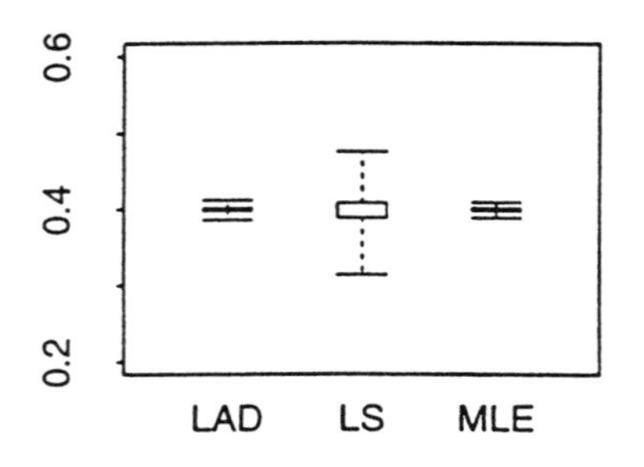

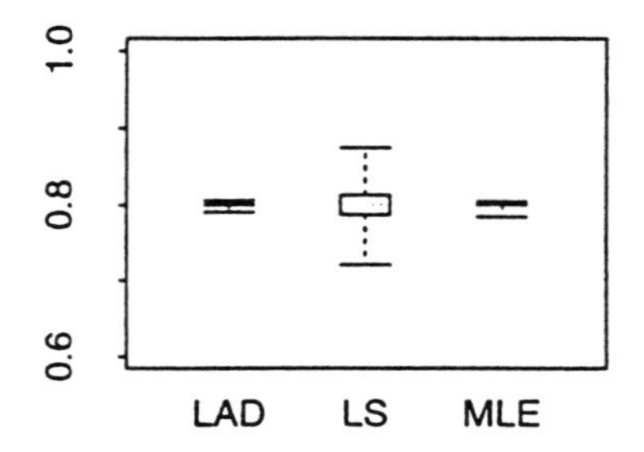

Figure 3: Distribution of estimates $\hat{\beta}_{LAD}$, $\hat{\beta}_{LS}$, and $\hat{\beta}_{MLE}$ for model (M.1) $X_t = 0.4X_{t-1} + Z_t$ (left), and (M.2) $X_t = Z_t + 0.8Z_{t-1}$ (right) with $\alpha = 0.75$.

The MLE estimator also estimated the parameters α and σ. Histograms of the estimates of α appear in Figure 7, and those of σ in Figure 8. The estimates shown are those found for model M.1, the results for the other models were similar. The accuracy is reasonably good and lead one to suspect that the MLE estimates based on estimated α's will be nearly as good as those based upon the true α.

Simulations estimating the parameter ϕ of the AR(1) model M.1 by maximum likelihood based upon the true α appear in Figure 9 and Table 4. Additionally, a maximum likelihood estimate of ϕ based upon a Cauchy density ($\alpha = 1$) was made. These simulations show that there is no real penalty for estimating the additional parameters. It is apparent that the maximum likelihood estimate is robust to the parameter specification of the stable density. This can also be seen in the results for the maximum likelihood estimates based upon the Cauchy density. The Cauchy based likelihood is seen to be quite respectable, at least for the cases $\alpha = 0.75$, and (of course) $\alpha = 1.0$. However, the performance degenerates for α farther from 1.0, that is the case $\alpha = 1.75$.

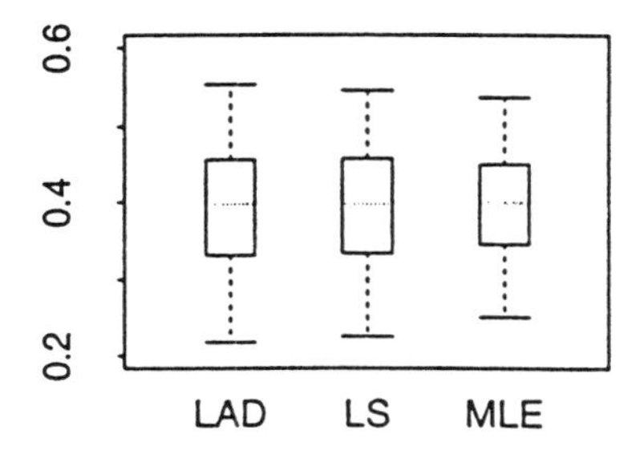

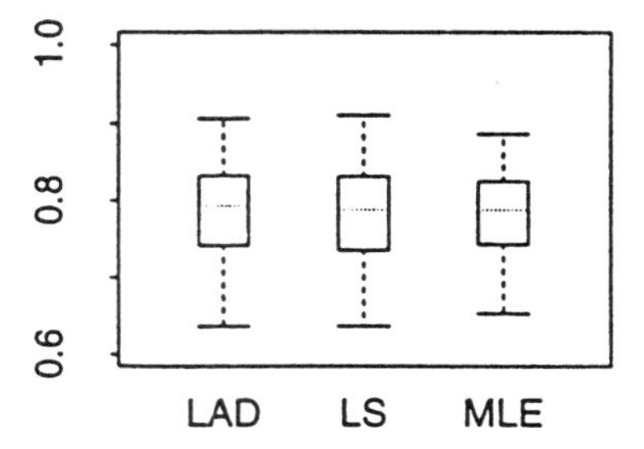

Figure 4: Distribution of estimates $\hat{\beta}_{LAD}$, $\hat{\beta}_{LS}$, and $\hat{\beta}_{MLE}$ for the model (M.3) $X_t = 0.4X_{t-1} + Z_t + 0.8Z_{t-1}$, with $\alpha = 1.75$ $\phi = 0.4$ (left), $\theta = 0.8$ (right).

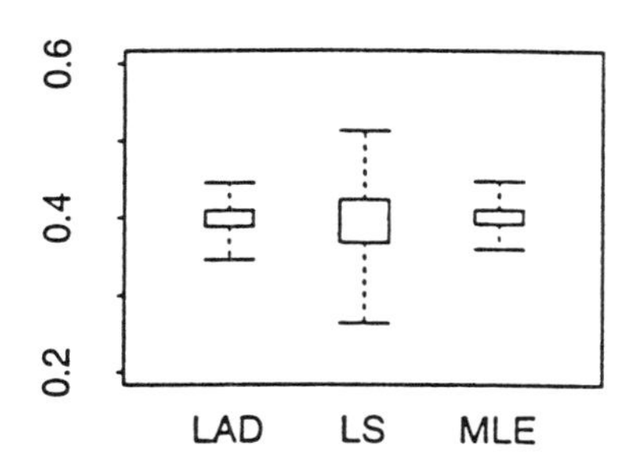

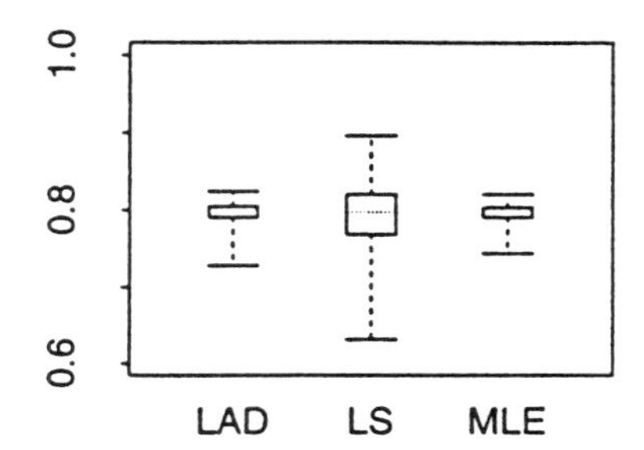

Figure 5: Distribution of estimates $\hat{\beta}_{LAD}$, $\hat{\beta}_{LS}$, and $\hat{\beta}_{MLE}$ for the model (M.3) $X_t = 0.4X_{t-1} + Z_t + 0.8Z_{t-1}$, with $\alpha = 1.0$, $\phi = 0.4$ (left), $\theta = 0.8$ (right).

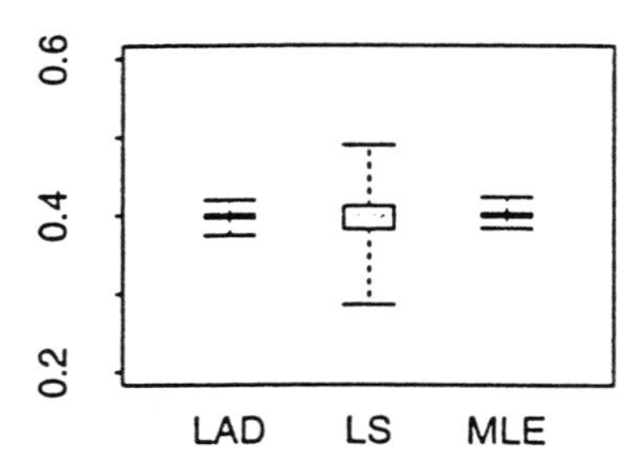

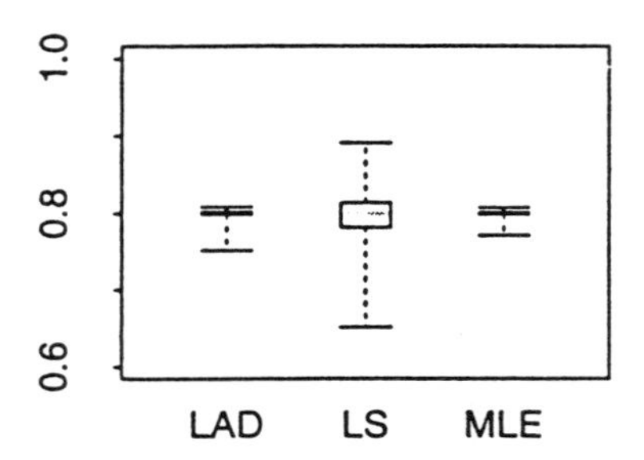

Figure 6: Distribution of estimates $\hat{\beta}_{LAD}$, $\hat{\beta}_{LS}$, and $\hat{\beta}_{MLE}$ for the model (M.3) $X_t = 0.4X_{t-1} + Z_t + 0.8Z_{t-1}$, with $\alpha = 1.75$ $\phi = 0.4$ (left), $\theta = 0.8$ (right).

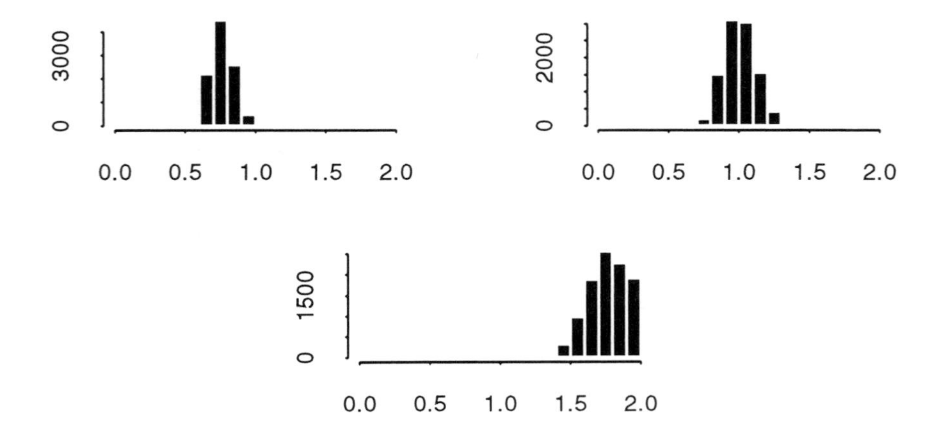

Figure 7: Distribution of estimates of α, for the model (M.1), with α = 0.75, 1.00, and 1.75.

In all of the previous simulations the innovations have had symmetric stable distributions, only the value of α was varied. In the final simulation a non-stable distribution is used to generate the innovations. Consider a random variable Y having a Pareto distribution such that,

$$P(Y > y) = \frac{k}{y^\alpha},$$

where $Y \in (1, \infty)$ and k is a constant. Now by subtracting 1 from Y and multiplying (with equal probability) ± 1, one obtains a symmetric random variable with polynomial tails, call it a symmetric Pareto. This distribution is similar to the symmetric α-stable distribution in that the tail behavior of the two distributions is the same for a suitably chosen k. It differs from the symmetric stable distribution in two important respects.

First, the density has a different shape near zero, being similar to a double-exponential rather than the familiar bell shape of the symmetric stable densities. Second, the value of α is not limited to $(0, 2)$, it can be any positive value.

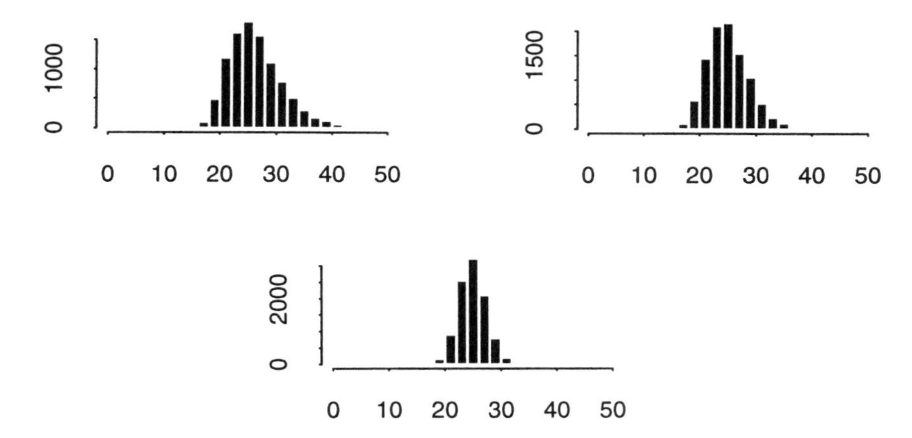

Figure 8: Distribution of estimates of σ, for the model (M.1), with α = 0.75, 1.00, and 1.75.

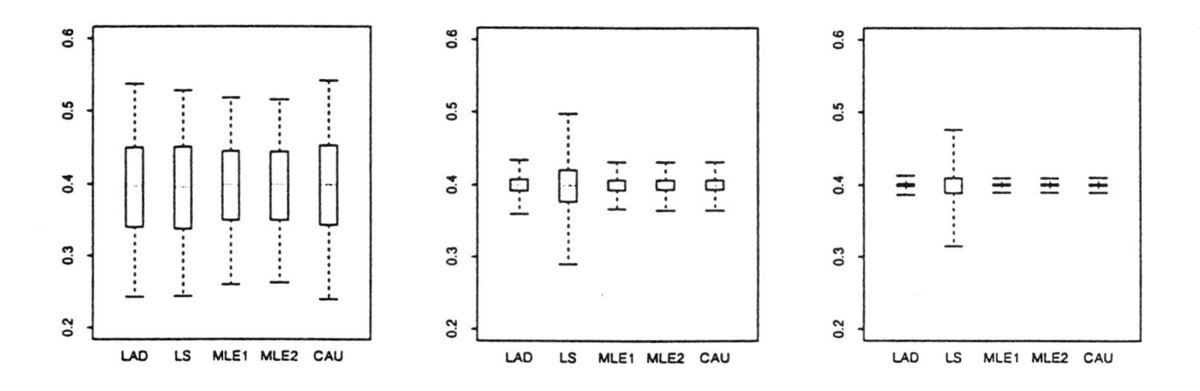

Figure 9: Distribution of estimates $\hat{\beta}_{LAD}$, $\hat{\beta}_{LS}$, $\hat{\beta}_{MLE1}$ (estimated parameters), $\hat{\beta}_{MLE2}$ (given parameters), and $\hat{\beta}_{Cauchy}$ for model (M.1) $X_t = 0.4X_{t-1} + Z_t$ with $\alpha = 1.75$ (left), $\alpha = 1.00$ (middle), $\alpha = 0.75$ (right).

Two values of α were used in the simulations, $\alpha = 1.0$ and 1.75. Only model $M.1$ was simulated from, and three estimators were computed, $\hat{\beta}_{LAD}$, $\hat{\beta}_{LS}$, and $\hat{\beta}_{MLE}$. Again, for the MLE estimator, the parameters α and σ were estimated. Note that the MLE estimator is based upon the stable density even though the innovations have a symmetric Pareto distribution.

α	$\hat{\beta}_{MLE}$	$\hat{\beta}_{Cauchy}$
$\alpha = 1.75$	0.39471 (0.05896)	0.39986 (0.06963)
$\alpha = 1.00$	0.39941 (0.01279)	0.39941 (0.01279)
$\alpha = 0.75$	0.39986 (0.00374)	0.40090 (0.00374)

Table 4:
Mean and mean-absolute deviations (in parentheses) for $\hat{\beta}_{MLE}$ with known parameters and $\hat{\beta}_{Cauchy}$, for model $M.1$, with $\alpha = 1.75$, $\alpha = 1.00$, and $\alpha = 0.75$.

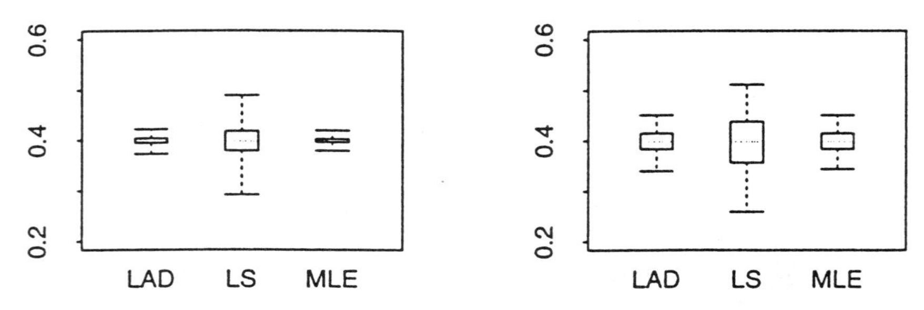

Figure 10: Distribution of estimates $\hat{\beta}_{LAD}$, $\hat{\beta}_{MLE}$, and $\hat{\beta}_{LS}$, for model (M.1) $X_t = 0.4X_{t-1} + Z_t$, with symmetric Pareto innovations $\alpha = 1.0$ (left) and $\alpha = 1.75$ (right).

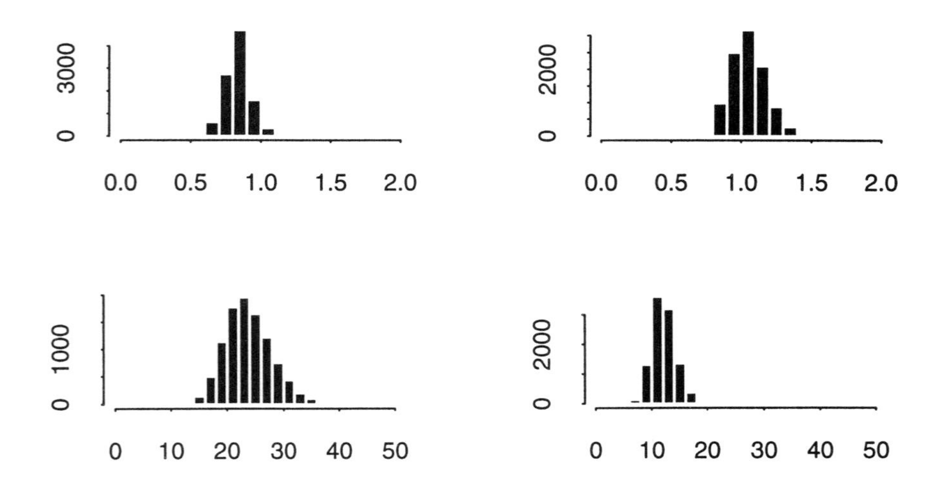

Figure 11: Histogram of estimates of α (top) and σ (bottom), for the model (M.1) $X_t = 0.4X_{t-1} + Z_t$, with symmetric Pareto innovations having α = 0.75 and σ = 25.0 (left), and α = 1.75 and $\sigma = 10.5$ (right).

The results of the simulation appear in Figure 10, ands are similar to what was found when the innovations had a stable distribution. The LAD and MLE estimates are superior to the LS estimator, and the spread of the distribution of the estimates is similar.

Histograms of the MLE estimates of α and σ are shown in Figure 10. The MLE estimate based upon the stable density underestimates α when the innovations have a Pareto distribution. This shows that estimating α can be quite sensitive to the underlying distribution. The estimates of σ are acceptable. The true scale parameter for the $\alpha = 1.0$ case is 25 as before, in the $\alpha = 1.75$ case, is 10.5.

5. Conclusion

Estimation of the parameters of ARMA models with heavy-tailed innovations can be done with very high accuracy. However care must be taken when choosing the estimation procedure. The class of M-estimates encompasses many of the commonly used estimation procedures, and the asymptotic properties of these estimators for heavy-tailed ARMA models give clues to which estimators will perform well.

The M-estimator based upon the absolute value of the errors performs well when the innovations are heavy-tailed. LAD estimation has long been used for such problems in regression analysis and it is natural to use them for ARMA modelling also. The estimator has good asymptotic properties and is relatively simple to compute. The simulations show quite conclusively that LAD estimation is an excellent choice for estimating ARMA parameters with heavy-tailed innovations.

The least-squares estimator is more accurate for ARMA models with heavy-tailed innovations than it is in the finite variance case, however, when compared to the LAD or MLE estimators it is inferior. The difference can be quite striking (see Figure 3). The least squares estimate, and second order based methods in general, should be avoided whenever possible under conditions of heavy-tailed innovations.

The MLE estimator has good asymptotic properties and has been shown in the simulations to perform as well as or slightly better than the LAD estimator. The obvious drawback to using the MLE estimator is the numerical difficulty of computing arbitrary α-stable densities, and the additional parameters that must be estimated. Moreover, the maximum likelihood procedure implicitly assumes that the innovations have a known distribution, which can be difficult to verify in practice.

References

[1] Brockwell, P.J. and Davis, R.A. (1991). *Time Series: Theory and Methods, 2nd edition*, Springer, New York.

[2] Davis, R.A. (1996) Gauss-Newton and M-Estimation for ARMA processes with infinite variance. *Stochastic Processes and their Applications* **63**, p. 75–95.

[3] Davis, R.A., Knight K., and Liu, J. (1992). M-estimation for autoregressions with infinite variance. *Stochastic Processes and their Applications* **40** p. 145-180.

[4] Davis, R.A. and Resnick, S. (1986). Limit theory for the sample covariance and correlation functions of moving averages. *Annals of Statistics* **14**, p. 533–558.

[5] Davis, R.A. and Wu, W. (1994) Bootstrapping M-estimates in regression and autoregression with infinite variance. *Preprint, Colorado State Univ., Dept. of Statistics.*

[6] McCulloch, J.H. (1994). Numerical approximation of the symmetric stable distribution and density. *Preprint, Ohio State University, Dept. of Economics.*

[7] Mikosch, Gadrich, Klüppelberg, and Adler (1995) Parameter estimation for ARMA models with infinite variance innovations. *The Annals of Statistics* **v. 23 n. 1** p. 305–326.

Department of Statistics, Colorado State University, Ft. Collins, CO 80523

On Estimating the Intensity of Long-Range Dependence in Finite and Infinite Variance Time Series

Murad S. Taqqu and Vadim Teverovsky [1] [2] [3]

Abstract

The goal of this paper is to provide benchmarks to the practitioner for measuring the intensity of long-range dependence in time series. It provides a detailed comparison of eight estimators for long-range dependence, using simulated FARIMA(p, d, q) time series with different finite and infinite variance innovations. FARIMA time series model both long-range dependence (through the parameter d) and short-range dependence (through the parameters p and q). We evaluate the biases and standard deviations of several estimators of d and compare them for each type of series used. We consider Gaussian, exponential, lognormal, Pareto, symmetric and skewed stable innovations. Detailed tables and graphs have been included. We find that the estimators tend to perform less well when p and q are not zero, that is, when there is additional short-range dependence structure. For most of the estimators, however, the use of infinite variance instead of finite variance innovations does not cause a great decline in performance.

1. Introduction

Long-range dependence has been observed in many time series, e.g., network traffic and finance (see [LTWW94], [Bai96], and also the special issue [JoE96]). Such dependence is also known as "long memory" or "$1/f$ noise." It is characterized, in the finite variance case, by slowly decaying covariances, a spectral density that tends to infinity as the frequencies tend to zero, and by the self-similarity of aggregated summands. The intensity of these phenomena is measured by a parameter d. Various estimators of d have been proposed. Even though some have known large sample properties, it is important to test their accuracy by using simulated series. Such a study was started in Taqqu, Teverovsky and Willinger [TTW95], using "ideal" models that display long-range dependence, namely Fractional Gaussian Noise (FGN) or Gaussian Fractional

---

[1] Research partially supported by NSF grants NCR-9404931 and DMS-9404093 at Boston University.

[2] *AMS Subject classification:* 60G18, 62-09, 60E07.

[3] *Keywords:* Long-range dependence, long memory, robustness, infinite variance, time series, periodogram, Whittle.

ARIMA$(0, d, 0)$. Both of these models are Gaussian and have a well-controlled short-range dependence structure. Since real-world data possess a more complex structure, the *robustness* of the estimators must also be examined. In this study we will analyze the performance of various estimators of long-range dependence when there is additional short-range dependence structure and/or infinite variance.

The FARIMA(p, d, q) family of models is widely used in the modeling of time series with long-range dependence (see, for example, Brockwell and Davis [BD91]). These are moving averages $X_n = \sum_{i=-\infty}^{n} c_{n-i}\epsilon_i$, where c_k behaves like k^{d-1} for large k and the ϵ_i's are independent, identically distributed random variables (innovations). Here we will focus specifically on FARIMA$(1, d, 1)$ series, with both finite and infinite variance innovations.

The estimation methods we consider in this paper are : 1) The Absolute Value Method 2) Regression on the Periodogram 3) R/S 4) Aggregated Variance 5) Variance of Residuals 6) Whittle Estimator 7) Aggregated Whittle 8) Local Whittle. (In addition we will briefly discuss three other estimators without presenting detailed results. These are: Higuchi's Fractal Dimension Method, Differenced Variance, Modified Periodogram.) These estimators are also described in Taqqu, Teverovsky, Willinger [TTW95]. All are graphical except the Whittle-based ones. The latter are related to Whittle's approximate MLE and discussed at greater length in Taqqu and Teverovsky [TT97a] and Taqqu and Teverovsky [TT96]. We do not use exact (Gaussian) MLE methods because these strongly depend on the series being Gaussian, which is often not the case in practice. The exact MLE is, moreover, much more computationally intensive than the approximate MLE (Whittle), which is, itself, much more computationally intensive than the graphical methods.

The innovations that appear in the FARIMA$(1, d, 1)$ series have either finite or infinite variance. We will use Gaussian, exponential and lognormal innovations in the finite variance case and Pareto and stable innovations in the infinite variance case. Both Pareto and stable distributions are characterized by a parameter α which describes the heaviness of the distribution tails. If ϵ is either Pareto or stable with parameter α, then $P(\epsilon > x) \sim Cx^{-\alpha}$ as $x \to \infty$, that is, the probability tails decrease slowly, like a power function. Moreover, $\mathrm{Var}(\epsilon) = \infty$ if $\alpha < 2$ and $E|\epsilon| = \infty$ if $0 < \alpha \leq 1$.

The inclusion of infinite variance innovations forces us to carefully differentiate between the two parameters d and H that are used to characterize long-range dependence. The parameter d, which plays the role of a differencing parameter in the FARIMA model, is described in Section 2. The parameter H is a scaling parameter that can be characterized as follows: given a time series X_i, of length N, define the corresponding aggregated series,

$$X^{(m)}(k) := \frac{1}{m} \sum_{i=(k-1)m+1}^{km} X_i, \quad k = 1, 2, ..., [N/m]. \tag{1.1}$$

Then, for large enough N/m and m,

$$X^{(m)} \overset{d}{\sim} m^{H-1} S \tag{1.2}$$

(where $\overset{d}{\sim}$ means "asymptotically equal to" in the sense of distributions). Here S is a process which depends on the distribution of X but does not depend on m (the term *asymptotically self-similar* is used to describe property (1.2)). In the finite variance case S is Fractional Gaussian Noise (FGN) and in the infinite variance case it is Linear Fractional Stable Noise (LFSN). These processes are exactly self-similar in the sense that relation (1.2) holds for all $m \geq 1$ and not just asymptotically. FGN and LFSN are introduced in the next section.

The scaling relation (1.2) defines the parameter H. For finite variance processes,

$$H = d + 1/2,$$

and for infinite variance processes,

$$H = d + 1/\alpha.$$

It is important therefore to know whether an estimator is estimating H or d.

The paper is structured as follows. The time series are introduced in Section 2 and the experimental set-up is described in Section 3. The estimators are defined in Section 4 and the results presented in Section 5. Several tables and graphs are included for reference. Our conclusions are stated in Section 6.

2. The Time Series

In this section we present the various types of series which are used in this paper. These include fractional Gaussian noise and FARIMA with innovations that are either Gaussian, non-Gaussian with finite variance or that have infinite variance, including stable innovations.

2.1 Fractional Gaussian Noise (FGN)

We start with the FGN series, which serves as a reference point for the FARIMA series introduced below. The importance of FGN stems from its exact self-similarity. If

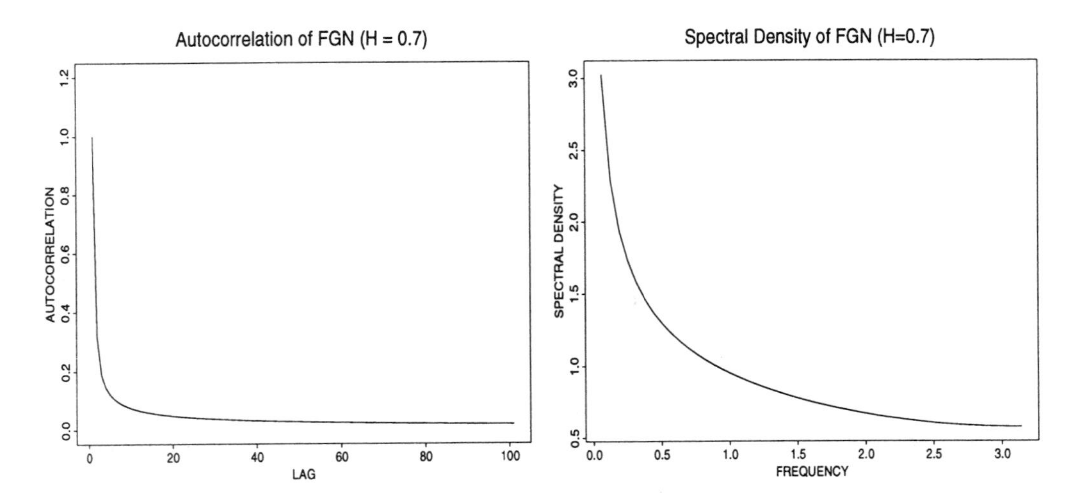

Figure 1: Autocorrelation function and spectral density of FGN with $d = 0.2\,(H = 0.7)$.

$X = \{X_i, i \geq 1\}$ is FGN and if $X^{(m)}$ is defined as in (1.1), then $X^{(m)} \stackrel{d}{=} m^{H-1}X$ ($\stackrel{d}{=}$ denotes equality in distribution). This relation is much stronger than the asymptotic relation (1.2).

The FGN series $\{X_i,\ i \geq 1\}$ is a zero mean, stationary Gaussian time series whose autocovariance function at lag h is:

$$\gamma(h) = EX_iX_{i+h} = \frac{1}{2}\{(h+1)^{2d+1} - 2h^{2d+1} + |h-1|^{2d+1}\},\ h \geq 0,\ -1/2 < d < 1/2. \tag{2.3}$$

For $d \neq 0$, the autocovariance satisfies

$$\gamma(h) \sim d(2d+1)h^{2d-1} \quad \text{as } h \to \infty \tag{2.4}$$

($\sim$ means here "asymptotic to"), that is, for large lags, it decreases to zero like a power function. d measures the intensity of the long-range dependence of the series X and it is related to the scaling exponent H by the relation $H = d + 1/2$. In addition, the spectral density (Fourier transform of γ) is given by:

$$f(\nu) = C_d\Big(2\sin\frac{\nu}{2}\Big)^2 \sum_{k=-\infty}^{\infty} \frac{1}{|\nu + 2\pi k|^{2d+2}} \sim C_d|\nu|^{-2d} \text{ as } \nu \to 0, \tag{2.5}$$

where C_d is a constant. The spectral density behaves like a power function at the origin. Long-range dependence corresponds to $d > 0$ ($H > 1/2$), because in this case, the series spectral density diverges to infinity at the origin. For graphs of the autocorrelation function and the spectral density for FGN with $d = 0.2$ ($H = 0.7$) see Figure 1.

2.2 Gaussian FARIMA

Let us now turn to the Gaussian FARIMA(p, d, q) processes. A Gaussian FARIMA$(0, d, 0)$ process is defined by $X_i = \Delta^{-d}\epsilon_i$, $i \geq 1$, where the ϵ_i are independent, identically distributed Gaussian random variables with mean 0, and where Δ is the differencing operator $\Delta\epsilon_i = \epsilon_i - \epsilon_{i-1}$. For fractional d we interpret Δ^{-d} by using formal power series expansion, as follows:

$$\Delta^{-d} = \sum_{i=0}^{\infty} b_i(-d)B^i, \tag{2.6}$$

where B is the backward operator, $B\epsilon_i = \epsilon_{i-1}$, and

$$b_i(-d) = \frac{\Gamma(i+d)}{\Gamma(d)\Gamma(i+1)}, \quad i = 1, 2, \ldots.$$

Γ denotes the gamma function. The autocovariance function of this process satisfies, for $-1/2 < d < 1/2$,

$$\gamma(h) \sim C_d h^{2d-1} \quad \text{as } h \to \infty \tag{2.7}$$

where $C_d = \pi^{-1}\Gamma(1-2d)\sin\pi d$. Thus, for large lags d, the autocovariance (2.7) of a Gaussian FARIMA$(0, d, 0)$ process has the same power decay as the autocovariance (2.4) of FGN. Correspondingly, the spectral density of a Gaussian FARIMA$(0, d, 0)$ process behaves in the same way as that of a FGN process, i.e. it is like a power function at the origin.

A Gaussian FARIMA$(0, d, 0)$ is a particular case of Gaussian FARIMA(p, d, q) processes. These are defined through the equation

$$X_i - \phi_1 X_{i-1} - \cdots - \phi_p X_{i-p} = \Delta^{-d}\epsilon_i - \theta_1\Delta^{-d}\epsilon_{i-1} - \cdots - \theta_q\Delta^{-d}\epsilon_{i-q}, \tag{2.8}$$

where the $\Delta^{-d}\epsilon_i$ are as above and the ϕ_i and the θ_i are the autoregressive and the moving average coefficients respectively. The Gaussian FARIMA(p, d, q) series follow the same asymptotic relations for their autocovariance and spectral density as do the Gaussian FARIMA$(0, d, 0)$ processes, although the actual functions are more complicated, and have additional short-term components. These short-term components can influence significantly the estimates of d. While some estimators are relatively insensitive to deviations from "ideal" long-range dependence, others are extremely sensitive to them, and thus lack robustness. The goal of this study is to determine which estimators are robust and which ones are not.

2.3 FARIMA with other finite variance innovations

In addition to Gaussian innovations ϵ_i, we also consider innovations ϵ_i which have *exponential* and *lognormal* distributions. An exponential random variable ϵ has the cumulative distribution function,

$$F(x) = P(\epsilon \le x) = \begin{cases} 0 & x < 0, \\ 1 - e^{-\lambda x} & x \ge 0. \end{cases} \tag{2.9}$$

We use $\lambda = 1$ in our series. A lognormal random variable ϵ has the cumulative distribution function (c.d.f.),

$$F(x) = \Phi\left(\frac{\ln(x) - \mu}{\sigma}\right), \tag{2.10}$$

where Φ is the c.d.f. of a standard Normal random variable. Here we use $\mu = 0, \sigma = 1$. We only analyze FARIMA$(0, d, 0)$ series for these types of innovations.

2.4 FARIMA with infinite variance innovations

For infinite variance innovations ϵ, we consider symmetric and skewed stable distributions and Pareto distributions. They are characterized by the parameter α and satisfy

$$P(\epsilon > x) \sim Cx^{-\alpha}, \quad \text{as } x \to \infty, \tag{2.11}$$

implying that the variance is infinite if $\alpha < 2$. The *symmetric α-stable* distribution, denoted $S_\alpha(\sigma)$ has the characteristic function

$$E[\exp(i\theta\epsilon)] = \exp(-\sigma^\alpha|\theta|^\alpha),$$

where σ is the scale parameter. We will concentrate on the case $1 < \alpha \le 2$, with the understanding that for $\alpha = 2$ the process is Gaussian. A stable random variable which is totally skewed to the right has the characteristic function,

$$E[\exp(i\theta\epsilon)] = \exp\left[-\sigma^\alpha|\theta|^\alpha(1 - i(\text{sign}\,\theta)\tan\frac{\pi\alpha}{2})\right],$$

where $i = \sqrt{-1}$, and sign $\theta = -1, 1, 0$ if $\theta < 0, > 0, = 0$ respectively. We use $\sigma = 1$ in all of the simulated series. Two of the important properties of stable distributions are: (1) If ϵ is α-stable with scale parameter σ then $a\epsilon$ is also α-stable with scale parameter $|a|\sigma$. (2) If ϵ_1 and ϵ_2 are independent α-stable with σ_1 and σ_2 respectively, then $\epsilon_1 + \epsilon_2$ is α-stable with parameter $(\sigma_1^\alpha + \sigma_2^\alpha)^{1/\alpha}$.

Finally, a Pareto random variable ϵ has the c.d.f.,

$$F(x) = 1 - \frac{1}{x^\alpha}, \quad x \ge 1. \tag{2.12}$$

Pareto random variables also have infinite variance, and in fact satisfy (2.11) exactly for all $x \ge 1$.

When generating series with infinite variance, we will use independent symmetric α-stable variables as innovations in FARIMA(p, d, q) series (see (2.8)), and skewed stable and *Pareto* distributions as the innovations in a FARIMA$(0, d, 0)$ series. The parameter d is restricted to the interval $[0, 1 - 1/\alpha)$. The resulting process X will be long-range dependent and α-stable if the innovations are α-stable, and asymptotically α-stable if the innovations are Pareto.

2.5 Linear Fractional Stable Noise (LFSN)

FGN is the only exactly self-similar Gaussian process, i.e. such that

$$X^{(m)} \stackrel{d}{=} m^{H-1} X. \tag{2.13}$$

If X is α-stable with $\alpha < 2$, then there are many different processes satisfying (2.13), and LFSN is one of them. It is usually defined through an integral representation, but, since this representation is not used here, we refer the reader to Chapter 7 of [ST94]. The only property of LFSN used in this paper is relation (2.13) (exact self-similarity).

For more information on Fractional Gaussian Noise or Gaussian FARIMA series, see Brockwell and Davis [BD91] or Beran [Ber94] and, for stable processes, Samorodnitsky and Taqqu [ST94]. The FARIMA is a useful family of models because any short-range dependence can be modeled by choosing the appropriate autoregressive and moving average coefficients in (2.8). Unfortunately these coefficients can also affect the estimates of long-range dependence. These effects may be considerable, even for the series of length 10, 000 used in this study.

3. Set-Up

In [TTW95] we analyzed

a) Fractional Gaussian Noise,
b) Gaussian FARIMA$(0, d, 0)$.

The results are summarized in Table 1 and the top row of Figure 5. These series provide a baseline for the present research. Here we focus on situations where there is additional short-range dependence and/or finite or infinite variance innovations. The ten types of series examined are:

1) Gaussian FARIMA$(1, d, 0)$, with $\phi_1 = 0.5$.
2) Gaussian FARIMA$(0, d, 1)$ with $\theta_1 = 0.5$.
3) Gaussian FARIMA$(1, d, 1)$ with $\phi_1 = -0.3,\ \theta_1 = -0.7$.
4) Gaussian FARIMA$(1, d, 1)$ with $\phi_1 = 0.3,\ \theta_1 = 0.7$.
5) FARIMA$(0, d, 0)$ with exponential innovations.
6) FARIMA$(0, d, 0)$ with lognormal innovations.
7) Symmetric Stable FARIMA$(0, d, 0)$ with $\alpha = 1.5,\ 1.2$.

8) Symmetric Stable FARIMA$(1, d, 1)$ with $\alpha = 1.5,\ 1.2, \phi_1 = 0.3,\ \theta_1 = 0.7$.
9) FARIMA$(0, d, 0)$ with Pareto innovations, $\alpha = 1.5$.
10) FARIMA$(0, d, 0)$ with skewed stable innovations, $\alpha = 1.5$.

The parameters ϕ_1 and θ_1 generate short-range dependence. Extreme values of these parameters (close to 1) can create very large deviations from pure long-range dependence. Our choices of values for ϕ_1 and θ_1 are a compromise between showing what can go wrong and totally obscuring the long-range dependence (for more information on the role of ϕ_1 and θ_1 see [TT96]).

For each type of series, we take d in the range $[0, 1 - 1/\alpha)$. This results in an $H < 1$, which is necessary for X to be a stationary series. Specifically,

a) for Gaussian series we chose $d = 0.0(0.1)0.4$,
b) for $\alpha = 1.5$ stable and Pareto, we chose $d = 0.0(0.1)0.3$,
c) for $\alpha = 1.2$ stable, we chose $d = 0.0, 0.1$.

The Gaussian FARIMA series were produced by the *arima.fracdiff.sim* command built into S-Plus Version 3.2 (S-Plus is a statistical package marketed by StatSci), except for the case $d = 0$ $(H = 0.5)$ when there is no long-range dependence. The method of generation is based on the Durbin-Levinson algorithm (see for example Brockwell and Davis [BD91], Chapter 8.2) and is described in Haslett and Raftery [HR89], p. 12-13. For $d = 0$, we wrote our own routine to generate the processes. The $d = 0$ case is very simple because no fractional differencing is necessary.

The stable innovation series were generated using the *rstab* command in S-Plus (in the skewed case this was modified to produce non-negative random variables). The stable FARIMA(p, d, q) processes were then produced by using Eqs. (2.6) and (2.8). The exponential and lognormal distributions were produced by the S-Plus commands *rexp* and *rlnorm* respectively. The Pareto innovations were produced by using the formula $U^{-1/\alpha}$ where U is a Uniform$(0, 1)$ random variable. The FARIMA$(0, d, 0)$ series were then produced by using Eq. (2.6). The Durbin-Levinson algorithm can not be used for infinite variance processes since it depends heavily on the existence of the covariance function. We used instead the infinite sums in (2.6), truncated at $i = 10,000$.

For each type of finite variance series and each value of d, 50 series of length 10,000 were simulated. Because of the potential for greater variability, 1000 independent series of length 10,000 were produced for the infinite variance series. We used the estimation methods described in Section 4 below to estimate d or H, tabulated the average biases of the estimates as well as the in-sample standard deviations (standard errors) and the square roots of the sample MSEs. The following notation is used in the tables. If d_i is the estimate for sample i out of I total samples, then,

$$\overline{d} = \frac{1}{I}\sum_{i=1}^{I} d_i, \quad BIAS = d_0 - \hat{d}, \quad \hat{\sigma}^2 = \frac{1}{I-1}\sum_{i=1}^{I}(d_i - \overline{d})^2, \quad \text{MSE} = \frac{1}{I}\sum_{i=1}^{I}(d_i - d_0)^2,$$

where d_0 is the nominal value of d. When H is estimated, we computed all of the statistics on the basis of $H_0 = d_0 + 1/\alpha$, and proceeded accordingly.

The results are presented in Tables 1 - 4 for the finite variance series and in Tables 6 - 8 for the infinite variance series. Detailed boxplots for the finite variance series are given in Figures 7 and 8. In these boxplots, the vertical axis shows the deviations from the nominal values of d or H. Each boxplot has the following components: (1) a box representing the middle 50% of the data. That box contains a) a shaded region representing an approximate 95% confidence interval about the median (see McGill, Tukey and Larsen [MTL78]); b) the median represented by the unshaded line in the middle of the shaded region; (2) "Whiskers" encompassing approximately 95% of the data, and designated by dashed lines; (3) Outliers that fall beyond the whiskers.

The greater the number of realizations, the more outliers are present. In the infinite variance case, we generated 1000 realizations for each type of series. The boxplots in this case are displayed **without** the outliers, because otherwise they would be too crowded. To give an idea of the variability, we present histograms (Figure 9) and detailed boxplots (Figure 10) for the Local Whittle Estimator applied to FARIMA(0, d, 0) series with the following innovations: Gaussian, stable with $\alpha = 1.5$ and stable with $\alpha = 1.2$.

In Table 5 we show how this estimator behaves for these time series as the number of independent realizations grows from 50 to 1000. The table shows that the mean and standard deviation of the estimates do not change significantly as the number of realizations grows. However, with infinite variance innovations, there are instances of very extreme outliers in the distribution of the estimates. Thus, when $\alpha = 1.2, d = 0.1$, there was one extreme instance out of 1000 when the estimate was biased by 0.33. (We note that this particular realization caused problems for other periodogram based estimators as well. Both the Whittle method and the periodogram method had biases of at least 0.3, much larger than for any other realization.) To see the effect such an outlier can have on the mean and standard deviation, we divided the 1000 estimates obtained for each type of series into 20 blocks of 50 estimates each. When $\alpha = 1.2$, we found that $\hat{\sigma}$ was approximately 0.02 for most of the 20 blocks, but it was .059 for the block where the outlier was included. The mean was only slightly affected. In the Gaussian case, $\hat{\sigma}$ did not vary by more than 0.005 from its average value of 0.030 over the 20 blocks. Therefore, whereas 50 realizations is enough in the finite variance case, more are needed in the infinite variance case. Our studies suggest using I of at least 300 or 400 if infinite variance innovations are used.

4. Estimators

The various estimators used here are described briefly below (see also [TT97a], [TTW95]). A number of them are obtained through graphical methods. They are illustrated in Figures 2 - 6. In most of the figures, a line indicating the best fit to the data, whose slope provides the estimates of d or H, has been included. The choice of cut-offs is somewhat arbitrary and reflects our experience. Slight changes in the values of the cut-offs tend not to change the estimates significantly. The figures include also a second line whose slope is the correct "nominal" slope (the intercept of this second line was chosen so as not to interfere with the data points themselves).

4.1 Absolute Value Method

Consider the aggregated series (1.1), obtained by dividing a given series of length N into blocks of length m, and averaging the series over each block. We take the first absolute moment of this series,

$$AM^{(m)} = \frac{1}{N/m} \sum_{k=1}^{N/m} \left| X^{(m)}(k) - \overline{X} \right|, \tag{4.14}$$

where $X^{(m)}$ is defined in (1.1) and $\overline{X}$ is the overall series mean. The centered aggregated series $X^{(m)} - EX^{(m)}$ is asymptotically equal to $m^{H-1}S_{\alpha,d}$ where $S_{\alpha,d}$ is an α-stable random sequence (in fact, it is LFSN). In particular, if X has finite variance, we get $S_{2,d}$ which is FGN.

Thus, $AM^{(m)}$ behaves like m^{H-1} for large m. Since $H = d + 1/2$ for finite variance series, $m^{H-1} = m^{d-1/2}$. For infinite variance series, $H = d + 1/\alpha$ (see Samorodnitsky and Taqqu [ST94]) and hence $m^{H-1} = m^{d+1/\alpha-1}$. This estimation method gives H and not d.

For successive values of m, the sample absolute first moment of the aggregated series is plotted versus m on a log-log plot. The result should be a straight line with a slope of $H - 1$. In practice, the slope is estimated by fitting a least-squares line to the points of the plot. We assume that both N and N/m are large. This ensures that both the length of each block and the number of blocks is large. If the series has no long-range dependence and finite variance, then $H = 0.5$ and the slope of the fitted line should be $-1/2$. In practice, the points at the very low and high ends of the plots are not used to fit the least-squares line. Indeed, short-range effects can distort the estimates of H if the low end of the plot is used, and at the very high end of the plot there are too few blocks to get reliable estimates of AM. In this study, the cut-offs are $10^{0.7}$ and $10^{2.5}$.

A technique suggested by Higuchi [Hig88] is very similar to the Absolute Value

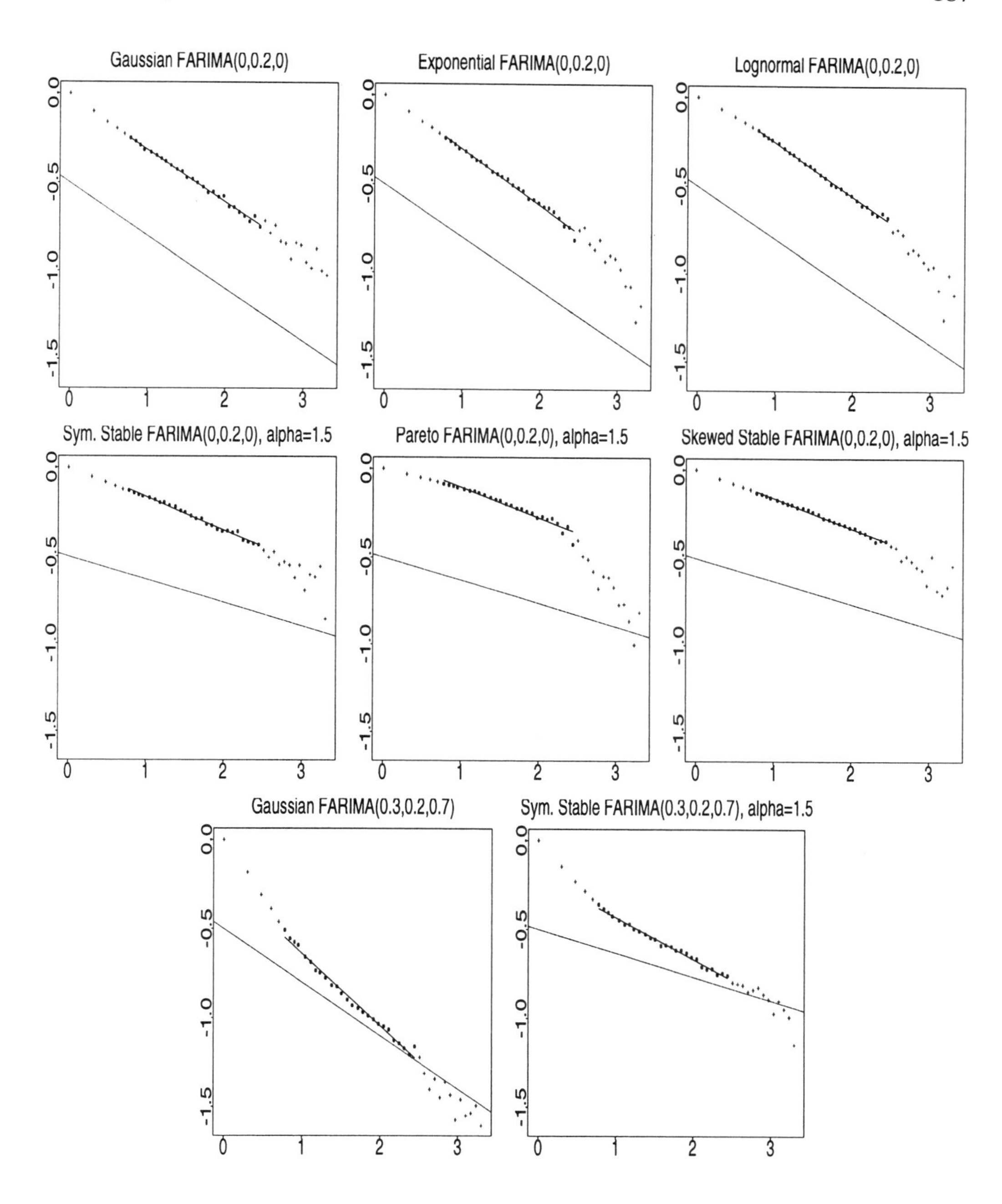

Figure 2: Absolute Value Method. $d = 0.2$. Vertical scale, $\log_{10}(AM^{(m)})$. Horizontal scale, $\log_{10}(m)$. *Top row*: Finite variance, FARIMA$(0, d, 0)$ series. *Second row*: Infinite variance, FARIMA$(0, d, 0)$ series. *Third row*: FARIMA$(1, d, 1)$ series. Reference lines with the theoretical slope $d + 1/\alpha - 1 = H - 1$ are included. The innovations in the infinite variance series (stable, Pareto) have $\alpha = 1.5$.

method. It uses

$$L(m) = \frac{N-1}{m^3} \sum_{i=1}^{m} \left[\frac{N-1}{m}\right]^{-1} \sum_{k=1}^{[(N-i)/m]} \left| \sum_{j=i+(k-1)m+1}^{i+km} X(j) \right| . \tag{4.15}$$

Here, N is the length of the time series, m is the equivalent of a block size, and [ ] denotes the greatest integer function. For long-range dependent series, $EL(m) \sim Cm^{H-2}$. Essentially, the difference with the absolute value method lies in using a sliding window to compute the aggregated series, instead of using non-intersecting blocks. Because of this, the method is much more computationally intensive. While this modification may result in increased accuracy in shorter time series, there seems to be no advantage in using it for the series length of 10, 000. Hence, the results of this method are not presented here.

4.2 Variance Method

We again consider the aggregated series $X^{(m)}$defined in (1.1), but instead of the first absolute moment, we compute its sample variance,

$$\widehat{\mathrm{Var}} X^{(m)} = \frac{1}{N/m} \sum_{k=1}^{N/m} (X^{(m)}(k) - \overline{X})^2. \tag{4.16}$$

As stated earlier, the series $X^{(m)} - EX^{(m)}$ scales like m^{H-1}, thus if the series is Gaussian, or at least finite variance, the sample variance will be asymptotically proportional to $m^{2H-2} = m^{2d-1}$ for large N/m and m.

Deriving the asymptotic behavior of $\widehat{\mathrm{Var}} X^{(m)}$ is more complicated when the series has infinite variance. Here is a sketch. In view of (1.2),

$$\begin{aligned} \widehat{\mathrm{Var}} X^{(m)} &\overset{d}{\sim} \widehat{\mathrm{Var}}(m^{H-1} S_{\alpha,d}) \\ &= m^{2H-2}\left[\frac{1}{N/m} \sum_{k=1}^{N/m} S_{\alpha,d}^2(k) - \left(\frac{1}{N/m} \sum_{k=1}^{N/m} S_{\alpha,d}(k)\right)^2\right], \end{aligned} \tag{4.17}$$

where $S_{\alpha,d}$ is an α-stable sequence (as stated earlier, it is LFSN). Therefore,

$$\widehat{\mathrm{Var}} X^{(m)} \overset{d}{\sim} m^{2H-2}\left[\left(\frac{N}{m}\right)^{2/\alpha-1} U_{\alpha/2} - C^2\right] \tag{4.18}$$

where $U_{\alpha/2}$ is a positive $\alpha/2$-stable random variable and C is a constant (see Kokoszka and Taqqu [KT96]).

Thus, for large N/m and large m,

$$\widehat{\mathrm{Var}} X^{(m)} \overset{d}{\sim} C(N) m^{2H-2+1-2/\alpha} U_{\alpha/2} = C(N) m^{2d-1} U_{\alpha/2}, \tag{4.19}$$

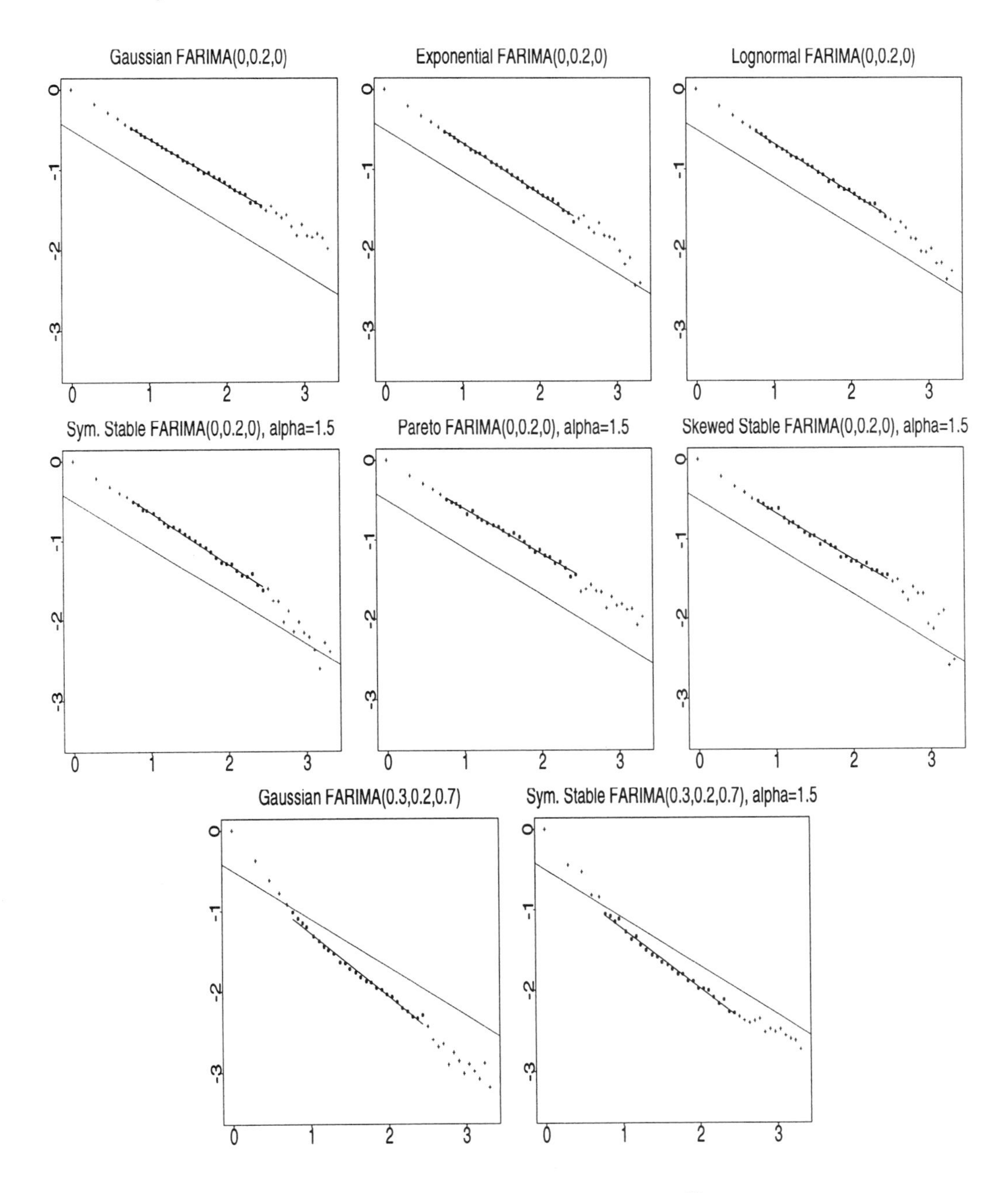

Figure 3: Variance Method. $d = 0.2$. Vertical scale, $\log_{10} \widehat{\mathrm{Var}} X^{(m)})$. Horizontal scale, $\log_{10}(m)$. *Top row*: Finite variance, FARIMA$(0, d, 0)$ series. *Second row*: Infinite variance, FARIMA$(0, d, 0)$ series. *Third row*: FARIMA$(1, d, 1)$ series. Reference lines with the theoretical slope $2d - 1$ are included. The innovations in the infinite variance series (stable, Pareto) have $\alpha = 1.5$.

where $C(N)$ is a constant dependent on N. For fixed sample size $N >> m$, the Variance method thus produces a random variable proportional to m^{2d-1}, in the finite variance as well as in the infinite variance case, providing an estimate for d.

As for the Absolute Value method, a log-log plot versus m should give a straight line, this time with a slope of $2d - 1$. The cut-offs of $10^{0.7}$ and $10^{2.5}$ were used in this study.

Jumps in the mean and slowly decaying trends are two types of non-stationarity which can occur. To distinguish these from long-range dependence, we can difference the variance (see Teverovsky and Taqqu [TT97b] for details), and study

$$\widehat{\mathrm{Var}}X^{(m_i+1)} - \widehat{\mathrm{Var}}X^{(m_i)} \tag{4.20}$$

where the m_i's are the successive values of m as defined above. Differencing introduces more fluctuation, but it can provide a way of detecting the presence of the two types of non-stationarity mentioned above. The method should be used together with the original aggregated variance method. For stable FARIMA(1, d, 1) series, the scattering of the estimates produced by this method is so great that the estimator does not produce very useful results. Since we do not focus here on non-stationarity, results for the differenced variance method are not included.

4.3 Variance of Residuals Method

The Variance of Residuals method was introduced by Peng et al. [PBS+94]. First the series is divided into blocks of size m. Then, within each block, the partial sums of the series are calculated, $Y(t) = \sum_{i=1}^{t} X_i$. A least-squares line, $a + bt$, is fitted to the partial sums within each block, and the sample variance of the residuals is computed,

$$\frac{1}{m}\sum_{t=1}^{m}(Y(t) - a - bt)^2 .$$

The following is a sketch of how this estimator behaves in the presence of long-range dependence. Define two random variables,

$$A_1 = \int_0^1 L_{\alpha,H}(t)dt, \quad A_2 = \int_0^1 tL_{\alpha,H}(t)dt,$$

with $L_{\alpha,H}(t)$, $H = d + 1/\alpha$, denoting α-stable Fractional Stable Motion (see [ST94] for a definition). Since, for large m,

$$Y(mt) \overset{d}{\sim} m^H L_{\alpha,H}(t),$$

we can use the following approximations, where all of the sums are from 1 to m:

$$\sum Y(t) \sim \int_0^m Y(t)dt \overset{d}{\sim} m^{H+1}A_1,$$

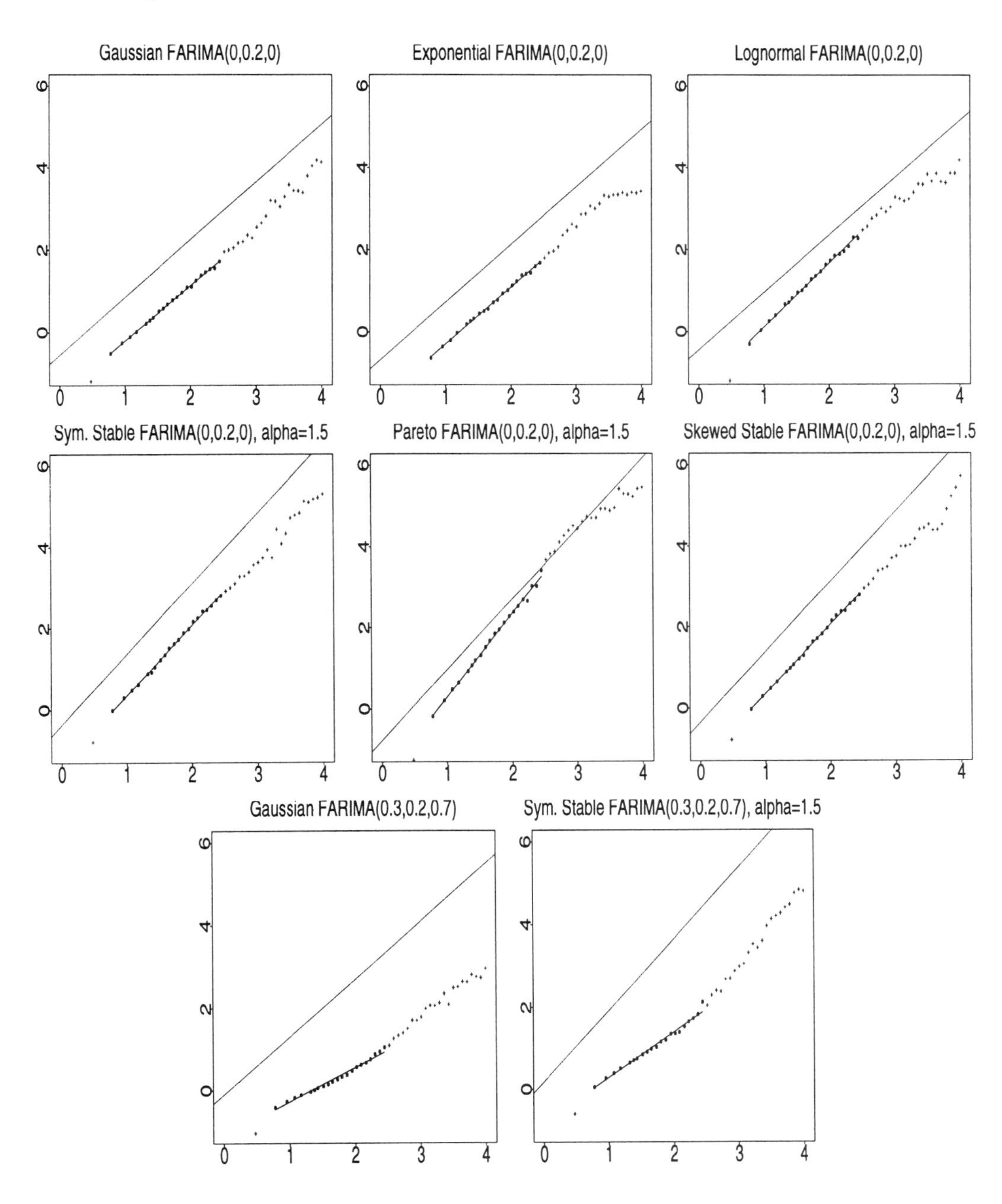

Figure 4: Variance of Residuals Method. $d = 0.2$. Vertical scale, $\log_{10}(statistic)$. Horizontal scale, $\log_{10}(m)$. *Top row*: Finite variance, FARIMA$(0, d, 0)$ series. *Second row*: Infinite variance, FARIMA$(0, d, 0)$ series. *Third row*: FARIMA$(1, d, 1)$ series. Reference lines with the theoretical slope $2(d + 1/\alpha) = 2H$ are included. The innovations in the infinite variance series (stable, Pareto) have $\alpha = 1.5$.

$$\sum tY(t) \sim \int_0^m tY(t)dt \overset{d}{\sim} m^{H+2}A_2.$$

Then, the least-squares coefficients can be calculated as:

$$b = \frac{\sum Y(t)t - \frac{1}{m}\sum Y(t)\sum t}{\sum t^2 - \frac{1}{m}(\sum t)^2} \overset{d}{\sim} \frac{m^{H+2}A_2 - m^{H+2}A_1/2}{m^3/3 - m^3/4} = m^{H-1}[12A_2 - 6A_1],$$

$$a = \frac{1}{m}\sum Y(t) - \frac{1}{m}\sum bt \overset{d}{\sim} m^H A_1 - m^H[12A_2 - 6A_1]/2 = m^H[4A_1 - 6A_2],$$

where $\overset{d}{\sim}$ means approximately equal in distribution. Therefore,

$$\frac{1}{m}\sum_{t=1}^{m}(Y(t) - a - bt)^2 \overset{d}{\sim} m^{2H}\int_0^1 \left(L_{\alpha,H}(t) - (4A_1 - 6A_2) - 6t(2A_2 - A_1)\right)^2 dt. \tag{4.21}$$

Since the integral in (4.21) does not depend on m, the sample variance of residuals within each block scales like m^{2H}. The above calculations apply for both the infinite and finite variance case. For a related more rigorous proof in the Gaussian case, see [TTW95].

The sample variance of residuals is computed for each block, and its median is then obtained over the N/m blocks. If we assume, as a first approximation, that the median also behaves like m^{2H}, then a log-log plot versus m should follow a straight line with a slope of $2H$. The cut-offs used here are $(10^{0.7}, 10^{2.5})$.

If, instead of the median, the average over the blocks is computed, some complications arise. The problem stems from the fact that the integral in (4.21) behaves like an $\alpha/2$-stable random variable (hence has infinite mean) if the original series has infinite variance with parameter α. Thus, if the average of an $\alpha/2$-stable random variable over N/m blocks is computed, the result is proportional to $m^{1-2/\alpha}$, and the average of the corresponding variance of residuals is proportional to $m^{2H+1-2/\alpha} = m^{2d+1}$. Therefore, taking the average produces an estimator of d, not H. If d is the parameter of interest, taking the average over the blocks would be a possible procedure. In practice, this is not recommended because the scatter is too large for the infinite variance series.

4.4 R/S Method

The R/S method is one of the oldest methods for estimating H and is discussed in detail in Mandelbrot and Wallis [MW69], Mandelbrot [Man75] and Mandelbrot and Taqqu [MT79]. Given the partial sums of a time series $Y(t) = \sum_{i=1}^{t} X(i)$, and its sample variance $S^2(n)$, the R/S statistic is defined:

$$\frac{R}{S}(n) := \frac{1}{S(n)}\left[\max_{0\le t\le n}\left(Y(t) - \frac{t}{n}Y(n)\right) - \min_{0\le t\le n}\left(Y(t) - \frac{t}{n}Y(n)\right)\right]. \tag{4.22}$$

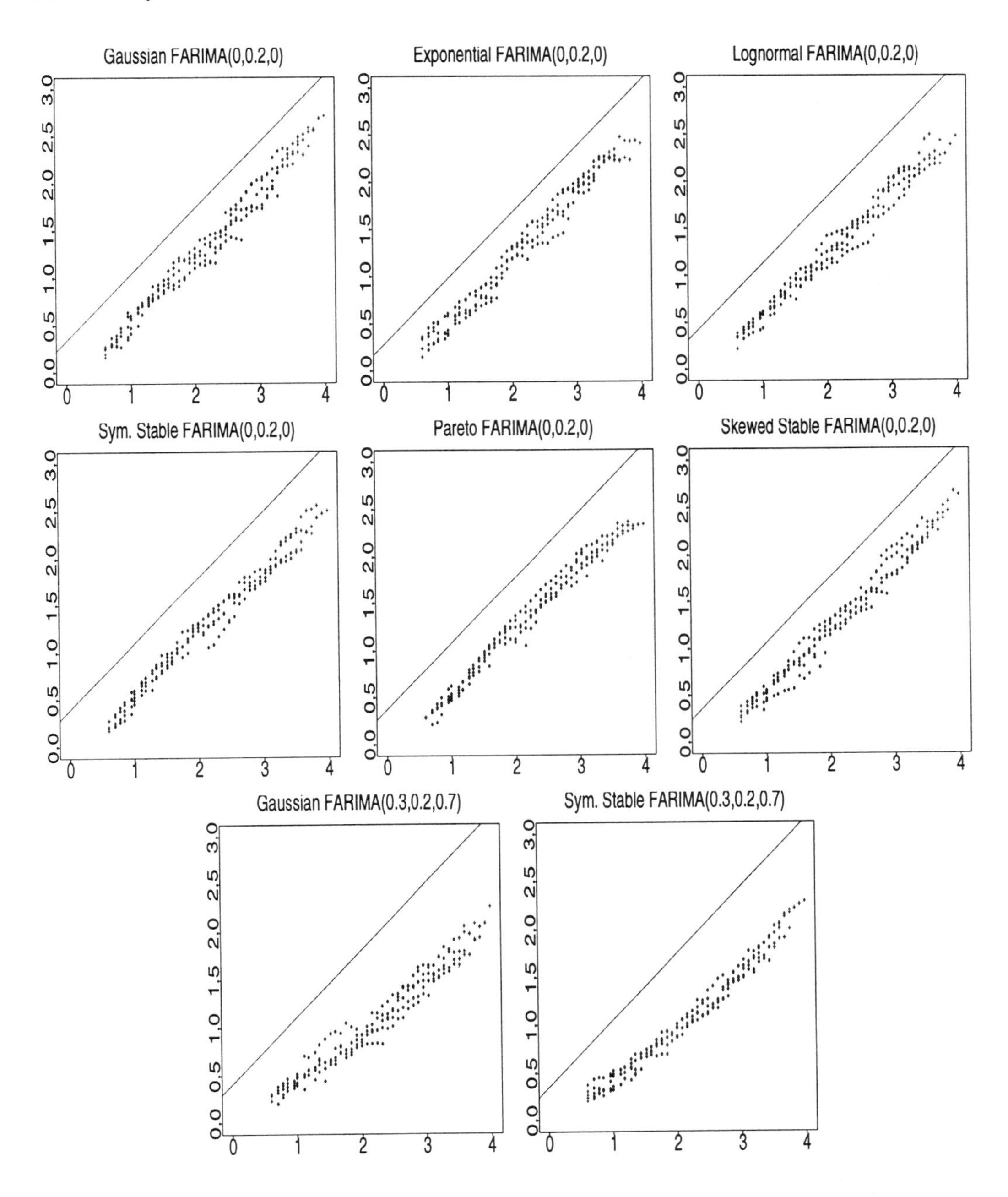

Figure 5: R/S Method. $d = 0.2$. Vertical scale, $\log_{10}(statistic)$. Horizontal scale, $\log_{10}(m)$. *Top row*: Finite variance, FARIMA$(0, d, 0)$ series. *Second row*: Infinite variance, FARIMA$(0, d, 0)$ series. *Third row*: FARIMA$(1, d, 1)$ series. Reference lines with the theoretical slope $d + 1/2$ are included. The innovations in the infinite variance series (stable, Pareto) have $\alpha = 1.5$.

Because of the asymptotic scaling property of $Y(t)$, $R(n)$behaves like $n^H = n^{d+1/\alpha}$. On the other hand, $S(n)$ is just the square root of the sample variance, which, as in (4.18), is proportional to $n^{2/\alpha-1}$. Thus, $S(n)$ behaves like $n^{1/\alpha-1/2}$. Since, in fact, one has joint convergence of $(R(n), S(n))$, $R/S(n)$ behaves like $n^{d+1/2}$, as $n \to \infty$. This provides a way of estimating d, whatever the value of α. A rigorous proof, based on the results of Mandelbrot [Man75], can be found in Avram [Avr86].

The actual method divides the original time series into K blocks, each of size N/K. Then, for each lag n, we compute $R(k_i, n)/S(k_i, n)$, starting at points $k_i = iN/K + 1, i = 1, 2, \ldots$, such that $k_i + n \leq N$.

We plot R/S versus the lag n on a log-log plot. A line is fitted to the plot, and its slope should equal $d + 1/2$. The cut-offs are $10^{0.7}$ and $10^{3.5}$.

4.5 Periodogram Method

The periodogram is defined as

$$I(\nu) = \frac{1}{2\pi N} \left| \sum_{j=1}^{N} X(j) e^{ij\nu} \right|^2, \tag{4.23}$$

where ν is the frequency, N is the length of the series, and X is the time series. In the finite variance case, $I(\nu)$ is an estimator of the spectral density of X, and a series with long-range dependence will have a spectral density proportional to $|\nu|^{-2d}$ close to the origin. A log-log regression thus provides an estimate of d. In the infinite variance case the problem is significantly more complicated. Although extensive work has been done in this general area (see Klüppelberg and Mikosch [KM96a], [KM96b] and Kokoszka and Mikosch [KM]), there are no theoretical results for the periodogram regression method. The proportionality to $|\nu|^{-2d}$ as $\nu \to 0$, however, seems to hold empirically in the infinite variance case as well, suggesting the need for theoretical research in this direction.

We thus expect a log-log plot of the periodogram versus the frequency to display a straight line with a slope of $-2d$. In practice we use only the lowest 10% of the frequencies for the calculation, since this behavior holds only for frequencies close to zero.

We do not include here the "smoothed" periodogram method discussed in [TTW95], because in our studies the original periodogram has consistently performed better.

4.6 Whittle Method

The Whittle estimator is also based on the periodogram. It involves the function

$$Q(\eta) := \int_{-\pi}^{\pi} \frac{I(\nu)}{f(\nu; \eta)} d\nu + \int_{-\pi}^{\pi} \log f(\nu; \eta) d\nu, \tag{4.24}$$

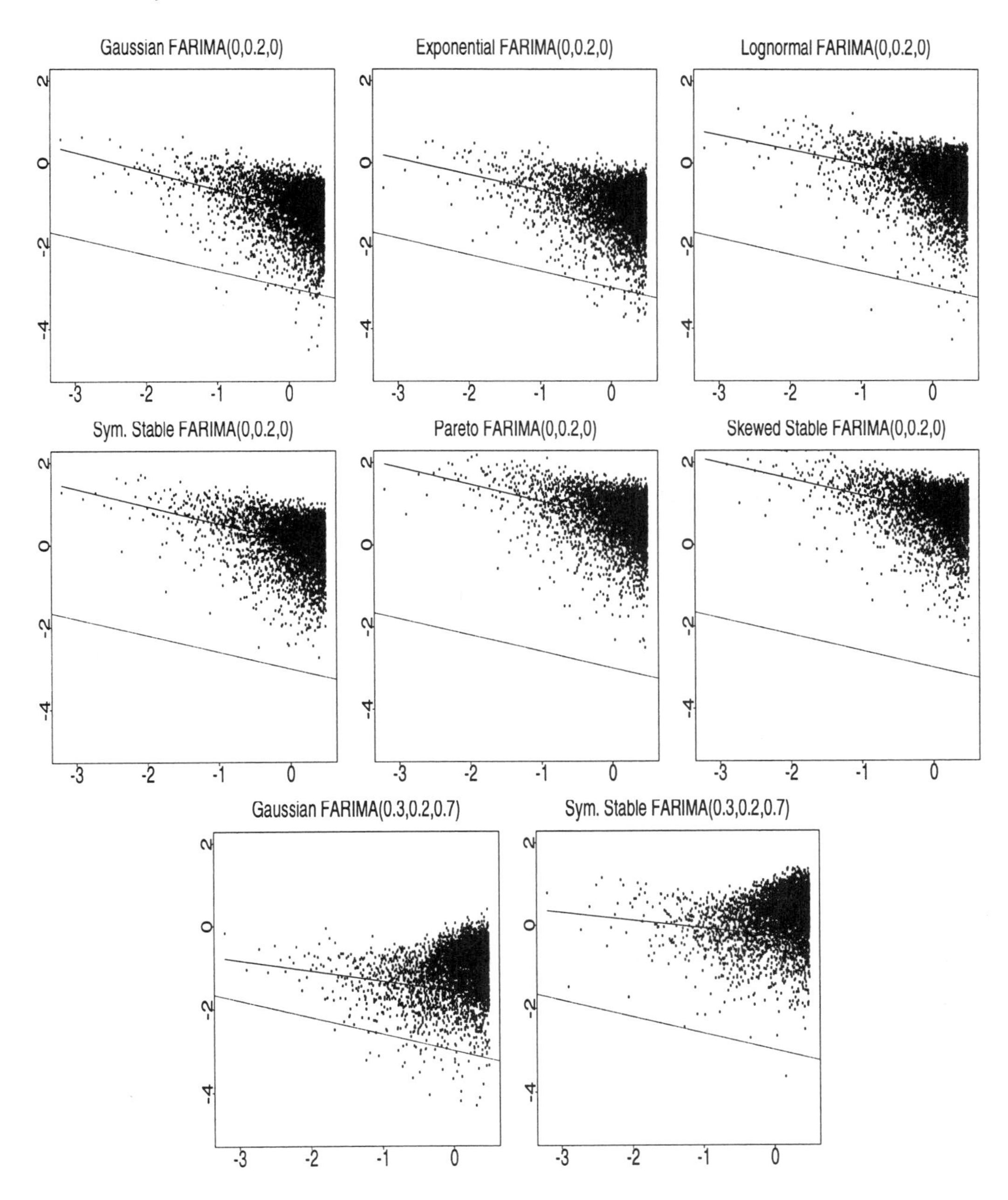

Figure 6: Periodogram Method. $d = 0.2$. Vertical scale, $\log_{10}(periodogram)$. Horizontal scale, $\log_{10}(frequency)$. *Top row*: Finite variance, FARIMA$(0, d, 0)$ series. *Second row*: Infinite variance, FARIMA$(0, d, 0)$ series. *Third row*: FARIMA$(1, d, 1)$ series. Reference lines with the theoretical slope $-2d$ are included. The innovations in the infinite variance series (stable, Pareto) have $\alpha = 1.5$.

where η is the vector of unknown parameters, $I(\nu)$ is the periodogram (see (4.23)) and $f(\nu;\eta)$ is the spectral density at frequency ν. The term $\int_{-\pi}^{\pi}\log f(\nu;\eta)d\nu$ can be set equal to 0 by renormalizing $f(\nu;\eta)$. The normalization only depends on a scale parameter, and not on the rest of the components of η. Thus, we replace f by f^*, such that $f^* = cf$, and $\int_{-\pi}^{\pi}\log f^*(\nu;\eta)d\nu = 0$.

The Whittle estimator is defined as the value of η which minimizes the function Q. In practice, (4.24) is replaced by a sum over Fourier frequencies. η is the parameter d if the series is FARIMA$(0,d,0)$, and it also includes the autoregressive and moving average parts if the series is FARIMA(p,d,q) (see (2.8)). It is d and not H which is estimated even in the infinite variance case.

The main disadvantages of the Whittle estimator are: (1) the need to know the parametric form of the spectral density and (2) a greater computational effort than for the graphical methods listed above. If the exact form of the spectral density is not known, the estimator can become very biased. Moreover, one may not notice that bias exists because the output is non-graphical. Thus, this estimator is not very robust. For details see Fox and Taqqu [FT86] and Beran [Ber94] in the Gaussian case and Klüppelberg and Mikosch [KM96b] and Kokoszka and Taqqu [KT96] in the infinite variance case. In this study we assume for all of the series that the true model is FARIMA(p,d,q) with $p=q=1$. This assumption is reasonable because we do not generate series with p or q higher than 1. The estimator would become biased if the generated series had a p or q larger than 1 because of underspecification.

4.7 Aggregated Whittle Method

The Aggregated Whittle estimator provides a way of robustifying the Whittle estimator in the absence of exact parametric information about the spectral density of the given data series. It can be used if the time series is long enough. The idea is to aggregate the data, which creates a new, shorter series

$$X_k^{(m)} := \frac{1}{m}\sum_{i=m(k-1)+1}^{mk} X_i. \tag{4.25}$$

Of course the shortened length of the aggregated series increases the standard deviation of the estimator, but, on the other hand, for high enough aggregation level m, the centered aggregated series $X^{(m)} - EX^{(m)}$ will be close to Fractional Gaussian Noise and the bias can be reduced by using the Whittle estimator with an underlying FGN model, in the finite variance case. The limit is LFSN in the infinite variance case, but the results of [KT96] suggest that it is reasonable to still use the Whittle estimator with an underlying FGN model.

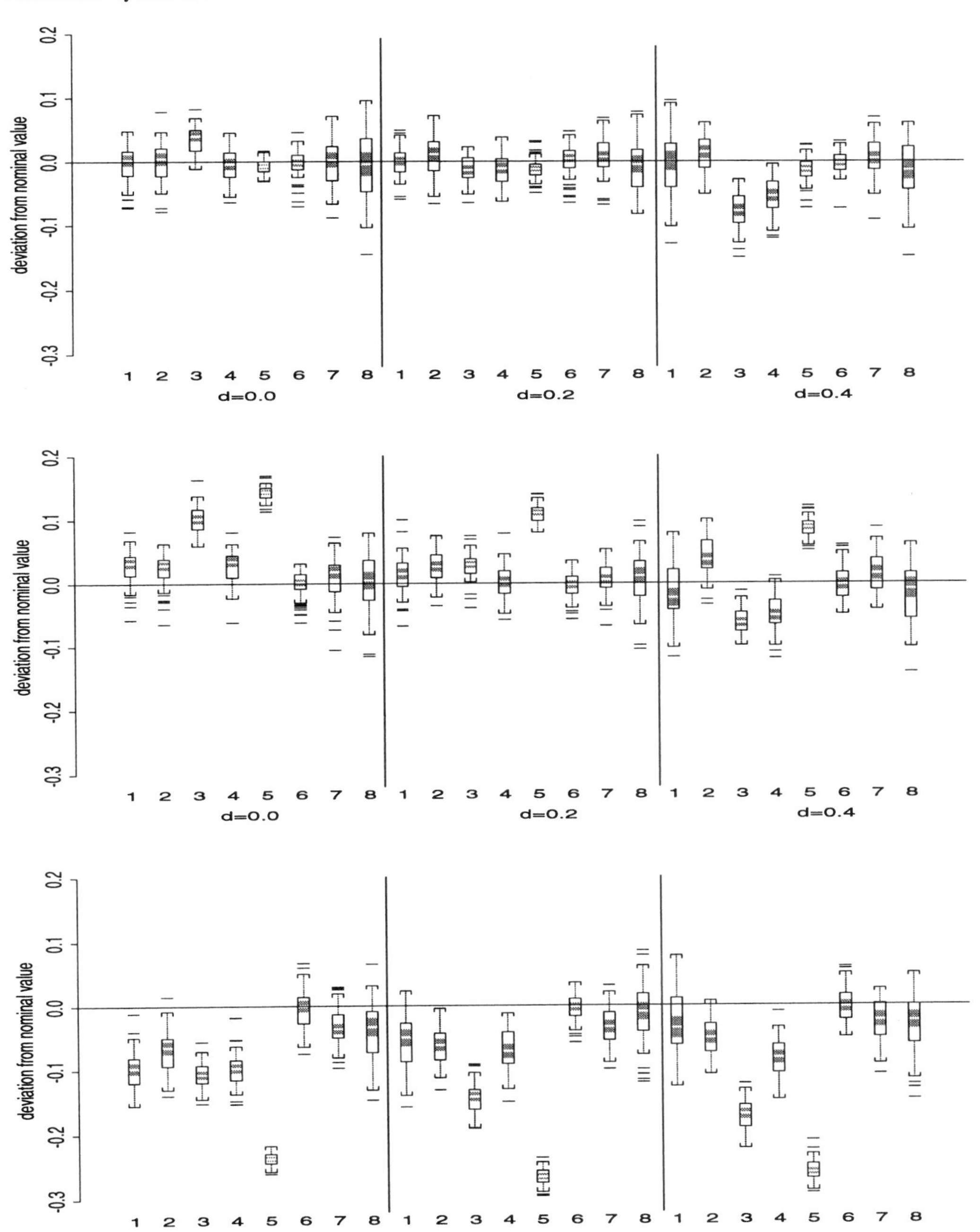

Figure 7: *Top row*: FARIMA$(0, d, 0)$, *Middle row*: FARIMA$(0, d, 0.5)$, *Bottom row*: FARIMA$(0.3, d, 0.7)$. Gaussian innovations. *Boxplots:* 1 – Absolute Value. 2 – Periodogram. 3 – R/S. 4 – Variance. 5 – Variance of Residuals. 6 – Whittle. 7 – Local Whittle ($N/M = 32$). 8 – Aggregated Whittle ($m = 50$).

4.8 Local Whittle Method

Unlike the Whittle estimator, the Local Whittle estimator is a semi-parametric estimator, in that it only specifies the parametric form of the spectral density when ν is close to zero, namely, it assumes that

$$f(\nu) \sim G(d)|\nu|^{-2d} \quad \text{as } \nu \to 0, \tag{4.26}$$

where $G(d)$ is some function of d. This estimator is essentially equivalent to the Whittle, except that it only assumes the behaviour of the spectral density up to some frequency $2\pi M/N$. One estimates d by minimizing

$$R(d) = \log\left(\frac{1}{M}\sum_{j=1}^{M}\frac{I(\nu_j)}{\nu_j^{-2d}}\right) - 2d\frac{1}{M}\sum_{j=1}^{M}\log\nu_j. \tag{4.27}$$

The Local Whittle estimator was developed by Robinson [Rob95] and was further studied, along with the Whittle and Aggregated Whittle estimators, by Taqqu and Teverovsky [TT97a] in the Gaussian case. As we will show here, it works well in the infinite variance case, although there are as yet no corresponding theoretical results.

5. Results

We will first present some general features of our results, and then comment on the behavior of each estimator separately.

5.1 General Observations

As can be seen in Table 1 and Figure 7 (pg. 197), most estimators perform well when FGN or Gaussian FARIMA(0, d, 0) series are used. This provides a comparative benchmark for what happens when one adds deviations from "pure" Gaussian long-range dependence.

When Gaussian FARIMA(p, d, q) series are used the results depend very strongly on the values of the AR and MA parameters [4]. If the parameters are negative, as in the left part of Table 3, most estimators are essentially unbiased, and the results are comparable to those for FARIMA(0, d, 0) series. On the other hand, if the parameters are positive, as in Table 2 and on the right-hand side of Table 3, as well as the bottom two-thirds of Figure 7, then most estimators are quite biased. The exception is the Whittle estimator when assuming a FARIMA(1, d, 1) model because the series we generate have p and q no higher than 1. The next best estimator is the Local Whittle.

[4] When p and/or q are 1, the notation FARIMA(ϕ_1, d, θ_1) is used in the tables in order to display the values of ϕ_1 and/or θ_1.

Using exponential and lognormal innovations instead of Gaussian ones does not affect most of the estimators (see Table 4 and the boxplots in Figure 8). The exceptions are the Absolute Value Method which is biased downward for high d, and the Variance of Residuals method which is very biased for the lognormal-based series.

Most of the estimators performed well for symmetric stable FARIMA$(0, d, 0)$ series (see the left sides of Tables 6 and 8 and the top halves of Figures 11 and **??**). The exception again was the Absolute Value estimator which appeared both more scattered and more biased. In addition, while the standard deviations and confidence intervals for most of the methods are similar to the Gaussian cases, more outliers appear. This is because the series have infinite variance. When stable FARIMA$(1, d, 1)$ are used, the results are very similar to the Gaussian $(1, d, 1)$ case with the same values of the parameters $(\phi = 0.3,\ \theta = 0.7)$ (see the right sides of Tables 6 and 8 and the bottom halves of Figures 11 and **??**). The exception is the Absolute Value method which is affected by the infinite variance of the sample series. We note again that while most of the estimators estimate the parameter d, the Variance of Residuals and the Absolute Value methods estimate H. Also, the Whittle method tends to have infrequent but very large outliers in the FARIMA$(1, 0, 1)$ cases. Whereas the estimate of d should be 0, approximately 1% of the time the boundary value of 0.5 is returned. (The Whittle estimator software restricts the values of d to the interval $(-0.5, 0.5)$ corresponding to the stationary range for finite variance series.) Because of these outliers, the standard errors for the Whittle estimator are larger in the FARIMA$(1, 0, 1)$ cases, both for $\alpha = 1.5$ and $\alpha = 1.2$.

Using Pareto and skewed stable innovations for the FARIMA$(0, d, 0)$ series gives results very similar to the symmetric stable FARIMA$(0, d, 0)$, with $\alpha = 1.5$ (see Table 7 and Figure 12). The exception is the Variance of Residuals method, which is very biased for Pareto innovations.

5.2 Graphical Methods

The *Absolute Value method* is one of the better graphical methods for Gaussian series. Unfortunately its performance declines for infinite variance series, particularly as α gets closer to 1 (and as d grows for non-Gaussian finite variance series) (see Figure 2). An intuitive explanation is that as α gets closer to 1, the series become more and more heavy-tailed, and even the mean diverges when $\alpha = 1$. Thus, outliers in the original series can have an inordinate effect on the estimate of H, and therefore the estimates are very scattered, as seen in the boxplots. Therefore, while this method can be used if the data series is known to be Gaussian or has finite variance, it is not recommended if the series is likely to be infinite variance.

The *Variance method* is biased downward for high values of d even if Gaussian FARIMA$(0, d, 0)$ series are used. This bias is not especially large unless $d > 0.3$, so this should not be an overriding consideration. For Gaussian FARIMA$(1, d, 1)$ series the performance declines, as with the other methods, but the bias is never greater than 0.1. The biases and standard deviations remain approximately the same for Gaussian and non-Gaussian series. Although the method is never as accurate as the Whittle-type methods, it appears quite robust (Figure 3).

The *Periodogram method* is perhaps the best of the graphical methods examined here. While it is not extremely accurate for the "ideal" series, it suffers the smallest losses in efficiency as deviations from the "ideal" are introduced. For the FARIMA$(1, d, 1)$ series it has some of the smallest MSEs along with the Absolute Value method, but unlike the latter, its efficiency does not suffer when non-Gaussian series are introduced. Even for stable FARIMA$(1, d, 1)$ its $\sqrt{\text{MSE}}$ is never larger than .095 (see Figure 6).

The *R/S method* has long been known to be biased towards $d = 0.2$. It is biased upward for small values of d and downward for large values of d. The method is also very biased when FARIMA$(1, d, 1)$ series are used, and while not suffering a loss of efficiency for non-Gaussian series, it is already so inaccurate that it should be used only to get a very rough idea of the intensity of long-range dependence (see Figure 5).

The *Variance of Residuals method* is one of only two methods which give H, the scaling parameter, instead of d. It works very well for both Gaussian and stable FARIMA$(0, d, 0)$ series, giving the smallest MSEs of any of the graphical methods, and is better than the Whittle estimator for several parameter values. Unfortunately, while its standard error remains small, its bias becomes quite large for lognormal, Pareto and also FARIMA$(1, d, 1)$ series. Apparently the effects of the short-range dependence persist to a greater degree here than for the other graphical estimators (see Figure 4). Its accuracy for "ideal" series makes it desirable to attempt to make this method more robust. A possible solution is to only use the larger values of m for our estimates by moving the lower cutoff to the right. This does result in decreased bias, but since the interval where H is estimated becomes shorter, the estimates are more scattered and the standard error increases. Another way in which this method can be robustified is by again aggregating the original series, just as in the Aggregated Whittle, and for the same reasons. The aggregated series should become closer and closer to FGN (LFSN) as the aggregation level increases, and the estimator is quite good for those kinds of series. Clearly the level of aggregation is very important. If it is not large enough, the estimates will be very biased. If it is too large, the estimates will be very scattered, and the standard error will be large. To use the Variance of Residuals effectively, one would need to run the estimator at several cutoff values or aggregation levels. The techniques proposed for the Whittle-type estimators in [TT96] would also be reasonable here.

5.3 Whittle-type estimators

The *Whittle estimator* with the assumed model of FARIMA$(1, d, 1)$ gave the best performance for the series used in this study. Recall that all of the series used here fall within the FARIMA$(1, d, 1)$ family of models. Certainly, if the parametric form of a time series is known, then the Whittle estimator is to be recommended. Even if the exact form is not known, but the maximum order (p, q) is known, this estimator can give good results. There is no problem with symmetric stable series, as the estimator's performance is essentially unaffected. When other infinite variance series are introduced, the estimator gives very incorrect estimates for a few realizations. For the majority of realizations, however, it gives good results.

We note that in addition to d, the Whittle estimator provides values for ϕ and θ, the AR and MA coefficients. In general, the estimates of these coefficients are approximately as accurate as the estimates of d. But a problem can arise if the actual series is FARIMA$(0, d, 0)$ or FGN and if a FARIMA$(1, d, 1)$ model is assumed. For ϕ and θ the estimator will often give non-zero values which are usually almost equal, and thus cancel each other. They can be quite large and this can affect, somewhat, the estimation of d. It is, however, better to overspecify the order of the model than to underspecify it. In the latter case, the biases will be very large.

Aggregated Whittle is in general fairly unbiased and although its standard error is larger than that of the Whittle estimator, it is smaller than that of the graphical methods for the FARIMA$(1, d, 1)$ series. Just as for other methods, the level of aggregation can be an important factor in the performance of the estimator. We used an aggregation level of $m = 50$, which seemed to provide reasonable results for the time series of length $10,000$ considered here. An aggregation level which is much smaller would not eliminate the bias, and a much larger aggregation level would introduce too much uncertainty.

The *Local Whittle estimator* seems to perform quite well. Overall, it is better than the other methods, except the original Whittle (under the correct underlying model). Unlike the Whittle it does not estimate the AR and MA coefficients. The Local Whittle may be the preferred estimator to use when the parametric form of a series is completely unknown, except for its long memory. Just as for the Whittle method, some realizations yield very incorrect estimates, but, in the vast majority of realizations this problem does not arise. As with the other methods, the choice of M in (4.27) is an important one. $M = N/32$ seems to be a good choice. For more details about the choice of M (Local Whittle) or m (Aggregated Whittle), see Taqqu and Teverovsky [TT96].

As mentioned at the end of Section 3, we analyzed the Local Whittle in particular detail. The results for series with Gaussian, stable $\alpha = 1.5$ and stable $\alpha = 1.2$ innovations are illustrated in Figures 9 and 10. These figures clearly show the difference between finite and infinite variance series. Although the estimates for the stable series have more outliers, they are in fact more concentrated in the center of the distribution. This is best seen in the boxplots (Figure 10) for the stable series, as the box representing the middle 50% of the estimates is narrower, but more outliers are present. While the standard deviation of the estimates is smaller for the stable series (see Table 5), one needs a larger number of realizations, I, to account for the greater variability.

6. Conclusions

We studied the robustness of several estimators. Finite and infinite variance FARIMA series were considered, using Gaussian, stable, exponential, lognormal, and Pareto distributions, as well as different combinations of $p, q = 0, 1$. Since most of the estimators were developed for Gaussian (or at most finite variance) series, testing whether they still work in the infinite variance case is important. We sketched the theoretical behavior of some estimators in the infinite variance case, and used simulations to test their robustness. The inclusion of infinite variance processes requires making a sharp distinction between the two parameters H and d which are used almost indistinguishably in the finite variance case. We have found that most of the estimators estimate d (or $d + 1/2$) and not $H = d + 1/\alpha$. It is best, in practice, to use several different methods to estimate long-range dependence since this can provide a better perspective on the structure of the time series.

Two main conclusions can be drawn from our simulations. First, many of the estimators are relatively robust with respect to deviations from Gaussian series. There tends to be somewhat more scattering of the estimates in the infinite variance case, but the standard errors and MSEs are not very different. In fact, in many cases the standard errors are smaller in the infinite variance case, as discussed for the Local Whittle estimator.

The exceptions are the Absolute Value and Variance of Residuals methods, which tend to be quite sensitive to the distribution of the innovations. However, we have found that for FARIMA$(0, d, 0)$ series, the Variance of Residuals method still works well (for estimating H), while the Absolute Value method loses efficiency, especially for values of α close to 1. Note that these methods estimate $H = d + 1/\alpha$ and not d.

Our second general conclusion is that the estimators are strongly affected by non-zero AR and MA components. Even for the non-extreme values used in this study, the biases of most of the estimators are considerable. The best of the methods, Whittle, only worked because its assigned model was correct (or overspecified). If the model is underspecified, the Whittle estimator becomes more biased than any of the others used here. The Local Whittle and the Periodogram are the two methods which seem the most robust and accurate. One can improve them further by using several different cut-off values, as described in Taqqu and Teverovsky [TT96].

	Gaussian Innovations Nominal d								
	FGN					FARIMA(0,d,0)			
	0.0	0.1	0.2	0.3	0.4	0.1	0.2	0.3	0.4
	ABSOLUTE								
BIAS	-.003	-.005	.000	-.005	-.004	-.006	-.002	-.003	-.012
$\hat{\sigma}$	.028	.028	.024	.028	.049	.033	.040	.037	.040
$\sqrt{\text{MSE}}$	.028	.028	.023	.028	.049	.033	.039	.037	.041
	PERIODOGRAM								
BIAS	.001	.001	.009	.012	.011	.009	.015	.016	.005
$\hat{\sigma}$	.032	.029	.033	.025	.028	.039	.033	.031	.028
$\sqrt{\text{MSE}}$	.032	.029	.034	.028	.030	.039	.034	.035	.028
	R/S								
BIAS	.035	.009	-.013	-.034	-.079	.014	-.012	-.040	-.077
$\hat{\sigma}$	.023	.024	.020	.024	.027	.023	.022	.022	.026
$\sqrt{\text{MSE}}$	.042	.025	.024	.042	.083	.027	.025	.046	.081
	VARIANCE								
BIAS	-.005	-.012	-.013	-.028	-.056	-.009	-.014	-.027	-.060
$\hat{\sigma}$	.026	.027	.024	.022	.031	.031	.036	.028	.032
$\sqrt{\text{MSE}}$	.026	.029	.027	.036	.063	.031	.038	.039	.068
	VARIANCE OF RESIDUALS								
BIAS	-.009	-.011	-.014	-.018	-.016	-.017	-.023	-.028	-.035
$\hat{\sigma}$	.012	.015	.014	.018	.016	.016	.016	.017	.017
$\sqrt{\text{MSE}}$	.015	.019	.020	.026	.022	.023	.028	.033	.039
	WHITTLE (FARIMA(1, d, 1))								
BIAS	-.004	-.006	-.002	-.005	-.002	-.002	.000	-.001	-.004
$\hat{\sigma}$	.022	.015	.017	.019	.015	.022	.025	.020	.018
$\sqrt{\text{MSE}}$	.022	.016	.017	.019	.015	.022	.025	.020	.018
	AGGREGATED WHITTLE ($m = 50$)								
BIAS	-.007	-.017	-.005	-.002	-.017	-.008	-.005	-.011	-.024
$\hat{\sigma}$	.054	.048	.041	.042	.047	.052	.056	.046	.054
$\sqrt{\text{MSE}}$	.054	.050	.041	.041	.050	.052	.056	.047	.058
	LOCAL WHITTLE ($N/m = 32$)								
BIAS	-.003	-.005	.006	.002	.006	.003	.000	.004	-.004
$\hat{\sigma}$	.034	.028	.030	.026	.031	.031	.033	.022	.025
$\sqrt{\text{MSE}}$	.034	.028	.031	.026	.031	.031	.033	.022	.025

Table 1: Estimation results for d using 50 independent copies of time series of length 10, 000. These are the "ideal" results for FGN and for Gaussian FARIMA(0,d,0).

	Gaussian Innovations Nominal d									
	FARIMA(.5, d, 0)					FARIMA(0, d, .5)				
	0.0	0.1	0.2	0.3	0.4	0.0	0.1	0.2	0.3	0.4
	ABSOLUTE									
BIAS	.026	.017	.013	.012	-.014	-.061	-.040	-.034	-.022	-.029
$\hat{\sigma}$	.028	.029	.031	.044	.045	.017	.031	.033	.041	.052
$\sqrt{MSE}$	.038	.033	.033	.045	.047	.064	.051	.047	.046	.060
	PERIODOGRAM									
BIAS	.022	.029	.027	.031	.040	-.029	-.015	-.028	-.014	-.016
$\hat{\sigma}$	.027	.026	.025	.032	.033	.028	.029	.035	.033	.029
$\sqrt{MSE}$	.035	.039	.037	.044	.052	.040	.033	.044	.036	.032
	R/S									
BIAS	.103	.074	.027	-.016	-.064	-.076	-.092	-.113	-.122	-.141
$\hat{\sigma}$	.020	.022	.021	.024	.021	.017	.022	.021	.026	.021
$\sqrt{MSE}$	.105	.078	.034	.029	.067	.077	.095	.115	.124	.143
	VARIANCE									
BIAS	.025	.013	.001	-.022	-.052	-.063	-.044	-.051	-.050	-.073
$\hat{\sigma}$	.027	.024	.026	.032	.019	.031	.029	.029	.034	.033
$\sqrt{MSE}$	.036	.028	.026	.039	.059	.065	.053	.058	.060	.080
	VARIANCE OF RESIDUALS									
BIAS	.121	.101	.087	.077	.070	-.192	-.195	-.198	-.187	-.177
$\hat{\sigma}$	.011	.013	.014	.017	.019	.011	.013	.013	.016	.016
$\sqrt{MSE}$	.121	.102	.088	.079	.073	.192	.195	.198	.188	.177
	WHITTLE (FARIMA(1, d, 1))									
BIAS	-.002	-.002	-.005	-.004	.000	-.006	.002	-.012	-.001	-.004
$\hat{\sigma}$	.023	.022	.021	.026	.026	.020	.021	.023	.026	.022
$\sqrt{MSE}$	.023	.022	.021	.026	.026	.021	.021	.026	.026	.022
	AGGREGATED WHITTLE ($m = 50$)									
BIAS	.001	.007	.005	-.009	-.021	-.032	-.007	-.024	-.024	-.027
$\hat{\sigma}$	.045	.057	.044	.050	.046	.034	.054	.044	.048	.047
$\sqrt{MSE}$	.044	.057	.044	.051	.050	.047	.054	.049	.053	.053
	LOCAL WHITTLE ($N/m = 32$)									
BIAS	.010	.013	.007	.010	.016	-.010	-.008	-.015	-.012	-.014
$\hat{\sigma}$	.036	.027	.025	.034	.031	.025	.030	.031	.032	.032
$\sqrt{MSE}$	.037	.030	.026	.035	.035	.027	.030	.034	.033	.035

Table 2: Estimation results for d using 50 independent copies of FARIMA time series of length 10, 000. The series are Gaussian FARIMA(0.5, d, 0) and (0, d, 0.5).

	Gaussian Innovations, Nominal d									
	FARIMA($-.3, d, -.7$)					FARIMA($.3, d, .7$)				
	0.0	0.1	0.2	0.3	0.4	0.0	0.1	0.2	0.3	0.4
	ABSOLUTE									
BIAS	.001	.006	.011	-.006	-.010	-.098	-.077	-.055	-.043	-.029
$\hat{\sigma}$	.031	.029	.029	.044	.039	.028	.029	.040	.046	.052
$\sqrt{\text{MSE}}$	.031	.029	.031	.044	.040	.102	.082	.068	.062	.059
	PERIODOGRAM									
BIAS	-.003	.004	.017	.013	.017	-.068	-.067	-.063	-.063	-.051
$\hat{\sigma}$	.032	.033	.032	.032	.031	.033	.030	.028	.033	.028
$\sqrt{\text{MSE}}$	.032	.033	.036	.034	.035	.076	.073	.069	.071	.058
	R/S									
BIAS	.062	.028	-.010	-.040	-.091	-.106	-.123	-.145	-.157	-.173
$\hat{\sigma}$	.023	.024	.024	.028	.034	.018	.024	.024	.022	.024
$\sqrt{\text{MSE}}$	.066	.036	.025	.049	.097	.108	.126	.147	.158	.175
	VARIANCE									
BIAS	.000	-.002	-.008	-.034	-.054	-.099	-.082	-.069	-.074	-.086
$\hat{\sigma}$	.028	.029	.025	.032	.028	.027	.029	.031	.030	.031
$\sqrt{\text{MSE}}$	.028	.029	.026	.027	.060	.102	.087	.076	.079	.091
	VARIANCE OF RESIDUALS									
BIAS	.014	.002	-.002	-.008	-.010	-.235	-.246	-.253	-.249	-.237
$\hat{\sigma}$	.014	.014	.015	.018	.017	.011	.012	.015	.017	.016
$\sqrt{\text{MSE}}$	.020	.014	.015	.020	.019	.235	.246	.254	.249	.238
	WHITTLE (FARIMA($1, d, 1$))									
BIAS	-.003	-.005	-.002	-.002	.001	-.004	-.004	-.007	-.009	-.006
$\hat{\sigma}$	.012	.013	.012	.012	.010	.033	.032	.029	.031	.033
$\sqrt{\text{MSE}}$	.013	.013	.012	.012	.010	.033	.032	.029	.030	.033
	AGGREGATED WHITTLE ($m = 50$)									
BIAS	.002	.003	.000	-.016	-.020	-.040	-.024	-.015	-.026	-.031
$\hat{\sigma}$	.051	.051	.043	.055	.043	.046	.047	.048	.050	.046
$\sqrt{\text{MSE}}$	.050	.050	.042	.057	.047	.061	.053	.050	.056	.055
	LOCAL WHITTLE ($N/M = 32$)									
BIAS	-.003	-.001	.008	-.002	.007	-.032	-.024	-.033	-.031	-.030
$\hat{\sigma}$	.029	.030	.029	.31	.031	.031	.032	.030	.030	.032
$\sqrt{\text{MSE}}$	.029	.029	.030	.031	.031	.044	.039	.044	.043	.044

Table 3: Estimation results for d using 50 independent copies of Gaussian FARIMA$(1, d, 1)$ time series of length 10,000.

	Nominal d									
	Exponential Innovations FARIMA(0, d, 0)					Lognormal Innovations FARIMA(0, d, 0)				
	0.0	0.1	0.2	0.3	0.4	0.0	0.1	0.2	0.3	0.4
	ABSOLUTE									
BIAS	-.001	-.008	-.013	-.029	-.050	.027	.012	.005	-.010	-.040
$\hat{\sigma}$	.024	.024	.033	.027	.036	.026	.028	.027	.029	.031
$\sqrt{\text{MSE}}$	.023	.025	.035	.039	.061	.037	.030	.027	.030	.050
	PERIODOGRAM									
BIAS	.006	-.003	-.003	-.005	-.001	.004	-.004	-.002	.003	.001
$\hat{\sigma}$	.026	.027	.031	.031	.032	.031	.032	.030	.029	.027
$\sqrt{\text{MSE}}$	.026	.027	.031	.031	.032	.031	.032	.030	.029	.027
	R/S									
BIAS	.036	.007	-.014	-.040	-.079	.036	.005	-.018	-.047	-.085
$\hat{\sigma}$	.020	.019	.021	.022	.024	.019	.019	.020	.022	.019
$\sqrt{\text{MSE}}$	.041	.020	.025	.045	.083	.040	.019	.027	.051	.087
	VARIANCE									
BIAS	-.004	-.008	-.014	-.030	-.051	.003	-.013	-.013	-.027	-.052
$\hat{\sigma}$	.025	.022	.031	.025	.033	.025	.026	.029	.026	.029
$\sqrt{\text{MSE}}$	.025	.024	.034	.039	.060	.025	.029	.031	.037	.059
	VARIANCE OF RESIDUALS									
BIAS	.035	.021	.010	.007	-.002	.125	.105	.097	.086	.081
$\hat{\sigma}$	.015	.014	.015	.014	.015	.015	.016	.015	.019	.022
$\sqrt{\text{MSE}}$	.038	.026	.018	.016	.015	.126	.106	.098	.088	.084
	WHITTLE (FARIMA(1, d, 1))									
BIAS	.000	-.004	-.001	-.003	-.003	.005	-.005	-.005	.001	-.005
$\hat{\sigma}$	.017	.017	.023	.024	.028	.021	.020	.023	.021	.025
$\sqrt{\text{MSE}}$	.017	.018	.023	.024	.028	.021	.021	.023	.021	.026
	AGGREGATED WHITTLE ($m = 50$)									
BIAS	-.003	-.005	-.004	-.009	-.015	.000	-.012	-.001	-.011	-.018
$\hat{\sigma}$	.051	.040	.047	.047	.048	.043	.051	.049	.039	.044
$\sqrt{\text{MSE}}$	.051	.040	.046	.047	.050	.043	.052	.049	.041	.047
	LOCAL WHITTLE ($N/M = 32$)									
BIAS	.000	-.003	-.001	.003	.002	.004	-.010	.003	.003	-.002
$\hat{\sigma}$	.031	.024	.033	.031	.029	.028	.027	.030	.031	.028
$\sqrt{\text{MSE}}$	.031	.024	.032	.031	.029	.028	.029	.030	.031	.028

Table 4: Estimation results for d using 50 independent copies of FARIMA(0, d, 0) time series of length 10, 000. The innovations are exponential and lognormal.

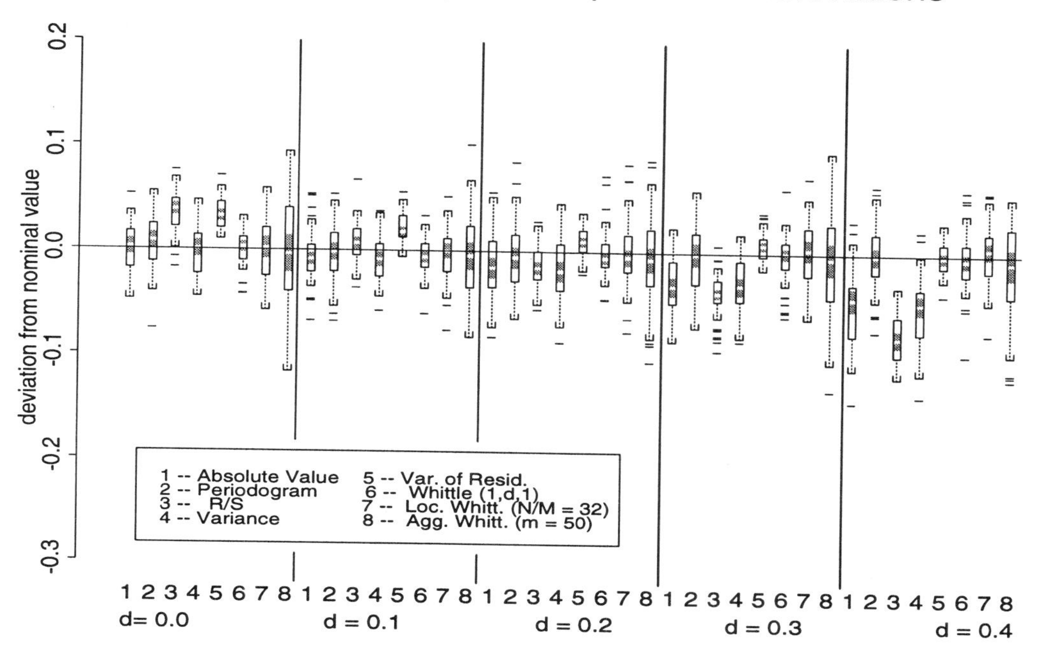

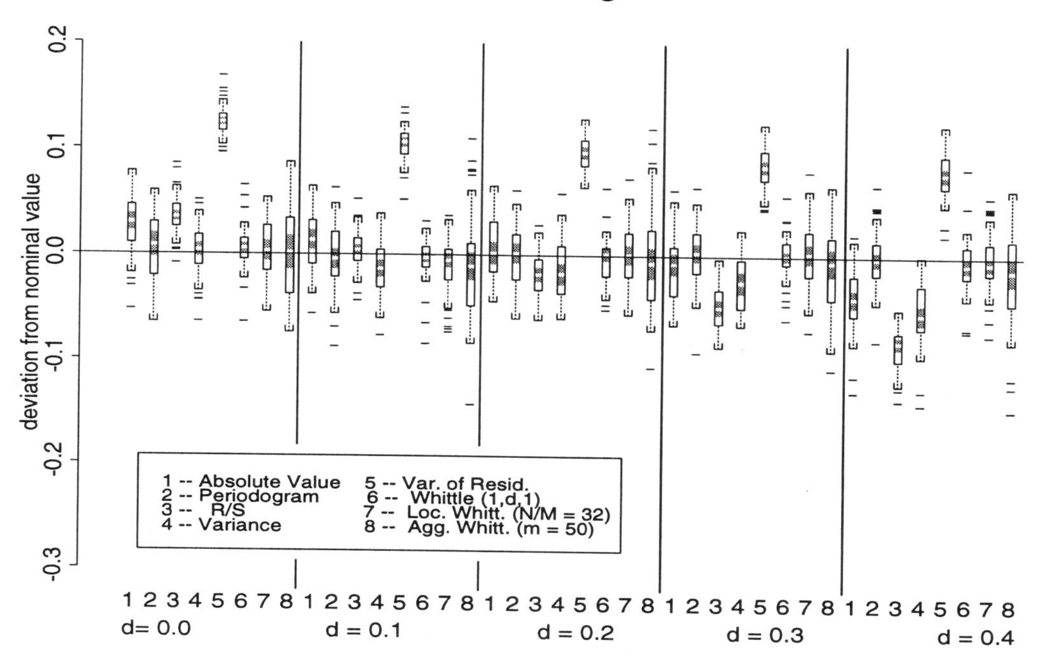

Figure 8: Boxplots for FARIMA(0, d, 0) series with finite variance innovations.

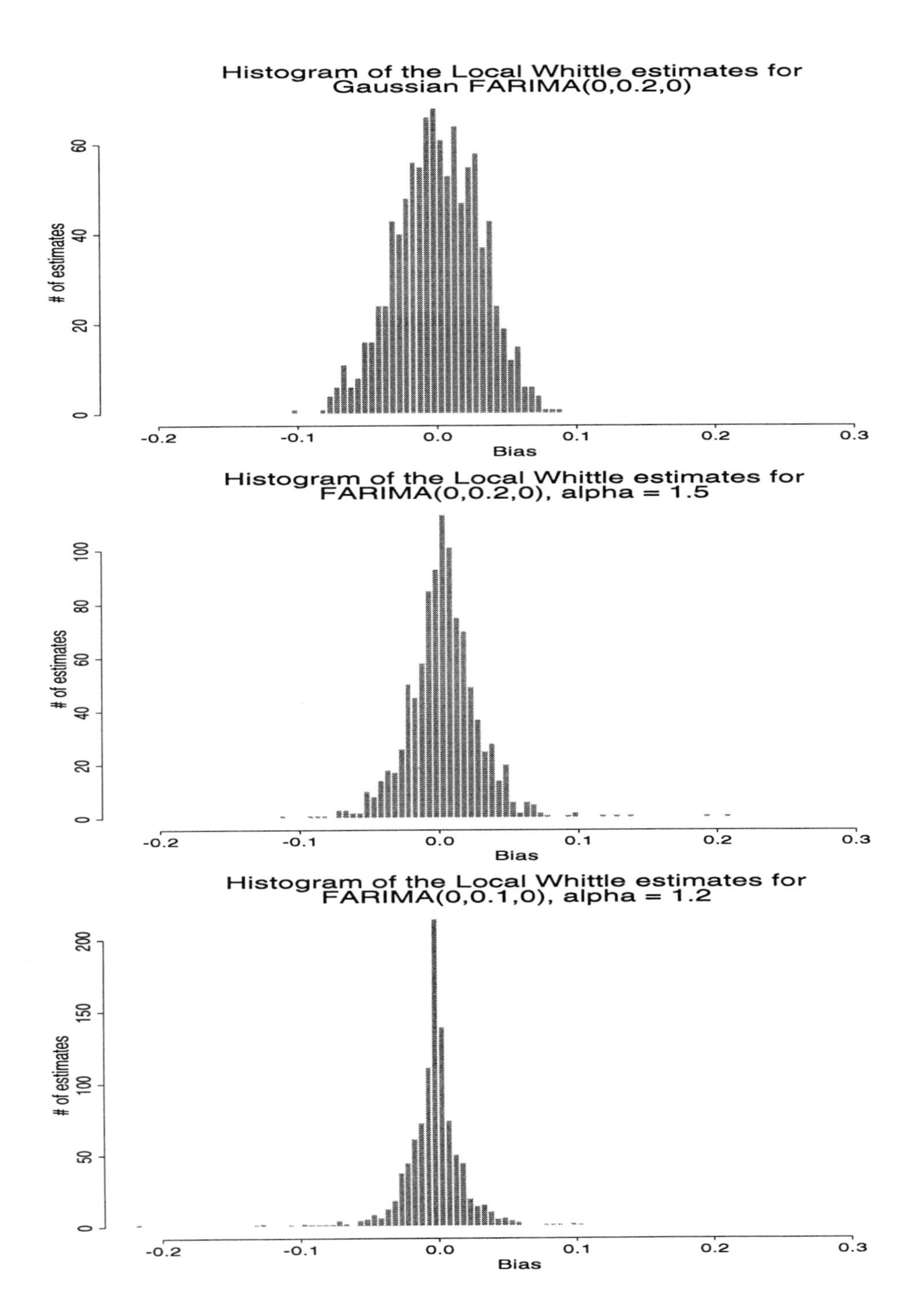

Figure 9: Histograms of the Local Whittle estimates for series with different α's, using $I = 1000$ realizations. The width of each bar is 0.005.

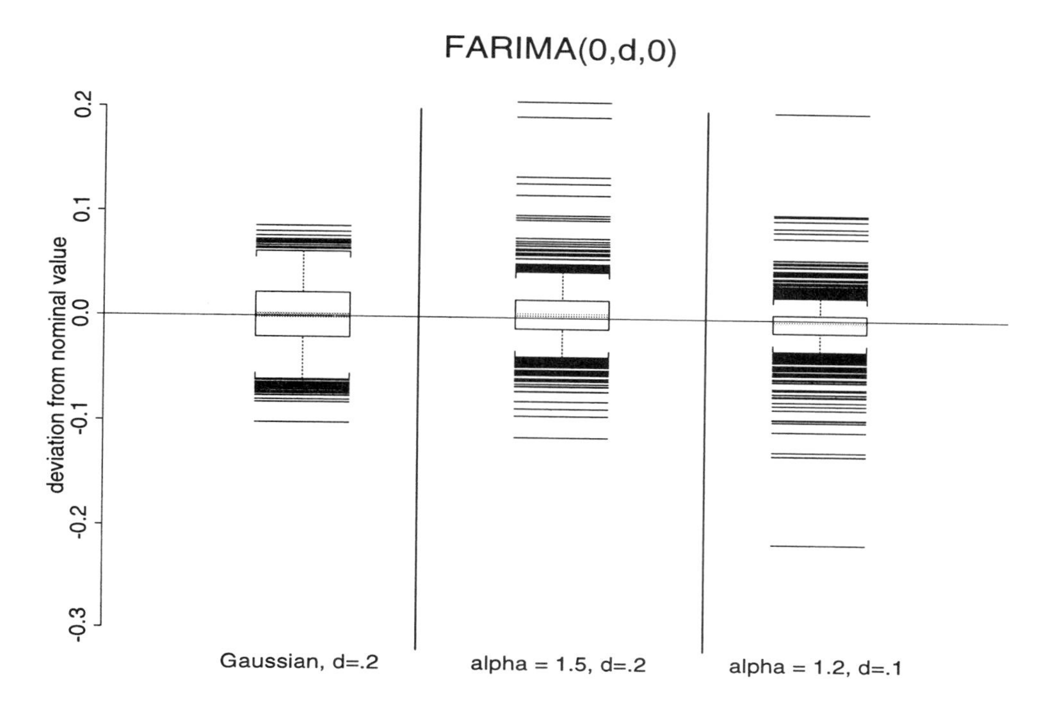

Figure 10: Detailed boxplots for the Local Whittle estimates corresponding to the histogram in Figure 10. Outliers are included.

I	Gaussian		$\alpha = 1.5$		$\alpha = 1.2$	
	BIAS	$\hat{\sigma}$	BIAS	$\hat{\sigma}$	BIAS	$\hat{\sigma}$
50	-.007	.030	-.001	.027	-.005	.020
100	-.001	.032	.002	.027	-.008	.021
200	.000	.031	.002	.025	-.007	.020
300	.002	.031	.003	.024	-.006	.022
400	.002	.030	.004	.024	-.005	.021
500	.002	.031	.004	.026	-.005	.021
600	.002	.030	.004	.026	-.005	.022
700	.002	.030	.004	.027	-.005	.023
800	.001	.030	.004	.027	-.004	.023
900	.001	.030	.003	.027	-.004	.023
1000	.001	.030	.003	.027	-.003	.026

Table 5: Mean and standard deviation of the Local Whittle estimator applied to FARIMA$(0, d, 0)$ series with various α's, using various numbers of independent realizations, I. The d's are as in Figure 11.

	Symmetric Stable Innovations with $\alpha = 1.5$							
	Nominal d							
	FARIMA(0, d, 0)				FARIMA(.3, d, .7)			
	0.0	0.1	0.2	0.3	0.0	0.1	0.2	0.3
	ABSOLUTE							
BIAS	-.015	-.041	-.056	-.089	-.111	-.110	-.100	-.110
$\hat{\sigma}$	.064	.056	.048	.045	.071	.066	.059	.049
$\sqrt{\text{MSE}}$	.066	.070	.073	.100	.132	.129	.116	.120
	PERIODOGRAM							
BIAS	.000	-.003	.003	.003	-.063	-.065	-.062	-.061
$\hat{\sigma}$	.026	.029	.027	.033	.027	.027	.026	.028
$\sqrt{\text{MSE}}$	.026	.029	.027	.033	.069	.071	.067	.067
	R/S							
BIAS	.023	-.001	-.014	-.039	-.041	-.117	-.149	-.156
$\hat{\sigma}$	.017	.018	.021	.021	.017	.015	.020	.023
$\sqrt{\text{MSE}}$	.029	.018	.025	.044	.044	.118	.150	.157
	VARIANCE							
BIAS	-.002	-.014	-.012	-.024	-.087	-.085	-.070	-.063
$\hat{\sigma}$	.024	.025	.028	.038	.037	.030	.029	.032
$\sqrt{\text{MSE}}$	.024	.029	.031	.045	.095	.090	.075	.070
	VARIANCE OF RESIDUALS							
BIAS	-.001	-.010	-.017	-.023	-.241	-.257	-.263	-263
$\hat{\sigma}$	.023	.024	.025	.026	.021	.022	.024	.025
$\sqrt{\text{MSE}}$	.023	.026	.030	.035	.242	.258	.264	.264
	WHITTLE (FARIMA(1, d, 1))							
BIAS	-.003	-.004	-.001	.000	.001	-.010	-.004	.000
$\hat{\sigma}$	.021	.022	.023	.026	.063	.026	.030	.029
$\sqrt{\text{MSE}}$	.021	.023	.023	.026	.063	.027	.030	.029
	AGGREGATED WHITTLE ($m = 50$)							
BIAS	.001	-.014	.005	.005	-.030	-.041	-.017	-.008
$\hat{\sigma}$	.045	.049	.060	.067	.053	.053	.052	.062
$\sqrt{\text{MSE}}$	.045	.051	.060	.068	.061	.067	.055	.062
	LOCAL WHITTLE ($N/M = 32$)							
BIAS	-.001	-.003	.003	.004	-.031	-.034	-.027	-.025
$\hat{\sigma}$	.027	.027	.027	.032	.025	.026	.026	.029
$\sqrt{\text{MSE}}$	.027	.027	.027	.032	.040	.042	.037	.038

Table 6: Estimation results for d using 1000 independent copies of FARIMA time series of length 10, 000. The innovations are symmetric Stable with $\alpha = 1.5$.

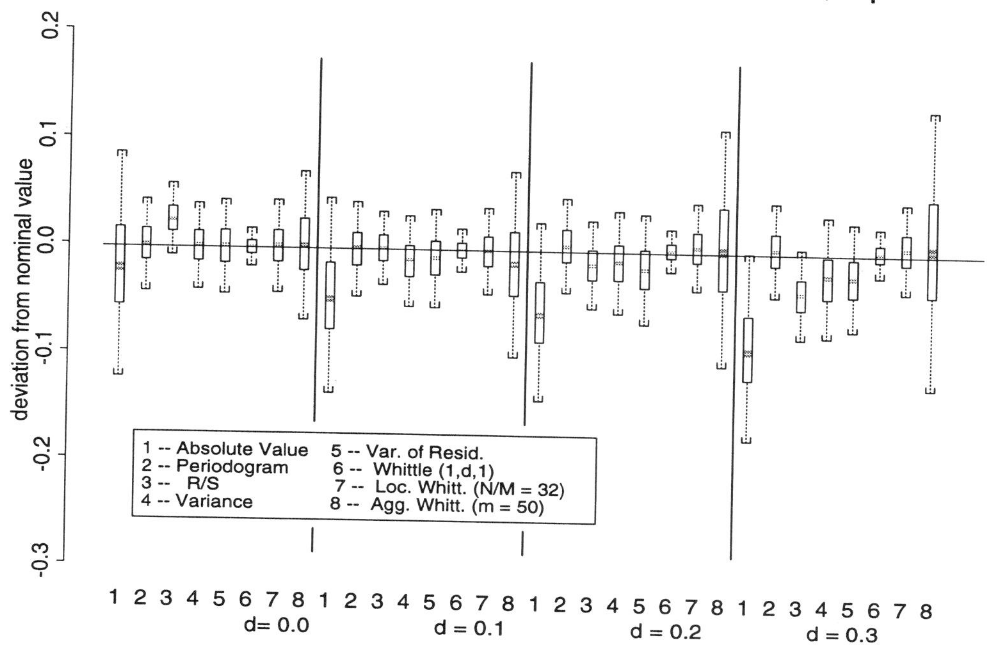

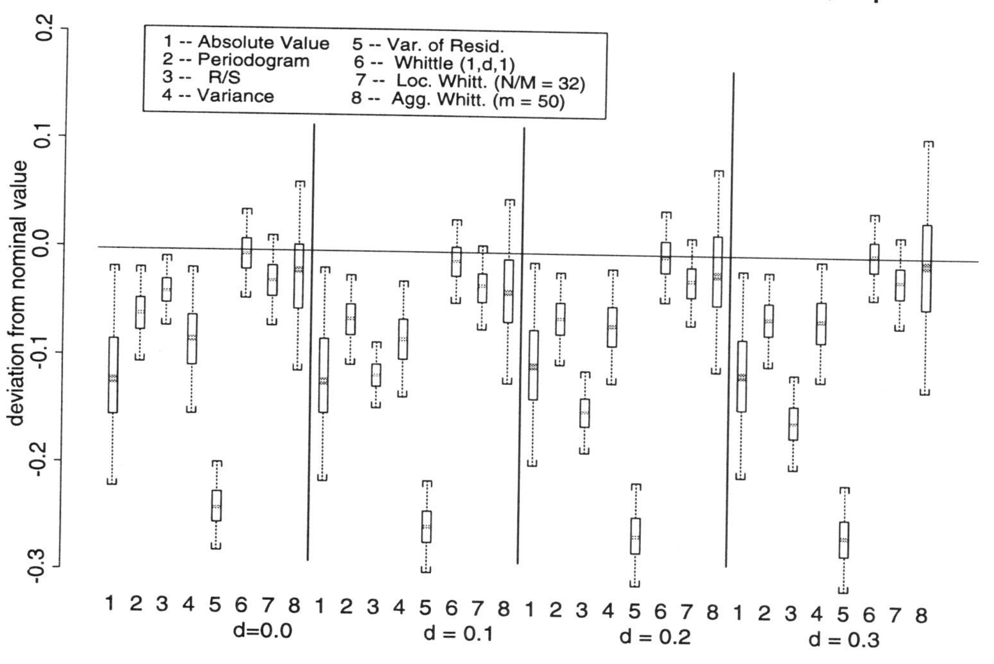

Figure 11: Boxplots for FARIMA$(0, d, 0)$ and FARIMA$(1, d, 1)$ $(\phi_1 = 0.3, \theta_1 = 0.7)$ series with symmetric Stable innovations, $\alpha = 1.5$.

	Heavy tailed Innovations with $\alpha = 1.5$							
	Nominal d							
	Pareto				Skewed Stable			
	0.0	0.1	0.2	0.3	0.0	0.1	0.2	0.3
	ABSOLUTE							
BIAS	.019	-.014	-.037	-.079	-.026	-.051	-.062	-.094
$\hat{\sigma}$	.067	.055	.046	.040	.065	.059	.051	.045
$\sqrt{\text{MSE}}$	.070	.057	.059	.088	.070	.078	.080	.104
	PERIODOGRAM							
BIAS	.001	-.003	.002	.004	-.001	-.004	.002	.004
$\hat{\sigma}$	.026	.024	.031	.027	.028	.027	.028	.027
$\sqrt{\text{MSE}}$	.026	.024	.031	.028	.028	.027	.028	.027
	R/S							
BIAS	.020	-.005	-.015	-.041	.025	-.001	-.014	-.037
$\hat{\sigma}$	.016	.018	.021	.021	.018	.019	.021	.023
$\sqrt{\text{MSE}}$	.026	.018	.025	.046	.031	.019	.025	.044
	VARIANCE							
BIAS	-.001	-.014	-.011	-.022	-.002	-.015	-.012	-.022
$\hat{\sigma}$	.021	.022	.028	.029	.022	.022	.026	.030
$\sqrt{\text{MSE}}$	.021	.026	.031	.037	.022	.027	.028	.037
	VARIANCE OF RESIDUALS							
BIAS	.172	.160	.145	.131	-.003	-.009	-.019	-.023
$\hat{\sigma}$	.030	.030	.030	.030	.022	.023	.024	.026
$\sqrt{\text{MSE}}$	.175	.163	.148	.134	.022	.026	.030	.034
	WHITTLE (FARIMA$(1, d, 1)$)							
BIAS	-.001	-.003	.000	.002	-.002	-.003	.000	.001
$\hat{\sigma}$	.020	.018	.023	.024	.019	.019	.022	.023
$\sqrt{\text{MSE}}$	.020	.018	.023	.024	.020	.020	.022	.023
	AGGREGATED WHITTLE ($m = 50$)							
BIAS	.000	-.017	.004	.006	-.001	-.018	.004	.006
$\hat{\sigma}$	.039	.048	.061	.065	.042	.048	.061	.064
$\sqrt{\text{MSE}}$	.039	.048	.061	.065	.042	.052	.061	.064
	LOCAL WHITTLE ($N/M = 32$)							
BIAS	-.001	-.004	.004	.005	-.001	-.004	.003	.006
$\hat{\sigma}$	.025	.025	.029	.028	.026	.027	.027	.030
$\sqrt{\text{MSE}}$	.025	.026	.030	.029	.026	.027	.028	.031

Table 7: Estimation results for d using 1000 independent copies of FARIMA$(0, d, 0)$ time series of length $10,000$. The innovations are Pareto and skewed Stable, with $\alpha = 1.5$.

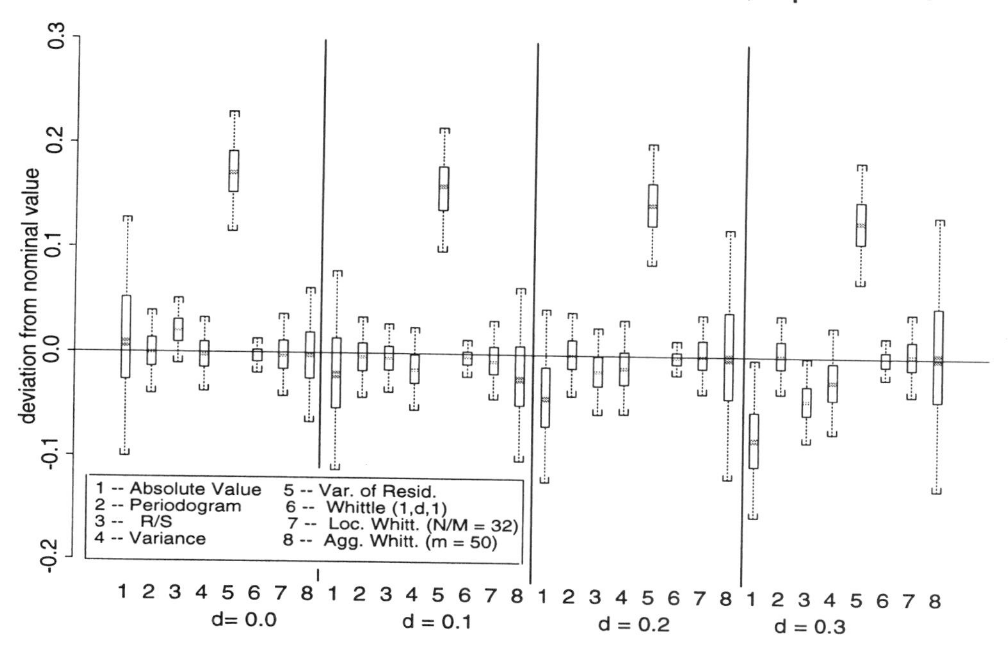

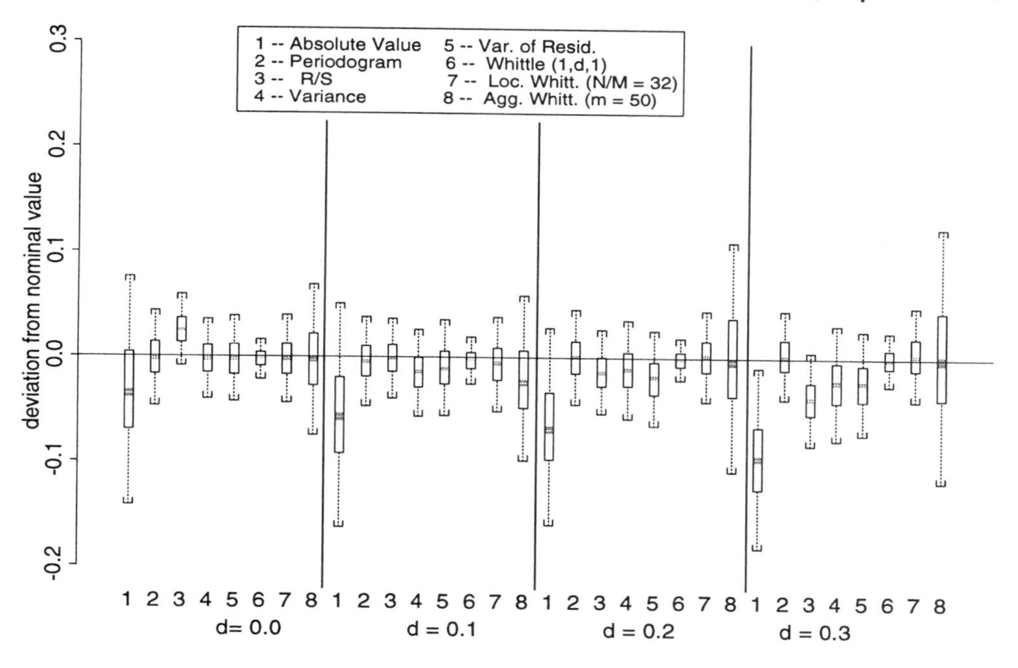

Figure 12: Boxplots for FARIMA(0, d, 0) series with Pareto and skewed Stable innovations, $\alpha = 1.5$.

	Symmetric Stable Innovations with $\alpha = 1.2$ Nominal d			
	FARIMA$(0, d, 0)$		FARIMA$(.3, d, .7)$	
	0.0	0.1	0.0	0.1
	ABSOLUTE			
BIAS	-.058	-.101	-.156	-.163
$\hat{\sigma}$	.074	.059	.079	.071
$\sqrt{\text{MSE}}$	.093	.117	.175	.178
	PERIODOGRAM			
BIAS	.000	-.002	-.063	-.067
$\hat{\sigma}$	.024	.027	.027	.026
$\sqrt{\text{MSE}}$	.024	.027	.069	.072
	R/S			
BIAS	.014	-.008	-.025	-.113
$\hat{\sigma}$	.015	.016	.014	.013
$\sqrt{\text{MSE}}$	.020	.018	.029	.114
	VARIANCE			
BIAS	-.002	-.013	-.080	-.084
$\hat{\sigma}$	.022	.026	.037	.031
$\sqrt{\text{MSE}}$	.022	.029	.088	.089
	VARIANCE OF RESIDUALS			
BIAS	.007	-.003	-.239	-.251
$\hat{\sigma}$	.034	.036	.033	.032
$\sqrt{\text{MSE}}$	.034	.036	.241	.253
	WHITTLE (FARIMA$(1, d, 1)$)			
BIAS	-.003	-.003	.001	-.010
$\hat{\sigma}$	.020	.025	.058	.031
$\sqrt{\text{MSE}}$	.021	.025	.058	.032
	AGGREGATED WHITTLE ($m = 50$)			
BIAS	-.002	-.011	-.022	-.041
$\hat{\sigma}$	.039	.054	.056	.053
$\sqrt{\text{MSE}}$	.039	.055	.061	.067
	LOCAL WHITTLE ($N/M = 32$)			
BIAS	-.001	-.003	-.030	-.034
$\hat{\sigma}$	.024	.026	.025	.024
$\sqrt{\text{MSE}}$	.024	.026	.039	.042

Table 8: Estimation results for d using 1000 independent copies of FARIMA time series of length 10, 000. The innovations are symmetric Stable with $\alpha = 1.2$.

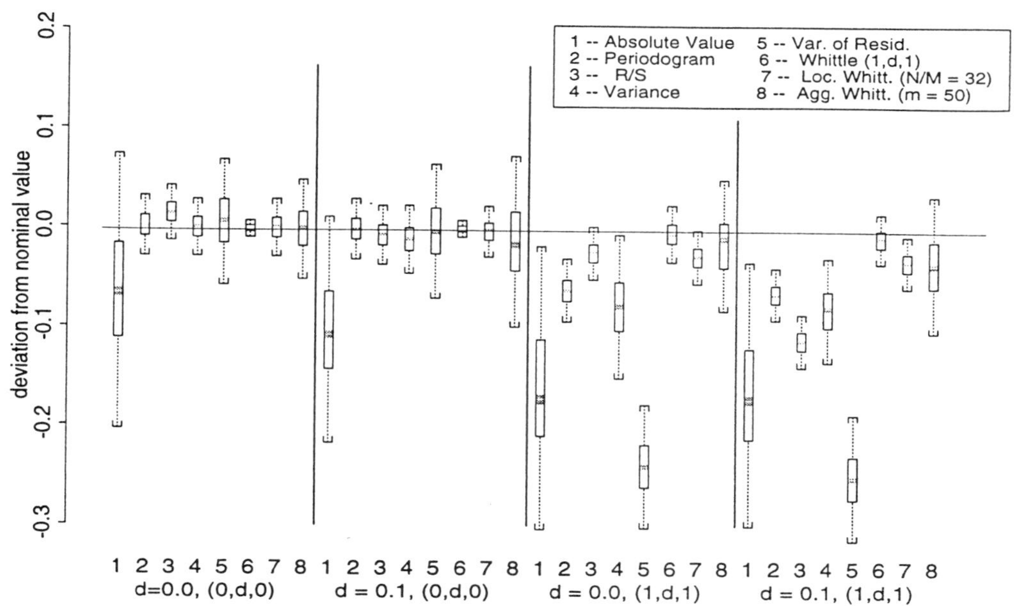

Figure 13: Boxplots for FARIMA(0, d, 0) and FARIMA(1, d, 1) ($\phi_1 = 0.3, \theta_1 = 0.7$) series with symmetric Stable innovations, $\alpha = 1.2$.

References

[Avr86] F. Avram. *Some limit theorems for stationary sequences with infinite or finite variance.* PhD thesis, Cornell University, Ithaca, New York, 1986.

[Bai96] R. T. Baillie. Long memory processes and fractional integration in econometrics. *Journal of Econometrics*, 73:5–59, 1996.

[BD91] P. J. Brockwell and R. A. Davis. *Time Series: Theory and Methods.* Springer-Verlag, New York, 2nd edition, 1991.

[Ber94] J. Beran. *Statistics for Long-Memory Processes.* Chapman & Hall, New York, 1994.

[FT86] R. Fox and M. S. Taqqu. Large-sample properties of parameter estimates for strongly dependent stationary Gaussian time series. *The Annals of Statistics*, 14:517–532, 1986.

[Hig88] T. Higuchi. Approach to an irregular time series on the basis of the fractal theory. *Physica D*, 31:277–283, 1988.

[HR89] J. Haslett and A. E. Raftery. Space-time modelling with long-memory dependence: assessing Ireland's wind power resource. *Applied Statistics*, 38:1–50, 1989. Includes discussion.

[JoE96] Fractional differencing and long memory processes, R.T. Baillie and M.L.King, editors. Special issue of the *Journal of Econometrics*, Vol. 73, No. 1, July 1996.

[KM] P. Kokoszka and T. Mikosch. The integrated periodogram for long-memory processes with finite or infinite variance. Preprint 1995. To appear in *Stochastic Processes and their Applications*.

[KM96a] C. Klüppelberg and T. Mikosch. The integrated periodogram for stable processes. *The Annals of Statistics*, 24:1855-1879, 1996.

[KM96b] C. Klüppelberg and T. Mikosch. Self-normalized and randomly centered spectral estimates. In P. M. Robinson and M. Rosenblatt, editors, *Athens Conference on Applied Probability and Time Series Analysis. Volume II: Time Series Analysis in Memory of E. J. Hannan*, pages 259–271, New York, 1996. Springer-Verlag. Lecture Notes in Statistics, **115**.

[KT96] P. S. Kokoszka and M. S. Taqqu. Parameter estimation for infinite variance fractional ARIMA. To appear in *The Annals of Statistics*, 1996.

[LTWW94] W. E. Leland, M. S. Taqqu, W. Willinger, and D. V. Wilson. On the self-similar nature of Ethernet traffic (Extended version). *IEEE/ACM Transactions on Networking*, 2:1–15, 1994.

[Man75] B. B. Mandelbrot. Limit theorems on the self-normalized range for weakly and strongly dependent processes. *Zeitschrift für Wahrscheinlichkeitstheorie und verwandte Gebiete*, 31:271–285, 1975.

[MT79] B. B. Mandelbrot and M. S. Taqqu. Robust R/S analysis of long-run serial correlation. In *Proceedings of the 42nd Session of the International Statistical Institute*, Manila, 1979. Bulletin of the International Statistical Institute. Vol.48, Book 2, pp. 69-104.

[MTL78] R. McGill, J. W. Tukey, and W. A. Larsen. Variation of box plots. *The American Statistician*, 32:12–16, 1978.

[MW69] B .B. Mandelbrot and J. R. Wallis. Computer experiments with fractional Gaussian noises, Parts 1,2,3˙ *Water Resources Research*, 5:228–267, 1969.

[PBS+94] C. K. Peng, S. V. Buldyrev, M. Simons, H. E. Stanley, and A. L. Goldberger. Mosaic organization of DNA nucleotides. *Physical Review E*, 49:1685–1689, 1994.

[Rob95] P. M. Robinson. Gaussian semiparametric estimation of long range dependence. *The Annals of Statistics*, 23:1630–1661, 1995.

[ST94] G. Samorodnitsky and M. S. Taqqu. *Stable Non-Gaussian Processes: Stochastic Models with Infinite Variance*. Chapman and Hall, New York, London, 1994.

[TT97a] M. S. Taqqu and V. Teverovsky. Robustness of Whittle-type estimates for time series with long-range dependence. *Stochastic Models*, 13:723-757, 1997.

[TT97b] V. Teverovsky and M. S. Taqqu. Testing for long-range dependence in the presence of shifting means or a slowly declining trend using a variance-type estimator. *Journal of Time Series Analysis*, 18:279-304, 1997.

[TT96] M. S. Taqqu and V. Teverovsky. Semi-parametric graphical estimation techniques for long-memory data. In P. M. Robinson and M. Rosenblatt, editors, *Athens Conference on Applied Probability and Time Series Analysis. Volume II: Time Series Analysis in Memory of E. J. Hannan*, pages 420–432, New York, 1996. Springer-Verlag. Lecture Notes in Statistics, **115**.

[TTW95] M. S. Taqqu, V. Teverovsky, and W. Willinger. Estimators for long-range dependence: an empirical study. *Fractals*, 3(4):785–798, 1995. Reprinted in *Fractal Geometry and Analysis*, C.J.G. Evertsz, H-O Peitgen and R.F. Voss, editors. World Scientific Publishing Co., Singapore, 1996.

Mathematics Department, Boston University, Boston, MA 02215, USA
Email: murad@math.bu.edu, vteverovsky@alumn.mit.edu

Why Non-Linearities can Ruin the Heavy-Tailed Modeler's Day

Sidney I. Resnick

ABSTRACT. A heavy-tailed time series that can be expressed as an infinite order moving average has the property that the sample autocorrelation function (acf) at lag h, converges in probability to a constant $\rho(h)$ despite the fact that the mathematical correlation typically does not exist. A simple bilinear model considered by Davis and Resnick (1996) has the property that the sample autocorrelation function at lag h converges in distribution to a non-degenerate random variable. Examination of various data sets exhibiting heavy-tailed behavior reveals that the sample correlation function typically does not behave like a constant. Usually, the sample acf of the first half of the data set looks considerably different than the sample acf of the second half. A possible explanation for this acf behavior is the presence of nonlinear components in the underlying model. This seems to imply that infinite order moving average models and, in particular, ARMA models do not adequately capture dependency structure in the presence of heavy tails. Some additional results about the simple nonlinear model are discussed, and in particular, we consider how to estimate coefficients.

1. Introduction

There are now numerous data sets from the fields of telecommunications, finance and economics which appear to be compatible with the assumption of heavy-tailed marginal distributions. Examples include file lengths, cpu time to complete a job, call holding times, inter-arrival times between packets in a network and lengths of on/off cycles (Duffy, et al., 1993, 1994; Meier–Hellstern et al., 1991; Willinger, Taqqu, Sherman and Wilson, 1995; Crovella and Bestavros, 1995; Cunha, Bestavros and Crovella, 1995).

A key question of course is how to fit models to data which require heavy-tailed marginal distributions. In the traditional setting of a stationary time series with finite variance, every purely non-deterministic process can be expressed as a linear process driven by an uncorrelated input sequence. For such time series, the autocorrelation function can be well approximated by that of an finite order ARMA(p, q) model. In particular, one can choose an autoregressive model of order p (AR(p)) such that the acf of the two models agree for lags $1, \ldots, p$ (see Brockwell and Davis (1991), p. 240).

Key words and phrases. heavy tails, regular variation, Hill estimator, Poisson processes, linear programming, autoregressive processes, parameter estimation, weak convergence, consistency, time series analysis, estimation, independence.

Sidney Resnick was partially supported by NSF Grant DMS-9400535 at Cornell University and also received some support from NSA Grant MDA904-95-H-1036.

So when finite variance models are considered from a second order point of view, linear models are sufficient for data analysis.

In the infinite variance case, we have no such confidence that linear models are sufficiently flexible and rich enough for modeling purposes. Yet theoretical attempts to date to study heavy tailed time series models have concentrated effort on ARMA models or infinite order moving averages despite little evidence that such models would actually fit heavy-tailed data. Understandably, these attempts were motivated by the desire to see how well classical ARMA models perform in the heavy-tailed world. However, the point which this paper emphasizes is that the class of infinite order moving averages is unlikely to provide a sufficiently broad class which is capable of accurately capturing the dependency structure of a variety of heavy-tailed data.

Some theoretical perspective on this issue is provided in the interesting work of Rosiński (1995) who decomposes a general symmetric α-stable process $\{X(t), t \geq 0\}$ into an independent sum of 3 processes

$$X(t) = X_1(t) + X_2(t) + X_3(t),$$

where X_1 is a superposition of moving average processes, X_2 is a Harmonizable α-stable process and X_3 is a process of 'third' type. There is no reason to suppose the moving average processes are in any way dense within the class of symmetric α-stable processes and thus no reason to suspect that moving averages should be given a prominent role in data analysis.

The challenge, of course, is to find flexible parametric families of heavy-tailed models. One possible family is the class of bilinear models which has received attention in the finite variance world (Gabr and Subba Rao, 1984). Davis and Resnick (1996) study a simple bilinear model and show that the behavior of the sample correlation function is strikingly different for such models than for linear heavy-tailed time series. This has the important implication that any heavy-tailed inference such as Yule-Walker estimation (Brockwell and Davis, 1991; Resnick, 1997) based on the sample acf, will be dramatically misleading if the analyst fails to adequately account for nonlinearities. This is discussed further in Section 2. Section 3 gives examples of several data sets whose sample correlation function behaves like that of the bilinear process. It is doubtful if such data can be fit by linear heavy-tailed models. Further consideration is given in Section 4 to linear programming estimators (Feigin and Resnick, 1992, 1994, 1997; Feigin, Resnick, Stărică, 1995; Feigin, Kratz, Resnick, 1996; Davis and McCormick, 1989) applied to a simple bilinear process. Generalizations of this simple process will be necessary to achieve the flexibility needed of a desirable parametric family.

2. Sample correlations of linear and bilinear processes

The sample correlation function is a basic tool in classical time series for not only assessing dependence but also for estimation purposes since, for example, the Yule-Walker estimators of autoregressive coefficients in an autoregressive model depend on sample correlations. (See Brockwell and Davis, 1991; Resnick, 1997.) For a stationary sequence $\{X_n, n = 0, \pm 1, \pm 2, \dots\}$ the classical definition of the sample correlation function at lag h ($h = 0, 1, \dots$) is

$$\hat{\rho}(h) = \frac{\sum_{t=1}^{n-h}(X_t - \bar{X})(X_{t+h} - \bar{X})}{\sum_{t=1}^{n}(X_t - \bar{X})^2}.$$

When heavy tails are present, and especially when the data is positive as is frequently the case, it makes little sense to center at $\bar{X}$ and the following heavy-tailed version is used:

$$\hat{\rho}_H(h) = \frac{\sum_{t=1}^{n-h} X_t X_{t+h}}{\sum_{t=1}^{n} X_t^2}.$$

Consider an infinite order moving average

$$X_t = \sum_{j=0}^{\infty} \psi_j Z_{t-j}, \tag{2.1}$$

where $\{Z_t\}$ is an iid sequence of heavy-tailed random variables satisfying

$$\begin{aligned} P[|Z_1| > x] &= x^{-\alpha} L(x), \quad (x \to \infty), \\ \frac{P[Z_1 > x]}{P[|Z_1| > x]} &\to p, \quad (x \to \infty), \end{aligned} \tag{2.2}$$

with $\alpha > 0$, L slowly varying, $0 \leq p \leq 1$, and with the ψ's satisfying mild summability conditions. Note that if $Z_1 \geq 0$, then $p = 1$. If $\alpha < 2$, there is no finite variance and hence the mathematical correlations of the Z's and presumably the X's do not exist. However, Davis and Resnick (1985a,b; 1986) proved that $(\hat{\rho}_H(h), h = 1, \dots, q)$ still has nice asymptotic properties which can be used for assessing dependence and for Yule-Walker estimation. Define

$$\rho(h) = \sum_{j=0}^{\infty} \psi_j \psi_{j+h} / \sum_{j=0}^{\infty} \psi_j^2.$$

Then for the linear model (2.1) we have the consistency result

$$\hat{\rho}_H(h) \xrightarrow{P} \rho(h),$$

and a reasonably fast rate of convergence also ensues. This leads to consistency of Yule-Walker estimates of autoregressive coefficients in an AR(p) model and allows for computation of a limit distribution for these estimates.

For contrasting behavior we consider a simple bilinear process satisfying the recursion

$$X_t = cX_{t-1}Z_{t-1} + Z_t, \tag{2.3}$$

where the Z's satisfy (2.2) and

$$|c|^{\alpha/2} E|Z_1|^{\alpha/2} < 1. \tag{2.4}$$

Under this condition (see Liu (1989)), there exists a unique stationary solution to the equations (2.3) given by

$$X_t = \sum_{j=0}^{\infty} c^j Y_t^{(j)},$$

where

$$Y_t^{(j)} = \begin{cases} Z_t, & \text{if } j = 0, \\ \left(\prod_{i=1}^{j-1} Z_{t-i}\right) Z_{t-j}^2, & \text{if } j \geq 1. \end{cases} \tag{2.5}$$

Define b_n to be the $1 - n^{-1}$ quantile of $|Z_1|$, i.e.

$$b_n = \inf\{x : P[|Z_1| > x] < n^{-1}\}. \tag{2.6}$$

In order to describe the basic result about acf's of bilinear processes and also for later work on estimation, we review rapidly some notation and concepts about point processes. For a locally compact Hausdorff topological space $\mathbb{E}$, we let $M_p(\mathbb{E})$ be the space of Radon point measures on $\mathbb{E}$. This means $m \in M_p(\mathbb{E})$ is of the form

$$m = \sum_{i=1}^{\infty} \epsilon_{x_i},$$

where $x_i \in \mathbb{E}$ are the locations of the point masses of m and ϵ_{x_i} denotes the point measure defined by

$$\epsilon_{x_i}(A) = \begin{cases} 1, & \text{if } x \in A, \\ 0, & \text{if } x \notin A. \end{cases}$$

We emphasize that we assume that all measures in $M_p(\mathbb{E})$ are Radon which means that for any $m \in M_p(\mathbb{E})$ and any compact $K \subset \mathbb{E}$, $m(K) < \infty$. On the space $M_p(\mathbb{E})$ we use the vague metric $\rho(\cdot,\cdot)$. Its properties are discussed for example in Resnick (1987, Section 3.4) and Kallenberg (1983). Note that a sequence of measures $m_n \in M_p(\mathbb{E})$ converge vaguely to $m_0 \in M_p(\mathbb{E})$ if for any continuous function $f : \mathbb{E} \mapsto [0,\infty)$ with compact support we have $m_n(f) \to m_0(f)$ where $m_n(f) = \int_{\mathbb{E}} f\, dm_n$. The non-negative continuous functions with compact support will be denoted by $C_K^+(\mathbb{E})$.

A Poisson process on $\mathbb{E}$ with mean measure μ will be denoted by PRM(μ). The primary example of interest in our applications is the case when $\mathbb{E}_m = [-\infty,\infty]^m \setminus \{\mathbf{0}\}$, where compact sets are closed subsets of $[-\infty,\infty]^m$ which are bounded away from $\mathbf{0}$.

Here is the result from Davis and Resnick (1996) describing the behavior of the simple bilinear process.

Theorem 2.1. *Suppose $\{X_t\}$ is the bilinear process (2.3) where the marginal distribution F of the iid noise $\{Z_t\}$ satisfies (2.2), the constant c satisfies (2.4) and b_n is given by (2.6). Suppose further that $\sum_{s=1}^\infty \epsilon_{j_s}$ is PRM(μ) with μ given by*

$$\mu(dx) = p\alpha x^{-\alpha-1}dx 1_{(0,\infty)}(x) + q|x|^{-\alpha-1}dx 1_{(-\infty,0)}(x),$$

and $\{U_{s,k},\ s \geq 1, k \geq 1\}$ are iid with distribution F.

(i) In $M_p(\mathbb{E}_1)$,

$$\sum_{t=1}^n \epsilon_{b_n^{-2}X_t} \Rightarrow \sum_{k=1}^\infty \sum_{s=1}^\infty \epsilon_{j_s^2 c^k W_{s,k}},$$

where

$$W_{s,k} = \begin{cases} \prod_{i=1}^{k-1} U_{s,i}, & \text{if } k > 1, \\ 1, & \text{if } k = 1, \\ 0, & \text{if } k < 1. \end{cases}$$

(ii) In $M_p(\mathbb{E}_{h+1})$,

$$\sum_{t=1}^n \epsilon_{b_n^{-2}(X_t, X_{t-1},\dots,X_{t-h})} \Rightarrow \sum_{k=1}^\infty \sum_{s=1}^\infty \epsilon_{j_s^2(c^k W_{s,k}, c^{k-1}W_{s,k-1},\dots,c^{k-h}W_{s,k-h})}.$$

Furthermore, if $0 < \alpha < 4$, we have for any $h = 1, 2, \dots$ that

$$(\hat{\rho}_H(l), l = 1,\dots,h) \Rightarrow (L_i, i = 1,\dots,h)$$

in $\mathbb{R}^h$, where

$$L_i = \frac{\sum_{s=1}^\infty \sum_{k=1}^\infty j_s^4 c^{2k-i} W_{s,k} W_{s,k-i}}{\sum_{s=1}^\infty \sum_{k=1}^\infty j_s^4 c^{2k} W_{s,k}^2}, \qquad i = 1,\dots,h.$$

Contrast the random limits for $(\hat{\rho}_H(l), l = 1, \ldots, h)$ when $\{X_t\}$ is the bilinear process (2.3) with the nonrandom limits obtained when $\{X_t\}$ is the linear process (2.1). This difference can lead to dramatic errors if one models heavy-tailed nonlinear data with a linear model. This contrast is demonstrated clearly with simulated data. In Section 3 we present simple analyses of several real heavy-tailed data sets to illustrate the likelihood that linear models are unsuitable.

We simulated three independent samples ($test_i$, $i = 1, 2, 3$) of size 5000 from the bilinear process

$$X_t = .1Z_{t-1}X_{t-1} + Z_t, \quad t = 0, \pm 1, \pm 2, \ldots, \tag{2.7}$$

where $\{Z_t\}$ are iid Pareto random variables,

$$P[Z_1 > x] = 1/x, \quad x > 1.$$

For contrast, we also simulated three independent samples of size 1500 of AR(2) data. The AR(2) is

$$X_t = 1.3X_{t-1} - 0.7X_{t-2} + Z_t, t = 0, \pm 1, \pm 2, \ldots$$

and the innovations have a Pareto distribution as for the bilinear example. The AR data sets were called $testar_i$, $i = 1, 2, 3$.

The erratic nature of the behavior of $\hat{\rho}_H$ for the bilinear model is illustrated in Figure 2.1 which graphs the heavy tail acf for $test_i$, $i = 1, 2, 3$. The graphs look rather different reflecting the fact that we are basically sampling independently three times from the non-degenerate limit distribution of the heavy-tailed acf. If one were not aware of the non-linearity in the data, one would be tempted to model with a low order moving average, based for example on the left hand plot. Furthermore, partial autocorrelation plots and plots of the AIC statistic as a function of the order of the model all show similar erratic behavior as one moves from independent sample to independent sample. So failure to account for non-linearity means there is great potential to be misled in the sorts of models one tries to fit.

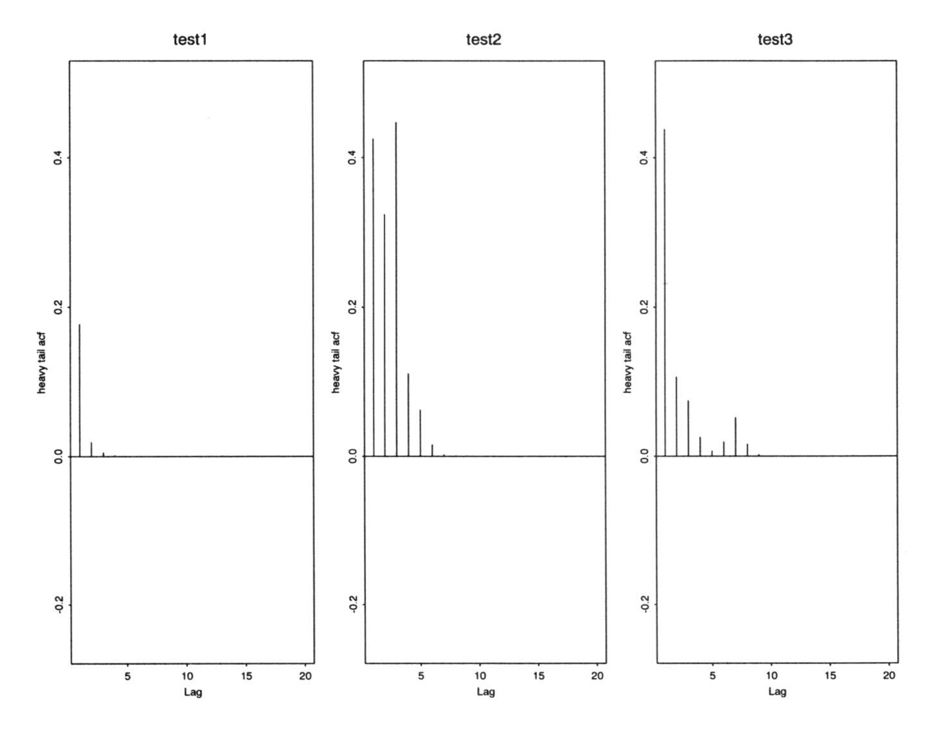

Figure 2.1. Heavy-tailed ACF for 3 bilinear samples

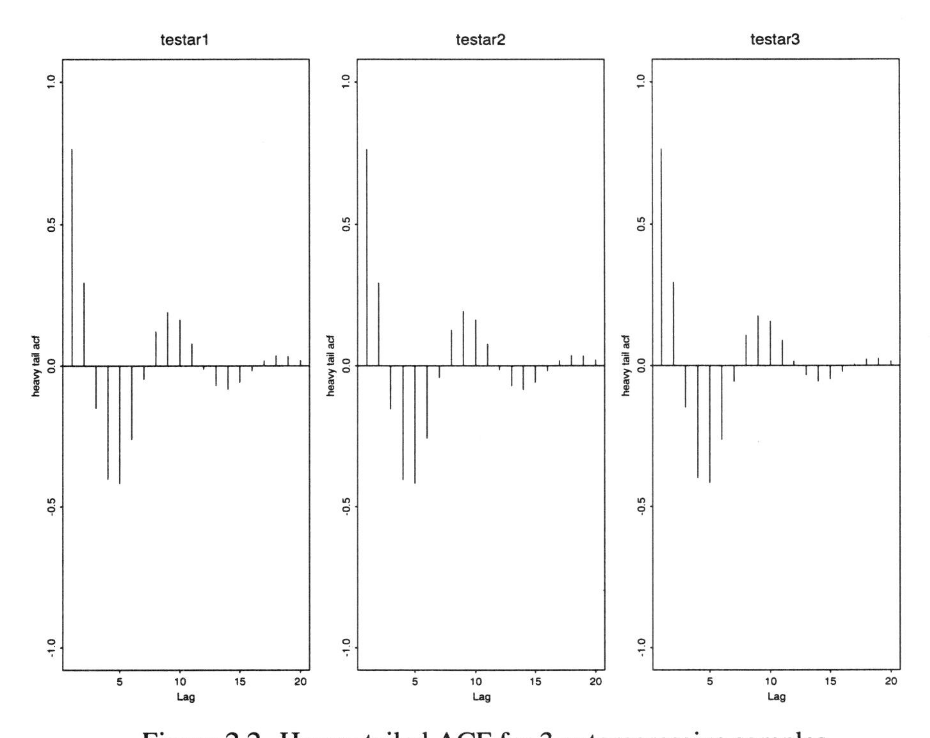

Figure 2.2. Heavy-tailed ACF for 3 autoregressive samples

As a contrast, Figure 2.2 above presents the comparable heavy-tailed acf plots for the three independent AR samples. Here, the pictures look identical reflecting the fact that we are sampling from an essentially degenerate distribution.

3. Analysis of heavy-tailed data sets

We now present several examples of real data and make the argument that it is unlikely that the data can be modelled as a linear model of the form (2.1). For each data set we note why we believe a heavy-tailed model is appropriate and why any sort of infinite order moving average is likely to be an inadequate model.

Given a particular data set, there are various methods of checking that a heavy-tailed model is appropriate. Such methods are reviewed in Resnick (1997). Suppose $\{X_n, n \geq 1\}$ is a stationary sequence and that

$$P[X_1 > x] = x^{-\alpha} L(x), \quad x \to \infty \tag{3.1}$$

where L is slowly varying and $\alpha > 0$. Consider the following techniques:

(1) The Hill plot. Let

$$X_{(1)} > X_{(2)} > \cdots > X_{(n)}$$

be the order statistics of the sample $X_1, \ldots, X_n$. We pick $k < n$ and define the Hill estimator (Hill, 1975) to be

$$H_{k,n} = \frac{1}{k} \sum_{i=1}^{k} \log \frac{X_{(i)}}{X_{(k+1)}}.$$

Note that k is the number of upper order statistics used in the estimation. The Hill plot is the plot of

$$((k, H_{k,n}^{-1}), 1 \leq k < n)$$

and, if the process is linear or satisfies mixing conditions, then since $H_{k,n} \xrightarrow{P} \alpha^{-1}$ as $n \to \infty$, $k/n \to 0$ the Hill plot should have a stable regime sitting at height roughly α. See Mason (1982), Hsing (1991), Resnick and Stărică (1995), Rootzen et al. (1990), Rootzen (1996). In the iid case, under a second order regular variation condition, $H_{k,n}$ is asymptotically normal with asymptotic variance $1/\alpha^2$. (Cf. de Haan and Resnick, 1996).

(2) The smooHill plot. The Hill plot often exhibits extreme volatility which makes finding a stable regime in the plot more guesswork than science. To counteract this,

Resnick and Stărică (1997) developed a smoothing technique yielding the smooHill plot. Pick an integer u (usually 2 or 3) and define

$$smooH_{k,n} = \frac{1}{(u-1)k} \sum_{j=k+1}^{uk} H_{j,n}.$$

In the iid case, when a second order regular variation condition holds, the asymptotic variance of $smooH_{k,n}$ is less than that of the Hill estimator, namely,

$$\frac{1}{\alpha^2}\frac{2}{u}(1-\frac{\log u}{u}).$$

(3) Alt plotting; Changing the scale. As an alternative to the Hill plot, it is sometimes useful to display the information provided by the Hill or smooHill estimation as

$$\{(\theta, H^{-1}_{\lceil n^\theta \rceil, n}), 0 \leq \theta \leq 1, \},$$

and similarly for the smooHill plot, where we write $\lceil y \rceil$ for the smallest integer greater or equal to $y \geq 0$. We call such plots the *alternative Hill plot* abbreviated AltHill and the *alternative smoothed Hill plot* abbreviated AltsmooHill. The alternative display is sometimes revealing since the initial order statistics get shown more clearly and cover a bigger portion of the displayed space.

(4) Dynamic and static qq plots. As we did for the Hill plots, pick k upper order statistics

$$X_{(1)} > X_{(2)} > \dots X_{(k)}$$

and neglect the rest. Plot

$$\{(-\log(1-\frac{j}{k+1}), \log X_{(j)}), 1 \leq j \leq k\}. \tag{3.2}$$

If the data is approximately Pareto or even if the marginal tail is only regularly varying, this should be approximately a straight line with slope$=1/\alpha$. The slope of the least squares line through the points is an estimator called the qq-estimator (Kratz and Resnick, 1996). Computing the slope we find that the qq–estimator is given by

$$\widehat{\alpha^{-1}}_{k,n} = \frac{\frac{1}{k}\sum_{i=1}^{k}(-\log(\frac{i}{k+1}))\log(\frac{X_{(i)}}{X_{(k+1)}}) - \frac{1}{k}\sum_{i=1}^{k}(-\log(\frac{i}{k+1}))H_{k,n}}{\frac{1}{k}\sum_{i=1}^{k}(-\log(\frac{i}{k+1}))^2 - (\frac{1}{k}\sum_{i=1}^{k}(-\log(\frac{i}{k+1}))^2} \tag{3.4}$$

There are two different plots one can make based on the qq–estimator. There is the dynamic qq–plot obtained from plotting $\{(k, 1/\widehat{\alpha^{-1}}_{k,n}, 1 \leq k \leq n\}$, which is similar to the Hill plot. Another plot, the static qq–plot, is obtained by choosing and fixing k, plotting the points in (3.2) and putting the least squares line through the points while computing the slope as the estimate of α^{-1}.

The qq-estimator is consistent for the iid model if $k \to \infty$ and $k/n \to 0$. Under a second order regular variation condition and further restriction on $k(n)$, it is asymptotically normal with asymptotic variance $2/\alpha^2$. This is larger than the asymptotic variance of the Hill estimator The volatility of the qq–plot always seems to be less than that of the Hill estimator.

We now consider several data sets and illustrate some features and describe problems encountered when trying to fit MA(∞) models. Where appropriate we give sample acf plots. (These are drawn using the Splus routine which draws a 95% confidence window based on Bartlett's formula assuming asymptotic normality. So these bounds should be viewed with a big grain of salt.)

(i) ISDN2. This dataset consists of 4868 interarrival times of ISDN D-channel packets. The time series plot and qq–plot giving evidence of heavy tails are shown in Fig. 3.1.

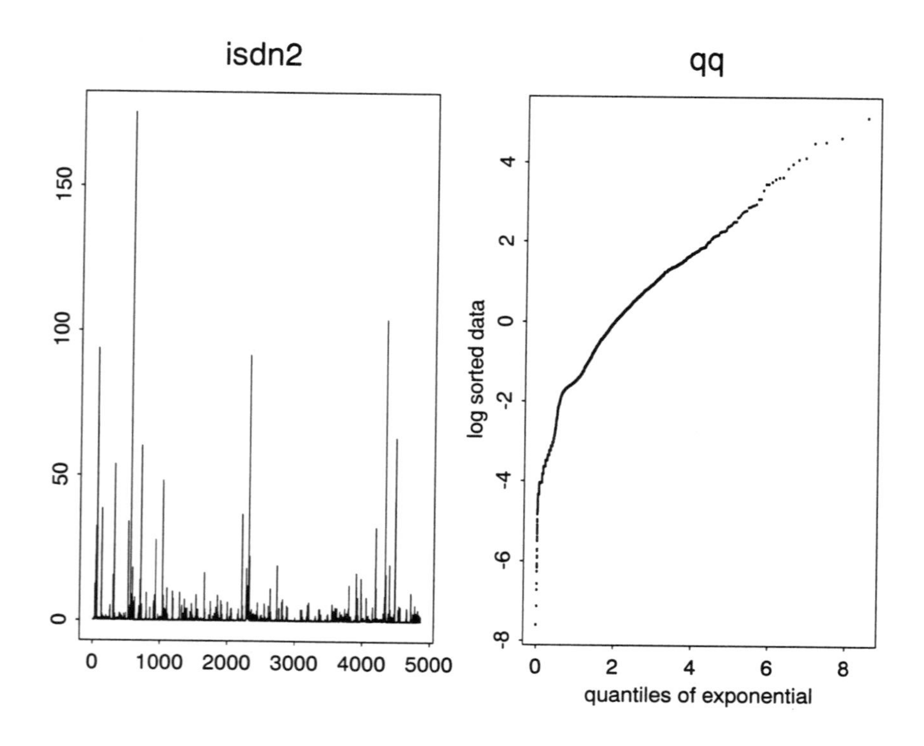

Figure 3.1. Tsplot of ISDN2 and qq-plot

Hill plots given in Figure 3.2 indicate an α in the neighborhood of 1.2. The static qq-plot based on 1000 upper order statistics given in Figure 3.3 yields an α value of 1.136.

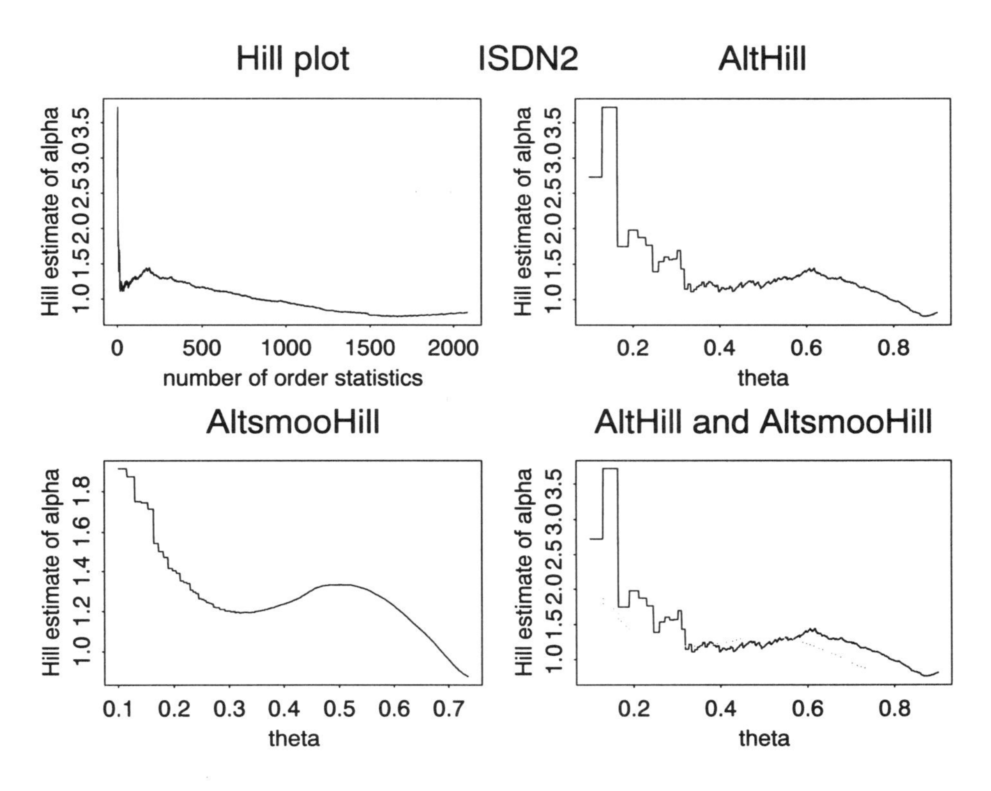

Figure 3.2. Hill plots of ISDN2

The last two plots in Figure 3.3 give the acf of the first 1500 observations and the acf of the last 1500 observations in the data set. Note the two graphs are quite different, which seems to rule out any sort of ARMA model as a potential candidate to be fit to the data.

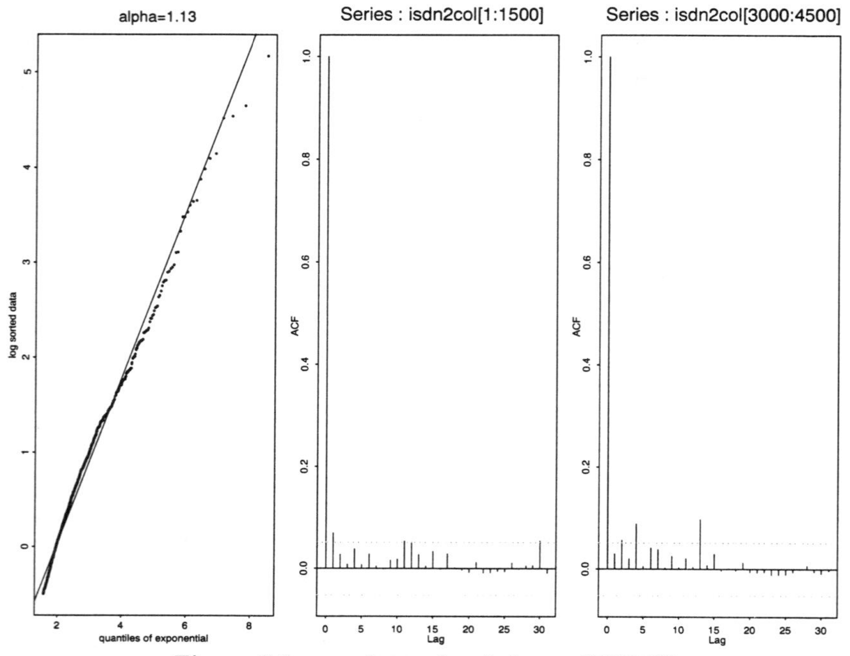

Figure 3.3. qq-plot and acf plots of ISDN2

(ii) Interar. This data set represents 176834 interarrivals of externally generated TCP packets to a server. The recording period was one hour. Figure 3.4 gives the time series plot and the qq-plot. Both show clear evidence of heavy tails.

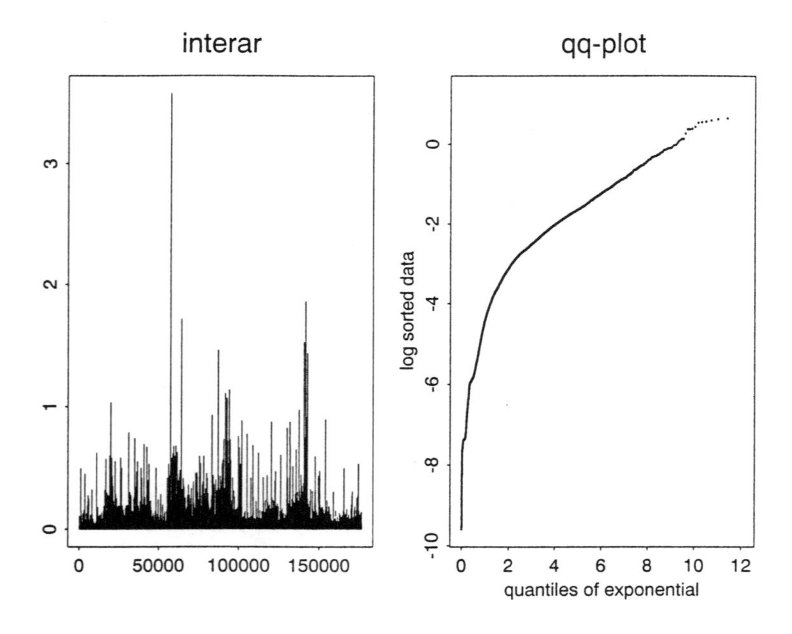

Figure 3.4. Tsplot and qq-plot for interar

The Hill plots given in Figure 3.5 show a value of α in the neighborhood of 2.5 and the static qq-plot in Figure 3.6 gives a value of $\alpha = 2.28$.

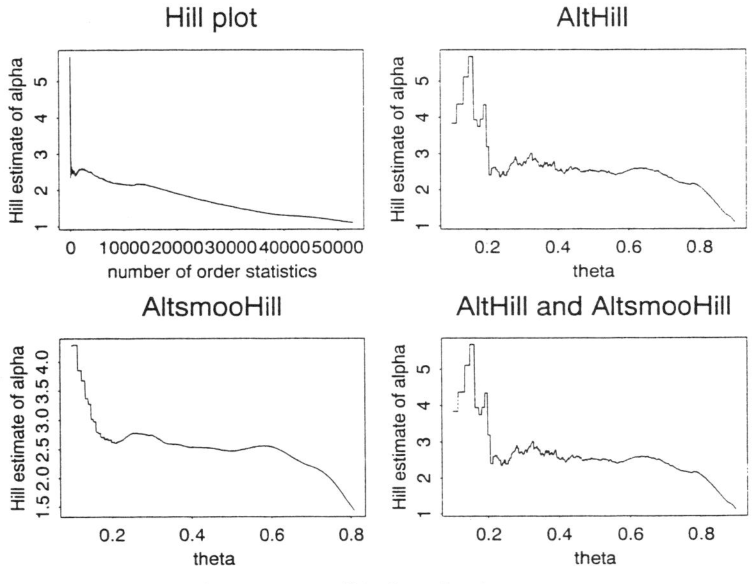

Figure 3.5. Hill plots for interar

The acf of the first 10,000 values does not remotely resemble the plot for 10,000 values taken in the middle of the time series. These are the last two plots in Figure 3.6.

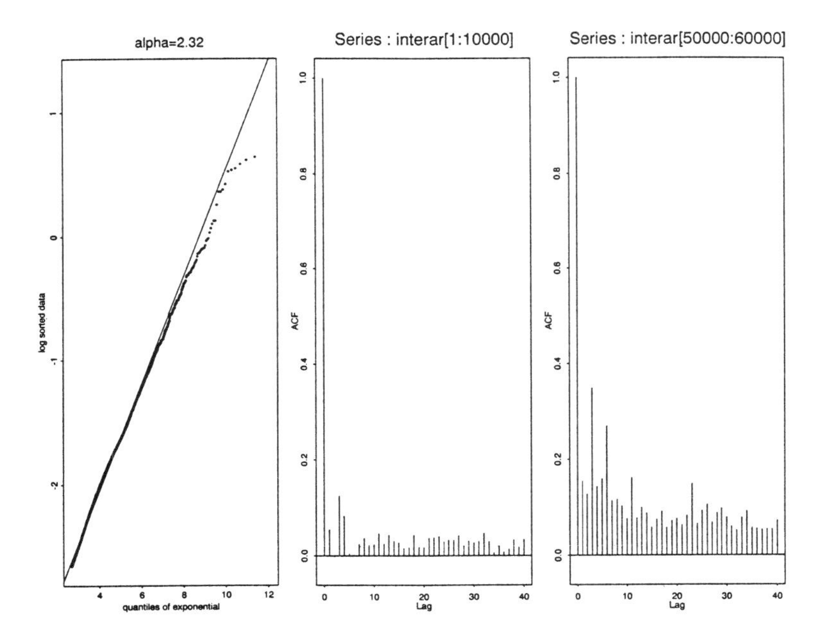

Figure 3.6. Static qq-plot and acf plots for interar

(iii) SILENCE. Consider a time series of length 1027 shown in Figure 3.7 which represents the off periods between transmission of packets generated by a terminal during a logged-on session. The left graph in Figure 3.7 is the time series plot and the right graph is the static qq-plot using 500 upper order statistics giving ample evidence of heavy tails. The estimate of α given by this plot is 0.6696.

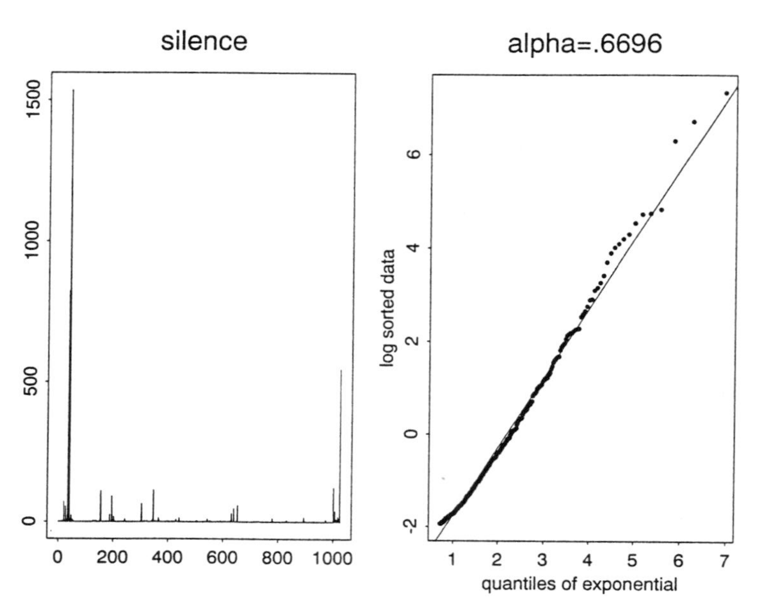

Figure 3.7. Tsplot and static qq-plot of SILENCE

The Hill plots confirm the estimate of α given by the static qq-plot and give $\alpha \approx .64$.

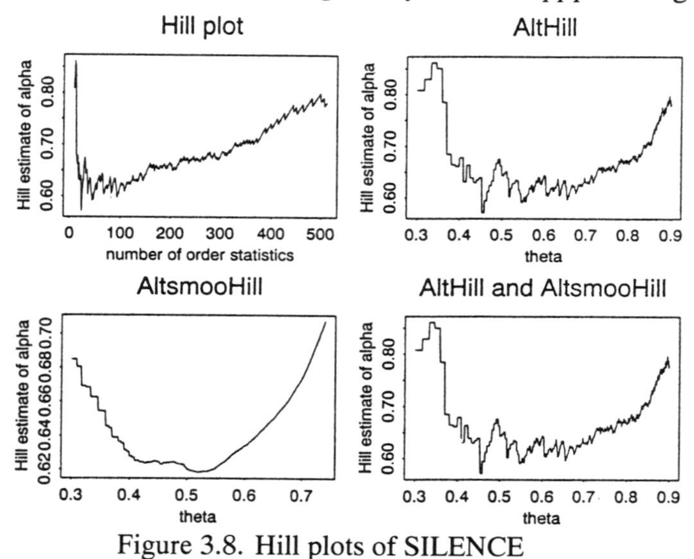

Figure 3.8. Hill plots of SILENCE

For the last graph, we split the data set into thirds and graphed the acf for each piece separately. The pictures are obviously quite different.

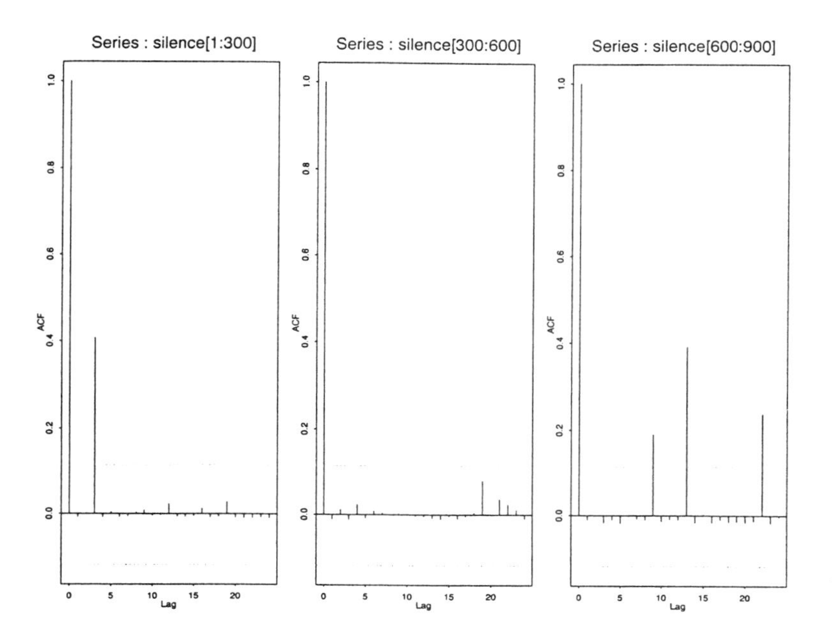

Figure 3.9. Acf plots of SILENCE

4. Estimation for the simple bilinear process. We now consider the simple bilinear process given in (2.3)

$$X_t = cX_{t-1}Z_{t-1} + Z_t \tag{4.1}$$

where in this section we assume $Z_t \geq 0$ and the distribution of Z_t has left endpoint

$$l = \inf\{x > 0 : P[Z_t \leq x] > 0\}. \tag{4.2}$$

Further, we assume $c > 0$ and (2.4) assumes the form

$$c^{\alpha/2} E Z_1^{\alpha/2} < 1. \tag{4.3}$$

In particular, this implies $c^{\alpha/2} l^{\alpha/2} < 1$ or $cl < 1$.

The next proposition gives consistent estimators for (c, l).

Proposition 4.1. *Suppose $\{X_t\}$ is the bilinear process given in (4.1) and that (c, l) satisfy (4.2) and (4.3). Suppose in addition that $l > 0$ and $0 < \alpha < 4$. Define*

$$\hat{m} = \bigwedge_{t=1}^{t} X_t \text{ and } \hat{r} = \bigwedge_{t=2}^{n} \frac{X_t}{X_{t-1}}.$$

Then consistent estimators of (c, l) *are given by*

$$\hat{l} = \hat{m}(1 - \hat{r}), \quad \hat{c} = \frac{\hat{r}}{\hat{m}(1 - \hat{r})}.$$

Note that $\hat{r}$ is the linear programming estimator studied by Davis and McCormick (1989) and Feigin and Resnick (1992, 1994) for linear autoregressions.

Proof. Since $Z_t \geq l$ with probability 1, we have from (2.5) that a.s. for each t

$$\begin{aligned} X_t \geq & l + \sum_{j=1}^{\infty} c^j l^{j-1} l^2 \\ = & l + \frac{cl^2}{1 - cl} \\ = & \frac{l}{1 - cl} := l^{\#}. \end{aligned}$$

Now we apply the Davis and Resnick (1996) result quoted as Theorem 2.1 in this paper. We have in $M_p[l^{\#}, \infty]^2$ (and therefore by Theorem 3.3 of Feigin, Kratz and Resnick (1996) in $M_p[l^{\#}, \infty)^2$) that

$$\sum_{t=2}^{n} \epsilon_{b_n^{-2}(X_t, X_{t-1})} \Rightarrow \sum_{k=1}^{\infty} \sum_{s=1}^{\infty} \epsilon_{j_s^2(c^k W_{s,k}, c^{k-1} W_{s,k-1})}. \tag{4.4}$$

We seek to apply a map

$$(x, y) \mapsto x/y$$

but in order to do this we must compactify the state space of the converging point processes. From (4.4) we get for any large $K > 0$

(4.5)

$$\sum_{t=2}^{n} 1_{[l^{\#} \leq b_n^{-2} X_{t-i} \leq K,\ i=0,1]} \epsilon_{b_n^{-2}(X_t, X_{t-1})} \Rightarrow \sum_{k=1}^{\infty} \sum_{s=1}^{\infty} 1_{[l^{\#} \leq j_s^2 (c^{k-i} W_{s,k-i} \leq K,\ i=0,1]} \epsilon_{j_s^2(c^k W_{s,k}, c^{k-1} W_{s,k-1})}.$$

and now applying the division map we get in $M_p[l^{\#}, \infty)$

$$\sum_{t=2}^{n} 1_{[l^{\#} \leq b_n^{-2} X_{t-i} \leq K,\ i=0,1]} \epsilon_{X_t / X_{t-1}} \Rightarrow \sum_{k=1}^{\infty} \sum_{s=1}^{\infty} 1_{l^{\#} \leq j_s^2 (c^{k-i} W_{s,k-i} \leq K,\ i=0,1]} \epsilon_{c U_{s,k-1}}. \tag{4.6}$$

We now argue we can remove the truncation level K. In order to do this we must show by Billingsley (1968), Theorem 4.2, that for any $\eta > 0$ and any continuous function f with compact support in $[0, \infty)$

$$\lim_{K \to \infty} \limsup_{n \to \infty} P[|\sum_{t=2}^{n} 1_{[l^{\#} \leq b_n^{-2} X_{t-i} \leq K,\ i=0,1]} f(\frac{X_t}{X_{t-1}}) - \sum_{t=2}^{n} f(\frac{X_t}{X_{t-1}})| > \eta] = 0. \tag{4.7}$$

Suppose the compact support of f is contained in $[0, \theta]$. The probability in (4.7) is bounded by

$$
\begin{aligned}
&nP[\frac{X_t}{b_n^2} > K, \frac{X_t}{X_{t-1}} \le \theta] + nP[\frac{X_{t-1}}{b_n^2} > K, \frac{X_t}{X_{t-1}} \le \theta] \\
&\le 2nP[\frac{X_t}{b_n^2} > K] \\
&\to (const)K^{-\alpha},
\end{aligned}
$$

as $n \to \infty$ by Corollary 2.4 of Davis and Resnick (1996) and as $K \to \infty$ the above

$$\to 0.$$

This shows that (4.6) holds with the truncation level K replaced by ∞.

We thus conclude that in $M_p[0, \infty)$

$$\sum_{t=2}^{n} \epsilon_{X_t/X_{t-1}} \Rightarrow \sum_{k=1}^{\infty}\sum_{s=1}^{\infty} \epsilon_{cU_{s,k-1}}$$

and applying the a.s. continuous function that maps the point measure into the minimum of the points yields

$$\hat{r} = \bigwedge_{t=2}^{n} \frac{X_t}{X_{t-1}} \Rightarrow c \bigwedge_{s,k} U_{s,k} = cl.$$

Since it is also clear that

$$\hat{m} \Rightarrow l^{\#},$$

the desired consistency result follows. □

In a simulation experiment, we simulated a time series of length 5000 from the simple bilinear process given in (4.1) with $c = .3$ and $l = .1$. Our estimators yielded values of

$$(\hat{c}, \hat{l}) = (.317023, .1011468).$$

5. Concluding remarks.

The estimator $\hat{c}$ proposed in Proposition 4.1 for the simple bilinear process given in (4.1) is consistent but it is not at all clear that there is an asymptotic distribution or even what the rate of convergence might be. It seems likely that if an asymptotic distribution exists, it will depend on the unknown parameters c and α. Other estimators need to be explored.

However the obvious priority must be to find a flexible parametric family which is large enough to fit the abundance of heavy-tailed data that exists but is tractable enough to yield excellent model selection and estimation techniques. The point of emphasis of this paper is that any parametric family of stationary processes which can be expressed as infinite order moving averages is not likely to satisfy the requirements of adequately fitting existing heavy-tailed data. The general bilinear model is one possible family of processes that merits exploration.

The Hill estimator has been proven to be consistent for observations coming from a process which is iid (Mason, 1982) or MA(∞) (Resnick and Stărică, 1995) or which satisfy mixing conditions (Rootzen, et al, 1990; Rootzen, 1995). The Hill estimator appears to work just fine for the simple bilinear process in (4.1). Figure 5.1 displays the Hill plots for 5000 observations coming from (4.1) with $c = 0.1$ and $l = 1$. Since the tail of X_t satisfies

$$P[X_t > x] \sim (const)P[Z_1^2 > x]$$

the correct answer that the Hill plots seek is 0.5.

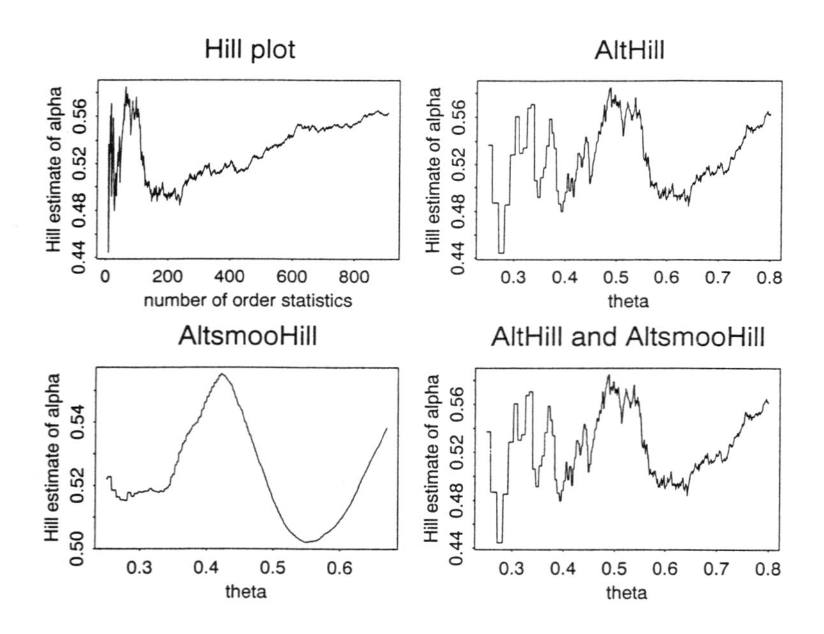

Figure 5.1. Hill plots for the simple bilinear process

We intend to give some thought to showing directly that the Hill estimator can be applied successfully to estimating the shape parameter when the underlying model is nonlinear.

References

1. Billingsley, P., *Convergence of Probability Measures*, Wiley, NY, 1968.
2. Brockwell, P. and Davis, R., *Time Series: Theory and Methods, 2nd edition*, Springer–Verlag, New York, 1991.
3. Crovella, M and Bestavros, A., *Explaining world wide web traffic self–similarity*, Preprint available as TR-95-015 from {crovella, best}@cs.bu.edu, 1995
4. Cunha, C., Bestavros, A. and Crovella, M., *Characteristics of www client–based traces*, Preprint available as BU-CS-95-010 from {crovella,best}@cs.bu.edu
5. Davis, R. and McCormick, W., *Estimation for first-order autoregressive processes with positive or bounded innovations*, Stochastic Processes and their Applic. **31** (1989), 237–250.
6. Davis, R. and Resnick, S., *Limit theory for moving averages of random variables with regularly varying tail probabilities*, Ann. Probability **13** (1985a), 179–195.
7. Davis, R. and Resnick, S., *More limit theory for the sample correlation function of moving averages*, Stochastic Processes and their Applications **20** (1985b), 257–279.
8. Davis, R. and Resnick, S., *Limit theory for the sample covariance and correlation functions of moving averages*, Ann. Statist. **14** (1986), 533–558.
9. Davis, R. and Resnick, S. *Limit theory for bilinear processes with heavy-tailed noise*, Annals of Applied Probability, **6** (1996), 1191–1210.
10. Duffy, D., McIntosh, A., Rosenstein, M., and Willinger, W. *Statistical analysis of CCSN/SS7 traffic data from working CCS subnetworks*, IEEE Journal on Selected Areas in Communications **12** (1994 544–551.
11. Duffy, D., McIntosh, A., Rosenstein, M., and Willinger, W. *Analyzing telecommunications traffic data from working common channel signaling subnetworks*, Proceedings of the 25th Interface, San Diego CA, 1993.
12. Feigin, P. and Resnick, S., *Estimation for autoregressive processes with positive innovations*, Stochastic Models **8** (1992), 479–498.
13. Feigin, P. and Resnick, S., *Limit distributions for linear programming time series estimators*, Stochastic Processes and their Applications **51** (1994), 135–166.
14. Feigin, P. and Resnick, S., *Linear programming estimators and bootstrapping for heavy-tailed phenomena*, Advances in Applied Probability **29** (1997), 759–805.

15. Feigin, P., Kratz, M. and Resnick, S., *Parameter estimation for moving averages with positive innovations*, Annals of Applied Probability **6** (1996), 1157–1190.

16. Feigin, P. Resnick, S. and Stărică, Cătălin, *Testing for independence in heavy-tailed and positive innovation time series*, Stochastic Models **11** (1995), 587–612.

17. Gabr, M. and Subba Rao, T., *An Introduction to Bispectral Analysis and Bilinear Time Series Models*, Lecture Notes in Statistics, v. 24, Springer-Verlag, New York, 1984.

18. Haan, L. de and Resnick, S., *On asymptotic normality of the Hill estimator*, TR1155. ps.Z available at line http://www.orie.cornell.edu/trlist/trlist.html; to appear, Stochastic Models (1996).

19. Hill, B., *A simple approach to inference about the tail of a distribution*, Ann. Statist. **3** (1975), 1163–1174.

20. Hsing, T., *On tail estimation using dependent data*, Ann. Statist. **19** (1991), 1547–1569.

21. Kallenberg, O., *Random Measures*, Third edition, Akademie-Verlag, Berlin, 1983.

22. Kratz, M. and Resnick, S., *The qq–estimator and heavy tails Stochastic Models* **12** (1996), 699–724.

23. Liu, J., *On the existence of a general multiple bilinear time series*, J. Time Series Analysis **10** (1989), 341–355.

24. Mason, D., *Laws of large numbers for sums of extreme values*, Ann. Probability **10** (1982), 754–764.

25. Meier-Hellstern, K., Wirth, P., Yan, Y., Hoeflin, D., Traffic models for ISDN data users: office automation application, *Teletraffic and Datatraffic in a Period of Change.* Proceedings of the 13th ITC, A. Jensen and V.B. Iversen (eds), North Holland, Amsterdam, (1991), 167–192.

26. Resnick, Sidney, *Extreme Values, Regular Variation, and Point Processes*, Springer-Verlag, New York, 1987.

27. Resnick, S., *Heavy tail modelling and teletraffic data*, Ann. Statistics **25** (1997), 1805–1869.

28. Resnick, S. and Stărică, C., *Consistency of Hill's estimator for dependent data*, J. Applied Probability **32** (1995), 139–167.

29. Resnick, S. and Stărică, Cătălin *Smoothing the Hill estimator*, Advances in Applied Probability **29** (1997), 271–293.

30. Rootzen, H., Leadbetter, M. and de Haan, L., *Tail and quantile estimation for strongly mixing stationary sequences*, Technical Report 292, Center for Stochastic Processes, Department of Statistics, University of North Carolina, Chapel Hill, NC 27599-3260, (1990).

31. Rootzen, H. *The tail empirical process for stationary sequences*, Preprint 1995:9 ISSN 1100-2255, Studies in Statistical Quality Control and Reliability, Chalmers University of Technology (1995).

32. Rosiński, Jan *On the structure of stationary stable processes*, Ann. Probab. **23** (1995), 1163–1187.

33. Willinger, W., Taqqu, M., Sherman, R. and Wilson, D., *Self–similarity through high–variability: Statistical analysis of ethernet LAN traffic at the source level* (1995), Preprint.

Sidney I. Resnick, Cornell University
School of Operation Research and Industrial Engineering
Rhodes Hall 223
Ithaca, NY 14853 USA

email: sid@orie.cornell.edu

Periodogram Estimates from Heavy-Tailed Data

T. Mikosch

Abstract

We give an overview of asymptotic results for the periodogram and for the integrated periodogram of a linear process. We contrast the theory in the case of finite/infinite variance processes. Then we conclude how periodogram–based statistical methods can be modified when large values in the observations are present.

1. Introduction

We consider throughout the time series model

$$X_t = \sum_{j=-\infty}^{\infty} \psi_j Z_{t-j}\,, \quad t \in \mathbb{Z}\,, \tag{1.1}$$

i.e. (X_t) is a *linear process* driven by iid noise (Z_t) with real coefficients (ψ_j). Linear processes, with finite variance noise, are attractive models (see for instance [4, 31]) because we can easily fit many real data, calculate asymptotic confidence bands for parameter estimates, control the goodness of fit, make predictions, etc. The theory crucially depends on the assumption of a *finite variance of Z* (equivalently of X) in order to be able to apply Hilbert space (L^2) techniques. The linear structure of L^2 and that of (1.1) fit nicely.

Recently, there has been an increasing interest in the analysis of time series exhibiting large fluctuations (see for instance [4], Section 13.3, [12, 23]). One way to model such behaviour is by *infinite variance linear processes*. The question we want to answer is the following:

What happens when classical time series methods are applied to data with large fluctuations caused by "outliers" in the noise? Can we modify the standard methods (*spectral estimates, parameter estimation, goodness of fit tests, etc.*) *in order to achieve results analogous to the classical theory?*

The word "outlier" is used here only in a heuristic sense; we choose the underlying distribution in such a way that large values of Z_t are inherent to the model. For example, Z_t could have a stable or Pareto distribution with infinite variance.

It is in general very hard to get statistical evidence about the finiteness of *any* moment of X (see for instance [19, 33] for some results on tail estimation for dependent data). Numerous real data sets from finance, insurance, meteorology, hydrology, the earth sciences, etc. do exhibit large fluctuations, and so the question arises what sort of

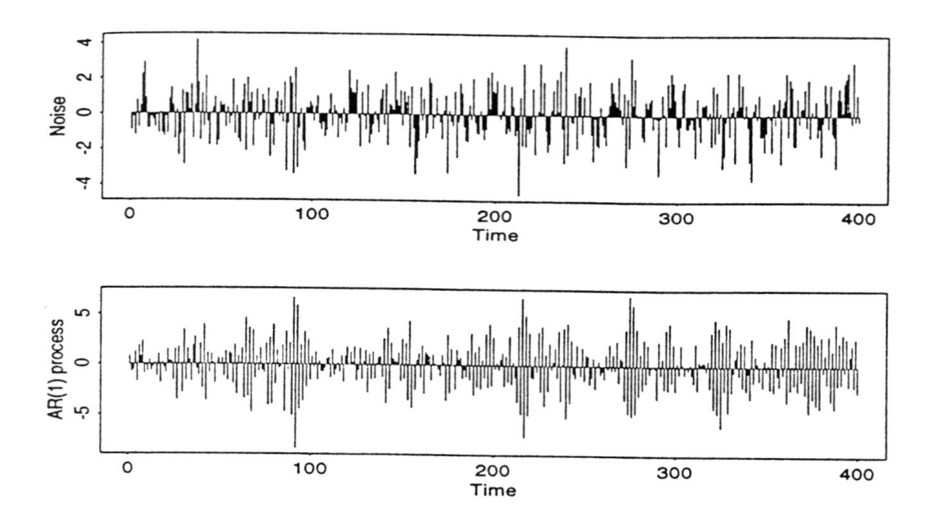

Figure 1: Realisation of iid $N(0,2)$ noise (Z_t) (top) and the corresponding AR(1) process $X_t = -0.8X_{t-1} + Z_t$ (bottom). □

model should be fitted. Various suggestions have been made. In finance, for example, SDE with stochastic volatility or Gaussian mixture models (including the (G)ARCH family) have been introduced. It has been pointed out that real financial data have a flat spectrum (i.e. they have the spectrum of white noise) and therefore linear processes are not appropriate models; see for instance the discussion in [37]. However, there are exceptions for which such a general statement does not hold. Our world is certainly "first order linear", and so the use of linear processes as first order approximations seems appropriate for any data set with or without large fluctuations. Below we collect some facts which might convince the reader that linear processes with large (infinite variance) Z–values allow for a reasonable theory comparable with the classical L^2 case. We consider modifications of various well–known estimators and testing procedures that are based on the periodogram as an empirical spectral estimate. At this point we anticipate that these methods work equally well under a finite or infinite variance regime.

We proceed as follows: in Section 2 we contrast estimation techniques for the spectral density of (X_t) under heavy– or light–tailed noise (Z_t), in Section 3 we continue with a discussion of the corresponding estimators of the spectral distribution function. In Section 4 we consider the weighted integrated periodogram and its applications to goodness of fit. In Section 5 we discuss parameter estimation based on the periodogram.

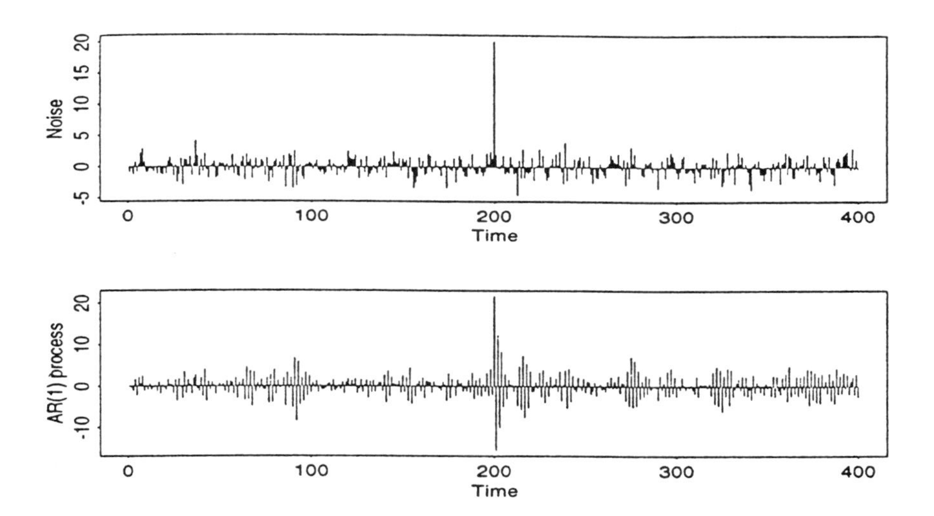

Figure 2: Realisation of iid $N(0, 2)$ noise (Z_t) (top) and the corresponding AR(1) process $X_t = -0.8X_{t-1} + Z_t$ (bottom). The noise is the same as in Figure 1, but it is perturbed by one large value $Z_{200} = 20$. Notice the different order of magnitude in the time series before and after $t = 200$. □

2. Estimation of the spectral density

In classical time series analysis we study linear processes (1.1) under the assumptions

$$EZ = 0\,, \quad \sigma^2 = \text{var}(Z) < \infty \quad \text{and} \quad \Psi^2 = \sum_{j=-\infty}^{\infty} \psi_j^2 < \infty\,. \tag{2.1}$$

Then (X_t) constitutes a stationary process which has spectral representation

$$X_t = \int_{-\pi}^{\pi} e^{it\lambda} dZ(\lambda)\,, \quad t \in \mathbb{Z}\,, \tag{2.2}$$

with respect to a complex–valued process $Z(\cdot)$ with orthogonal increments. The *spectral distribution function* F is given by

$$F(\lambda) - F(\lambda') = E|Z(\lambda) - Z(\lambda')|^2\,, \quad \lambda > \lambda'\,, \quad \text{and}\, F(-\pi) = 0\,.$$

If (X_t) is linear, then, under quite general conditions on (ψ_j), F is absolutely continuous with spectral density

$$f(\lambda) = \frac{\sigma^2}{2\pi}\, |\psi(\lambda)|^2\,, \quad \psi(\lambda) = \sum_{j=-\infty}^{\infty} \psi_j e^{-i\lambda j}\,, \quad \lambda \in [-\pi, \pi]\,. \tag{2.3}$$

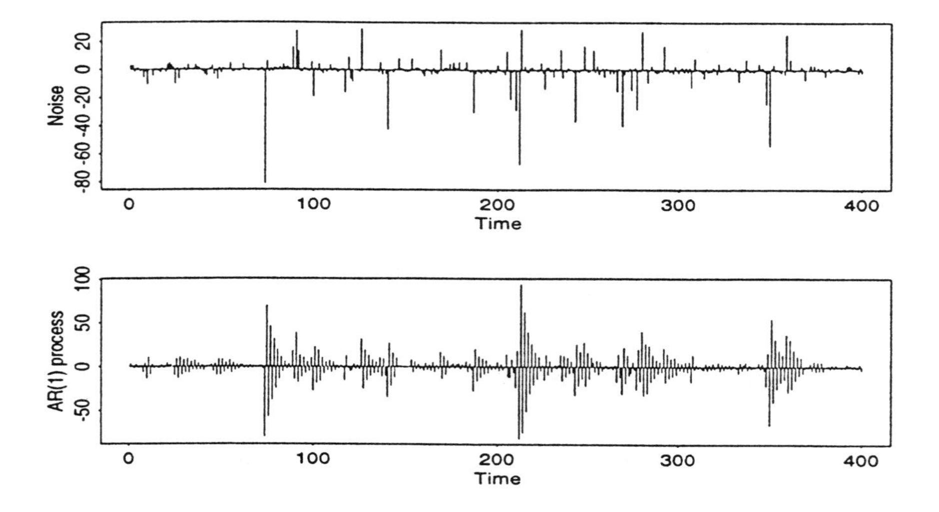

Figure 3: Realisation of iid Cauchy noise (Z_t) (top) and the corresponding AR(1) process $X_t = -0.8X_{t-1} + Z_t$ (bottom). □

The *power transfer function* $|\psi(\lambda)|^2$ is of crucial importance for the spectral analysis of a time series. The classical estimator of $2\pi f(\lambda)$ is the *periodogram*

$$I_{n,X}(\lambda) = \left| \frac{1}{\sqrt{n}} \sum_{t=1}^{n} X_t e^{-i\lambda t} \right|^2 = \sum_{|t|<n} \gamma_{n,X}(t) e^{-i\lambda t}, \quad \lambda \in [-\pi, \pi], \tag{2.4}$$

where

$$\gamma_{n,X}(t) = n^{-1} \sum_{s=1}^{n-|t|} X_s X_{s+|t|}$$

is the sample version of the autocovariance $\gamma_X(t) = EX_0X_t$. The periodogram is the naive empirical estimator of

$$2\pi f(\lambda) = \sum_{t=-\infty}^{\infty} \gamma_X(t) e^{-i\lambda t}, \quad \lambda \in [-\pi, \pi].$$

Following the basic aim of this paper we may ask:

What happens to the classical periodogram estimation procedures for $f(\lambda)$ and $F(\lambda)$ when there are large values among the Z_t?

In Figure 4 we consider the logarithmic smoothed periodogram from 400 realisations of the AR(1) process $X_t = -0.8X_{t-1} + Z_t$ corresponding to Figures 1–3. The estimated curves almost have the same shape. Since we consider the log–periodogram, the deviation of the periodogram estimates from the true spectrum, when large values in the

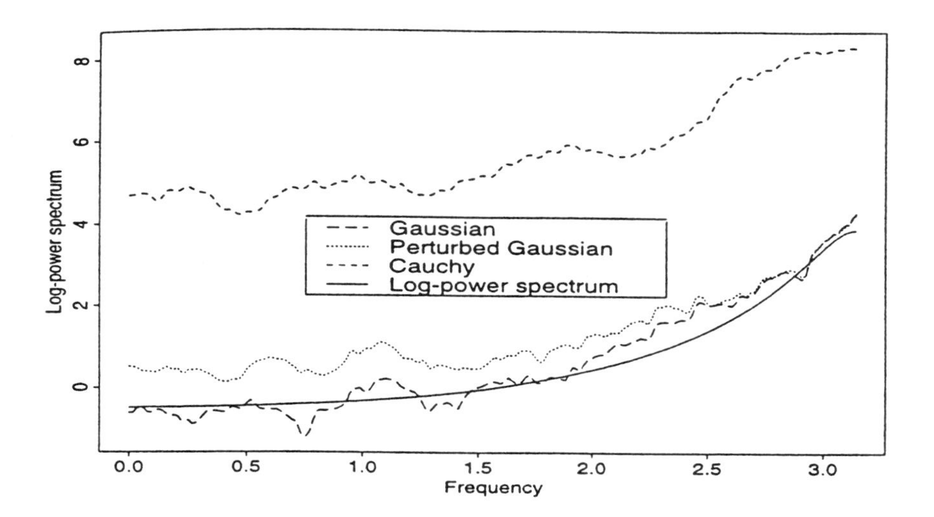

Figure 4: Logarithmic smoothed periodogram from 400 realisations of the AR(1) process $X_t = -0.8X_{t-1} + Z_t$ corresponding to Figures 1–3 with Gaussian, perturbed Gaussian or Cauchy noise. The solid line represents the true logarithmised power transfer function; see (2.3). □

noise are present, must be due to the normalisation of the periodogram. We first give a heuristic argument explaining this phenomenon: under general assumptions, for σ^2 either finite or infinite, one can show that

$$I_{n,X}(\lambda) = I_{n,Z}(\lambda)\,|\psi(\lambda)|^2 + o_P\left(\gamma_{n,Z}(0)\right), \quad \lambda \in [-\pi, \pi]. \tag{2.5}$$

Except for the remainder term this is the formal empirical analogue to the left–hand side of (2.3). By the strong law of large numbers, $\gamma_{n,Z}(0) \stackrel{\text{a.s.}}{\to} \sigma^2 \leq \infty$. Thus the remainder term in (2.5) is negligible, provided that $\sigma^2 < \infty$, and $I_{n,Z}(\lambda)$ converges to an exponential random variable with variance σ^2. Moreover, a vector of periodogram ordinates at distinct frequencies $\lambda_i \in (0, \pi)$ converges in distribution to iid exponential random variables. But what happens if $\sigma^2 = \infty$? In this case the remainder in (2.5) cannot be neglected. Notice that $\gamma_{n,Z}(0)$ is a sample version of the variance σ^2 (finite or infinite), but as a function of the data, $\gamma_{n,Z}(0)$ is always finite. The idea, then, is to re–normalise the periodogram by means of $\gamma_{n,Z}(0)$:

$$\widetilde{I}_{n,Z}(\lambda) = \frac{I_{n,Z}(\lambda)}{\gamma_{n,Z}(0)} = \frac{\left|\sum_{t=1}^{n} Z_t e^{-i\lambda t}\right|^2}{\sum_{t=1}^{n} Z_t^2}, \quad \lambda \in [-\pi, \pi].$$

The random variables $Z_t/\sqrt{n}$ have now been replaced by the quantities $Z_t/\sqrt{n\gamma_{n,Z}(0)}$, $t = 1, \ldots, n$, whose absolute value does not exceed 1.

This idea is the key to the theory given below!

Self–normalisation or *studentisation* of the data makes the classical spectral estimates behave *as if they came from a time series with finite variance although the theoretical variance σ^2 may be infinite.*

Similarly, we shall replace $I_{n,X}$ by $\widetilde{I}_{n,X} = I_{n,X}/\gamma_{n,X}(0)$. In this context, the following observation is of great help:

$$\gamma_{n,X}(0) = \gamma_{n,Z}(0)\Psi^2 + o_P\left(\gamma_{n,Z}(0)\right), \tag{2.6}$$

where Ψ^2 is defined in (2.1). Hence, by (2.5) and (2.6),

$$\widetilde{I}_{n,X}(\lambda) = \frac{I_{n,X}(\lambda)}{\gamma_{n,X}(0)} = \widetilde{I}_{n,Z}(\lambda)\frac{|\psi(\lambda)|^2}{\Psi^2} + o_P(1), \quad \lambda \in [-\pi, \pi]. \tag{2.7}$$

As in the finite variance case, the asymptotic properties of the right–hand side depend only on $\widetilde{I}_{n,Z}(\lambda)$. Convergence in distribution for $n \to \infty$ cannot be expected without additional assumptions. The general limit theory for sums of independent random variables suggests (see for instance [28]) that conditions on the distribution tails of Z are needed. For instance, if we assume that Z is symmetric with Pareto–like tails

$$P(Z \le -x) = P(Z > x) \sim cx^{-\alpha}, \quad x \to \infty, \quad \alpha < 2, c > 0, \tag{2.8}$$

then a limit distribution of $\widetilde{I}_{n,Z}(\lambda)$ exists. Condition (2.8) means that Z is in the domain of normal attraction of a symmetric α–stable distribution (see [13]); (2.8) is in particular satisfied for symmetric α–stable random variables.

The limit distribution of $\widetilde{I}_{n,Z}(\lambda)$ is very complicated. It depends on whether λ is a rational or an irrational multiple of π. Moreover, periodogram ordinates at distinct frequencies are asymptotically dependent. Fortunately, the limits of $\widetilde{I}_{n,Z}(\lambda)$ have mean 1, exponentially decreasing tails, and they are uncorrelated for distinct $\lambda \in (0, \pi)$. Thus the situation is not so much different from the finite variance case. In the latter we know that smoothing techniques yield consistent estimates of the spectral density (cf. [4], Chapter 10; [31], Chapter 6). Introduce non–negative symmetric weights $(W_n(k))$ that satisfy the usual conditions (cf. [4], p. 351)

$$\sum_{|k|\le m} W_n(k) = 1, \quad \sum_{|k|\le m} W_n^2 = o(1)$$

as $m = m_n \to \infty$ and $m/n \to 0$. For example, $W_n(k) = 1/(2m+1)$ with $m = [n^\delta]$, for some $\delta \in (0, 1)$, is such a weight sequence (it corresponds to the Daniell window). In the infinite variance case it can be shown that discrete weighted average estimators are consistent: let $\lambda_k = 2\pi k/n$ denote the Fourier frequencies. Then

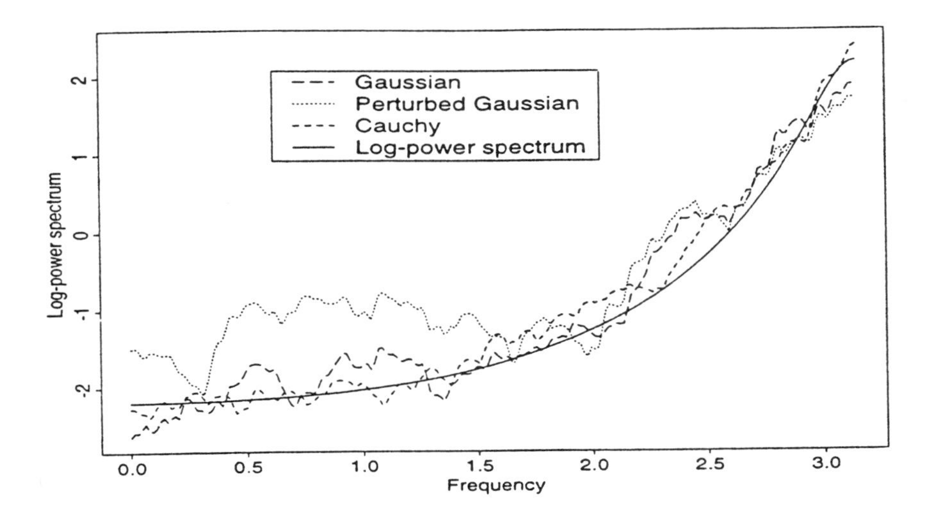

Figure 5: Logarithmic smoothed self–normalised periodogram from 400 realisations of the AR(1) process $X_t = -0.8X_{t-1} + Z_t$ corresponding to Figures 1–3. The solid line represents the true logarithmic power transfer function; see (2.3), cf. Figure 4. □

$$\sum_{|k|\le m} W_n(k)\widetilde{I}_{n,X}(\lambda+\lambda_k) \quad \xrightarrow{P} \quad \Psi^{-2}\,|\psi(\lambda)|^2\,, \quad \lambda \in (0,\pi)\,. \tag{2.9}$$

Figures 4 and 5 have been obtained by applying the weights $W_n(k) = 1/(2m+1)$ with $m = 11$.

Remarks. 1) The interpretation of $|\psi(\lambda)|^2$ as spectral density of (X_t) is not possible in the infinite variance case. However, results of type (2.9) show that classical estimators are applicable when they are self–normalised or studentised. A heuristic explanation is that, if there occur large X–values, then division of X_t by $\sqrt{n\gamma_{n,X}(0)}$ "corrects" the "too big" empirical variance of the observations. The vector $(X_t/\sqrt{n\gamma_{n,X}(0)})_{t=1,\dots,n}$ behaves for large n as if it came from a bounded stationary sequence with spectral density function $\Psi^{-2}\,|\psi(\lambda)|^2$. Notice that (2.9) remains valid if $\sigma^2 < \infty$ (see [4], Chapter 7). In this case, $\gamma_{n,X}(0) \xrightarrow{P} \mathrm{var}(X) = \sigma^2\Psi^2$, and therefore self–normalisation is not really needed.

2) The arguments for the infinite variance case given above are proved in [20, 21]. In the first reference the convergence in distribution of a vector of periodogram ordinates is studied under (2.8). In this case,

$$n^{-2/\alpha+1}\gamma_{n,Z}(0) \xrightarrow{d} Y$$

for an $\alpha/2$–stable positive random variable Y. This allows one to switch from the self–normalised periodogram $\widetilde{I}_{n,X}$ to the periodogram with deterministic normalisation $n^{2/\alpha}$.

Weighted average spectral estimates are studied in [21]. The methods proposed there can be extended to kernel–type estimators of $|\psi(\lambda)|^2$. More sophisticated methods (uniform (in λ) convergence, rates of convergence) do not seem to exist for infinite variance linear processes. As for the theory given below, one may expect that the rates are better than in the finite variance case.

3) Recall the spectral representation (2.2) of a (second order) stationary process (X_t). One may replace the orthogonal increment process $Z(\cdot)$ by an α–stable process with independent increments, yielding an α–stable harmonizable process. This class of stationary processes is disjoint from the class of the linear processes (1.1) with iid α–stable noise (Z_t) (see [34, 35]). However, harmonizable processes allow one to introduce a "pseudo–spectral density" (see [26] for details).

4) In the infinite variance case the condition $\Psi^2 < \infty$ is not sufficient to ensure the a.s. convergence of the infinite series (1.1). Conditions of type $\sum_j |\psi_j|^\delta < \infty$ for some $\delta \in (0, 1 \wedge \alpha)$ are required under the tail assumption (2.8). This condition is satisfied for causal invertible ARMA processes in which case $|\psi_j| < a^j$ for some $a < 1$ and large j. However, it rules out long memory processes such as fractional ARIMA(p, d, q) for some $d \in (0, 1 - 1/\alpha)$ with α–stable noise, for $1 < \alpha < 2$. In this case, $\psi_j \sim j^{d-1} L(j)$ for a slowly varying function L (see [35], Section 7.13). □

3. Estimation of the spectral distribution function

Now we turn to the estimation of the spectral distribution function F. A natural estimator of $F(\lambda) - F(0)$ is the *empirical spectral distribution function* (integrated periodogram)

$$J_{n,X}(\lambda) = \int_0^\lambda I_{n,X}(x) dx\,, \quad \lambda \in [0, \pi]\,.$$

In Figure 6, $\ln J_{n,Z}$ is calculated from 400 realisations of Gaussian or α–stable noise ($\alpha < 2$). These graphs again show that the normalisation of $J_{n,X}$ is not correct for infinite variance X. We mention that relation (2.5) holds uniformly for $\lambda \in [0, \pi]$, and therefore

$$J_{n,X}(\lambda) = \int_0^\lambda I_{n,Z}(x)\, |\psi(x)|^2\, dx + o_P(\gamma_{n,Z}(0))\,, \quad \lambda \in [0, \pi]\,.$$

Similar arguments as those for (2.7) show that, uniformly for $\lambda \in [0, \pi]$,

$$\begin{aligned} \widetilde{J}_{n,X}(\lambda) &= \int_0^\lambda \widetilde{I}_{n,Z}(x)\, |\psi(x)|^2\, dx + o_P(1) \\ &\xrightarrow{P} \Psi^{-2} \int_0^\lambda |\psi(x)|^2\, dx = \widetilde{F}(\lambda)\,. \end{aligned} \tag{3.1}$$

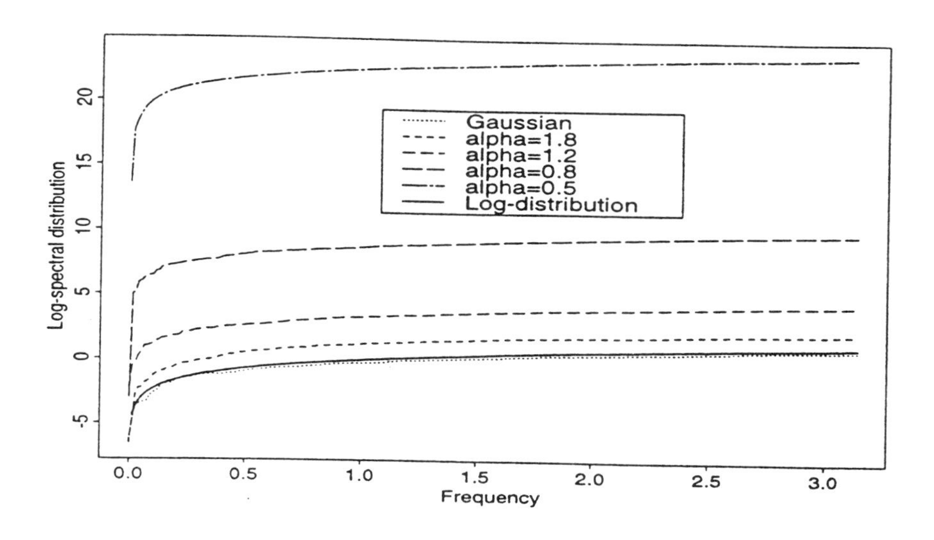

Figure 6: The logarithmic integrated periodogram from 400 iid realisations of (Z_t). Z is either standard Gaussian or symmetric α–stable ($\alpha < 2$), see (2.8). For comparison, the log–spectral distribution function of (Z_t) in the Gaussian case (solid line). □

Remark. 1) In the L^2 case, (3.1) follows for instance from the results in [17], see also [31], Chapter 6. It remains valid for certain infinite variance processes (see [22], the assumptions given there can be relaxed substantially). □

Relation (3.1) can be refined to a functional central limit theorem. With normalisation $a_n = \sqrt{n}$ in the L^2 case and with $a_n = (n/\ln n)^{1/\alpha}$ under (2.8), $(\widetilde{J}_{n,X} - \widetilde{F})$ has a Gaussian, respectively infinite variance limit process in $\mathbb{C}[0, \pi]$. For example, if $X = Z$, then

$$\widetilde{J}_{n,Z}(\lambda) = \widetilde{F}(\lambda) + 2\sum_{t=1}^{n-1} \frac{\sin(\lambda t)}{t} \widetilde{\gamma}_{n,Z}(t), \quad \lambda \in [0, \pi], \tag{3.2}$$

where $\widetilde{\gamma}_{n,Z}(t) = \gamma_{n,Z}(t)/\gamma_{n,Z}(0)$ are the sample autocorrelations of (Z_t). The normalised sample autocorrelations $(\widetilde{\gamma}_{n,Z}(t))_{t=1,\ldots,m}$ converge in distribution to a vector $(Y_t)_{t=1,\ldots,m}$ either with iid $N(0, 1)$ (see [4], Chapter 7) or with iid α–stable components divided by an independent $\alpha/2$–stable positive random variable (see [8, 9, 10]). The latter facts and a Slutsky argument applied to (3.2) finally yield in $\mathbb{C}[0, \pi]$ that

$$a_n(\widetilde{J}_{n,Z}(\lambda) - \widetilde{F}(\lambda)) \quad \stackrel{d}{\rightarrow} \quad B_\lambda = 2\sum_{t=1}^{\infty} \frac{\sin(\lambda t)}{t} Y_t . \tag{3.3}$$

For Gaussian noise (Z_t), the limit B is a Brownian bridge on $[0, \pi]$ (cf. [1]), and under (2.8) with $\alpha \in (1, 2)$, it is an infinite variance process with a.s. continuous sample paths (see Figure 7, cf. [22]).

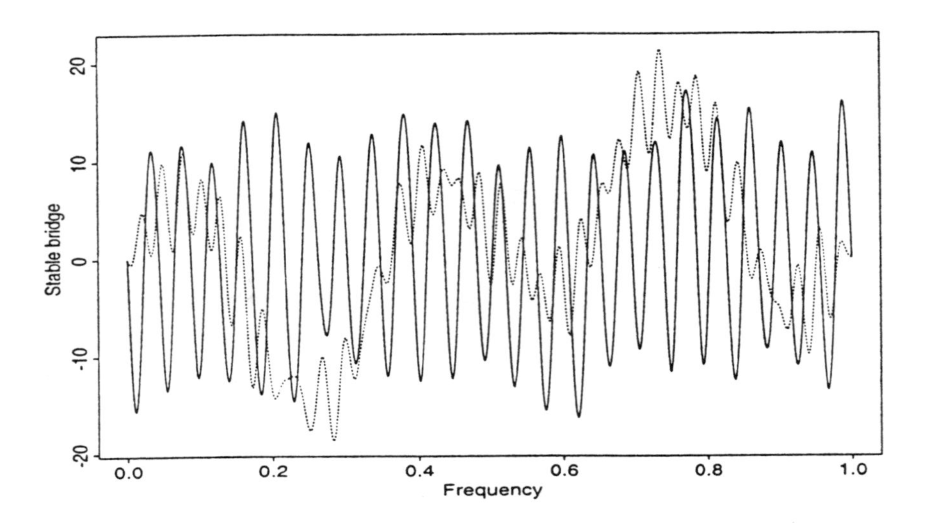

Figure 7: Two sample paths of the α–stable analogue to the Brownian bridge from (3.3), $\alpha = 1.5$. □

Remarks. 2) For general linear processes, the limit of $(a_n(\widetilde{J}_{n,X} - \widetilde{F}))$ is of the form

$$2\Psi^{-2}\sum_{t=1}^{\infty}\left(\int_0^{\lambda}\cos(xt)|\psi(x)|^2dx\right)Y_t\,, \tag{3.4}$$

see [24] for the finite/infinite variance long memory case, [22] for the infinite variance short memory case and [1] for the short memory L^2 case. In the L^2 case, this is *not* a Brownian bridge; see [1].

3) Results such as (3.1), (3.3) and (3.4) show that a weak limit of $(\widetilde{J}_{n,X} - \widetilde{F})$ exists under quite general assumptions. When $\sigma^2 = \infty$, the (infinite variance) limit process is rather unfamiliar. However, the rate of convergence $(n/\ln n)^{1/\alpha}$ compares favourably with $\sqrt{n}$ in the finite variance case. □

4. Goodness of fit tests

A glance at Figures 4 and 5 gives us a first impression on the goodness of fit of the time series data from model (1.1). A natural question arises:

How can we test the goodness of fit of the model (1.1) *when there are large values in the noise?*

In Section 3 it is also suggested how to do that: one may proceed analogously to the classical Kolmogorov–Smirnov or Cramér–von Mises tests known from non–parametric

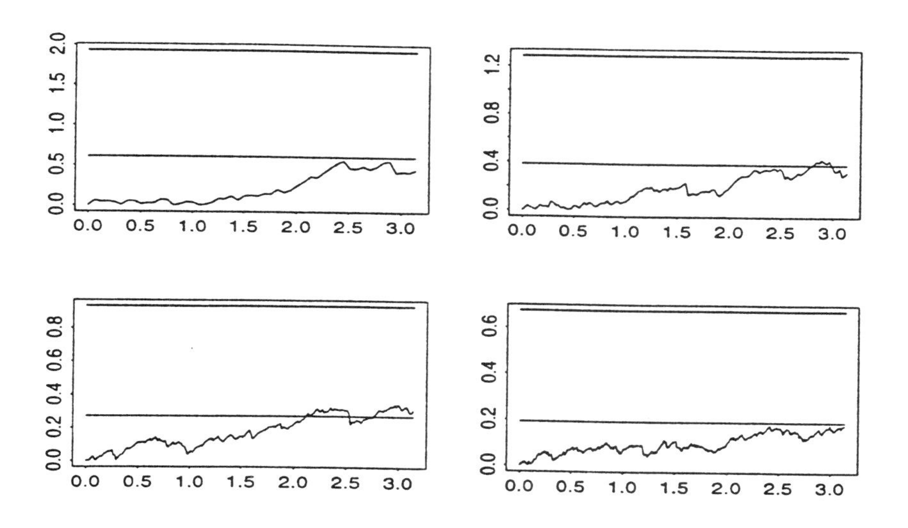

Figure 8: The statistic $|\widetilde{J}_{n,Z}(\lambda) - \widetilde{F}(\lambda)|$, $\lambda \in [0, \pi]$, for iid 1.8–stable noise for $n = 100$ (left, top), $n = 250$ (right,top), $n = 500$, (left, bottom) and $n = 1000$ (right,bottom). Note the different vertical scale. The solid line on top of the figures and the x–axis constitute a 95% asymptotic confidence band based on the limit relation (4.1); the solid line at the bottom of the figures and the x–axis constitute a 95% confidence band for the normalised absolute supremum functional of the Brownian bridge. □

statistics (for instance [30, 36]). For example, from (3.3) and by the continuous mapping theorem we conclude for iid (Z_t) that

$$a_n \sup_{\lambda \in [0,\pi]} \left| \widetilde{J}_{n,Z}(\lambda) - \widetilde{F}(\lambda) \right| \xrightarrow{d} \sup_{\lambda \in [0,\pi]} |B_\lambda|, \tag{4.1}$$

$$a_n^2 \int_0^\pi (\widetilde{J}_{n,Z}(\lambda) - \widetilde{F}(\lambda))^2 d\lambda \xrightarrow{d} \int_0^\pi B_\lambda^2 d\lambda. \tag{4.2}$$

In the L^2 case, the quantiles of the limit distributions in (4.1) and (4.2), as functionals of the Brownian bridge, are well known (for instance [1, 36]). In the infinite variance case, no explicit formulae for the quantiles exist, so one has to rely on simulations (see [22]). We mention that *the asymptotic quantiles from the L^2 case can be used to construct conservative confidence bands and test regions in the infinite variance case.* For large n, the wider confidence bands caused by the infinite variance limit process get compensated by the better rate of convergence; see also the graphs in Figure 8. They show that the goodness of fit test for noise works convincingly well in the heavy–tailed case, even for small and medium sample sizes.

For general linear processes, the limit process of the goodness of fit test statistics involves the power transfer function (see (3.4)), which is rather inconvenient (see [1] for details). However, goodness of fit tests may also be based on weighted empirical spectral distribution functions of the form $\int_0^\lambda g(x)\widetilde{I}_{n,X}(x)dx$, for appropriate weight functions g. Above we considered the case $g \equiv 1$. A particularly attractive choice is $g(\lambda) = 1/|\psi(\lambda)|^2$ (cf. [2, 3]). The same arguments as in Section 2 yield in $\mathbb{C}[0, \pi]$

$$\begin{aligned}
\widehat{J}_{n,X}(\lambda) &= \int_0^\lambda \frac{\widetilde{I}_{n,X}(x)}{|\psi(x)|^2} dx \\
&= \int_0^\lambda \frac{\widetilde{I}_{n,Z}(x)|\psi(x)|^2}{|\psi(x)|^2} dx + o_P(1) \\
&= \widetilde{J}_{n,Z}(\lambda) + o_P(1) \quad \stackrel{P}{\to} \quad \lambda\,,
\end{aligned}$$

and

$$a_n(\widehat{J}_{n,X}(\lambda) \quad - \quad \widetilde{F}(\lambda)) \quad \stackrel{d}{\to} \quad B_\lambda\,,$$

where B is defined in (3.3). *Thus the integrated periodogram $\widehat{J}_{n,X}$ with weight function $1/|\psi(\lambda)|^2$ has the same limit as $\widetilde{J}_{n,Z}$ for iid noise. In particular, we may $\widetilde{J}_{n,Z}$ in* (4.1), (4.2) *replace by $\widehat{J}_{n,X}$.* This reduces the problem of finding the limit distribution of a whole class of goodness of fit test statistics for linear processes to the iid case as discussed above. This parametric method works indeed for a wide class of coefficients, as proved in [22] for the short memory, infinite variance case, and in [24] for the long memory, infinite/finite variance case.

4. Parameter estimates based on the periodogram

Parameter estimates for ARMA and FARIMA processes (for a definition of these processes see [4, 35]) can be constructed from the periodogram. One of them was suggested by Whittle [38]. For pure AR processes, it coincides with the Yule–Walker estimator. In the L^2–case, the Whittle estimator is asymptotically equivalent to Gaussian maximum likelihood and least sqares estimators (cf. [4], Section 10.8).

We describe the Whittle estimation procedure for ARMA(p, q) processes: let

$$\phi(z) = 1 - \phi_1 z - \cdots - \phi_p z^p\,, \quad \theta(z) = 1 + \theta_1 z + \cdots + \theta_q z^q\,,$$

be polynomials with coefficents $\beta = (\phi_1, \ldots, \phi_p, \theta_1, \ldots, \theta_q)^\top$ which belong to the natural parameter space

$$\begin{aligned}
C \quad = \quad \{\beta \in \mathbb{R}^{p+q} : \phi_p\theta_q \neq 0\,,\ \phi(z)\theta(z) \neq 0 \quad \text{for} \quad |z| \leq 1 \quad \text{and} \\
\phi(z) \text{ and } \theta(z) \text{ have no common zeros}\}\,.
\end{aligned}$$

If B denotes the backshift operator, i.e. $BX_t = X_{t-1}$, and $\beta \in C$, then the difference equations $\phi(B)X_t = \theta(B)Z_t$, $t \in \mathbb{Z}$, define a causal, invertible ARMA(p,q) process. Assume that the true parameter value β_0 belongs to C. Define

$$\sigma_n^2(\beta) = \int_{-\pi}^{\pi} I_{n,X}(\lambda)\,|\psi(\lambda,\beta)|^{-2}\,d\lambda\,. \tag{4.1}$$

Here $|\psi(\lambda,\beta)|^2$ denotes the power transfer function of the ARMA process corresponding to the parameter $\beta \in C$. If we formally replace $I_{n,X}(\lambda)$ in $\sigma_n^2(\beta)$ by the true density $|\psi(\lambda,\beta_0)|^2$, then this new function of β assumes its absolute minimum at $\beta = \beta_0$. This suggests estimating β_0 by

$$\beta_n = \operatorname{argmin}_{\beta \in C}\, \sigma_n^2(\beta)\,.$$

Under general conditions, β_n is a consistent estimator of β_0 (see [27]). In the L^2 case, it is asymptotically normal (see [4], Section 10.8) with rate $\sqrt{n}$; under (2.8) it has an infinite variance limit and rate $(n/\ln n)^{1/\alpha}$. It is proved in Gadrich [15] that the integral in (4.1) may be replaced by a Riemann sum evaluated at the Fourier frequencies $\lambda_k = 2\pi k/n$:

$$\widehat{\sigma}_n^2(\beta) = \frac{2\pi}{n} \sum_{-\pi<\lambda_k\le\pi} I_{n,X}(\lambda_k)\,|\psi(\lambda_k,\beta)|^{-2} \approx \sigma_n^2(\beta)\,.$$

The Whittle estimate works for large classes of models. It can be extended to FARIMA(p,d,q) processes which are long memory processes for $d < 0.5$ in the L^2 case and for $0 < d < 1 - 1/\alpha$ under (2.8), $\alpha \in (1,2)$. The parameter estimation for infinite variance ARMA is treated in [27], for finite variance FARIMA in [14, 16] and for infinite variance FARIMA in [25]. There exist various parameter estimates for infinite variance ARMA processes (see for instance [5, 6, 7]) which are not based on the periodogram.

5. Concluding remarks

The above outline shows that a general theory can be built up for the periodogram of infinite variance linear processes. Such a theory shows how one has to modify estimates and tests when large Z–values are present. A general suggestion (which is supported by the theoretical results) is always to use the self–normalised periodogram. In this case, various classical estimates and tests work equally well under general assumptions such as finite/infinite variance, long/short memory.

There exist two significant differences between the finite and the infinite variance case:

1. In the L^2 case the limit distributions/processes are Gaussian. In the infinite variance case they are infinite variance distributions/processes which are closely related to

α–stable distributions/processes for some $\alpha < 2$. The latter limits are less familiar and their theoretical properties are difficult to establish.

2. In the L^2 case the rates of convergence in the weak limit theory are of the order $\sqrt{n}$. In the infinite variance case they are of the order $n^{-1/\alpha}L(n)$ for some $\alpha < 2$ and a slowly varying function L.

Thus "bad" (i.e. infinite variance) limits get compensated by a "good" (faster than $\sqrt{n}$) rate of convergence. At the moment no measure is available which describes the interrelationship of these contradictory phenomena. In particular, it is unclear how large n must be chosen in order to use the faster rate of convergence for the construction of conservative test regions or confidence bands based on the quantiles of the classical (Gaussian) limits. However, simulation studies show that the proposed procedures work equally well in the finite/infinite variance cases, even for medium sample sizes $n \in [100, 200]$. Simulation experience also gives some evidence that the faster rates in the infinite case can be exploited to construct conservative confidence bands and test regions as described above. The quantiles of the limiting infinite variance quantities cannot be calculated explicitly at the moment. They typically also depend on unknown parameters (for instance the index α for stable distributions/processes) which have to be estimated. The accuracy of those estimates is known to be poor (see e.g. [29, 33]).

Finally, we summarise our experience in two rules of thumb:

1. *Always use the self–normalised periodogram.*

2. *You may use the classical confidence bands and test regions based on the L^2–theory in a conservative sense.*

If one follows these two hints one may also trust the standard computer packages on statistics, even when there are large values in the noise of the data.

The above theory depends crucially on the assumption that (X_t) is a linear process. It can be expected that there is not such an analogy between the finite/infinite variance case for non–linear time series models. An indication of this is given by recent work on sample autocorrelations for bilinear processes (see [11, 32]).

References

[1] T.W. Anderson, Goodness of fit tests for spectral distributions, *Ann. Statist.* **21** (1993), 830–847.

[2] M.S. Bartlett, Problemes de l'analyse spectrale des séries temporelles stationnaires, *Publ. Inst. Statist. Univ. Paris.* **III-3** (1954), 119–134.

[3] M.S. Bartlett, *An Introduction to Stochastic Processes with Special Reference to Methods and Applications*, 3rd edition, Cambridge University Press, Cambridge (UK), 1978.

[4] P.J. Brockwell and R.A. Davis, *Time Series: Theory and Methods,* 2nd edition, Springer, New York, 1991.

[5] M. Calder and R.A. Davis, Inference for linear processes with stable noise, (1997). This Volume.

[6] R.A. Davis, Gauss–Newton and M–estimation for ARMA processes with infinite variance, *Stoch. Proc. Appl.* **63** (1996), 75–95.

[7] R.A. Davis, K. Knight and J. Liu, M–estimation for autoregressions with infinite variance, *Stoch. Proc. Appl.* **40** (1992), 145–180.

[8] R.A. Davis and S.I. Resnick, Limit theory for moving averages of random variables with regularly varying tail probabilities, *Ann. Probab.* **13** (1985), 179–195.

[9] R.A. Davis and S.I. Resnick, More limit theory for the sample correlation function of moving averages, *Stoch. Proc. Appl.* **20** (1985), 257–279.

[10] R.A. Davis and S.I. Resnick, Limit theory for the sample covariance and correlation functions of moving averages, *Ann. Statist.* **14** (1986), 533–558.

[11] R.A. Davis and S.I. Resnick, Limit theory for bilinear processes with heavy-tailed noise, *Ann. Appl. Probab.* **6** (1996), 1191–1210.

[12] P. Embrechts, C. Klüppelberg, and T. Mikosch, *Modelling Extremal Events Towards Insurance and Finance,* Springer, Berlin, 1997.

[13] W. Feller, *An Introduction to Probability Theory and Its Applications II,* Wiley, New York, 1971.

[14] R. Fox and M.S. Taqqu, Large sample properties of parameter estimates for strongly dependent stationary Gaussian time series, *Ann. Statist.* **14** (1986), 517–532.

[15] T. Gadrich, Parameter estimation for ARMA processes with symmetric stable innovations (in Hebrew), D.Sc. Thesis, Technion, Haifa, 1993.

[16] L. Giraitis and D. Surgailis, A central limit theorem for quadratic forms in strongly dependent linear variables and application to asymptotic normality of Whittle's estimate, *Probab. Theor. Rel. Fields.* **86** (1990), 87–104.

[17] U. Grenander and M. Rosenblatt, *Statistical Analysis of Stationary Time Series,* 2nd edition, Chelsea Publishing Co., New York, 1984.

[18] C.C. Heyde and R. Gay, Smoothed periodogram asymptotics and estimation for processes and fields with possible long–range dependence, *Stoch. Proc. Appl.* **45** (1993), 169–182.

[19] T. Hsing, On tail index estimation using dependent data, *Ann. Statist.* **19** (1991), 1547–1569.

[20] C. Klüppelberg and T. Mikosch, Spectral estimates and stable processes, *Stoch. Proc. Appl.* **47** (1993), 323–344.

[21] C. Klüppelberg and T. Mikosch, Some limit theory for the self–normalised periodogram of stable processes, *Scand. J. Statist.* **21** (1994), 485–491.

[22] C. Klüppelberg and T. Mikosch, The integrated periodogram for stable processes, *Ann. Statist.* **24** (1996) 1855–1879.

[23] C. Klüppelberg and T. Mikosch, Self–normalised and randomly centred spectral estimates, in: *Proceedings of the Athens Conference on Applied Probability and Time Series Analysis.* Time series volume in honour of E.J. Hannan, Lect. Notes in Statistics. Springer, New York, 1996, pp. 259–271.

[24] P. Kokoszka and T. Mikosch, The integrated periodogram for long–memory processes with finite or infinite variance, *Stoch. Proc. Appl.* **66** (1997), 55–78.

[25] P. Kokoszka and M.S. Taqqu, Parameter estimation for infinite variance fractional ARIMA, *Ann. Statist.* **24** (1996), 1880–1913.

[26] E. Masry and S. Cambanis, Spectral density estimation for stationary stable processes. *Stoch. Proc. Appl.* **18** (1984), 1–31.

[27] T. Mikosch, T. Gadrich, C. Klüppelberg and R.J. Adler, Parameter estimation for ARMA models with infinite variance innovations. *Ann. Statist.* **23** (1995), 305–326.

[28] V.V. Petrov, *Limit Theorems of Probability Theory,* Oxford University Press, Oxford, 1995.

[29] O.V. Pictet, H.M. Dacorogna and U.A. Müller, Behaviour of tail estimation under various distributions, (1997). This Volume.

[30] D. Pollard. *Convergence of Stochastic Processes,* Springer, Berlin, New York, 1984.

[31] M.B. Priestley, *Spectral Analysis and Time Series I,II,* Academic Press, New York, 1981.

[32] S.I. Resnick, Why non–linearities can ruin the heavy-tailed modeller's day, (1997). This Volume.

[33] S.I. Resnick and C. Stărică, Consistency of Hill's estimator for dependent data, *J. Appl. Prob.* **32** (1995), 239–267.

[34] J. Rosinski, On the structure of stationary stable processes, *Ann. Probab.* **23** (1995), 1163–1187.

[35] G. Samorodnitsky and M.S. Taqqu, *Stable Non–Gaussian Random Processes. Stochastic Models with Infinite Variance,* Chapman and Hall, London, 1994.

[36] G.R. Shorack and J.A. Wellner, *Empirical Processes with Applications to Statistics,* Wiley, New York, 1986.

[37] S.J. Taylor, *Modelling Financial Time Series,* Wiley, Chichester, 1986

[38] P. Whittle, Estimation and information in stationary time series. *Ark. Mat.* **2** (1953), 423-434.

University of Groningen
Department of Mathematics
P.O. Box 800
NL–9700 AV Groningen
The Netherlands

Bayesian Inference for Time Series with Infinite Variance Stable Innovations

Nalini Ravishanker and Zuqiang Qiou

Abstract

This article describes the use of sampling based Bayesian inference for infinite variance stable distributions and for time series with infinite variance stable innovations. For time series, an advantage of the Bayesian approach is that it enables the simultaneous estimation of the parameters characterizing the stable law, together with the parameters of the univariate or multivariate linear ARMA model. Our approach uses a Metropolis-Hastings algorithm to generate samples from the joint posterior distribution of all the parameters and is an extension to univariate and multivariate time series processes of the approach in [Bu] for independent observations.

1. Introduction

A random variable X has a stable distribution $S(\alpha, \beta, \delta, \sigma)$ if there are parameters $0 < \alpha \leq 2, -1 \leq \beta \leq 1, \sigma > 0$ and $-\infty < \delta < \infty$ such that its characteristic function has the form ([GK])

$$E(e^{itx}) = \begin{cases} \exp(-|\sigma t|^{\alpha}(1 - i\beta \text{sign}(t) \tan(\pi\alpha/2) + i\delta t) & \text{if } \alpha \neq 1 \\ \exp(-|\sigma t|(1 + 2i\beta \ln|t| \text{sign}(t)/\pi) + i\delta t) & \text{if } \alpha = 1, \end{cases} \tag{1.1}$$

where t is a real number, $i = \sqrt{-1}$ and $\text{sign}(t) = 1$ if $t > 0$, $\text{sign}(t) = 0$ if $t = 0$ and $\text{sign}(t) = -1$ if $t < 0$. In (1.1), α is the stability parameter, β is the skewness parameter, σ is the scale parameter and δ is the location parameter. The special cases $\alpha = 2, \alpha = 1$ and $\alpha = 0.5$ correspond respectively to the Gaussian, Cauchy and inverse Gaussian distributions which are cases for which there exists a closed form expression for the density function of X. [ST] provides an excellent discussion of the probabilistic properties of stable distributions, which have proved useful in modeling independent observations that admit occasional large values ([Z]). In time series modeling, the assumption of stable innovations has found use in modeling series with infinite variance; references in this area include [Bh], [CT], [CB], [F], [GO], [HK], [K1],[K2], [N], [Ro],

[SK] and [YM]. The role of positive stable random variables, (where $0 < \alpha < 1$ and $\beta = 1$) in survival analysis and longitudinal data modeling has been explored (see [Ho] and [CD]).

Several methods for estimating the stable law parameters based on independent observations have been discussed in the literature. See [A], [Du1], [Du2], [Du3], [F], [FR1], [FR2], [He], [Hi], [K1], [K2], [Ko1], [Ko2], [Ma], [Mc], [PHL], [PD], [P1] and [QR1]. [Du1] proved the asymptotic normality of maximum likelihood estimators of the parameters of the stable distribution and pointed out the possibility of a Bayesian approach through the incorporation of prior specifications. However, the unavailability of a closed form expression for the density function of X for all α values makes direct evaluation of the likelihood function computationally very intensive. For this reason, until recently, Bayesian inference in the context of stable distributions was impractical.

[Bu] provided an expression for the joint density of n *iid* observations from a stable distribution through the incorporation of an augmented variable, which permits computationally feasible sampling based Bayesian inference. In general, the sampling based Bayesian approach is extremely useful for conducting inference on an analytically intractable posterior distribution. Samples are drawn from the required joint posterior distribution via complete conditional distributions. Subsequent marginal and joint inference on the parameters may be carried out on the basis of these samples. For instance, posterior moments and quantiles may be obtained as well as estimated marginal and joint posterior densities.

Bayesian model determination involves model choice (selection among models) and model adequacy (performance of a particular model). Predictive distributions are useful in assessing model adequacy in order to compare what the model predicts and what was observed (see [GD]). Formal pairwise choice between models requires the calculation of the Bayes factor (ratio of marginal predictive distributions), which however, loses interpretability when there are improper priors on the parameters.

Alternatively, EDA- style diagnostics may be used ([MRGP]). [Bu] used Gibbs sampling ([GS]) to generate samples from the joint posterior distribution of the stable law parameters. An alternate approach ([QR1]) involves an approximation of the posterior moments through a fully exponential Laplace approximation ([TK]) and of the posterior density by a mixture of suitable densities (such as normal densities, see [W]) and yields results similar to the approach in [Bu]. However, [Bu]'s approach is attractive since it avoids the necessity of numerical integration to obtain the stable likelihood. The usefulness of the Bayesian framework for univariate and multivariate Gaussian time series modeling has been explored by [MRGP], [PR] and [RR].

For time series with infinite variance stable innovations which are useful in applications in diverse areas such as telecommunications, hydrology, physics of condensed matter, economics and finance, we present sampling based Bayesian inference. The outline of the paper is as follows. In Section 2, we give a brief review of the sampling based Bayesian framework. In Section 3.1, we provide a summary of inference for stable distributions ([Bu]), while in Section 3.2, we describe inference for univariate time series modeled by an ARMA(p, q) process with stable innovations. In Section 4, we show how to carry out inference for multiple time series modeled by a VARMA(p, q) process with symmetric stable innovations. In each case, we present illustrations of our approach.

2. Sampling based Bayesian Framework

Given data $X_n = (x_1, \ldots, x_n)$ along with model parameters $\Psi = (\psi_1, \cdots, \psi_k)$, the Bayesian specification requires a likelihood $L(X_n|\Psi)$ and a prior $\pi(\Psi)$, from which, using Bayes theorem, the posterior distribution for Ψ is obtained as $\pi(\Psi|X_n) \propto L(X_n|\Psi).\pi(\Psi)$. In situations where $L(X_n|\Psi)$ is not available in closed form, but the density $f(X_n, Y_n|\Psi)$, jointly with some extra random variables $Y_n = (y_1, \ldots, y_n)$ is available, then the posterior may be obtained by integrating out the unwanted variables, *viz.* $\pi(\Psi|X_n) \propto \int f(X_n, Y_n|\Psi)\pi(\Psi)dY_n$. When the data are drawn from a stable distribution, there is no analytical form for $\pi(\Psi|X_n)$. Numerical integration is a possibility, but this may also pose problems in high dimensions of the parameter space and for large n.

A recent approach to Bayesian inference is through the use of Markov chain Monte Carlo methods ([T]) whereby samples are drawn from the required posterior distributions. In particular, the Gibbs sampling algorithm, introduced by [GS] as a tool for Bayesian calculations, is a Markovian updating scheme developed earlier for use in image processing ([GG]); it requires sampling from the complete conditional distributions associated with Ψ in some systematic order. Note that Ψ might include, apart from model parameters, other unobservables such as auxiliary variables and missing values as well. Given starting values $\Psi^{(0)} = (\psi_1^{(0)}, \cdots, \psi_k^{(0)})$, we generate $\psi_1^{(1)}$ from $\pi(\psi_1|\psi_2^{(0)}, \cdots, \psi_k^{(0)}, X_n)$, $\psi_2^{(1)}$ from $\pi(\psi_2|\psi_1^{(1)}, \psi_3^{(0)}, \cdots, \psi_k^{(0)}, X_n)$, and so on up to $\psi_k^{(1)}$ from $\pi(\psi_k|\psi_1^{(1)}, \cdots, \psi_{k-1}^{(1)}, X_n)$, thus completing one iteration and yielding the sample $\Psi^{(1)} = (\psi_1^{(1)}, \cdots, \psi_k^{(1)})$. Under mild regularity conditions (see [Ti]), it can be shown that $\Psi^{(L)}$, the sample obtained after L such iterations tends, in distribution, to a sample from the joint posterior $\pi(\Psi|X_n)$ as L tends to infinity and that ergodic averages of suitable functions of these samples provide consistent estimates of features of interest of $\pi(\Psi|X_n)$. In some situations, it might be appropriate to group elements of Ψ so that $\Psi = (\Psi_{(1)}, \cdots, \Psi_{(K)})$ and then run the Gibbs sampler to sample elements within a group together.

When the complete conditional distributions have standard forms, sampling is straightforward using the customary Gibbs sampler. In more challenging cases, these emerge only as nonstandard, nonnormalized densities, in which case, the Metropolis-Hastings algorithm, which uses a Metropolis step ([MRRTT]) within each iteration of the Gibbs sampling algorithm ([CG] and [Ha]) may be used. For each draw, the Metropolis algorithm runs a conditional Markov chain whose stationary distribution is the required posterior density. Let U denote the current value of the ith group of parameters, say $\Psi_{(i)}$. A new draw V is made from a suitable candidate proposal density, centered at U. $\delta(U, V)$, the ratio of the posterior evaluated at V to that evaluated at U, holding all other parameters fixed, is computed. If $\delta(U, V) \geq 1$, we move to V; if $\delta(U, V) < 1$, we move to V with probability $\delta(U, V)$. Similar Metropolis schemes are run for each group of parameters for an appropriate number of subtrajectories.

Maximum likelihood estimates of the parameters, when available cheaply, provide reasonable initial starting values for the sampler, although in well behaved problems, convergence is usually achieved with arbitrary initial starting values that are relatively distant from the true values. Given the initial values, a single Markov chain is usually run for a fairly large number of iterations and convergence monitored. After the chain has converged, say at the jth iteration, a set of samples (of size N) is obtained by choosing every h^{th} sample, where the sample autocorrelations of the chain at lags $l \geq h$ are very small. Alternatively, the initial parameter estimates may be perturbed based on an overdispersion criterion (see [GR]) to start independent parallel Markov chains. The convergence of the iterative simulation may be monitored (see [GR] for details) and then samples collected from the converged chains to yield independent samples from the posterior density based on which various summary features of interest are obtained.

3. Univariate Time Series with Stable Innovations

3.1 Gibbs sampling in the *iid* case

[Bu] described sampling based Bayesian inference for the parameters of a stable distribution, based on n *iid* observations, motivated by problems in portfolio analysis. Given α and β, let $f(z, y)$ be a bivariate function such that it projects $(-\infty, 0) \times (-1/2, l_{\alpha,\beta}) \cup (0, \infty) \times (l_{\alpha,\beta}, 1/2)$ to $(0, \infty)$:

$$f(z, y|\alpha, \beta) = \frac{\alpha}{|\alpha - 1|} \exp[-|\frac{z}{t_{\alpha,\beta}(y)}|^{\alpha/(\alpha-1)}]|\frac{z}{t_{\alpha,\beta}(y)}|^{\alpha/(\alpha-1)} \frac{1}{|z|} \tag{3.2}$$

where $t_{\alpha,\beta}(y) = (\sin[\pi\alpha y + \eta_{\alpha,\beta}]/\cos\pi y)(\cos\pi y/\cos[\pi(\alpha - 1)y + \eta_{\alpha,\beta}])^{(\alpha-1)/\alpha}$, $\alpha \in (0, 1) \cup (1, 2]$, $\beta \in [-1, 1]$, $z \in (-\infty, \infty)$, $y \in (-1/2, 1/2)$, $\eta_{\alpha,\beta} = \beta \min(\alpha, 2 - \alpha)\pi/2$ and $l_{\alpha,\beta} = -\eta_{\alpha,\beta}/\pi\alpha$. [Bu] showed that (3.2) is a proper bivariate probability density function for (z, y) conditional on α and β and moreover, $z|\alpha, \beta \sim S(\alpha, \beta, 0, 1)$ with density

$$f(z|\alpha, \beta) = \frac{\alpha|z|^{1/(\alpha-1)}}{|\alpha - 1|} \int_{-\frac{1}{2}}^{\frac{1}{2}} \exp[-|\frac{z}{t_{\alpha,\beta}(y)}|^{\alpha/(\alpha-1)}]|\frac{1}{t_{\alpha,\beta}(y)}|^{\alpha/(\alpha-1)} dy. \quad (3.3)$$

Let $X_n = (x_1, \ldots, x_n)$ be *iid* observations from $S(\alpha, \beta, \delta, \sigma)$ and $Y_n = (y_1, \ldots, y_n)$ denote a vector of augmented variables. From (3.2), it is easily seen that

$$\pi(\alpha, \beta, \delta, \sigma|X_n) \propto (\frac{\alpha}{|\alpha - 1|})^n$$

$$\times \int \exp[-\sum_{i=1}^{n} |\frac{x_i - \delta}{\sigma t_{\alpha,\beta}(y_i)}|^{\alpha/(\alpha-1)}] \prod_{i=1}^{n} |\frac{x_i - \delta}{\sigma t_{\alpha,\beta}(y_i)}|^{\alpha/(\alpha-1)} \frac{1}{|x_i - \delta|} \pi(\alpha, \beta, \delta, \sigma) dY_n. \quad (3.4)$$

Assuming noninformative priors for all parameters and independence among them, parameter samples were generated from the respective conditional distributions $\pi(Y_n|\alpha, \beta, \delta, \sigma, X_n)$, $\pi(\alpha|\beta, \delta, \sigma, Y_n, X_n)$,$\pi(\beta|\alpha, \delta, \sigma, Y_n, X_n)$, $\pi(\sigma|\alpha, \beta, \delta, Y_n, X_n)$ and $\pi(\delta|\alpha, \beta, \sigma, Y_n, X_n)$ which are all easily derived from (3.4). The rejection method ([De]) was used for generating Y_n while the Metropolis-Hastings algorithm ([Ha]) was used to generate posterior samples of α, β and δ since their conditional distributions do not have standard forms. The transformation $v = t_{\alpha,\beta}(y)$ was employed in generating α and β while the transformation of $v = t_{\alpha,\beta}(y)/(x - \delta)$ was used in generating δ; [Bu] pointed out that these transformations gave posterior densities that were more spread out and unimodal. The posterior of $\sigma^{\alpha/(\alpha-1)}$ is inverse gamma, from which a simple transformation gives σ.

As a validation example, for 1,000 observations generated from a $S(1.7, .9, 125, 4)$ distribution, the Gibbs sampler was run for 20,000 iterations with starting point $(1.2, 0, 124, 5)$, with uniform priors on all the parameters and restricting $\alpha \in [1.1, 2]$. Adaptive rejection sampling envelopes ([Gi]) were used as the candidate generators for the conditional distributions of α, β and σ. These envelopes avoid the necessity of the user selecting the candidate generator; they adaptively fit to the required density. Convergence was reported after 5,000 iterations; the posterior means (and standard deviations) of α, β, δ and σ were 1.69 (.045), .91 (.056), 125.06 (.265) and 3.91 (.118) respectively. Scatterplots of the generated samples showed some dependence between α and σ and between σ and δ. The second example consisted of daily price returns of Abbey National shares between 31/7/91 and 8/10/91. The return at time t defined as $\rho_t = (p_{t-1} - p_t)/p_{t-1}$, where p_t is the share price on day t, was modeled using the stable distribution. Running the Gibbs sampler for 10,000 iterations, with starting point $(1.5, 0, .00029, .0065)$, posterior mean estimates for α, β, δ and σ of $1.61, -.55, 0.00053$ and 0.0079 confirmed the skewness and heavy tailed behavior of the data.

3.2 Implementation of Bayesian Inference for Time Series

A time series z_t is generated by an autoregressive moving average (ARMA) process of order p and q, ARMA (p, q) with mean μ and underlying stable innovations if

$$\Phi(B)(z_t - \mu) = \Theta(B)\epsilon_t \tag{3.5}$$

where B is the backward shift operator, i.e. $B^k z_t = z_{t-k}$, ϵ_t are independent and identically distributed as symmetric stable random variables with stability parameter α, scale parameter σ, skewness parameter β and location parameter δ fixed at 0. In (3.5), $\Phi(B) = 1 - \phi_1 B - \ldots - \phi_p B^p$ and $\Theta(B) = 1 - \theta_1 B - \ldots - \theta_q B^q$ are polynomials in B of degrees p and q respectively; we assume that the process is stationary and invertible (i.e. the roots of $\Phi(z) = 0$ and $\Theta(z) = 0$ all lie outside the unit circle) and identifiable (there are no common roots). The sampling based Bayesian approach enables simultaneous inference for the ARMA model parameters together with the parameters of the stable law. Our approach employs the Metropolis-Hastings algorithm to generate samples from the joint posterior distribution of all the model parameters and is an extension to infinite variance stable time series of the approach in [Bu] for independent observations.

Let $Z_n = (z_1, \cdots, z_n)$ denote n observations from an ARMA(p, q) process with stable innovations. Let $\Psi = (\alpha, \beta, \sigma)$, $\Lambda = (\mu, \Phi, \Theta)$, $Z_0 = (z_0, z_{-1}, \ldots, z_{1-p})$, $E_0 = (\epsilon_0, \epsilon_{-1}, \ldots, \epsilon_{1-q})$ and $Z_t = (z_1, \ldots, z_t)$. Expressing ϵ_t as $f_t(Z_0, E_0, Z_t, \Lambda)$ through (3.5) and since $\{\epsilon_t/\sigma\}_{t=1}^n$ are *iid* from $S(\alpha, \beta, 0, 1)$, the likelihood is obtained as a product of n marginal densities which have the form given in (3.3):

$$L(Z_n|\Psi, \Lambda) = \frac{\alpha^n}{|\alpha - 1|^n} \prod_{j=1}^{n} |\frac{f_j(Z_0, E_0, Z_j, \Lambda)}{\sigma^\alpha}|^{1/(\alpha-1)}$$

$$\times \int_{-\frac{1}{2}}^{\frac{1}{2}} \exp[-|\frac{f_j(Z_0, E_0, Z_j, \Lambda)}{\sigma t_{\alpha,\beta}(y_j)}|^{\alpha/(\alpha-1)}]|\frac{1}{t_{\alpha,\beta}(y_j)}|^{\alpha/(\alpha-1)} dy_j \tag{3.6}$$

where Z_0 and E_0 are set equal to the sample median and zero respectively.

Let $\pi(\Psi, \Lambda)$ denote the prior distribution on (Ψ, Λ). We assume that the prior distributions of all the parameters are independent and choose improper priors, i.e. $\pi(\alpha) = 1, 0 < \alpha \leq 2$, $\pi(\beta) = 1, -1 \leq \beta \leq 1$, $\pi(\mu) = 1, -\infty < \mu < \infty$, $\pi(\sigma) = 1/\sigma, 0 < \sigma < \infty$ and $\pi(\Phi, \Theta) = 1$ on the region of identifiability, stationarity and invertibility so that $\pi(\Psi, \Lambda) = \pi(\alpha)\pi(\beta)\pi(\sigma)\pi(\mu)\pi(\Phi, \Theta) = 1/\sigma$. By Bayes theorem, the posterior distribution of the parameters is

$$\pi(\Psi, \Lambda|Z_n) \propto \frac{\alpha^n}{\sigma|\alpha - 1|^n} \prod_{j=1}^{n} |\frac{f_j(Z_0, E_0, Z_j, \Lambda)}{\sigma^\alpha}|^{1/(\alpha-1)} \times$$

$$\int_{-\frac{1}{2}}^{\frac{1}{2}} \exp[-|\frac{f_j(Z_0, E_0, Z_j, \Lambda)}{\sigma t_{\alpha,\beta}(y_j)}|^{\alpha/(\alpha-1)}]|\frac{1}{t_{\alpha,\beta}(y_j)}|^{\alpha/(\alpha-1)} dy_j. \tag{3.7}$$

[QR2] verified that (3.7) is a proper posterior density. We generate samples from $\pi(\Psi, \Lambda | Z_n)$ through the complete conditional distributions associated with each component (or groups of components) of (Ψ, Λ). We use approximate maximum likelihood estimates, denoted by $(\alpha^{(0)}, \beta^{(0)}, \sigma^{(0)}, \mu^{(0)}, \Phi^{(0)}, \Theta^{(0)})$ as initial values for the Metropolis-Hastings algorithm, with initial covariances obtained from the corresponding Hessian matrix. For this, we obtain the approximate likelihood function directly by inverting the stable characteristic function (and not involving the auxiliary variable y) as

$$L^*(Z_n|\Psi, \Lambda) \propto$$

$$\begin{cases} \prod_{j=1}^{n} \int_0^\infty \exp[-(\sigma t)^\alpha] \cos(f_j(Z_0, E_0, Z_j, \Lambda)t - (\sigma t)^\alpha \beta \tan(\pi\alpha/2))dt & \text{if } \alpha \neq 1 \\ \prod_{j=1}^{n} \int_0^\infty \exp[-\sigma t] \cos(f_j(Z_0, E_0, Z_j, \Lambda) + 2\alpha\beta \ln t/(\pi))t dt & \text{if } \alpha = 1. \end{cases}$$

For the symmetric case when $\beta = 0$, the two expressions in the above equation coincide and simplify, so that

$$L^*(Z_n|\Psi, \Lambda) \propto \frac{1}{\sigma} \prod_{j=1}^{n} \int_0^\infty \exp[-(\sigma t)^\alpha] \cos(f_j(Z_0, E_0, Z_j, \Lambda)t)dt. \tag{3.8}$$

The logarithm of the likelihood function is absolutely integrable when $\alpha \in (\epsilon, 2]$ and is continuous when $\alpha \in (\epsilon, 1) \cup (1, 2]$, where ϵ is small and positive. Then it is straightforward to verify that the log likelihood function has continuous second partial derivatives at the stationary values so that Newton's method may be used to obtain approximate maximum likelihood estimates as well as the observed Fisher information.

Each iteration of the Gibbs sampler proceeds in the following way:
(i) Generate y_j from its complete conditional distribution $\pi(y|\alpha, (\beta, \mu, \Phi, \Theta), \sigma, Z_j)$, $j = 1, 2, \ldots, n$ using a rejection algorithm ([De]). Let $Y_n = (y_1, y_2, \ldots, y_n)$.
(ii) Generate α from its complete conditional distribution $\pi(\alpha|(\beta, \mu, \Phi, \Theta), \sigma, Z_n, Y_n)$, which has a nonstandard form, using the Metropolis-Hastings algorithm with a Gaussian proposal.
(iii) From the complete conditional distribution $\pi(\beta, \mu, \Phi, \Theta|\alpha, \sigma, Z_n, Y_n)$, again generate $(\beta, \mu, \Phi, \Theta)$ employing the Metropolis-Hastings algorithm with a Gaussian proposal.
(iv) Generate $\sigma^{\alpha/(\alpha-1)}$ from its complete conditional distribution $\pi(\sigma|\alpha, (\beta, \mu, \Phi, \Theta), Z_n, Y_n)$, which is an inverse Gamma distribution, from which σ is obtained by a simple transformation.

For details on the actual forms of these complete conditional distributions, the reader is referred to [QR2]. We mention a few points of interest here. First, in the time series case, $f_j(Z_0, E_0, Z_j, \Lambda)$ is used in place of z in equation (3.2). A useful transformation suggested by [Bu] *viz.* $v_j = t_{\alpha,\beta}(y_j)$ gives a more manageable form of the conditional posterior for α by supporting the parameter range more evenly and resulting in the density under the transformation being unimodal; however, as Buckle pointed out, it has the additional requirement of solving for $t_{\alpha,\beta}(y_j) = v_j$ for y_j , for $j = 1, \ldots, n$. Fortunately, due to the monotonicity of $t_{\alpha,\beta}(y_j)$, solving for each y_j value corresponding to v_j using an efficient algorithm such as Newton's method is not computationally very intensive. The conditional posterior for $(\beta, \mu, \Phi, \Theta)$ is also simplified and unimodal after the transformation $v_j = t_{\alpha,\beta}(y_j)/f_j(Z_0, E_0, Z_j, \Lambda)$.

3.3 Example

To illustrate our approach, we consider a data set consisting of $n = 394$ observations on daily stock prices of a retail store. A time series plot in Figure 3.3(a) reveals the presence of several large observations.

It might be argued that the large deviations present in the data are regularly spaced and are all positive so that a more general model than a linear time series model with stable innovations may be reasonable. However, for purposes of illustration, we fit ARMA models with stable $S(\alpha, 0, 0, \sigma)$ innovations using the Bayesian approach. In general, selection between several alternative and competing models might be made using a variety of approaches including EDA-style diagnostics (see [MRGP] for use with Gaussian ARMA models) or the conditional predictive ordinate or Bayes factors (see [PR] for use with Gaussian fractionally differenced ARMA models). Here, we present estimation results for five ARMA models with stable innovations as well as a comparison of residuals from a Gaussian innovations assumption and a stable innovations assumption for the AR(2) model.

Table 1 below presents the approximate MLE's together with their standard errors (which are used as initial estimates) as well as the posterior means and standard deviations of the parameters for AR(1), AR(2), MA(1), MA(2) and ARMA(1,1) models, obtained from an implementation of the sampling algorithm with 5,000 iterations, 6 replications and 40 Metropolis sub-trajectories within each Gibbs iteration. The posterior mean of α is approximately the same for all the models, about 1.5. Note that the posterior means are quite close to the MLE's, which is not surprising with the use of noninformative priors. Further, our experience showed that starting at different initial values than the MLE's, the Gibbs sampler converged to the posterior estimates in Table 1. For each of the ARMA models that we fit, we compared the residuals under a Gaussian innovations assumption and a stable innovations assumption.

Table 1: Parameter Estimates for the stock prices data under AR(1),AR(2),MA(1),MA(2),ARMA(1,1) models

Param. MLE(s.e) Mean(s.d)	AR(1)	AR(2)	MA(1)	MA(2)	ARMA(1,1)
α	1.38(0.07) 1.49(0.22)	1.41(0.08) 1.40(0.07)	1.39(0.07) 1.55(0.11)	1.48(0.07) 1.48(0.25)	1.43(0.08) 1.50(0.21)
μ	44.0(1.85) 44.58(2.22)	42.44(1.92) 42.41(1.42)	44.02(1.22) 46.52(1.32)	44.00(1.43) 45.42(1.78)	43.91(2.53) 38.86(2.88)
σ	11.12(0.62) 11.98(0.85)	11.12(0.66) 11.29(0.57)	11.08(0.57) 13.07(0.83)	11.01(0.58) 12.48(0.75)	10.95(0.63) 11.29(0.82)
ϕ_1	0.48(0.04) 0.48(0.04)	0.29(0.04) 0.30(0.04)	- -	- -	0.86(0.03) 0.87(0.03)
ϕ_2	- -	0.22(0.04) 0.22(0.04)	- -	- -	- -
θ_1	- -	- -	-0.32(0.04) -0.33(0.05)	-0.32(0.05) -0.28(0.05)	0.61(0.06) 0.61(0.05)
θ_2	- -	- -	- -	-0.20(0.04) -0.19(0.05)	- -

Table 2: Posterior means and standard deviations for VAR(1) models with $\alpha = 1.5$ and $\alpha = 1.0$

Model	VAR(1) with α=1.5		VAR(1) with α=1.0	
True Parameter	Post. mean	Post. stdev	Post. mean	Post. stdev
α(1.5)(1.0)	1.528	0.214	1.031	0.413
μ_1(-2.0)	-1.990	0.048	-1.966	0.094
μ_2(2.0)	1.998	0.132	1.994	0.018
σ_{11}(2.0)	2.304	0.911	2.472	1.320
$\sigma_{12} = \sigma_{21}$(0.0)	0.007	0.059	-0.009	0.066
σ_{22}(1.0)	1.132	0.429	1.130	0.554
ϕ_{11}(0.6)	0.594	0.008	0.583	0.001
ϕ_{12}(-0.8)	-0.745	0.013	-0.765	0.001
ϕ_{21}(0.3)	0.309	0.006	0.303	0.001
ϕ_{22}(-0.4)	-0.419	0.016	-0.407	0.003

Figure 3.3(b) shows a histogram which represents the empirical distribution of the residuals from fitting a Gaussian AR(2) model to the data with estimated parameters $\hat{\phi_1} = 0.544$, $\hat{\phi_2} = 0.356$, estimated mean 55.336 and estimated standard deviation 32.304. The smooth curve represents the estimated normal density centered at 0 and with standard deviation 32.304. Figure 3.3(c) presents the empirical distribution of the residuals from fitting an AR(2) model with stable innovations.

The symmetric stable density for the corresponding smooth curve is obtained by inverting the characteristic function (1.1) with $\beta = 0$ to get

$$f(x) = \int_0^\infty \exp[-(\sigma t)^\alpha] \cos(x - \delta) t dt / \pi.$$

Using the estimated $\hat{\alpha}$ and $\hat{\sigma}$ corresponding to the AR(2) model from Table 2, this expression is evaluated by numerical integration and yields the smooth curve in Figure 3.3(c). Based on this comparison, the AR(2) model with stable innovations provides a better fit to the data than the Gaussian AR(2) model.

4. Multivariate Time Series with Symmetric Stable Innovations

Multivariate autoregressive moving average (VARMA) models are widely used to model multiple time series in order to characterize the stochastic dependence within series and between series. Let $\tilde{Z}_t$ denote a k-variate time series generated by a VARMA(p, q) process

$$\Phi(B)(\tilde{Z}_t - \tilde{\mu}) = \Theta(B)\tilde{\epsilon}_t, \tag{4.9}$$

where $\Phi(B) = \mathrm{I} - \Phi_1 B - \cdots - \Phi_p B^p$ and $\Theta(B) = \mathrm{I} - \Theta_1 B - \cdots \Theta_q B^q$ are matrix polynomials of degrees p and q respectively and $\tilde{\mu} = (\mu_1, \ldots, \mu_k)$ is the location vector. Here we assume that Φ and Θ obey the usual stationarity and invertibility conditions. In order to ensure model identifiability, we further assume that Φ and Θ are left coprime and rank $(\Phi_p, \Theta_q) = k$ (see, *e.g.*, [Re],p.36). Let $\tilde{\Phi} = Vec(\Phi_1, ..., \Phi_p)$ and $\tilde{\Theta} = Vec(\Theta_1, ..., \Theta_q)$. In (4.9), $\tilde{\epsilon}_t$ are k-variate *iid* symmetric stable random variables, a definition of which follows.

The multivariate stable distribution ([P1], [P2] and [ST]) is a direct generalization of the definition of a univariate stable distribution so that all linear combinations of elements of the multivariate stable random variable are univariate stable and all linear combinations of multivariate stable variables are also multivariate stable. A k-dimensional random vector $\tilde{X} = (X_1, X_2, \ldots, X_k)$ with characteristic function $\phi_{\tilde{X}}(\tilde{t})$ where $\tilde{t} = (t_1, t_2, \ldots, t_k)$ follows a multivariate stable distribution if, and only if

$$\phi_{\tilde{X}}(\tilde{t}) = \exp(i\mu(\tilde{t}) - \sigma(\tilde{t})[1 + i\beta(\tilde{t})\omega(1, \alpha)]) \tag{4.10}$$

where the functions $\mu(\tilde{t})$, $\sigma(\tilde{t})$ and $\beta(\tilde{t})$ are such that for every scalar s, they must satisfy

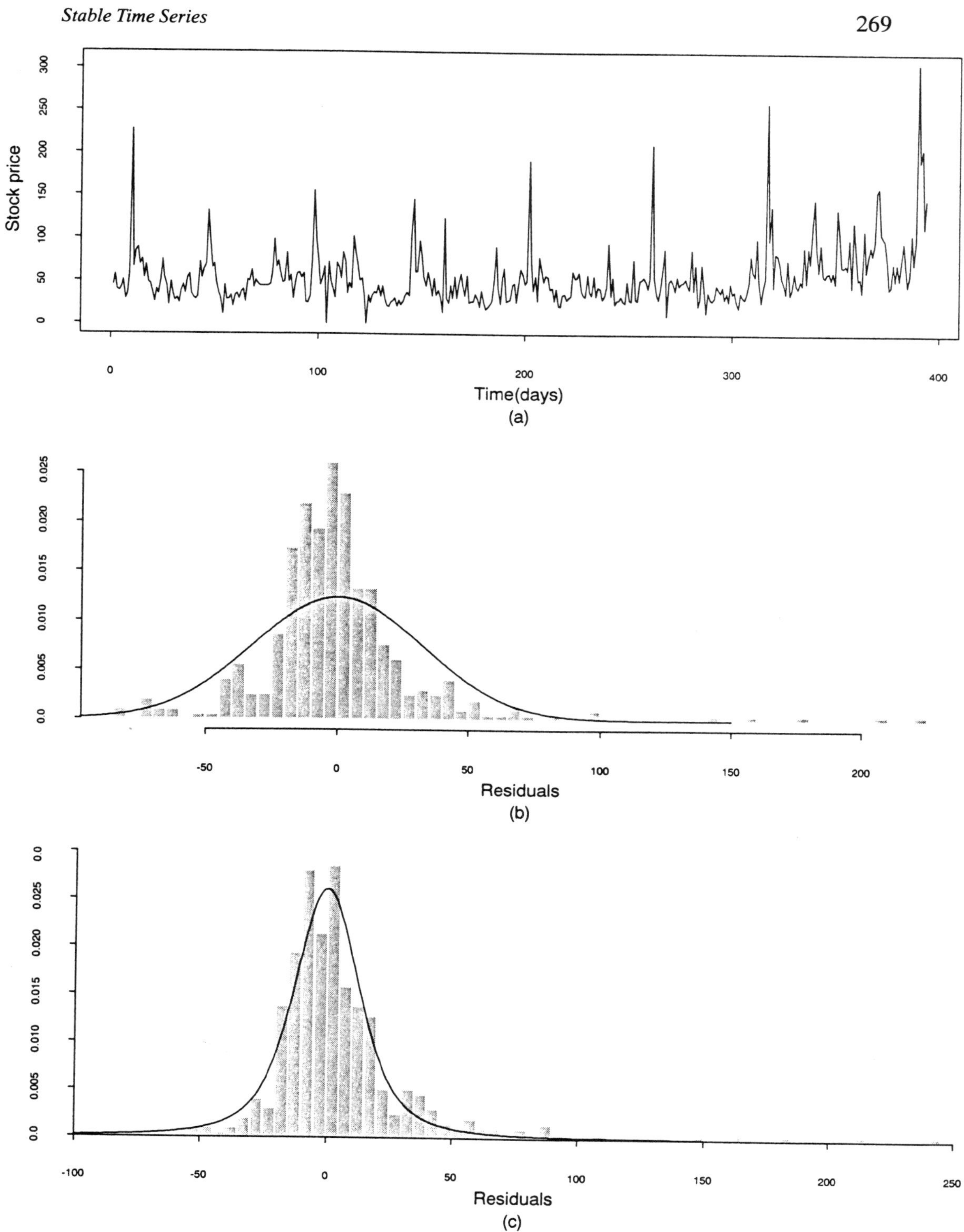

Figure 1: Modeling stock prices data.

$$\begin{array}{rcll} \mu(\tilde{t}s) & = & |s|^\alpha \mu(\tilde{t}) & \mu(\tilde{t}) \in (0, \infty) \\ \beta(\tilde{t}s) & = & \frac{s}{|s|}\beta(\tilde{t}) & \beta(\tilde{t}) \in [-1, 1] \\ \sigma(\tilde{t}s) & = & \sigma(\tilde{t})s - \mu(\tilde{t})\beta(\tilde{t})|s|^\alpha \frac{s}{|s|}[\omega(s,\alpha) - \omega(1,\alpha)] & \sigma(\tilde{t}) \in (-\infty, \infty) \end{array}$$

where $\omega = \tan(\alpha\pi/2)$ if $\alpha \neq 1$ and $\omega = 2\log|s|/\pi$ if $\alpha = 1$. Now consider the symmetric stable random vector with $\mu(\tilde{t})$ and $\sigma(\tilde{t})$ being defined as (see [P1]): $\mu(\tilde{t}) = \tilde{\mu}\tilde{t}^T$ and $\sigma(\tilde{t}) = \sum_{j=1}^m (\tilde{t}\Omega_j\tilde{t}^T)^{\alpha/2}/2$ where T denotes transpose, the Ω_i's are scale matrices and m is an unknown integer-valued scalar; its characteristic function has the very simple form

$$\phi_{\tilde{X}}(\tilde{t}) = \exp(i\tilde{\mu}(\tilde{t}^T) - \frac{\sum_{j=1}^m (\tilde{t}\Omega_j\tilde{t}^T)^{\alpha/2}}{2}). \tag{4.11}$$

When $m = 1$, the definition in [P2] of the multivariate symmetric stable random vector is equivalent to the definition of a sub-Gaussian random vector ([ST]). In what follows, we assume that the innovations from the VARMA process are multivariate symmetric stable with $m = 1$ (or equivalently sub-Gaussian) and describe how to carry out Bayesian inference using Markov chain Monte Carlo methods.

4.1 Form of the Multivariate Stable Density Function

Let A denote a univariate stable random variable with stability parameter $\alpha/2$, skewness parameter $\beta = 1$, scale parameter $\sigma_\alpha = [\cos(\pi\alpha/4)]^{2/\alpha}$ and location parameter $\delta = 0$. Let $\tilde{G} = (G_1, G_2, \ldots, G_k)$ be a zero mean Gaussian vector independent of A and let $\tilde{\mu} = (\mu_1, \mu_2, \ldots, \mu_k)$ be a fixed vector. Then the random vector $\tilde{X} = \tilde{\mu} + (A^{1/2}G_1, A^{1/2}G_2, \ldots, A^{1/2}G_k)$ has a k-dimensional symmetric stable distribution with characteristic function

$$E\exp\{i\sum_{j=1}^k t_jX_j\} = \exp\{i\sum_{l=1}^k \mu_l t_l - |\frac{1}{2}\sum_{j=1}^k\sum_{l=1}^k t_jt_l\sigma_{jl}|^{\alpha/2}\} \tag{4.12}$$

where $\sigma_{jl} = EG_jG_l$, $j, l = 1, \ldots, k$ are the covariances of the underlying Gaussian random vector $(G_1, G_2, \ldots, G_k)$; let $\Sigma = \{\sigma_{jl}\}$.

Consider a transformation $\tilde{G} = (\tilde{X} - \tilde{\mu})/A^{1/2}$ and $A = S$ with Jacobian $|J| = S^{-k/2}$. The joint density of $(\tilde{X}, S)$ is

$$f(\tilde{x}, s) = f(\tilde{g}, a)|J| = f_g(\frac{\tilde{x} - \tilde{\mu}}{s^{1/2}})f_s(s)s^{-k/2} \tag{4.13}$$

where

$$f_g(\frac{\tilde{x} - \tilde{\mu}}{s^{1/2}}) = \frac{1}{(2\pi)^{k/2}|\Sigma|^{1/2}}\exp[-\frac{1}{2}(\tilde{x} - \tilde{\mu})(\Sigma^{-1}/s)(\tilde{x} - \tilde{\mu})^T] \tag{4.14}$$

and

$$f_s(s) = \frac{\alpha}{|\alpha - 2|} \int \exp[-|\frac{s}{\sigma_\alpha t_\alpha(y)}|^{\alpha/(\alpha-2)}]|\frac{s}{\sigma_\alpha t_\alpha(y)}|^{\alpha/(\alpha-2)} \frac{1}{s} dy, \quad (4.15)$$

$t_\alpha(y) = t_{\alpha/2,\beta}(y)$ with $\beta = 1$ where $t_{\alpha,\beta}(y)$ is defined to be

$$t_{\alpha,\beta}(y) = (\sin[\pi\alpha y + \pi\beta \min(\alpha, 2-\alpha)/2])/(\cos(\pi y))$$
$$\times((\cos(\pi y)/(\cos[\pi(\alpha-1)y + \pi\beta\min(\alpha, 2-\alpha)/2]))^{(\alpha-1)/\alpha}$$

and $\sigma_\alpha = [\cos(\pi\alpha/4)]^{2/\alpha}$. If we further consider an augmented variable Y, we can write the joint density of $(\tilde{X}, S, Y)$ as

$$f(\tilde{x}, s, y) = \frac{1}{(2\pi)^{k/2}|\Sigma|^{1/2}} \frac{\alpha}{|\alpha-2|} |\frac{s}{\sigma_\alpha t_\alpha(y)}|^{\alpha/(\alpha-2)} \frac{1}{s^{k/2+1}}$$
$$\times \exp[-|\frac{s}{\sigma_\alpha t_\alpha(y)}|^{\alpha/(\alpha-2)} - \frac{1}{2}(\tilde{x} - \tilde{\mu})(\Sigma^{-1}/s)(\tilde{x} - \tilde{\mu})^T]. \quad (4.16)$$

4.2 Bayesian Inference through Gibbs Sampling

Let $\tilde{Z}_n = (\tilde{z}_1, \ldots, \tilde{z}_n)$ denote n observations from a k-variate VARMA process. Suppose we denote the prior on all the parameters by $\pi(\alpha, \tilde{\mu}, \Sigma, \tilde{\Phi}, \tilde{\Theta})$ and denote $\Theta^{-1}(B)\Phi(B)$ by $\Theta^{-1}\Phi(B)$; then from (4.16) and (4.9), the likelihood function

$$L(\tilde{Z}_n|\alpha, \tilde{\mu}, \Sigma, \tilde{\Phi}, \tilde{\Theta}) \propto (\frac{\alpha}{|\Sigma|^{1/2}|\alpha-2|})^n \prod_{j=1}^n \int\int |\frac{s_j}{\sigma_\alpha t_\alpha(y_j)}|^{\alpha/(\alpha-2)} \frac{1}{s_j^{k/2+1}}$$
$$\times \exp\{-(|\frac{s_j}{\sigma_\alpha t_\alpha(y_j)}|^{\alpha/(\alpha-2)} + \frac{1}{2}[\Theta^{-1}\Phi(B)(\tilde{z}_j - \tilde{\mu})](\Sigma^{-1}/s_j)[\Theta^{-1}\Phi(B)(\tilde{z}_j - \tilde{\mu})]^T)\}dy_j ds_j$$

where $Y_n = (y_1, \ldots, y_n)$ and $S_n = (s_1, \ldots, s_n)$. We assume a product prior specification, with $\pi(\alpha) = 1/2, 0 < \alpha \leq 2$, $\pi(\mu_i) = 1, -\infty < \mu_i < \infty, i = 1, 2, ..., k$, $\pi(\tilde{\mu}) = \prod_{i=1}^k \pi(\mu_i) = 1, \pi(\Sigma) = 1/|\Sigma|^{(k+1)/2}$ and $\pi(\tilde{\Phi}, \tilde{\Theta}) = \pi(\tilde{\Phi})\pi(\tilde{\Theta}) = 1$. Then, $\pi(\alpha, \tilde{\mu}, \Sigma, \tilde{\Phi}, \tilde{\Theta}) = \pi(\alpha)\pi(\tilde{\mu})\pi(\Sigma)\pi(\tilde{\Phi}, \tilde{\Theta}) = 1/(2|\Sigma|^{(k+1)/2})$. The joint posterior density of the parameters has the form

$$\pi(\alpha, \tilde{\mu}, \Sigma, \tilde{\Phi}, \tilde{\Theta}|\tilde{Z}_n) \propto \frac{1}{|\Sigma|^{(n+k+1)/2}}(\frac{\alpha}{|\alpha-2|})^n \times \prod_{j=1}^n \int\int |\frac{s_j}{\sigma_\alpha t_\alpha(y_j)}|^{\alpha/(\alpha-2)} \frac{1}{s_j^{k/2+1}} \exp\{$$
$$-(|\frac{s_j}{\sigma_\alpha t_\alpha(y_j)}|^{\alpha/(\alpha-2)} + \frac{1}{2}[\Theta^{-1}\Phi(B)(\tilde{z}_j - \tilde{\mu})]\frac{\Sigma^{-1}}{s_j}[\Theta^{-1}\Phi(B)(\tilde{z}_j - \tilde{\mu})]^T)\}dy_j ds_j. \quad (4.17)$$

[QR3] verified that this is a proper posterior density.

Although the above posterior density still involves a double integration, simulation through Markov chain Monte Carlo methods, together with the form of the joint density representation (4.16) will enable us to easily generate samples from the joint posterior density of the parameters in (4.17). The idea is to run the Gibbs sampler on an augmented vector of unknowns,viz. Y_n and S_n and generate samples from $\pi(\alpha, \tilde{\mu}, \Sigma, \tilde{\Phi}, \tilde{\Theta}|\tilde{Z}_n)$. This approach is an extension to multivariate time series of the idea in [Bu] for independent observations and the application of [QR2] to univariate time series. Given initial values $\alpha^{(0)}$, $\tilde{\mu}^{(0)}$, $\Sigma^{(0)}$, $\tilde{\Phi}^{(0)}$, $\tilde{\Theta}^{(0)}$ and $S_n^{(0)} = (s_1^{(0)}, \ldots, s_n^{(0)})$, we repeatedly generate samples of parameters α, $\tilde{\mu}$, Σ and $(\tilde{\Phi}, \tilde{\Theta})$ as well as the augmented vectors Y_n and S_n from their respective complete conditional posterior densities, viz., $\pi(Y_n|\alpha, \tilde{\mu}, \Sigma, \tilde{\Phi}, \tilde{\Theta}, \tilde{Z}_n, S_n)$, $\pi(\alpha|\tilde{\mu}, \Sigma, \tilde{\Phi}, \tilde{\Theta}, \tilde{Z}_n, Y_n, S_n)$, $\pi(S_n|\alpha, \tilde{\mu}, \Sigma, \tilde{\Phi}, \tilde{\Theta}, \tilde{Z}_n, Y_n), \pi(\Sigma|\alpha, \tilde{\mu}, \tilde{\Phi}, \tilde{\Theta}, \tilde{Z}_n, Y_n, S_n), \pi(\tilde{\mu}|\alpha, \Sigma, \tilde{\Phi}, \tilde{\Theta}, \tilde{Z}_n, Y_n, S_n)$ and $\pi((\tilde{\Phi}, \tilde{\Theta})|\alpha, \Sigma, \tilde{\mu}, \tilde{Z}_n, Y_n, S_n)$. The forms of these conditional densities are given in detail in [QR3].

Approximate maximum likelihood estimates, obtained by using the well-known EM algorithm (see [T], p.38-57) provide good initial values. Let $\tilde{\Psi}$ denote the vector of parameters. Using the EM algorithm, with (Y_n, S_n) being considered two latent vectors, the mode of the posterior distribution $\pi(\tilde{\Psi}|\tilde{Z}_n)$ (which is the approximate MLE under a noninformative prior specification) can be estimated by iterating the following two steps until convergence is achieved. For $k = 1, 2, \ldots$,
Step 1: Compute $Q(\tilde{\Psi}, \tilde{\Psi}^{\{k\}}) = \int_{Y_n, S_n} \log[\pi(\tilde{\Psi}|\tilde{Z}_n, Y_n, S_n)]\pi(Y_n, S_n|\tilde{\Psi}^{\{k\}}, \tilde{Z}_n)dY_n dS_n$.
Step 2: Find $\tilde{\Psi}^{\{k+1\}}$ such that $Q(\tilde{\Psi}^{\{k+1\}}, \tilde{\Psi}^{\{k\}}) = \max(Q(\tilde{\Psi}, \tilde{\Psi}^{\{k\}}))$ for all $\tilde{\Psi}$.
The first step involves a double integration with respect to each component of (Y_n, S_n). The computation of Q is carried out through a Monte-Carlo method (rather than numerical integration), since random variates from the conditional density of (Y_n, S_n) given $\tilde{\Psi}$ and $\tilde{Z}_n$ can be easily generated through Gibbs sampling. For details, see [QR3].

We use a Metropolis-Hastings algorithm with a linearly transformed Beta distribution as the proposal density since α is bounded both above and below and the density appears to be unimodal (see [Bu] for the *iid* case). The conditional posterior of s_j has a nonstandard form; the ratio of Uniforms method ([KM]), which requires the conditional density to be known only up to a proportionality constant and can besides easily deal with distributions with unbounded sample spaces, is appropriate. The conditional posterior density of Σ^{-1} is shown to be a Wishart distribution $W_k(B^{-1}, n+k+3)$ where $B = \sum_{j=1}^{n}[\Theta^{-1}\Phi(B)(\tilde{z}_j - \tilde{\mu})]^T[\Theta^{-1}\Phi(B)(\tilde{z}_j - \tilde{\mu})]/s_j$ from which the generation of samples of Σ is straightforward. It can also be shown that the conditional posterior of $\tilde{\mu}$ is normal with mean $(\sum_{j=1}^{n} \tilde{z}_j/s_j)\Phi^T(B)\Theta^{-T}(B)\Theta^T(1)\Phi^T(1)/(\sum_{j=1}^{n} 1/s_j)$ and variance $\Phi^{-1}(1)\Theta(1)\Sigma\Theta^T(1)\Phi^{-T}(1)/(\sum_{j=1}^{n} 1/s_j)$, from which the generation of $\tilde{\mu}$ is straightforward.

In general, the Metropolis-Hastings algorithm is used to generate $\tilde{\Phi}$ and $\tilde{\Theta}$ samples, using a Gaussian proposal with mean and covariance given by the mode and the Hessian at the mode of the likelihood function. The procedure simplifies considerably for the pure VAR(p) case with $q = 0$. Then the conditional density of $\phi_{j,l,m}$, the $\{l,m\}^{th}$ element of the jth coefficient matrix Φ_j, given all other parameters (denoted by *rest* below) is $\pi(\phi_{j,l,m}|rest) \sim N(\mu_{j,l,m}, \sigma_{j,l,m})$ where $\mu_{j,l,m} = [\sum_{i=1}^{n}(z_{i-j,m} - \mu_m)(a_i + b_i)/s_i]/[\sigma_{l,l}\sum_{i=1}^{n}(z_{i-j,m} - \mu_m)^2/s_i]$, $a_i = [(z_{i+1,l} - \mu_l - \sum_{u=1,u\neq m}^{k}\phi_{j,l,u}(z_{i-j,u} - \mu_u) - \sum_{u=1,u\neq j}^{p}\sum_{v=1}^{k}\phi_{u,l,v}(z_{i-u,v} - \mu_v)]\sigma_{l,l}$, $b_i = \sum_{u=1,u\neq l}^{k}[(z_{i+1,u} - \mu_u) - \sum_{v=1}^{p}\sum_{w=1}^{k}\phi_{v,u,w}(z_{i-v,w} - \mu_w)]\sigma_{l,u}$ $\sigma_{j,l,m} = 1/[\sigma_{l,l}\sum_{i=1}^{n}(z_{i-j,m} - \mu_m)^2/s_i]$, $(j = 1, \cdots, p)$, $(l = 1, \cdots, k)$, $(m = 1, \cdots, k)$.

4.3 Example

We illustrate our approach using bivariate time series data simulated from a VAR(1) model with multivariate sub-Gaussian innovations, generated as follows. We generate a univariate stable random variable $S(\alpha, \beta, \delta, \sigma)$ where $0 < \alpha < 1$, $\beta = 1$, $\delta = 0$ and $\sigma = [\cos(\alpha\pi/4)]^{2/\alpha}$ using the algorithm of [CMS]. Clearly, $S(\alpha, 1, 0, 1) = (a(U)/W)^{(1-\alpha)/\alpha}$ where $a(u) = \sin[(1-\alpha)u](\sin\alpha u)^{\alpha/(1-\alpha)}/[\sin(u)]^{1/(1-\alpha)}$, $0 < u < \pi$ and U is Uniform on $(0, \pi)$, W has a standard exponential density and U and W are mutually independent. The random variable we require, viz., $S(\alpha, 1, 0, \sigma)$ is then obtained as $\sigma S(\alpha, 1, 0, 1)$. Given these univariate stable innovations, we generate two sets of multivariate sub-Gaussian stable innovations following the definition in Section 4.1, corresponding to $\alpha = 1.5$ and $\alpha = 1.0$ respectively. In each case, $n = 1,000$, $\mu_1 = -2.0$, $\mu_2 = 2.0$,$\sigma_{11} = 2.0$,$\sigma_{12} = \sigma_{21} = 0.0$,$\sigma_{22} = 1.0$ where $\sigma_{11}, \sigma_{12}, \sigma_{21}, \sigma_{22}$ are the elements of the covariance matrix Σ and $\beta = 0$. Assuming that $\tilde{Z}_1 = \tilde{\epsilon}_1$, we generate $n = 1,000$ observations from a VAR(1) process with elements of Φ_1 given by $\phi_{11} = 0.6$,$\phi_{12} = -0.8$,$\phi_{21} = 0.3$ and $\phi_{22} = -0.4$ using the model function (4.9) with $p = 1, q = 0$. In each case, we generate samples from the joint posterior distribution of all the parameters, thus facilitating simultaneous inference in the Bayesian framework.

For $\alpha = 1.0$, the generated data has some very extreme observations compared to the $\alpha = 1.5$ case. The sampling algorithm is implemented with 10,000 iterations of a single chain. We select every fifth sample from the last 5,000 iterations to obtain a sample of 1,000 observations from the joint posterior distribution of the parameters. Table 2 presents the posterior means and standard deviations of the parameters. The posterior means are all quite close to the true parameter values used to simulate the data, which confirms the accuracy of our approach. The estimated marginal posterior densities of all the parameters except σ_{11} and σ_{22} appear to be fairly symmetric in both cases. Useful relationships between the parameters might be obtained from scatterplots of the posterior samples.

Based on 100 random posterior samples, Figure 2 and Figure 3 present selected scatter plots of α against the distinct elements of Σ, α versus the elements of Φ_1 and

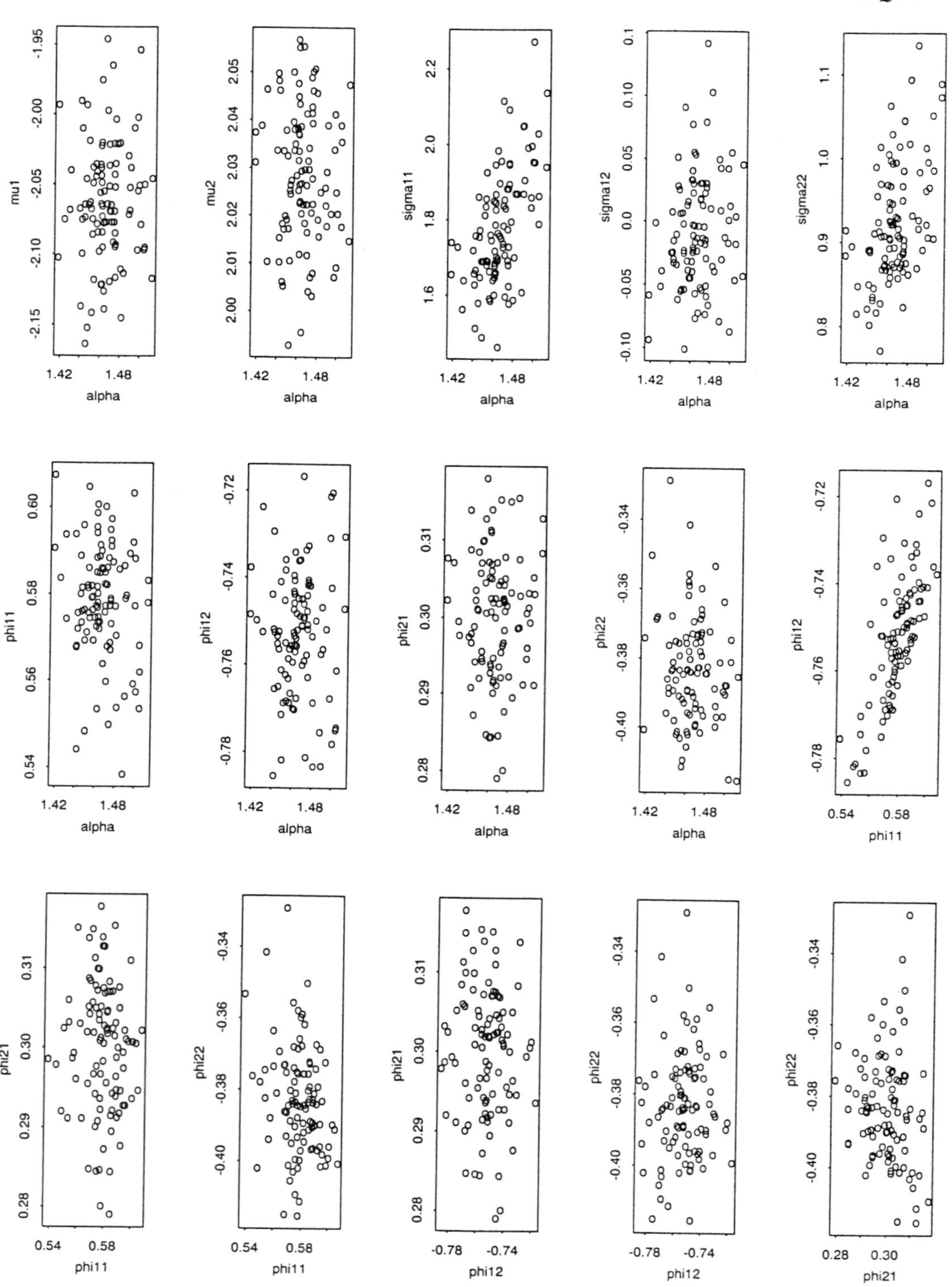

Figure 2: Scatter plots of parameter samples under VAR(1) model with $\alpha = 1.5$

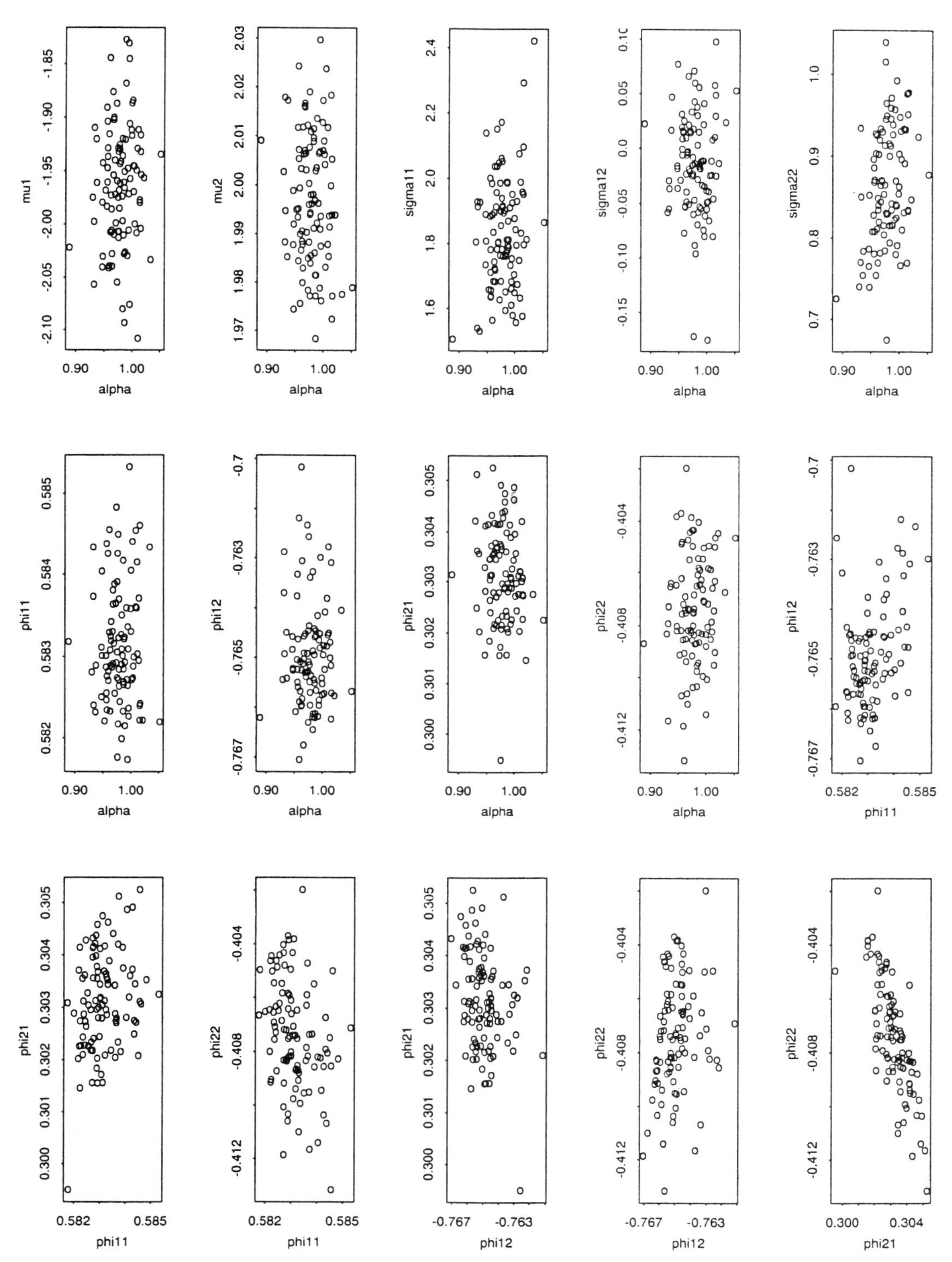

Figure 3: Scatter plots of parameter samples under VAR(1) model with $\alpha = 1.0$

the pairwise plots of the elements of Φ_1 with one another for $\alpha = 1.5$ and $\alpha = 1.0$ respectively. Note from Figure 2 that α is positively correlated with σ_{11} and σ_{22} (as expected), whereas there appears to be no correlation between α and the elements of Φ_1. Apart from a positive correlation between ϕ_{11} and ϕ_{12}, the scatterplots between the elements of Φ_1 show no useful information. Figure 3 presents similar information, with a positive correlation between ϕ_{11} and ϕ_{12} and a negative correlation between ϕ_{21} and ϕ_{22}.

Acknowledgements: This work was supported by an NSF Grant DMS 9510348. The authors are grateful to the editors and a referee for their very useful suggestions.

References

[A] R. W. Arad, Parameter estimation for symmetric stable distributions, *Int. Econ. Rev.*, **21** (1980), 209–220.

[Bh] R. J. Bhansali, Estimation of the Impulse Response Coefficients of a Linear process with infinite Variance, *J. Mult. Anal.*,**45** (1993), 274–290.

[Bu] D. J. Buckle, Bayesian Inference for Stable Distributions, *J. Amer. Statist. Assoc.*,**90** (1995), 605–613.

[CMS] J. M. Chambers, C. L. Mallows and B. W. Stuck, A Method for Simulating Stable Random Variables, *J. Amer. Statist. Assoc.*, **71** (1976), 340–344.

[CT] N. H. Chan and L. T. Tran, On the first order autoregressive process with infinite variance, *Econometric Theory,***5** (1989), 354-362.

[CD] M. H. Chen and D. K. Dey, Bayesian analysis of longitudinal binary data models using scale mixture of multivariate norml link functions, *Sankhya Ser. A*, (1998), to appear.

[CG] S. Chib and E. Greenberg, Understanding the Metropolis-Hastings algorithm, *Amer. Statist.*, **49** (1995), 327-335.

[CB] D. B. H. Cline and P. J. Brockwell, Linear prediction of ARMA processes with infinite variance, *Stoc. Proc. Appl.*,**19** (1985),281–296.

[De] L. Devroye, *Nonuniform Random Variate Generation*, New York: Springer-Verlag, 1986.

[Du1] W. H. DuMouchel, On the Asymptotic Normality of the Maximum Likelihood Estimate When Sampling from a Stable Distribution, *Ann. Statist.*,**1** (1973), 948–957.

[Du2] W. H. DuMouchel, Stable Distributions in Statistical inference:2. Information From Stably Distributed Samples, *J. Amer. Statist. Assoc.*,**70** (1975), 386–393.

[Du3] W. H. DuMouchel, Estimating the stable index α in order to measure tail thickness:a critique, *Ann. Statist.*, **11** (1983), 1019–1031.

[F] E. F. Fama,(1965). The Behavior of Stock Market Prices, *J. Business*,**38** (1965), 34–105.

[FR1] E. F. Fama and R. Roll, Some Properties of Symmetric Stable Distributions, *J. Amer. Statist. Assoc.*, **63** (1968), 817–837.

[FR2] E. F. Fama and R. Roll, Parameter estimates for Symmetric Stable Distributions, *J. Amer. Statist. Assoc.*, **66** (1971), 331–339.

[GS] A. E. Gelfand and A. F. M. Smith, Sampling based approaches to calculating marginal densities, *J. Amer. Statist. Assoc.*, **85** (1990), 398–409.

[GD] A. E. Gelfand and D. K. Dey, Bayesian model choice: Asymptotics and exact calculations, *J. Roy. Statist. Soc. B*, **55** (1993), 501-514.

[GR] A. Gelman and D. B. Rubin, Inference from iterative simulation using multiple sequences (with discussion), *Statist. Sci.*, **7** (1992), 457–511.

[GG] S. Geman and D. Geman, Stochastic Relaxation, Gibbs Distributions, and the Bayesian Restoration of Images, *IEEE Transactions on Pattern Analysis and Machine Intelligence*, PAMI-6 (1984), 721–741.

[Gi] W. R. Gilks, Derivative-Free Adaptive Rejection Sampling for Gibbs Sampling, in *Bayesian Statistics 4*, eds. J.M. Bernardo et al., Oxford, U.K.: Oxford University Press, (1992), 631–649.

[GK] B. V. Gnedenko and A. N. Kolmogorov, *Limit Distribution for Sums of Independent Random Variables*, trans. by K.L.Chung, Reading, Mass.: Addison-Wesley Publishing Co., 1954.

[GO] C. W. J. Granger and D. Orr, “ Infinite Variance ” and Research Strategy in Time Series Analysis, *J. Amer. Statist. Assoc.*, **67** (1972), 275–285.

[HK] E. Hannan and M. Kanter, Autoregressive processes with infinite variance, *J. Appl. Prob.*,**14** (1977), 411–415.

[Ha] W. K. Hastings, Monte Carlo sampling methods using Markov chains and their applications, *Biometrika*, **57** (1970), 97–109.

[He] C. R. Heathcote, The Integrated Squared Error Estimation of Parameters, *Biometrika*,**64** (1977), 255–264.

[Hi] B. M. Hill, A Simple General Approach to Inference about the Tail of a Distribution, *Ann. Statist.*, **3** (1975), 1163–1174.

[Ho] P. Houggard, A class of multivariate failure time distributions, *Biometrika*, **73** (1986), 671–678.

[KM] A. J. Kinderman and J. F. Monahan, Computer generation of random variables using the ratio of uniform deviates, *ACM Trans. of Math. Software*, **3** (1977), 257–260.

[K1] K. Knight, Rate of convergence of centered estimates of autoregressive parameters for infinite variance autoregressions, *J. Time Ser. Anal.*, **8** (1987), 51–60.

[K2] K. Knight, Limit Theory for Autoregressive-Parameter Estimates in an Infinite-Variance Random Walk, *Canad. J. Statist.*, **17** (1989), 261–278.

[Ko1] I. A. Koutrouvelis, Regression-type Estimation of the Parameters of Stable Laws, *J. Amer. Statist. Assoc.*, **75** (1980) 918–928.

[Ko2] I. A. Koutrouvelis, An Iterative Procedure for the Estimation of the Parameters of Stable Laws, *Commun. Statist.*, **B10** (1981), 17–28.

[Mc] J. H. McCulloch, Simple Consistent Estimators of Stable Distribution Parameters, *Commun. Statist.*, **B15**, (1986), 1109–1136.

[Ma] B. Mandelbrot, The Variation of Certain Speculative Prices, *J. Business*, **36** (1963), 394–419.

[MRGP] J. M. Marriott, N. Ravishanker, A. E. Gelfand and J. S. Pai, Bayesian analysis for ARMA processes: Complete sampling based inference under exact likelihoods, *Bayesian Statistics and Econometrics: Essays in honor of Arnold Zellner*, eds. D.Barry, K.Chaloner and J.Geweke, New York: John Wiley, 1996, 243–256.

[MRRTT] N. Metropolis, A. W. Rosenbluth, M. N. Rosenbluth, A. H. Teller and E. Teller, Equations of state calculations by fast computing machines, *J. Chem. Phy.*, **21** (1953), 1087–1091.

[N] D. K. Nassiuma, Symmetric stable sequences with missing observations, *J. Time Ser. Anal.*, **15** (1994), 313–323.

[PR] J. S. Pai and N. Ravishanker, Bayesian modeling of ARFIMA processes by Markov chain Monte Carlo methods, *J. Forecasting*, **15** (1996), 63–82.

[PHL] A. S. Paulson, E. W. Holcomb and R. A. Leitch, The Estimation of the Parameters of the Stable Laws, *Biometrika*,**62** (1975),163–170.

[PD] A. S. Paulson and T. A. Delahanty, Some Properties of Modified Integrated Squared Error Estimators for the Stable Laws, *Commun. Statist.*, **B13** (1984), 337–365.

[P1] S. J. Press, Estimation in Univariate and Multivariate Stable Distributions, *J. Amer. Statist. Assoc.*, **67** (1972), 842–846.

[P2] S. J. Press, Multivariate Stable Distributions, *J. Mult. Anal.*, **2** (1972), 444–462.

[QR1] Z. Qiou and N. Ravishanker, Bayesian inference for stable law parameters, Technical Report, University of Connecticut, (1995).

[QR2] Z. Qiou, and N. Ravishanker, Bayesian inference for time series with stable innovations, *J. Time Ser. Anal.*, **19** (1998), 235–249.

[QR3] Z. Qiou and N. Ravishanker, Bayesian inference for mutivariate time series with stable innovations, Technical Report, University of Connecticut, (1996).

[RR] N. Ravishanker and B. K. Ray, Bayesian analysis of Vector ARMA models using Gibbs sampling, *J. Forecasting*, **16** (1997), 177–194.

[Re] G. C. Reinsel, *Elements of Multivariate Time Series Analysis*, Springer Series in Statistics, New York: Springer-Verlag, 1993.

[Ro] G. Rosenfeld, Identification of Time Series with Infinite Variance, *Appl. Statist.*, **5** (1976), 147–153.

[ST] G. Samorodnitsky and M. S. Taqqu, *Stable Non- Gaussian Random Processes: Stochastic Models with Infinite Variance.* Chapman & Hall, New York, 1994.

[SK] B. W. Stuck and B. Kleiner, A statistical analysis of telephone noise, *Bell System Technical Journal*, **53** (1974), 1263–1312.

[T] M. Tanner, *Tools for statistical inference: Observed Data and Data Augmentation methods*, Lecture Notes in Statistics, New York: Springer-Verlag, 1993.

[TK] L. Tierney and J. B. Kadane, Accurate Approximations for Posterior Moments and Marginal Densities, *J. Amer. Statist. Assoc.*, **81** (1986), 82–86.

[Ti] L. Tierney, Markov chains for exploring posterior distributions (with discussion), *Ann. Statist.*, **22** (1994), 1701–1762.

[W] M. West, Approximating Posterior Distributions by Mixtures, *J. Roy. Statist. Soc.*, **B55** (1993), 409–422.

[YM] V. J. Yohai and R. A. Maronna, Asymptotic Behavior of Least-Squares Estimates for Autoregressive Processes with Infinite Variances, *Ann. Statist.*, **5** (1977), 554–560.

[Z] V. M. Zolotarev, *One- dimensional Stable Distributions*, Vol. 65 of "Translations of mathematical monographs", American Mathematical Society. Translation from the original 1983 Russian edition, 1986.

Nalini Ravishanker Department of Statistics
University of Connecticut
U-120, 196 Auditorium Road
Storrs, CT 06269, USA.
and
Zuqiang Qiou
Palisades Research Inc.
101 Bedford PI.
Morganville, NJ 07751, USA

III. Heavy-Tail Estimation

Hill, Bootstrap and Jackknife Estimators for Heavy Tails

Olivier V. Pictet, Michel M. Dacorogna and Ulrich A. Müller

Abstract

We perform an analysis of tail index estimation through Monte-Carlo simulations of synthetic data, in order to evaluate several tail estimators proposed in the literature. We derive and discuss the error of the Hill estimator under a general tail expansion of the distribution function. The analysis is extended to study the behavior of tail estimation under aggregation.

A detailed description is given of an algorithm designed to reduce the bias of the Hill estimator. This algorithm is based on a subsample bootstrap combined with the jackknife method. We show through simulations that this algorithm gives reasonable estimations of the tail index provided the number of observations is sufficiently large. The bias of the Hill estimator is successfully reduced. We also show that the estimation gives a constant tail index under aggregation up to an aggregation factor of 12. We recommend this method as a standard for tail estimation of empirical data.

1. Motivation

The purpose of this paper is to analyze and discuss the possibilities and the limitations of tail index estimation. We achieve this goal through two means: one, a careful formal derivation of the bias and the variance of the Hill estimator for a particular tail expansion of the distribution function; two, a set of Monte Carlo simulations. These simulations are performed under various distributions for which the theoretical tail index is known, and the results obtained for the estimator are compared to the theoretical ones.

This work originates from our interest in studying extreme risks in financial assets which are presented in a companion paper. Such studies are always faced with the paradoxical situation that extreme risks are by definition rare, whereas, significant statistical results can only be achieved if a sufficient number of these events can be analyzed. Unfortunately, the number of data in real cases is relatively limited. One possible solution to this problem is to use high frequency data for financial assets (Dacorogna et al., 1994) but the risk over 10 minutes is not relevant for a risk manager. He would like to be able to evaluate his risk over longer time periods. Therefore, it is important to be able to compute the behavior of the tail index under time aggregation if one wants to achieve a meaningful study of extreme risk in financial markets. In other words, we would like to find a way to extrapolate from the high-frequency data what is the risk at low frequency. There are only a few theoretical results on the problem of temporal aggregation related to tail behavior. Therefore, we turn to Monte Carlo simulations to explore this problem in detail.

In section 2, we present a Monte-Carlo study of a set of tail estimators as a function of their parameters under various theoretical distributions and conclude that the Hill estimator remains among the best estimators when it comes to fat-tailed distributions. In section 3, we concentrate on a careful derivation and discussion of the bias and the variance of the Hill estimator for a particular expansion of the distribution function of the tail. This derivation is presented from the point of view of a user discussing the approximations and their consequences on a possible application. In section 4, we develop a variation of Hall's bootstrap method (Hall, 1990) to determine the best parameters to compute the Hill estimator. In section 5, we examine the behavior of the Hill estimator for different types of fat-tailed distributions and as a function of the number of independent observations in the sample for the special case of temporal aggregation.

2. Tail Estimators

Before presenting in detail the theory of the Hill estimator, we present a simple analysis of a set of tail index estimators. It is is beyond the scope of this paper to derive tail estimators with full mathematical rigor[1]. We restrict ourselves to presenting their expressions as a function of the ordered values and refer the reader to the original papers[2].

Let $X_1, X_2, \ldots, X_n$ be a sequence of n observations drawn from a stationary i.i.d. process whose probability distribution function F is unknown. We assume that the distribution is fat-tailed, that is, the tail index α is finite. A good review of the definitions used in this paper can be found in (Leadbetter et al., 1983). Let us define $X_{(1)} \geq X_{(2)} \geq \ldots \geq X_{(n)}$ as the descending order statistics from $X_1, X_2, \ldots, X_n$.

Extreme value theory states that the extreme value distribution of the ordered data must belong to one of just three possible general families, regardless of the original distribution function F (Leadbetter et al., 1983). Besides, if the original distribution is fat-tailed there is only one general family it can belong to:

[1]For interested readers, good review articles on the subject are (de Haan, 1990; Hols and De Vries, 1991) who present the statistics and economic points of view respectively. A rigorous probability treatment may be found in (Leadbetter et al., 1983).

[2]In the presentation of these estimators, we follow the paper by Hols and de Vries (1991).

$$G(x) = \begin{cases} 0 & x \le 0 \\ \exp(-x^{-\alpha}) & x > 0, \quad \alpha > 0 \end{cases} \tag{1}$$

where $G(x)$ is the probability that $X_{(1)}$ exceeds x. There is only one parameter to estimate: α, which is also termed the tail index. The non-trivial stable distributions, the Student-t model and the unconditional distribution of the ARCH-process all fall in the domain of attraction of this type of distribution.

To give more intuition to the statements above, we plot the logarithm of the order statistics m as a function of the difference between the logarithms of the most extreme observation $\ln X_{(1)}$ and the other ordered observations $\ln X_{(m)}$. Such a plot is shown on Figure 1 for the case of a Student-t distribution with 4 degrees of freedom. Since we are in the domain of attraction of $\exp(-x^{-\alpha})$, it is trivial to see that the problem of estimating α becomes the problem of estimating the slope of the tangent at 0 of the curve shown in Figure 1. We see that a straight line with a slope equal to 4 is indeed a good tangent to the curve at 0, as it should be since the theoretical tail index of the Student-t distribution is equal to the number of degrees of freedom. Although the behavior of $\ln X_{(m)}$ is quite regular on Figure 1 because we took the average values over 10 replications of the data set, it is not always so and the problem of how to choose the number of points that are really in the tails is not trivial. One needs a more formal way to estimate the tail index. All the estimators we present below are different ways of estimating this slope.

We concentrate our efforts on the following estimators for $\gamma = 1/\alpha$:

1. The Pickands estimator:

$$\hat{\gamma}^P_{n,m} = \left[\ln \frac{X_{(m)} - X_{(2m)}}{X_{(2m)} - X_{(4m)}} \right] / \ln 2 \tag{2}$$

was first proposed in (Pickands III, 1975). The author also presents its weak consistency. When $m(n)$ increases suitably rapidly, Dekkers and De Haan (1989) obtain strong consistency and derive the asymptotic normality of $(\hat{\gamma} - \gamma)^{1/2}$ with zero mean and variance $\gamma^2(2^{2\gamma+1} + 1) \,/\, (2(2^\gamma - 1)\ln 2)^2$.

2. The Hill estimator:

$$\hat{\gamma}^H_{n,m} = \frac{1}{m-1} \sum_{i=1}^{m-1} \ln X_{(i)} - \ln X_{(m)} \quad \text{where } m > 1 \tag{3}$$

was first proposed in (Hill, 1975). It was proven to be a consistent estimator of $\gamma = 1/\alpha$ for fat-tailed distributions in (Mason, 1982). From (Hall, 1982) and (Goldie and Smith, 1987), it follows that $(\gamma_{n,m} - \gamma)m^{1/2}$ is asymptotically normal with mean zero and variance γ^2. For finite samples, however, the expectation value of the Hill estimator is biased, as shown in this paper. As long as this bias

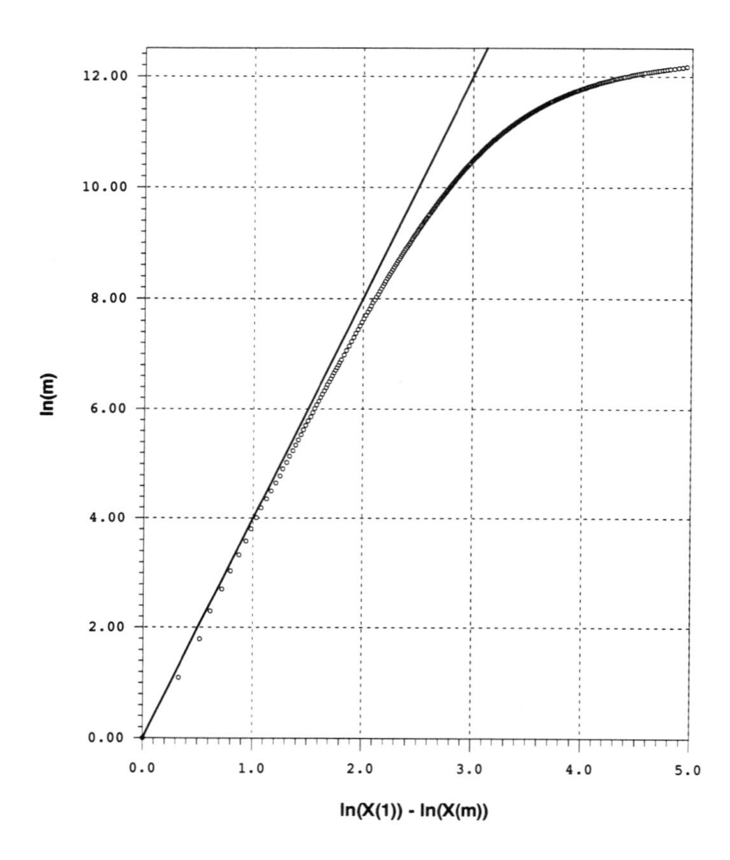

Figure 1: The logarithm of the order statistics m is plotted as a function of the difference between the logarithm of the most extreme observation and the logarithm of the ordered random observations. The data are drawn from a Student distribution with 4 degrees of freedom averaged over 10 replications. The straight line represents the theoretical tangent to this curve.

is unknown, the practical applicability of the Hill estimator for empirical samples is doubtful. This is motivation enough to study the bias of the Hill estimator (see the next section).

3. De Haan and Resnick estimator (de Haan and Resnick, 1980):

$$\hat{\gamma}^{R}_{n,m} = [\ \ln X_{(1)} - \ln X_{(m)}\ ]\ /\ \ln m \tag{4}$$

4. An extension of the Hill estimator proposed in (Dekkers et al., 1990):

$$\hat{\gamma}^{D}_{n,m} = \hat{\gamma}^{H\ (1)}_{n,m} + 1 - \frac{1}{2}\left\{1 - \frac{(\hat{\gamma}^{H\ (1)}_{n,m})^2}{\hat{\gamma}^{H\ (2)}_{n,m}}\right\}^{-1} \tag{5}$$

where $\hat{\gamma}^{H\ (1)}_{n,m}$ is the normal Hill estimator of eq. (3) and $\hat{\gamma}^{H\ (2)}_{n,m}$ is defined as the "second" moment:

$$\hat{\gamma}^{H\ (2)}_{n,m} = \frac{1}{m-1}\sum_{i=1}^{m-1}[\ \ln X_{(i)} - \ln X_{(m)}\ ]^2 \tag{6}$$

The consistency and normality of $\hat{\gamma}^{D}_{n,m}$ are proven in (Dekkers et al., 1990).

Except for the Pickands estimator, all the other estimators are only applicable in the case of fat-tailed distributions. They all require $m(n) \to \infty$, but it is not known how to choose m optimally for a finite sample. In section 4 we discuss a way of solving this problem. Here we compute these different estimators as a function of m, the number of tail observations for a set of distribution functions.

As noted above, this work was developed to explore the tail index of financial assets studied with high frequency data. Therefore, we simulate artificial time series corresponding to theoretical processes with well known tail indices. Each generated new random value is assumed to correspond to the change of the logarithmic price over the next ten minute interval[3]. We generate the random variable by inverting the cumulated distribution function for uniformly distributed probability values between 0 and 1. In this study the number of observations is 100, 000 per time series. For each distribution, we can compute many time series. The computation is done with 100 draws of 100, 000 observations. This means that we evaluate the tail estimators 100 times and we show on the plots the average results of these 100 simulations.

Note that the numbers that are generated by our numerical methods are precise up to 10^8. This will mostly affect values in the extreme tails. This limit forbids us to use a number of observations larger than a few hundred thousands without substantially changing the algorithms for evaluating the fractiles. Another remark of caution should

[3]We use the typical σ of 10 minutes returns of the foreign exchange rates.

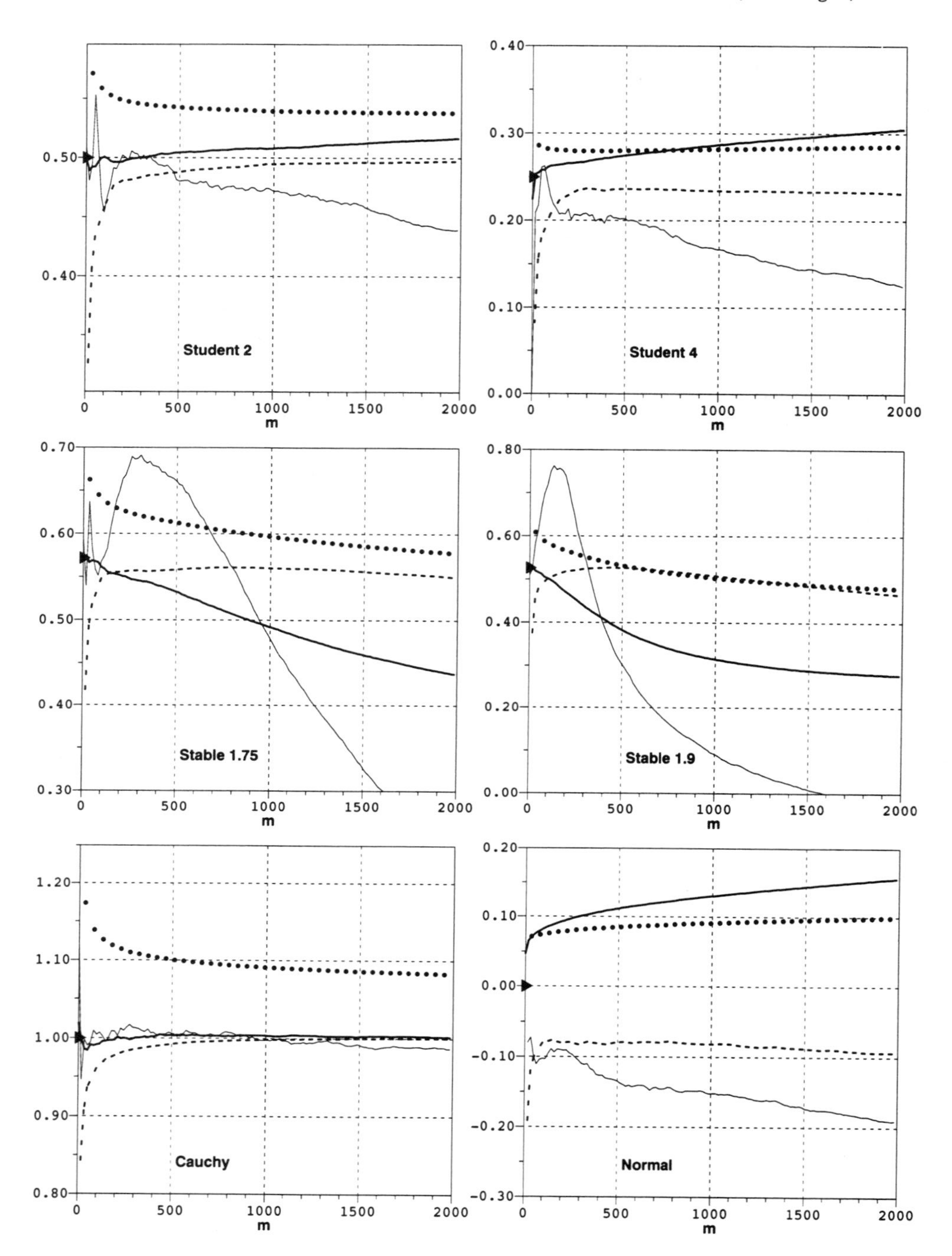

Figure 2: Comparison between the different tail estimators. The dashed line represents the Dekkers-Einmahl-De Haan estimator as a function of the maximum order statistics. The small black circles represent the De Haan and Resnick estimator. The thin line shows the Pickands estimator and the thick line the Hill estimator. The black triangles (▷) indicate the theoretical values of γ.

be made here: all these simulations require a very good random generator to obtain the probabilities between 0 and 1. Here we use Marsaglia's "Mother of all Pseudo-Random-Number-Generators". The code is available in the public domain and is claimed to have a period of 2^{250} and a precision of $0.4 \cdot 10^{-9}$ with 32 bits integers (Marsaglia and Zaman, 1994).

In the graphs of Figure 2, we compare the different indicators with two types of distributions: the stable distributions and the Student-t distributions with varying tail indices. We also present the case of the Gaussian distribution with $\alpha \to \infty$. The stable distributions have a tail index α varying from 1 to below 2 (chap 4.38 in (Kendall et al., 1987)). The limit $\alpha = 1$ corresponds to the Cauchy distribution and $\alpha \to 2$ corresponds to distributions approaching the normal distribution[4]. We compute the fractiles using the series expansions presented in (Bergström, 1952) and we choose $\alpha = 1.75,\ 1.9$ to see how the estimators vary when approaching a Gaussian. The tail index of a Student-t distribution is equal to the number of degrees of freedom ν. We choose to study Student-t distributions with $\nu = 2,\ 4$.

We show on the graphs the resulting values taken by $\gamma_{n,m}$ for the estimators of eqs. (2), (3), (4) and (5) using different numbers of tail observations m (maximum order statistics). The graphs are drawn with values of m up to 2000. This represents 2% of the total number of observations.

From these different graphs, it is clear that the De Haan-Resnick estimator is systematically biased upwards[5]. The Pickands estimator results in values oscillating around the theoretical value but it is very noisy and should not be used as such without some type of averaging. The Hill and the Dekkers-Einmahl-De Haan estimators give better results in all the theoretical cases examined. The Dekkers-Einmahl-De Haan estimator gives a reasonable answer for large m values although it generally presents a downward bias. The Hill estimator, for small values of m, also gives a reasonable answer. Its bias is, in this region, of the same order as the smallest bias given by the Dekkers-Einmahl-De Haan estimator. Moreover, there are many theoretical results on the Hill estimator that can be used for understanding its behavior. This is why we choose to concentrate on the Hill estimator in this study.

In Figure 3, we answer the question of how sensitive the Hill estimator is to the data. The variation of the Hill estimator over 100 simulations is shown by plotting the average values and their standard deviations. The typical behavior is apparent: very strong fluctuations when m approaches zero and a bias when m increases. This bias depends on the distributions. For the Cauchy distribution, the bias is very small. In

[4]In the case of the normal distribution, the index of stability has the value of 2, but it no longer equals the tail index α. The normal distribution has a tail index $\alpha = \infty$.

[5]In (Hall and Welsh, 1984), it is shown that this estimator does not achieve optimum rate of convergence unless certain restrictive conditions are imposed.

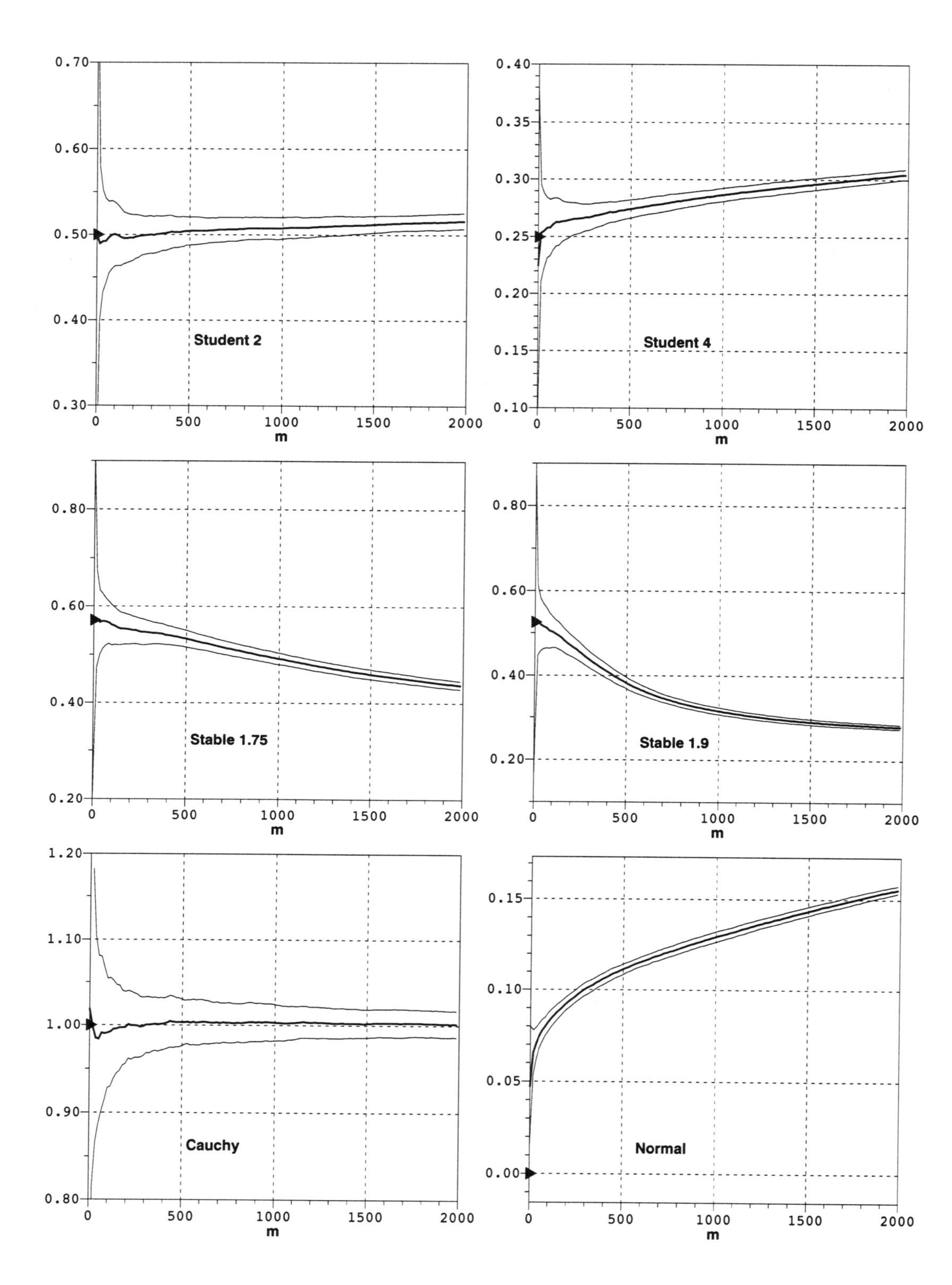

Figure 3: The Hill estimator is plotted as a function of the maximum order statistics. The thick curves represent the average over 100 simulations and the two thin curves the standard deviation around this average. The black triangles (▷) indicate the theoretical values of γ.

general, for distribution functions that are less fat-tailed in the same family the bias is stronger (Student with degree of freedom 4 and stable with $\alpha = 1.9$). For low m values, the estimator can no longer be trusted.

After this study, one question still remains: the choice of the maximum order statistic m to be used to compute the Hill estimator in order to optimally estimate the tail index. A first proposal was made of using the value m that would minimize the mean square error (MSE) between $\gamma_{n,m}$ and the theoretical value γ. Unfortunately, in most empirical studies of data, the theoretical value of the tail index is not known. To derive an optimal way of choosing m without *a priori* knowledge of the tail index, some assumptions on the underlying distribution function must be made. From Figure 3, it is clear that the best m value is a compromise between minimizing the variance of the estimator and the bias. In the next section, we discuss the behavior of the Hill estimator for a general tail expansion of the distribution function.

3. The Hill estimator

Let us assume n random observations X_i drawn from the same, unknown, fat-tailed distribution function. The center of the distribution is not too far from $X = 0$, which can always be obtained by subtracting the sample median from all X_i values. The Hill estimator (Hill, 1975) for the tail index of the distribution is defined in eq. (3) where the observations X_i are sorted in descending order:

$$X_{(i)} \geq X_{(i+1)} , \quad i = 1 \ldots n-1 . \tag{7}$$

The Hill estimator is based on the m largest observations. Eq. (3) is formulated as in (Hols and De Vries, 1991); an alternative formulation would have m replaced by $m+1$ everywhere. We write the estimator as $\gamma_{n,m}$ (dropping the superscript H for ease of notation).

In the following section, we compute and discuss the bias and the stochastic error variance of the Hill estimator for an asymptotic tail expansion of the distribution function. Similar results have been obtained by (Goldie and Smith, 1987) and (Hall and Welsh, 1985).

3.1 Expectation and variance of the Hill estimator

Let the cumulative distribution function, from which the observations X_i are drawn, be $F(x)$. Its density is $f(x) = \mathrm{d}F/\mathrm{d}x$. Let y be the value of the mth largest observation, $X_{(m)}$. The expectation value of the Hill estimator $\gamma_{n,m}$ can then be derived from eq. (3):

$$\mathrm{E}[\gamma_{n,m}|X_{(m)} = y] = \int_y^\infty \ln \frac{x}{y} \frac{f(x)}{1-F(y)} \,\mathrm{d}x \tag{8}$$

The unconditional expectation value $\mathrm{E}[\gamma_{n,m}]$ for a certain m is equal to the weighted integral over all values of y:

$$\mathrm{E}[\gamma_{n,m}] = \int_{-\infty}^{\infty} p_{n,m}(y)\, \mathrm{E}[\gamma_{n,m} | X_{(m)} = y]\, \mathrm{d}y \tag{9}$$

The probability $p_{n,m}(y)\mathrm{d}y$ of $X_{(m)}$ lying between y and $y + \mathrm{d}y$ is a well-known result of order statistics, see for example eq. (2.1.6) of (David, 1981). In our notation, this is

$$p_{n,m}(y) = \frac{n!}{(m-1)!\,(n-m)!}\, F^{n-m}(y)\, [1 - F(y)]^{m-1}\, f(y) \tag{10}$$

If we insert asymptotic expansions for $F(y)$ and $f(y)$, $p_{n,m}(y)$ will also have unknown high order terms. Therefore, the following form of eq. (9) will be more suitable:

$$\mathrm{E}[\gamma_{n,m}] = \frac{\int_{-\infty}^{\infty} p_{n,m}(y)\, \mathrm{E}[\gamma_{n,m} | X_{(m)} = y]\, \mathrm{d}y}{\int_{-\infty}^{\infty} p_{n,m}(y)\, \mathrm{d}y} \tag{11}$$

The expected variance of the Hill estimator about the mean value given by eq. (11) can be written as follows:

$$\mathrm{E}\{[\gamma_{n,m} - \mathrm{E}[\gamma_{n,m}]]^2\} = \frac{\int_{-\infty}^{\infty} p_{n,m}(y)\, \mathrm{E}[\gamma_{n,m}^2 | X_{(m)} = y]\, \mathrm{d}y}{\int_{-\infty}^{\infty} p_{n,m}(y)\, \mathrm{d}y} - \mathrm{E}^2[\gamma_{n,m}] \tag{12}$$

The expectation of the squared Hill estimator can be expressed with the help of eq. (3):

$$\mathrm{E}[\gamma_{n,m}^2 | X_{(m)} = y] = \frac{1}{(m-1)^2}\, \mathrm{E}[(\sum_{j=1}^{m-1} \ln \frac{X_{(j)}}{y})^2] \tag{13}$$

This is valid for $X_{(m)} = y$; there are $m - 1$ independent observations $\geq y$, randomly distributed with density $f(x)$. The square on the right hand side can be multiplied out and the square terms separated from the mixed product terms:

$$\mathrm{E}[\gamma_{n,m}^2 | X_{(m)} = y] =$$

$$\frac{1}{(m-1)^2}\, \{\mathrm{E}[\sum_{j=1}^{m-1} \ln^2 \frac{X_{(j)}}{y}] + \sum_{j=1}^{m-1} [\mathrm{E}[\ln \frac{X_{(j)}}{y}]\, \mathrm{E}[\sum_{j'=1, j' \neq j}^{m-1} \ln \frac{X_{(j')}}{y}]]\} \tag{14}$$

All expectation values can now be expressed by integrals, again because of the i. i. d. nature of the $m - 1$ observations exceeding y. The last sum of eq. (14) includes only $m - 2$ terms because the term with $j' = j$ is excluded. We obtain

$$\mathrm{E}[\gamma_{n,m}^2 | X_{(m)} = y] = \frac{1}{m-1} \int_y^{\infty} \ln^2 \frac{x}{y}\, \frac{f(x)}{1 - F(y)}\, \mathrm{d}x + $$

$$+ \frac{m-2}{m-1} [\int_y^{\infty} \ln \frac{x}{y}\, \frac{f(x)}{1 - F(y)}\, \mathrm{d}x]^2 \tag{15}$$

In order to compute the expectation values of the Hill estimator and its variance about the mean given by eqs. (11) and (12), we need to assume more about the distribution functions $F(x)$ and $f(x)$.

3.2 The tail expansion

The behavior of the positive, fat tails of many theoretical and empirical cumulative distributions can be approximated by an asymptotic series expansion. The center of the distribution can always be placed (by a constant shift of the x coordinates) not too remote from 0 so as to have $F(0) = \frac{1}{2}$, but the relation between the center of the distribution and the tail is unknown. The translation of x coordinates may have an effect on the tail estimation that should be studied. Therefore, we propose the following expansion,

$$F(x) = 1 - a\,(x-c)^{-\alpha}\,[1 + b\,(x-c)^{-\beta} + o((x-c)^{-\beta})] \tag{16}$$

where o is the Landau symbol "of order higher than"

$$a > 0\,,\quad \alpha > 0\,,\quad \beta > 0, \tag{17}$$

and the parameters b and c are real numbers. The parameter α is the tail index of this distribution.

The form of eq. (16) is similar to the one proposed in eq.(1) of (Hall, 1982) except for the shift parameter c which reflects the fact that we do not know where on the x-axis the tail starts and the expansion becomes valid.

The positive tail of the distribution has $x \gg |c|$. This leads to the following expansions:

$$(x-c)^{-\alpha} = x^{-\alpha}\,[1 + c\,\alpha\,x^{-1} + o(x^{-1})] \tag{18}$$

and

$$(x-c)^{-\beta} = x^{-\beta}\,[1 + c\,\beta\,x^{-1} + o(x^{-1})] = x^{-\beta} + o(x^{-\beta}) \tag{19}$$

Inserting this in eq. (16) yields

$$F(x) = 1 - a\,x^{-\alpha}\,[1 + b\,x^{-\beta} + c\,\alpha\,x^{-1} + o(x^{-\beta} + x^{-1})] \tag{20}$$

where the orders of the two terms with $x^{-\beta}$ and x^{-1} have to be compared. We can write

$$F(x) = 1 - a\,x^{-\alpha}\,[1 + b^*\,x^{-\beta^*} + o(x^{-\beta^*})] \tag{21}$$

where

$$\beta^* = \min(\beta,\ 1) \tag{22}$$

and

$$b^* = \begin{cases} b & \text{if}\ \ \beta < 1 \\ b + c\,\alpha & \text{if}\ \ \beta = 1 \\ c\,\alpha & \text{if}\ \ \beta > 1 \end{cases} \tag{23}$$

Eq. (21) is formally identical to eq. (1) of (Hall, 1982), but with the important restriction that β^* is now limited to 1, see eq. (22). A β^* larger than 1 is possible only in the case $c = 0$ which is very unlikely for empirical data. We expect $\beta^* = 1$ to be a frequently occurring case.

For the ease of notation, we shall write β instead of β^* and b instead of b^*:

$$F(x) \;=\; 1 - a\,x^{-\alpha}[1 + b\,x^{-\beta} + o(x^{-\beta})] \tag{24}$$

bearing in mind that β is restricted by eq. (22).

The following, even simpler formulation helps to simplify the notation of many intermediary results in some sections below:

$$F(x) \;=\; 1 - a\,x^{-\alpha}[1 + b'\,x^{-\beta}] \tag{25}$$

where

$$b' \;=\; b\,[1 + o(1)] \;=\; b\,[1 + o(x^0)] \tag{26}$$

3.3 The expectation of the Hill estimator in asymptotic expansion

The expectation of the Hill estimator for a fixed $X_{(m)} = y$ is given by eq. (8). This equation contains the probability density $f(x)$ which can be obtained by deriving eq. (24) for $F(x)$ against x. However, this derivative includes the derivative of the term $o(x^{-\beta})$ which may be arbitrarily large in some cases[6]. Therefore, eq. (8) has to be solved by other means than inserting an expansion of $f(x)$. We apply the product rule of integration to obtain

$$\mathrm{E}[\gamma_{n,m}|X_{(m)} = y] \;=\; \int_y^\infty \ln\frac{x}{y}\,\frac{f(x)}{1-F(y)}\,\mathrm{d}x \;=$$

$$\frac{1}{1-F(y)}\,\lim_{R\to\infty}\left[F(R)\,\log\frac{R}{y} \;-\; \int_y^R \frac{F(x)}{x}\,\mathrm{d}x\right] \tag{27}$$

Inserting eq. (25) yields

$$\mathrm{E}[\gamma_{n,m}|X_{(m)} = y] \;=\; \frac{1}{\alpha}\,(1 - \frac{\beta\,b'}{\alpha+\beta}\,y^{-\beta}) \tag{28}$$

The unconditional expectation of the Hill estimator is given by eq. (11) with the probability $p_{n,m}(y)$ from eq. (10). In this latter equation, one factor can be expanded straight away with the help of eqs. (25) and (26):

$$[1 - F(y)]^{m-1} \;=\; a\alpha y^{\alpha(m-1)}\,[1 + b'\,(m-1)\,y^{-\beta}] \tag{29}$$

[6] $o(x^{-\beta}) \;=\; x^{-\beta-0.01}\sin(x^4)$, for example

Another factor of eq. (10), $F^{n-m}(y)$, can be analyzed by inserting eq. (25) and applying the logarithm and then the exponential function to the result:

$$F^{n-m}(y) = \exp\{(n-m)\ \ln[1 - a\ y^{-\alpha}(1 + b'\ y^{-\beta})]\} \tag{30}$$

The logarithm can be expanded: $\ln(1-x) = -\sum_{k=1}^{\infty}(x^k/k)$, see e. g. (Gradshteyn and Ryzhik, 1980), eq. (1.511). We obtain

$$F^{n-m}(y) = \mathrm{e}^{-a(n-m)y^{-\alpha}} \times$$
$$\times \exp\{-a\ (n-m)\ y^{-\alpha}\ [\sum_{k=1}^{\infty} \frac{1}{k+1}\ y^{-k\alpha} + b'\ y^{-\beta}]\} \tag{31}$$

The exponential function of the second factor can be expanded again:

$$F^{n-m}(y) = \mathrm{e}^{-a(n-m)y^{-\alpha}}\ [1 - a\ (n-m)\ y^{-\alpha}\ (\sum_{k=1}^{\infty} c_k\ y^{-k\alpha} + b'\ y^{-\beta})] \tag{32}$$

The coefficients c_k might be explicitly computed from the expansion of the exponential function, but this is not necessary here. The higher order terms with exponents containing β are again included by the term with factor b', in line with eq. (26). Now, all the factors of eq. (10) can be inserted to give

$$p_{n,m}(y) = a^2\,\alpha\ \frac{n!}{(m-1)!\ (n-m)!}\ \mathrm{e}^{-a(n-m)y^{-\alpha}}\ y^{-m\alpha-1} \times$$
$$\times [1 - a\ (n-m) \sum_{k=1}^{\infty} c_k\ y^{-(k+1)\alpha} - a\ (n-m)\ b'\ y^{-\alpha-\beta} +$$
$$+ b'\ (m - 1 + \frac{\alpha+\beta}{\beta})\ y^{-\beta}] \tag{33}$$

The high order term with $y^{-\alpha-\beta}$ is retained here because it is multiplied by the factor $n-m$ which can be large. In eq. (11), the expectation of the Hill estimator is computed with integrals over $p_{n,m}(y)\mathrm{d}y$. Now, this integration is done with an asymptotic expansion of only the positive tail; the probability of y being negative is zero. Therefore, we replace the lower integration limit $-\infty$ by 0. This is the natural choice also because $p_{n,m}(y)$ as given by eq. (33) reaches the value 0 at that point: the exponential function dominates all the other factors when $y \to 0$.

The integration over $p_{n,m}(y)\mathrm{d}y$ is done by using the following integral from (Gradshteyn and Ryzhik, 1980), eq. (3.478.1):

$$\int_0^{\infty} x^{\nu-1}\ \mathrm{e}^{-\mu x^{\alpha}}\mathrm{d}x = \frac{1}{\alpha}\ \mu^{-\frac{\nu}{\alpha}}\ \Gamma(\frac{\nu}{\alpha}) \tag{34}$$

followed by a substitution $x = 1/y$. The validity of this integral is conditional to

$$\mathrm{Re}\,\mu > 0 \;, \quad \mathrm{Re}\,\nu > 0 \;, \quad \alpha > 0 \tag{35}$$

These conditions are fulfilled for the integrals of eq. (11). The denominator of that equation can be integrated by inserting eq. (33), with the help of eq. (34):

$$P = \int_0^\infty p_{n,m}(y)\,\mathrm{d}y = a^2 \frac{n!}{(m-1)!\,(n-m)!} \times$$
$$\times \{ [a\,(n-m)]^{-m}\,\Gamma(m) - \sum_{k=2}^{\infty} c_k [a\,(n-m)]^{-m-k}\,\Gamma(m+k) -$$
$$- b'\,[a\,(n-m)]^{-m-\frac{\beta}{\alpha}}\,\Gamma(m+\frac{\beta}{\alpha}) +$$
$$+ b'\,(m-1+\frac{\alpha+\beta}{\beta})\,[a\,(n-m)]^{-m-\frac{\beta}{\alpha}}\,\Gamma(m+\frac{\beta}{\alpha}) \} \tag{36}$$

This is still an asymptotic series expansion: a large sample expansion in the variable $a(n-m)$. The quantity b' defined by eq. (26) takes the additional interpretation

$$b' = b\,[1+o(1)] = b\,\{1+o[\,[a\,(n-m)]^0]\} \tag{37}$$

The numerator of eq. (11) is similar to the denominator, but has the additional factor $\mathrm{E}[\gamma_{n,m}|X_{(m)} = y]$ in the integrand. By inserting eq. (28), the integral becomes similar to eq. (36):

$$\int_0^\infty p_{n,m}(y)\,\mathrm{E}[\gamma_{n,m}|X_{(m)} = y]\,\mathrm{d}y =$$
$$= \frac{1}{\alpha}\,\{P - a^2 \frac{n!}{(m-1)!\,(n-m)!}\,\frac{\beta\,b'}{\alpha+\beta}\,[a(n-m)]^{-m-\frac{\beta}{\alpha}}\,\Gamma(m+\frac{\beta}{\alpha})\} \tag{38}$$

Inserting in eq. (11) yields

$$\mathrm{E}[\gamma_{n,m}] = \frac{1}{\alpha}\,\{1 - \frac{a^2}{P}\,\frac{n!}{(m-1)!\,(n-m)!}\,\frac{\beta\,b'}{\alpha+\beta} \times$$
$$\times [a\,(n-m)]^{-m-\frac{\beta}{\alpha}}\,\Gamma(m+\frac{\beta}{\alpha})\} \tag{39}$$

P can be expressed by the lowest order of eq. (36) as all higher order terms are already included by the factor b' here. We obtain the result

$$\mathrm{E}[\gamma_{n,m}] = \frac{1}{\alpha}\,\{1 - \frac{\beta\,b'}{\alpha+\beta}\,\frac{\Gamma(m+\frac{\beta}{\alpha})}{\Gamma(m)}\,[a\,(n-m)]^{-\frac{\beta}{\alpha}}\} = \frac{1}{\alpha} + B \tag{40}$$

This shows that the Hill estimator is a biased estimator for $1/\alpha$. The bias or systematic error B is defined as follows:

$$B = \mathrm{E}[\gamma_{n,m}] - \frac{1}{\alpha} = -\frac{1}{\alpha}\frac{\beta\, b'}{\alpha+\beta}\frac{\Gamma(m+\frac{\beta}{\alpha})}{\Gamma(m)}\,[a\,(n-m)]^{-\frac{\beta}{\alpha}} \tag{41}$$

This equation can be expanded in two variables simultaneously: m/n and $1/m$. For the quotient of Γ functions, eq. (6.1.47) of (Abramowitz and Stegun, 1970) gives the following expansion, derived from Stirling's formula:

$$\frac{\Gamma(m+\frac{\beta}{\alpha})}{\Gamma(m)} = m^{-\frac{\beta}{\alpha}}\,\{1+\frac{1}{2}\frac{\beta}{\alpha}\,(\frac{\beta}{\alpha}-1)\,\frac{1}{m}+O[(\frac{1}{m})^2]\} \tag{42}$$

with the Landau symbol O "term of this order", unlike o in eq. (16). The factor $[a(n-m)]^{-\frac{\beta}{\alpha}}$ is expanded in m/n; b' as given by eq. (37) is expanded in both m/n and $1/m$. All expansions are inserted in eq. (41):

$$B = -\frac{1}{\alpha}\frac{\beta\, b}{\alpha+\beta}\,a^{-\frac{\beta}{\alpha}}\,(\frac{m}{n})^{\frac{\beta}{\alpha}}\,\{1+O(\frac{1}{m})+O(\frac{m}{n})+o[(\frac{1}{m})^0\,(\frac{m}{n})^0]\} \tag{43}$$

We need all the three order terms. The mixed term $o[(1/m)^0\,(m/n)^0]$ is expected to dominate in most cases but for very small m, $O(1/m)$ may take over and for very large m, $O(m/n)$ may become dominant.

3.4 The variance of the Hill estimator in asymptotic expansion

The variance of the Hill estimator about its mean is given by eq. (12). One quantity needed to compute this is the expectation of the squared Hill estimator conditional to $X_{(m)} = y$ as given by eq. (15): a sum of two integrals. The first of these integrals can be solved by applying the product rule of integration twice, in the same manner as in eq. (27):

$$\int_y^\infty \ln^2\frac{x}{y}\,\frac{f(x)}{1-F(y)}\,\mathrm{d}x = 2\,a\,\frac{1}{\alpha^2}\,\frac{y^{-\alpha}}{1-F(y)}\,[1+\frac{\alpha^2\,b'}{(\alpha+\beta)^2}y^{-\beta}] \tag{44}$$

This equation and eqs. (25) and (28) (in squared form) can be inserted in eq. (15). After regrouping the terms, we obtain:

$$\mathrm{E}[\gamma_{n,m}^2|X_{(m)}=y] = \frac{1}{\alpha^2}\frac{m}{m-1}\,\{1+\frac{2\,[(\frac{\alpha+\beta}{\alpha})^2-1-(m-2)\,\frac{\beta}{\alpha+\beta}]}{m}\,b'\,y^{-\beta}\} \tag{45}$$

This result and eq. (33) can be inserted in eq. (12). By analogy to eqs. (36) and (38), the integral of the numerator can be written

$$\int_0^\infty p_{n,m}(y)\,\mathrm{E}[\gamma_{n,m}^2|X_{(m)}=y]\,\mathrm{d}y = \frac{1}{\alpha^2}\frac{m}{m-1}\,\{P-a^2\,\frac{n!}{(m-1)!\,(n-m)!}\,\times$$
$$\times\,\frac{2\,[(\frac{\alpha+\beta}{\alpha})^2-1-(m-2)\,\frac{\beta}{\alpha+\beta}]}{m}\,b'\,[a\,(n-m)]^{-m-\frac{\beta}{\alpha}}\,\Gamma(m+\frac{\beta}{\alpha})\} \tag{46}$$

P can again be expressed by the lowest order of eq. (36) as all higher order terms are already included by the factor b' here. We also insert eq. (43) in eq. (12) to obtain the variance of the stochastic error of the Hill estimator:

$$\mathrm{E}\{[\gamma_{n,m} - \mathrm{E}[\gamma_{n,m}]]^2\} = \frac{1}{\alpha^2} \frac{1}{m-1} \{1 + O[(\frac{m}{n})^{\frac{\beta}{\alpha}}]\} \tag{47}$$

The computation of the explicit $(m/n)^{\beta/\alpha}$ term is possible but not necessary for the following discussion of the error of the Hill estimator. We can expand the result also in terms of $1/m$:

$$\mathrm{E}\{[\gamma_{n,m} - \mathrm{E}[\gamma_{n,m}]]^2\} = \frac{1}{\alpha^2 m} \{1 + O(\frac{1}{m}) + O[(\frac{m}{n})^{\frac{\beta}{\alpha}}]\} \tag{48}$$

The variance of the Hill estimator as well as its bias in the lowest order of the expansion have been first derived by (Goldie and Smith, 1987) in a different context. The results of this paper are in line with (Goldie and Smith, 1987).

3.5 The error of the Hill estimator

The error of the Hill estimator as defined by eq. (3) has two components: the systematic bias of eq. (43) and a stochastic error with the variance given by eq. (48). The total error has the following variance:

$$\begin{aligned} \mathrm{E}\{[\gamma_{n,m} - \frac{1}{\alpha}]^2\} &= B^2 + \mathrm{E}\{[\gamma_{n,m} - \mathrm{E}[\gamma_{n,m}]]^2\} \\ &= \frac{1}{\alpha^2} \frac{\beta^2 b^2}{(\alpha+\beta)^2} a^{-\frac{2\beta}{\alpha}} (\frac{m}{n})^{\frac{2\beta}{\alpha}} + \frac{1}{\alpha^2 m} \end{aligned} \tag{49}$$

where the higher order terms of eqs. (43) and (48) have been dropped. This total error is large both for large m (where the first term, the squared bias, dominates) and very small m (where the second term dominates). The error becomes smaller in a middle region of moderately small m.

There are two different ways to apply this information on the error:

1. The total error of the Hill estimator can be minimized with respect to m. Such an optimal m has been also derived in (Hall and Welsh, 1985).

2. We can subtract an estimate of the bias, eq. (43), from the Hill estimator to obtain a bias-corrected estimator.

Both of these approaches have one thing in common: they rely on eq. (43) where the unknown parameters β, a and b have to be estimated. The second approach, the bias correction, may lead to a more precise tail estimation, but is difficult. Danielson and de Vries propose a way to estimate these parameters through a moment estimator approach (Danielson and de Vries, 1996).

The first method relies on minimizing the error. We can determine the zero of the derivative of eq. (49) against m. The resulting $\overline{m}$ minimizes the error:

$$\overline{m} = [\frac{\alpha\ (\alpha + \beta)^2}{2\ \beta^3\ b^2}]^{\frac{\alpha}{\alpha+2\beta}}\ (a\ n)^{\frac{2\beta}{\alpha+2\beta}} \tag{50}$$

The expected variance of the total error, $E\{[\gamma_{n,m} - 1/\alpha]^2\}$ as a function of m, has a horizontal tangent at $\overline{m}$ so neighboring integer m values are almost as good as $\overline{m}$. Absolute precision in the choice of m is hence not required, and neglecting the high order terms of eq. (49) is justified.

The resulting $\overline{m}$ can be re-inserted in eq. (49) to give the minimum error variance that can be achieved by the Hill estimator. A full formula for this might be given, but we are only interested in the dependence of the error variance on the sample size n:

$$E\{[\gamma_{n,m} - \frac{1}{\alpha}]^2\} \propto n^{-\frac{2\beta}{\alpha+2\beta}} \tag{51}$$

Increasing the sample size n indeed leads to a smaller error of the Hill estimator. This effect is especially strong if α is small and β is large.

In the next sections, we follow the approach developed in (Dacorogna et al., 1994) to find the best m.

4. The bootstrap method

Eq. (50) contains unknown parameters, so the best m cannot be directly computed from it. To remedy this unfortunate situation, Hall proposed in (Hall, 1990) using bootstrap resamples of a small size n_1 and small m_1 values which differ in order of magnitude from n and m. These new samples are then used for computing $\gamma^*_{n_1,m_1}$.

Hall (1990) suggested finding the optimal $\overline{m}_1$ for the subsamples from

$$\min_{m_1} E\left[\{\gamma^*_{n_1,m_1} - \gamma_0\}^2 \mid F_n\right], \tag{52}$$

where F_n is the empirical distribution function and $\gamma_0 = \gamma_{n,m_0}$ is some initial full sample estimate for γ with a reasonably chosen but non-optimal m_0. Eq. (52) is a new, approximate version of eq. (49) for the subsamples and has the advantage of being empirically computable. The quantity γ_0 is a good approximation of $1/\alpha$ for subsamples since we know from eq. (51) that the error in γ is much larger for n_1 than for n observations. The value $\overline{m}_1$ is found by recomputing $\gamma^*_{n_1,m_1}$ for different values of m_1 and then empirically evaluating and minimizing eq. (52). Given $\overline{m}_1$, the suitable full sample $\overline{m}$ value can be found by taking the ratio $\overline{m}/\overline{m}_1$ and using eq. (50):

$$\hat{m} = \overline{m}_1 \left(\frac{n}{n_1}\right)^{\frac{2\beta}{(2\beta+\alpha)}}, \tag{53}$$

where $\hat{m}$ is the bootstrap estimate of $\overline{m}$ and $\hat{\alpha} = 1/\hat{\gamma}_{n,\hat{m}}$ the final estimate of the tail index α.

For this procedure to work, one still needs to know the exponents α and β to compute $2\beta/(2\beta+\alpha)$. For α, the initial estimate by means of γ_0 can be used[7]. One may therefore try to estimate β as well, and use these estimates for calculating the exponent in eq. (53). Unfortunately, estimating β is not straightforward. In (Hall, 1990), a value of 2/3 is proposed for the exponent in eq. (53) as applying to "many cases of practical interest". Such an assumption, which would imply $\beta = \alpha$, becomes doubtful when eq. (22) and the discussion following that equation are taken into account. For example, in case of the Student-t distribution with ν degrees of freedom, $\alpha = \nu$ and $\beta = 2$, the exponent is $4/(4+\nu)$. Thus, the original bootstrap method of (Hall, 1990) cannot be used in practice for $\nu > 2$. A way out of this dilemma is to administer an artificial location shift c to the data such that $\beta \in]0, 1]$ is ensured. The conclusion is that one does not have to worry about $\beta > 1$. Large values of β compared to α were not a problem to begin with, as this makes the second order term in eq. (25) negligible.

In the next section, we cover the interval $]0, 1]$ by evaluating the exponent in eq. (53) at different β-values. At the upper end $\beta = 1$, and by using γ_0 to identify α, $\hat{m}$ is readily calculated from eq. (53). At the lower end β approaches 0. Recall that the distribution in eq. (25) has to be read as $F(x) = 1 - ax^{-\alpha}[1 + b \ln x]$, in this case. This implies that the error given by eq. (49) becomes $1/(\alpha^2 \ln s)^2 + 1/(\alpha^2 m)$. In finite samples, using $\ln X \approx (X^\beta - 1)/\beta$ where β is small, yields a good approximation. In the applied sections we take $\beta = 0.01$.The intermediate value $\beta = 1/2$ is considered as well. In addition we evaluate eq. (53) at $\beta = \alpha = 1/\gamma$, the case considered by Hall and $\beta = 2$ to cover the occurrence $c = 0$ (see also (Hall and Welsh, 1985) where the case $\beta = 2$ is discussed).

5. Empirical study of the tail bootstrap method

In this section, we present the results of simulations of the tail index estimators for some theoretical processes. For testing the quality of the method, it is important to apply it to various distributions for which eq. (25) has different accuracies. As in section 2, we explore the stable distribution function (with more α values), various Student-t distribution functions as well as a simple ARCH process[8]. ARCH processes are important in modeling financial asset prices. All these distributions are assumed to be centered at zero, therefore without any shift along the x-axis, but we administer a

---

[7] We do not use the value α^* computed from $\gamma^*_{n_1,\bar{m}_1}$ because experience showed us that the latter was too biased.

[8] Since we use a bootstrap method that destroys the time dependence by picking data randomly, the i. i. d. assumption is still valid for the bootstrapped data of an ARCH process.

	10m	30m	1h	2h	6h	1d
k	1	3	6	12	36	144
n	473'327	157'775	78'887	39'443	13'147	3'286
n_1	11'833	3'944	1'972	986	329	100
m_0	1'183	394	197	98	33	8

Table 1: The aggregation factor k, the full-sample size n, the sub-sample size n_1 and the m_0 values corresponding to the different time intervals.

location shift $c = 0.1$ in several cases and report the two configurations $c = 0, 0.1$. We follow the method indicated in (Hols and De Vries, 1991) for defining the theoretical tail index of the ARCH process.

As in section 2, we generate artificial time series. In our example, these can be regarded as simulated price changes of a financial market. We use regularly spaced data with intervals of 10 minutes and enough observations to cover nine full years. We also test the effect of aggregation by studying price changes over longer time intervals than the original data but on the same period which means that the sample size will become smaller with an increasing aggregation factor. Formally, an aggregated value Y_i is computed from the original values X_i as follows:

$$Y_i = \sum_{j=1}^{k} X_{k(i-1)+j} \tag{54}$$

where k is the aggregation factor. In Table 1, we display the number of data used in calculating each statistic corresponding to the different time intervals together with the aggregation factor. Once the aggregated data set is produced using eq. (54), the bootstrap procedure is applied on the aggregated set. We see that under aggregation the number of data is heavily reduced and, in the case of the one-day time interval, the total number of data becomes almost too small. This is why for the longest time intervals, 2 hours, 6 hours and 1 day, we generate additional artificial time series by using overlapping intervals, where the starting points of these time series are in an hourly sequence. Using the average results of these different time series can slightly improve the estimation as shown in (Müller, 1993). Another way of improving the estimation is to use both the positive and negative tails since we are mostly dealing with symmetric processes. Thus we report the averages of both the negative and the positive tail estimation. To obtain results independent of the particular choice of the initial seed of the random generator, we report, in this section, average results computed over 10 realizations for each distribution.

The results presented in Tables 2 and 4 are calculated with the bootstrap method under the following conditions:

- For the initial γ_0 in eq. (52), the m_0 value is chosen as 0.25% of the full sample size n.

- The resample size is always chosen as $n_1 = n/40$. The number of resamples is always 100. This is possible because the resamples are randomly picked from the full data set, so one data point can occur in several resamples.

  A question arises if one should use resamples on sequences of observations, as advocated in (Kunsch, 1989; Liu and Singh, 1992) or resamples without keeping sequences but with reshuffling. This question is particularly important in the case of financial time series because of the presence of heteroskedasticity (Dacorogna et al., 1993). We tried both approaches and the results are similar. Here we report results using the bootstrap with reshuffling the data because the whole theory assumes i.i.d. variables. By reshuffling the data we remove the heteroskedasticity present in financial time series and in the ARCH process we study here. Although there is no formal proof that the procedure proposed here should work for ARCH type processes, the empirical results show convincing evidence that indeed the theoretical index of an ARCH process is correctly evaluated both in this study and in (Hols and De Vries, 1991).

- The search space of m_1 for finding $\overline{m}_1$ through (52) is constrained to the tail (2% of subsample size n_1).

- To expand from the resamples to the full sample as in eq. (53), we input for α the initial guess $\alpha_0 = 1/\gamma_0$ and test for different values of β:

  1. $\beta = \alpha$, as is presupposed in (Hall, 1990)
  2. $\beta = 2$, which applies in case of the Student-t with $c = 0$ (Hall and Welsh, 1985),
  3. $\beta = 1$, a frequent case according to eq. (22),
  4. $\beta = 0.5$, to cover the range $\beta \in [0, 1]$,
  5. $\beta = 0.01$, to approximate the lower limit.

First, we examine the influence of different choices of β on the results of the bootstrap procedure. In Table 2, the results of the tail index for 10m intervals are shown as a function of β. The 95% confidence interval is computed from the stochastic variance given in eq. (48) i. e. $\pm 1.96[1/(\hat{\alpha}^2\hat{m})]^{1/2}$ (where we neglect all the smaller order terms).

distr.	c	α	$\beta = \alpha$	$\beta = 2.0$	$\beta = 1.0$	$\beta = 0.5$	$\beta = 0.01$
Cauchy	0.0	1.00	1.00 ±0.04	1.00 ±0.03	1.00 ±0.04	1.00 ±0.05	1.00 ±0.13
	0.1	1.00	1.26 ±0.12	1.22 ±0.11	1.12 ±0.13	1.05 ±0.17	1.02 ±0.33
stable	0.0	1.25	1.26 ±0.05	1.26 ±0.04	1.26 ±0.05	1.26 ±0.07	1.28 ±0.16
	0.0	1.50	1.52 ±0.07	1.53 ±0.06	1.51 ±0.08	1.51 ±0.12	1.50 ±0.23
	0.0	1.75	1.79 ±0.13	1.79 ±0.12	1.78 ±0.17	1.80 ±0.23	1.79 ±0.43
	0.0	1.90	1.98 ±0.19	1.97 ±0.20	1.92 ±0.26	1.87 ±0.35	1.92 ±0.63
	0.1	1.90	2.02 ±0.19	2.02 ±0.20	1.94 ±0.27	1.87 ±0.35	1.94 ±0.63
Student	0.0	2.00	2.00 ±0.08	2.00 ±0.08	2.00 ±0.12	2.00 ±0.16	1.98 ±0.28
	0.0	3.00	2.93 ±0.16	2.93 ±0.19	2.95 ±0.26	3.02 ±0.35	3.09 ±0.56
	0.0	4.00	3.78 ±0.25	3.86 ±0.34	3.91 ±0.47	3.97 ±0.61	4.05 ±0.91
	0.0	6.00	5.26 ±0.43	5.42 ±0.67	5.51 ±0.92	5.41 ±1.12	5.57 ±1.54
	0.1	6.00	5.34 ±0.42	5.50 ±0.66	5.58 ±0.90	5.59 ±1.11	5.60 ±1.50
ARCH	0.0	3.00	2.95 ±0.14	2.96 ±0.17	2.99 ±0.24	3.03 ±0.32	3.04 ±0.50
	0.1	3.00	3.01 ±0.13	3.01 ±0.16	3.04 ±0.22	3.07 ±0.30	3.06 ±0.47

Table 2: Estimated tail exponents for theoretical processes using the bootstrap method and varying the value of β for choosing the optimal m. We also present results for a location shift $c = 0.1$ in units of the scale of the distribution (here: symmetric shift for both tails, so taking $-c$ for the negative tail). Sample size $n = 473327$, the "10m" case of Table 1. The results are averages over 10 realizations for each distribution.

The results are quite insensitive to the choice of β. This is good news for the applicability of Hall's bootstrap method: even large errors in the assumed value of β do not lead to aberrant estimates of α.

It is interesting to compare the results of Table 2 with the case where the subsample bootstrap is not employed. This is the procedure commonly used in the published literature. Suppose that the initial estimate γ_0, which uses m_0 of the highest order statistics, were the final estimate. Using $m_0 = 1183$ as in Table 1, we obtain $\alpha_0 = 1/\gamma_0 = 5.06$ in the case of a Student-t variate with 6 degrees of freedom. This is clearly worse than the α values around 5.5 obtained in Table 2, no matter which β and c values were assumed. Hence, the bootstrap method helps to reduce the bias, in particular when the procedure is combined with a location shift, $c \neq 0$.

We also performed the simulation with a location shift, c, of the order of a tenth of the scale of the distribution. The shift is relatively large and, as a result, we see a slight increase of the value of the tail estimate. The reason is that the descent to zero of the density appears steeper than without a shift and this tends to increase the tail index. The

distr.	α	10m $\overline{m}_1$	10m $\hat{m}$	30m $\overline{m}_1$	30m $\hat{m}$	1h $\overline{m}_1$	1h $\hat{m}$	2h $\overline{m}_1$	2h $\hat{m}$	6h $\overline{m}_1$	6h $\hat{m}$	1d $\overline{m}_1$	1d $\hat{m}$
Cauchy	1.00	221	2588	76	897	36	429	18	221	8	105	8	62
stable	1.25	222	2146	76	727	37	350	18	182	8	83	8	61
stable	1.50	156	1277	68	555	35	290	18	149	8	71	8	58
stable	1.75	64	445	31	215	20	136	12	87	8	53	8	57
stable	1.90	35	205	14	86	9	57	7	42	7	44	8	45
Student	2.00	197	1251	73	456	37	230	18	114	8	54	8	43
Student	3.00	116	519	76	317	37	142	18	67	8	25	3	10
Student	4.00	76	277	49	155	28	81	13	36	4	11	2	4
Student	6.00	50	145	20	49	13	28	6	13	3	5	2	3
ARCH	3.00	152	684	62	280	36	163	18	81	8	53	7	41

Table 3: The resulting average values for $\overline{m}_1$ and $\hat{m}$ in the subsample bootstrap for $\beta = 1$ and $c = 0$. The total sample sizes n are indicated in Table 1. The results are averages over 10 realizations for each distribution.

tail index estimate for the shifted tails which is closest to the theoretical value is again obtained with $\beta = 1$, if one excepts the case $\beta = 0.01$ which is not very accurate as can be seen from the large errors in Table 2.

In conclusion, we recommend setting $\beta = 1$ in Hall's bootstrap method and testing the influence of a small location shift c.

6. The jackknife algorithm to compute an empirical variance

We evaluate the error on the tail index estimation using eq. (48) in Table 2. There is, however, an alternative to using the theoretical variance: an empirical variance computed through the jackknife algorithm (Yang and Robinson, 1986). To examine the quality of the theoretical error, we repeat the tail estimation (as outlined above) including the subsample bootstrap technique 10 times. Each time, we omit some randomly selected data points that amount to 10% of the original data set, making sure that every data point is omitted only once and used in the other 9 computations of $\hat{\gamma}_{0.9n}$. The new results of the jackknife method are based on the average of the ten $\hat{\gamma}_{0.9n}$ values denoted by $\hat{\gamma}_{(.)}$ for the tail index estimate $\hat{\alpha}$ and on the variance of $\hat{\gamma}_{(.)}$ for the stochastic error of this estimate:

distr.	α	10m	30m	1h	2h	6h	1d
Cauchy	1.00	1.00 ±0.04	1.00 ±0.07	1.00 ±0.10	1.00 ±0.14	0.99 ±0.19	1.00 ±0.26
stable	1.25	1.26 ±0.06	1.26 ±0.10	1.28 ±0.14	1.28 ±0.20	1.26 ±0.27	1.30 ±0.34
	1.50	1.51 ±0.09	1.52 ±0.13	1.54 ±0.18	1.52 ±0.26	1.54 ±0.36	1.60 ±0.43
	1.75	1.78 ±0.17	1.80 ±0.26	1.82 ±0.32	1.84 ±0.39	1.92 ±0.51	2.30 ±0.62
	1.90	1.93 ±0.27	1.93 ±0.42	1.96 ±0.53	2.03 ±0.64	2.42 ±0.70	3.35 ±1.03
Student	2.00	2.00 ±0.11	2.04 ±0.20	2.06 ±0.28	2.07 ±0.40	2.12 ±0.56	2.44 ±0.78
	3.00	2.94 ±0.26	3.23 ±0.38	3.46 ±0.60	3.75 ±0.95	4.62 ±1.88	6.96 ±5.45
	4.00	3.90 ±0.47	4.34 ±0.74	4.93 ±1.11	5.61 ±1.90	7.08 ±4.47	10.09 ±10.04
	6.00	5.48 ±0.90	6.60 ±1.92	7.75 ±2.94	9.36 ±5.24	12.68 ±11.12	13.17 ±13.99
ARCH	3.00	3.00 ±0.25	2.94 ±0.38	2.97 ±0.48	3.11 ±0.72	3.46 ±1.17	4.64 ±2.55

Table 4: Estimated tail exponent for theoretical processes using the bootstrap method and the jackknife algorithm. Here β is chosen to be 1 and $c = 0$. The results are the average over 10 realizations for each distribution.

$$\hat{\alpha}_{(.)} = 1/\hat{\gamma}_{(.)} \quad \text{and} \quad \text{error}(\hat{\alpha}) = 1.96\sqrt{\text{Var}[\hat{\gamma}_{(.)}]}\ \hat{\gamma}_{(.)}^{-2}. \tag{55}$$

for the error corresponding to 95% confidence which is reported in Table 4. Note that $\text{Var}[\hat{\gamma}_{(.)}]$ deviates from the usual variance definition and its exact expression is given in (Yang and Robinson, 1986), eq. (7.2.3) as the sum of all squared deviations of the $\hat{\gamma}_{0.9n}$ from $\hat{\gamma}_{(.)}$, multiplied by 9/10. Note that the numbers shown in Table 1 are all multipied by 0.9 if using this jackknife method (except, of course, the aggregation factors).

With the jackknife technique, we obtain computed standard errors reported in Table 4. They can be compared to the theoretical variance estimates shown in Table 2. On average, the jackknife errors are slightly higher. This fact can be seen by comparing the errors of the "10m" column of Table 4 to those of the column "$\beta = 1$" of Table 2. However, the errors obtained through the jackknife method better reflect the data set that is actually used. The theoretical error does not account for the degree of noise produced by the resampling while the jackknife method does. We thus recommend using this technique to obtain a good estimate of the errors[9].

The tail index under aggregation seems quite stable except for the long intervals, mainly the 6h and 1d intervals where the number of data becomes a problem to obtain

[9]The use of many realizations in the simple bootstrap method brings the results of the two methods closer to each other. In the case of one replication only, the jackknife method gives a better statistical estimate.

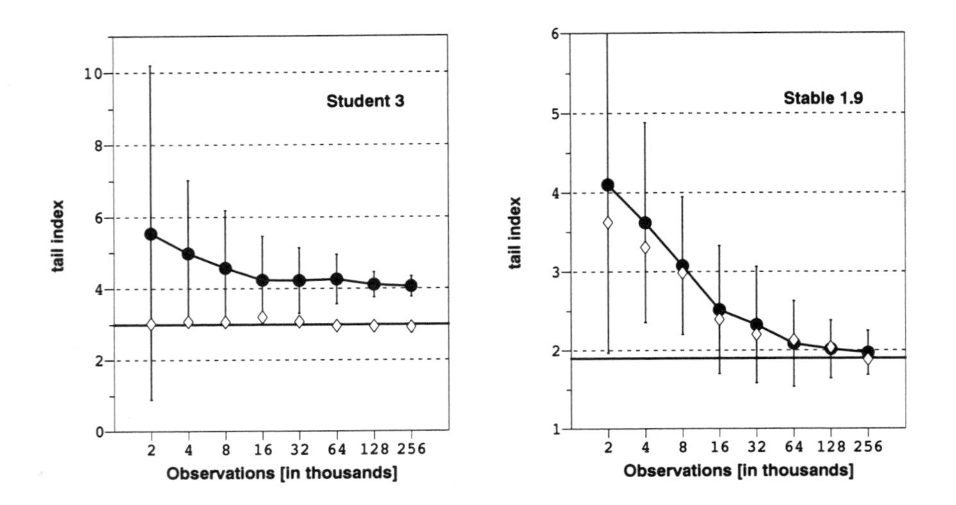

Figure 4: The estimated tail index is shown as a function of the number of observations. The filled circles (•) represent the aggregated time series and the diamonds (◇) the original series. The jackknife errors are reported only for the aggregated series. The horizontal line represents the theoretical value.

significant estimates. This is especially true when the theoretical value of α is big. The smaller number of data for large intervals forces the algorithm to use a larger fraction of this data, closer to the center of the distribution. There is also another reason for the instability of the estimator under aggregation: when small intervals are aggregated, the observed random variable values over long intervals are often aggregated from extreme as well as moderate values over small intervals. Thus the empirically measured tail properties become distorted by properties of the center of the distribution which, for $\alpha > 2$ and under aggregation, approaches the normal distribution (with $\alpha = \infty$) as a consequence of the central limit theorem. Indeed, the empirical $\hat{\alpha}$ values of the Student distributions and the ARCH process of Table 4 have a tendency to grow under aggregation. Going to time horizons longer than 6h or 1d therefore leads to unacceptable distortions. These results also show how important the extent of the data set is. A large size and a high frequency help to improve the quality of the tail estimates.

The results also demonstrate that large sample sizes (which can be achieved through high-frequency data) are important for the quality of the tail estimates. Eq. (51) has already shown this on a theoretical level. Now, we can compare empirical results from Table 4, for example the estimated tail exponents of stable distributions with $\alpha = 1.75$ in the columns "10m" and "6h". The difference between these results is solely due to the different sample sizes because the distribution function is stable under aggregation.

The "10m" column relies on 36 times as many independent observations as the "6h" column. Increasing the sample size by a factor 36 thus leads to a considerably improved estimate of α (= 1.75): 1.78 ± 0.17 as opposed to 1.92 ± 0.51, see Table 4. The empirical reduction of the error is of factor 3 which is close to the theoretically predicted reduction of 2.6 computed from eq. (51).

The advantage of synthetic time series is that the only limitation to the size of the data set is the available computing power. Thus, we examine how, in the case of the aggregated data, the tail index estimation varies when the number of data varies. We generate different sets of synthetic data representing 10 minute price changes and examine how the tail index estimation varies for 6 hour data as a function of the number of data obtained by aggregating the original 10 minute data 36 times following eq.(54).

In Figure 4, we show the tail index estimates with varying sample sizes going from 2,000 observations up to 256, 000. The estimates are computed both with synthetic 10 minute data (diamonds) and with observations aggregated 36 times to a total of 6 hour (filled circles). The largest number of observations represents a time series of almost 5 years for the 10 minute data and more than 175 years for the 6 hour data. We show the average results of 10 estimations with the combined bootstrap and jackknife algorithm with 10 different simulations. The points are the average estimate and the errors plotted around each point are the average error. For the 10 minute data, we do not draw the errors. For the Student-t distribution with $\nu = 3$, we see that the tail estimation on the original 10 minute data is very good already with the smallest number of observations while for the aggregated data the convergence of the estimator is slow and does not fully reach the theoretical value. It seems that with such a high aggregation factor part of the tail values are lost. In the case of the stable distribution with $\alpha = 1.9$, the aggregation does not play a role in the convergence of the tail estimation because of the very fact that this distribution is stable under aggregation. The important factor here is the number of observations. The errors as computed from the jackknife are decreasing with an increasing number of observations, as expected.

7. Conclusion

In this paper we deal with four interrelated issues: (i) the behavior of tail index estimators under different distributions for a very large data set; (ii) the problems and limitations of tail estimations; (iii) the estimation of the tail index semi-parametrically through a subsample bootstrap technique; (iv) the behavior of the estimation both with varying sample size and time aggregation.

The size of the data set is important when dealing with extreme observations. Extreme realizations are, by definition, rare; hence increasing the frequency and number of observations enhances the reliability of the estimate (recall eq. (49)). Up to now no technique other than this subsample bootstrap is known, and to the best of our knowledge we were the first to implement it. To this end, we have shown with the help of a location shift that the second order expansion has a natural restriction as formulated in eq. (22); this is well in line with the fact that assuming a second order exponent of $\beta = 1$ leads to optimal empirical results in the investigated cases. The subsample bootstrap technique necessarily requires large data sets as it requires sub-resamples which are of a different order of magnitude.

We have shown in this paper that, even without knowing in advance the distribution or the theoretical tail index, it is possible to achieve a reasonable tail index estimate for a wide set of distribution functions, with the help of a sophisticated subsample bootstrapping algorithm. In the case of time aggregation, the number of data is not always large enough for subsample bootstrapping, so one needs to rely on results obtained with high-frequency data although the tail index behavior under aggregation is not well understood. Our simulation results do not allow us to draw general conclusions on this subject. Although the tail index estimates move closer to the theoretical values when increasing the total number of observations, we still observe distinct differences between 10 minute data and 6 hour data even with very large samples corresponding to more than 175 years!

Acknowledgement: The authors would like to acknowledge the contribution of Casper G. de Vries. He introduced us to the wonderful world of extreme value theory.

References

Abramowitz M. and Stegun I. A., 1970, *Handbook of Mathematical Functions*, Dover Publications, Inc., New York, 9th edition.

Bergström H., 1952, *On some expansions of stable distribution functions*, Arkiv för Matematik, **2**(18), 375–378.

Dacorogna M. M., Müller U. A., Nagler R. J., Olsen R. B., and Pictet O. V., 1993, *A geographical model for the daily and weekly seasonal volatility in the FX market*, Journal of International Money and Finance, **12**(4), 413–438.

Dacorogna M. M., Pictet O. V., Müller U. A., and de Vries C. G., 1994, *The distribution of extremal foreign exchange rate returns in extremely large data sets*, Internal document UAM.1992-10-22, Olsen & Associates, Seefeldstrasse 233, 8008 Zürich, Switzerland.

Danielson J. and de Vries C. G., 1996, *Robust tail index and quantile estimation*, to be published in Journal of Empirical Finance.

David H. A., 1981, *Order Statistics*, John Wiley & Sons, New York.

de Haan L., 1990, *Fighting the arch-enemy with mathematics*, Statistica Neerlandica, **44**, 45–68.

de Haan L. and Resnick S. I., 1980, *A simple asymptotic estimate for the index of a stable distribution*, Journal of the Royal Statistics Society B, 83–87.

Dekkers A. L. M. and de Haan L., 1989, *On the estimation of the extreme-value index and large quantile estimation*, Annals of Statistics, **17**, 1795–1832.

Dekkers A. L. M., Einmahl J. H. J., and de Haan L., 1990, *A moment estimator for the index of an extreme value distribution*, Annals of Statistics, **17**, 1833–1855.

Goldie C. M. and Smith R. L., 1987, *Slow variation with remainder: Theory and applications*, Quarterly Journal of Mathematics, Oxford 2nd series, **38**, 45–71.

Gradshteyn I. S. and Ryzhik I. M., 1980, *Table of integrals, series, and products*, Academic Press, London.

Hall P., 1982, *On some simple estimates of an exponent of regular variation*, Journal of the Royal Statistical Society, Series B, **44**, 37–42.

Hall P., 1990, *Using the bootstrap to estimate mean square error and select smoothing parameter in nonparametric problem*, Journal of Multivariate Analysis, **32**, 177–203.

Hall P. and Welsh A. H., 1984, *Best attainable rates of convergence for estimates of parameters of regular variation*, The Annals of Statistics, **12**(3), 1079–1084.

Hall P. and Welsh A. H., 1985, *Adaptive estimates of parameters of regular variation*, The Annals of Statistics, **13**, 331–341.

Hill B. M., 1975, *A simple general approach to inference about the tail of a distribution*, Annals of Statistics, **3**(5), 1163–1173.

Hols M. C. and De Vries C. G., 1991, *The limiting distribution of extremal exchange rate returns*, Journal of Applied Econometrics, **6**, 287–302.

Kendall M., Stuart A., and Ord J. K., 1987, *Advanced Theory of Statistics*, volume 1, Charles Griffin & Company Limited, London, fifth edition.

Kunsch H. R., 1989, *The jacknife and the bootstrap for general stationary observations*, The Annals of Statistics, **17**, 1217–1241.

Leadbetter M. R., Lindgren G., and Rootzén H., 1983, *Extremes and related properties of random sequences and processes*, Springer Series in Statistics. Springer-Verlag, New York Berlin.

Liu R. Y. and Singh K., 1992, *Moving blocks jacknife and bootstrap capture weak dependence*, in Exploring the Limits of Bootstrap, R. LePage and Y. Billard (editors), Wiley, NewYork.

Marsaglia G. and Zaman A., 1994, *Some portable very-long-period random number generators*, Computers in Physics, **8**(1), 117–121.

Mason D. M., 1982, *Laws of large numbers for sums of extreme values*, the Annals of Probability, **10**(3), 754–764.

Müller U. A., 1993, *Statistics of variables observed over overlapping intervals*, Internal document UAM.1993-06-18, Olsen & Associates, Seefeldstrasse 233, 8008 Zürich, Switzerland.

Pickands III J., 1975, *Statistical inference using extreme order statistics*, The Annals of Statistics, **3**(1), 119–131.

Yang M. C. K. and Robinson D. H., 1986, *Understanding and Learning Statistics by Computer*, volume 4 of *Series in Computer Science*, World Scientific Publishing Co Pte Ltd, Singapore.

Authors address: Olsen & Associates, Seefeldstrasse 233, CH-8008 Zurich, Switzerland

Characteristic Function Based Estimation of Stable Distribution Parameters

Stephen M. Kogon and Douglas B. Williams

Abstract

This chapter is concerned with the estimation of stable distribution parameters using the characteristic function. The sample characteristic function is computed from the sample data and is subsequently fit to the stable distribution model. Of these methods, Koutrouvelis' has been shown to have the best performance. However, this technique involves a great deal of costly computations due to the numerous iterations required. A new method is proposed which greatly reduces the amount of computation by restricting the estimation to a common interval on the characteristic function. The new method is compared to Koutrouvelis', as well as other methods.

1. Introduction

The family of stable distributions has been shown to provide a useful model of data from long or heavy-tailed distributions [Fel71, Zol86, ST94]. Sample data from heavy-tailed distributions are characterized by the presence of observations with very large magnitudes, a phenomenon often referred to as high variability. Examples of such data have been reported in a wide variety of applications including economics [Man63], telephony [SK74], radar clutter modeling [JP76], and environmental sciences [KM96, Pai96].

The interest in stable distributions can be attributed to two very attractive theoretical properties of the model. First, the sum of independent, identically distributed (i.i.d.) stable random variables will retain the shape of the original distribution (the scale and center of the distribution may change), which is known as the stability property [Fel71]. Second, by the generalized central limit theorem, if the sum of i.i.d. random variables converges to a distribution, then that distribution must be a member of the family of stable distributions [Fel71]. Therefore, stable distributions constitute a domain of attraction for the sums of random variables, helping to justify their selection as a model for processes resulting from a large number of random events. However, the major drawback of the stable distribution model is the lack of a closed form expression for the probability density function (PDF) with only a few exceptions, namely the Gaussian, Cauchy, and Lévy distributions. For more information on stable distributions see [Fel71, Zol86, ST94, NS95].

The stable distribution model can be constructed via parameter estimation from the sample data by a number of methods that have been proposed. The various parameter estimators generally fall into three categories: maximum likelihood, quantile or fractile methods, and characteristic function based methods. DuMouchel [DuM73] proposed a maximum likelihood type algorithm which, although theoretically superior, requires a great deal of computation, making it impractical for many applications. Quantile methods, on the other hand, are empirically derived estimators based on order statistics which are computationally very inexpensive. Fama and Roll [FR71] proposed the first quantile technique for the estimation of the parameters of symmetric stable distributions. Later, McCulloch [McC86] presented an improved quantile method which was generalized to include asymmetric distributions. Parameter estimators based on the characteristic function, originally proposed by Press [Pre72], are attractive because they are based on the theoretical stable distribution model, yet can be implemented with a relatively small amount of computation.

A number of techniques have been proposed which fit the sample characteristic function (SCF) to the stable distribution model [Pre72, PHL75, Ara80, Kou80, Kou81, FM81b, BB81, PD85], of which the iterative weighted regression method of Koutrouvelis [Kou81] has been shown to have the best performance [AL89]. However, the numerous iterations designed to find the "best" interval on the characteristic function from lookup tables dramatically increase the complexity of this method and help to motivate the investigation into means of reducing the amount of computation while maintaining acceptable performance.

The purpose of this chapter is to discuss the important issues concerning the estimation of stable distribution parameters when using the characteristic function and to give practical guidelines for the implementation of these procedures. In the process, we give our own SCF based estimation technique which uses a common interval on the SCF, independent of parameter values, and does not require iterations. The chapter is organized as follows: first the general SCF based parameter estimator is derived along with Koutrouvelis' method. Next, the proposed SCF based method is given along with a discussion of the selection of the interval on the characteristic function. Results comparing the performance of the proposed estimator to Koutrouvelis' method, as well as some other techniques, are presented, followed by some concluding remarks.

2. Koutrouvelis' SCF Based Parameter Estimation

Because the characteristic function is the Fourier transform of the probability density function (PDF), there is a one-to-one correspondence between them [Pap91]. Since the PDF of a stable distribution is generally not known in closed form but the characteristic function is, the characteristic function is a natural vehicle for the parameter estimation problem. Although there are several different formulations, the most commonly used form of the characteristic function of a stable distribution is [DuM73, McC]

$$\Phi(\omega) = \mathrm{E}\{\exp(i\omega x)\} = \exp\left\{i\delta\omega - |c\omega|^{\alpha}(1 - i\beta f(\omega, \alpha, c))\right\} \qquad (2.1)$$

where the function $f(\omega, \alpha, c)$ is

$$f(\omega, \alpha, c) = \begin{cases} \frac{\omega}{|\omega|}\tan\frac{\pi\alpha}{2}, & \text{if } \alpha \neq 1 \\ -\frac{2}{\pi}\frac{\omega}{|\omega|}\log|c\omega|, & \text{if } \alpha = 1 \end{cases} \qquad (2.2)$$

and the stable distribution parameters are the characteristic exponent $\alpha \in (0, 2]$, the symmetry index $\beta \in [-1, 1]$, the spread parameter $c > 0$, and the location parameter δ. The characteristic exponent α determines the rate of decay, i.e., the heaviness, of the tails of the distribution, while the parameter β is an indication of the skewness of the distribution, with $\beta = 0$ corresponding to the symmetric case. The parameters δ and c simply translate and scale the distribution but have no effect on its shape. Note that the term $\beta f(\omega, \alpha, c)$ is discontinuous at $\alpha = 1$ for $\beta \neq 0$.

The derivation of Koutrouvelis' SCF (K-SCF) parameter estimator begins by taking the logarithm of the characteristic function, known as the log-characteristic function [Pap91]

$$\Psi(\omega) = \log\{\Phi(\omega)\} = -|c\omega|^{\alpha} + i\left(\delta\omega + |c\omega|^{\alpha}\beta\frac{\omega}{|\omega|}\tan\frac{\pi\alpha}{2}\right) \quad \alpha \neq 1\,. \qquad (2.3)$$

First, extracting the real part of (2.3)

$$Re[\Psi(\omega)] = -|c\omega|^{\alpha} = -c^{\alpha}|\omega|^{\alpha} \qquad (2.4)$$

and further manipulating it by taking another logarithm

$$\log\{-Re[\Psi(\omega)]\} = \alpha\log|\omega| + \alpha\log c\,. \qquad (2.5)$$

Clearly, this equation describes a linear relation between $\log|\omega|$ and $\log\{-Re[\Psi(\omega)]\}$ with a slope of α and a y-intercept determined by c. Similarly, the imaginary part of (2.3) for $\alpha \neq 1$ is

$$Im[\Psi(\omega)] = \delta\omega + |c\omega|^{\alpha}\beta\frac{\omega}{|\omega|}\tan\frac{\pi\alpha}{2} = \delta\omega + \beta|c\omega|^{\alpha}f(\omega, \alpha, c) \qquad (2.6)$$

where the function $f(\omega, \alpha, c)$ is from (2.2) and the only case considered is for $\alpha \neq 1$ since this is the case with probability one.

Now that the relations between the parameters and the log-characteristic function have been derived for the stable distribution model, the problem is how to extract the useful model information from the sample data. Since the characteristic function is generally unknown for a given set of data samples $\{x(n)\ ;\ n = 1, 2, \ldots, N\}$, it must be estimated. However, before proceeding with the computation of the SCF, the sample data is first normalized by initial estimates of the spread and location parameters in order to remove the effects of these parameters on the estimators, as originally proposed by Paulson et al. [PHL75]. Koutrouvelis proposed the use of the Fama and Roll (FR) quantile method [FR71] and the 25% truncated mean for initial estimates of $\hat{c}_0$ and $\hat{\delta}_0$, respectively [Kou80]. The data samples are then normalized

$$\tilde{x}(n) = \frac{x(n) - \hat{\delta}_0}{\hat{c}_0} \tag{2.7}$$

and the characteristic function of the normalized data is estimated using the SCF given by

$$\widehat{\Phi}(\omega) = \frac{1}{N} \sum_{n=1}^{N} \exp\{i\omega\tilde{x}(n)\} \tag{2.8}$$

where N is the number of data samples. Since the log-SCF is simply $\widehat{\Psi}(\omega) = \log\left\{\widehat{\Phi}(\omega)\right\}$, the real and imaginary parts of the log-SCF are obtained from the SCF:

$$Re\left[\widehat{\Psi}(\omega)\right] = \log\left|\widehat{\Phi}(\omega)\right|, \qquad Im\left[\widehat{\Psi}(\omega)\right] = Arctan\left(\frac{Im\left[\widehat{\Phi}(\omega)\right]}{Re\left[\widehat{\Phi}(\omega)\right]}\right) \tag{2.9}$$

where $Arctan(\cdot)$ is the principal value of the arctangent function. Using (2.5), the estimates of the stable characteristic exponent $\hat{\alpha}$ and an update of the spread parameter $\hat{c}_1$ are obtained from the regression equation [Shu88, Gol89]

$$\log\left\{-Re\left[\widehat{\Psi}(\omega_k)\right]\right\} - \left(\hat{\alpha}\log|\omega_k| + \hat{\alpha}\log\hat{c}_1\right) = e_k \tag{2.10}$$

for $k = 1, \ldots, K$ with the points $\{\omega_k\}$ chosen according to a lookup table [Kou80]. The terms e_k are the regression errors. Note that since $\hat{\alpha}$ is the slope of $\log\{-Re[\widehat{\Psi}(\omega)]\}$ versus $\log|\omega|$, it is trivial to compute $\hat{c}$ from the y-intercept.

Once the parameters $\hat{\alpha}$ and $\hat{c}_1$ have been found, the symmetry index β and an updated location parameter δ_1 are estimated by performing the regression of the imaginary part of the log-SCF onto (2.6)

$$Im\left[\widehat{\Psi}(\omega_l)\right] - \left(\hat{\delta}_1\omega_l + \hat{\beta}\left|\hat{c}_1\omega_l\right|^{\hat{\alpha}} f\left(\omega_l, \hat{\alpha}, \hat{c}_1\right)\right) = n_l \tag{2.11}$$

for $l = 1, \ldots, L$, again with the points $\{\omega_l\}$ selected using a lookup table [Kou80]. The terms n_l are the regression errors and the function $f(\omega, \hat{\alpha}, \hat{c}_1)$ is found by substituting the estimates $\hat{\alpha}$ and $\hat{c}_1$ into (2.2). The final estimates of the spread and location parameters are found using the current ($\hat{c}_1 \approx 1, \hat{\delta}_1 \approx 0$) and initial estimates

$$\hat{c} = \hat{c}_0\hat{c}_1 \tag{2.12}$$

$$\hat{\delta} = \hat{\delta}_0 + \hat{c}_0 \hat{\delta}_1 \,. \tag{2.13}$$

Note that the estimation of β and δ is dependent on the estimates of α and c. This coupling of the estimators for the different parameters is cited as one possible cause for the relatively poorer accuracy of the estimator of β [Kou80, AL89].

The regressions in (2.10) and (2.11) are performed using least squares (LS) methods [Shu88, Gol89]. The interval points for the regression, $\{\omega_k\}$ and $\{\omega_l\}$, are determined by independent lookup tables indexed by the sample size and initial parameter estimates where the intervals were found with a simulation study [Kou80]. Initially, the regressions were performed using ordinary least squares (OLS) regression [Kou80]. The performance of the estimators was improved by incorporating the covariances of the regression errors into a generalized least squares (GLS) regression [Kou81]. In this improved regression technique of Koutrouvelis, the GLS regression was iterated, while continually re-normalizing the sample data with updated spread and location parameter estimates, until the convergence criterion of the mth iteration, given by

$$\xi(m) = (\hat{\alpha}^{(m)} - \hat{\alpha}^{(m-1)})^2 + (\hat{\delta}_1^{(m)} - \hat{\delta}_1^{(m-1)})^2 \,, \tag{2.14}$$

falls below a prespecified threshold (recommended to be $\xi(m) < 0.01$) [Kou81]. The terms $\hat{\alpha}^{(m)}$ and $\hat{\delta}^{(m)}$ are the estimates of the mth iteration. The location parameter estimates in (2.14) do not have to be normalized since they were computed from normalized data samples. Note that Koutrouvelis found that the iterations did not always converge and, therefore, the process was automatically terminated after 10 iterations. A slight performance gain over the OLS regression was noted in most cases using the GLS regression, with a pronounced improvement for the estimation of β and δ for small values of α.

3. Fixed Interval SCF Based Parameter Estimation

In the derivation of our estimator, we use an alternate expression for the characteristic function of a stable distribution [Zol86]

$$\Phi(\omega) = \exp\left\{i\zeta\omega - |c\omega|^\alpha + i\beta g(\omega, \alpha, c)\right\} \tag{3.1}$$

where ζ is the modified location parameter and the function $g(\omega, \alpha, c)$ is

$$g(\omega, \alpha, c) = \begin{cases} c\omega \tan\frac{\pi\alpha}{2}\left(|c\omega|^{\alpha-1} - 1\right), & \text{if } \alpha \neq 1 \\ -\frac{2}{\pi} c\omega \log|c\omega|, & \text{if } \alpha = 1\,. \end{cases} \tag{3.2}$$

Note that in this formulation of the characteristic function, the term $\beta g(\omega, \alpha, c)$ is continuous in both α and β as opposed to $\beta f(\omega, \alpha, c)$ from (2.1) which is discontinuous for $\alpha = 1$ and $\beta \neq 0$. The modified location parameter ζ is related to the location parameter δ by

$$\zeta = \begin{cases} \delta + \beta c \tan \frac{\pi\alpha}{2} & \text{if } \alpha \neq 1 \\ \delta & \text{if } \alpha = 1 \, . \end{cases} \tag{3.3}$$

Manipulating the log-characteristic function $\Psi(\omega)$ from (2.3) yields the same equation for α and c in (2.5) and the following relation for β and ζ for $\alpha \neq 1$

$$Im\left[\Psi(\omega)\right] = \zeta\omega + \beta c\omega \tan\frac{\pi\alpha}{2}\left(|c\omega|^{\alpha-1} - 1\right) = \zeta\omega + \beta g(\omega, \alpha, c) \tag{3.4}$$

where the function $g(\omega, \alpha, c)$ is given by (3.2). Note that this formulation of the characteristic function is consistent with that used by Chambers et al. [CMS76] to generate stable random variables with $c = 1$ and $\zeta = 0$ (not $\delta = 0$).

The characteristic function $\Phi(\omega)$ is defined for all frequencies ω. However, the parameter estimation must be performed over some finite interval on the sample characteristic function (SCF), ideally where the SCF most accurately estimates the true characteristic function. The shape of the stable distribution characteristic function is completely determined by the two parameters α and β. The spread parameter c simply dilates or compresses the magnitude of the characteristic function and the modified location parameter ζ (or the normal location parameter δ) adds a linear phase term. Neither c nor ζ alters the shape of the characteristic function in any way, yet both strongly influence the frequency interval for parameter estimation. Our goal is to use a common estimation interval on the SCF, independent of the parameter values, to ensure the same level of performance over the entire parameter space and to avoid using lookup tables as is required for Koutrouvelis' method [Kou80, Kou81].

Before proceeding with the estimation of α and β, initial estimates of the spread and modified location parameters, c_0 and ζ_0, are computed and their effects removed. The requirement on the estimator is robustness over a wide range of the parameter space and computational simplicity. Koutrouvelis used the FR quantile method to find c and a truncated mean to find the location parameter. However, the FR quantile method is intended for symmetric distributions ($\beta = 0$) and contains a bias in its estimate of $\hat{c}$, even in the symmetric case [McC86]. In addition, the estimate of the location parameter using the truncated mean will also contain a bias for asymmetric distributions [Zol86]. Instead, we propose the use of the technique due to McCulloch [McC86] for the preliminary estimates of both c and ζ. This quantile method is intended for the general class of stable distributions, including the asymmetric case, where the estimators have been shown to be unbiased with better performance than the FR quantile method, even in the symmetric case. Note that the parameter ζ is also estimated with this method. The drawback with McCulloch's method is that it requires that the parameters α and β be estimated before c and ζ. However, the improvement in performance, particularly for asymmetric distributions, helps justify this slight increase in computational expense. Note that McCulloch's method is only intended for $\alpha > 0.5$ which corresponds to values typically found in practice. Refer to [McC86] for details.

Once the initial estimates, $\hat{c}_0$ and $\hat{\zeta}_0$, have been found, the data samples $\{x(n)\}$ are normalized

$$\tilde{x}(n) = \frac{x(n) - \hat{\zeta}_0}{\hat{c}_0} \tag{3.5}$$

where $\{\tilde{x}(n)\}$ are the normalized samples with spread and modified location parameters approximately $c_0 \approx 1$ and $\zeta_0 \approx 0$.

Using similar derivations to Koutrouvelis, the regression equations are found to be

$$\log\left\{-Re\left[\widehat{\Psi}(\omega_k)\right]\right\} - \left(\hat{\alpha}\log|\omega_k| + \hat{\alpha}\log\hat{c}_1\right) = e_k \tag{3.6}$$

$$Im\left[\widehat{\Psi}(\omega_k)\right] - \left(\hat{\zeta}_1\omega_k + \hat{\beta}g(\omega_k, \hat{\alpha}, \hat{c}_1)\right) = n_k \tag{3.7}$$

where again e_k and n_k are the regression errors for the respective estimators and $\{\omega_k\}$ is a set of frequency points on the SCF common to both regressions. The regression equations are solved using OLS rather than GLS [Shu88, Gol89] in order to avoid the increased computations and iterations necessary for Koutrouvelis' method. The generalization to GLS is straightforward using the covariances derived by Koutrouvelis [Kou81] for the $\log\left\{-Re\left[\widehat{\Psi}(\omega_k)\right]\right\}$ and $Im\left[\widehat{\Psi}(\omega_k)\right]$ quantities, but as will be shown in the next section, the performance improvement for the increased computations is minimal in most cases. The final estimate of the spread and modified location parameters are found to be

$$\hat{c} = \hat{c}_0\hat{c}_1 \tag{3.8}$$

$$\hat{\zeta} = \hat{\zeta}_0 + \hat{c}_0\hat{\zeta}_1 \tag{3.9}$$

and the final estimate of the location parameter is retrieved from the estimate of ζ by

$$\hat{\delta} = \begin{cases} \hat{\zeta} - \hat{\beta}\hat{c}\tan\frac{\pi\hat{\alpha}}{2} & \text{if } \hat{\alpha} \neq 1 \\ \hat{\zeta} & \text{if } \hat{\alpha} = 1\,. \end{cases} \tag{3.10}$$

Now that the data samples have been normalized and the SCF regression equations derived, a frequency interval on the SCF must be selected for the OLS regressions in (3.6) and (3.7), independent of the scale and location parameters. In order to choose the interval, we consider the statistical properties of the SCF. First note the SCF is a consistent estimate of the theoretical characteristic function that converges almost surely as $N \to \infty$ [FM81a]

$$\lim_{N\to\infty}\left\{\sup_{-\infty<\omega<\infty}|\widehat{\Phi}(\omega) - \Phi(\omega)|\right\} = 0\,. \tag{3.11}$$

Therefore, as expected, the estimate of the characteristic function will improve as the sample size is increased. Although general convergence of the SCF is a desirable property, interval selection is dependent upon its convergence on different intervals of ω. Consider the asymptotic statistics (as $\omega \to 0$ and $\omega \to \infty$) of the real and imaginary parts of the SCF, beginning with the means

$$\mathrm{E}\left\{Re\left[\widehat{\Phi}(\omega)\right]\right\} = \begin{cases} 1 & \omega \to 0 \\ 0 & \omega \to \infty \end{cases} \tag{3.12}$$

$$\mathrm{E}\left\{Im\left[\widehat{\Phi}(\omega)\right]\right\} = \begin{cases} 0 & \omega \to 0 \\ 0 & \omega \to \infty. \end{cases} \tag{3.13}$$

At these frequency points, the mean values are equal to the values of the true characteristic function indicating an unbiased estimate. On the other hand, the variances and covariance of the real and imaginary parts of the SCF can be shown to be [Kou80]

$$\mathrm{Var}\left\{Re\left[\widehat{\Phi}(\omega)\right]\right\} = \begin{cases} 0 & \omega \to 0 \\ \frac{1}{2N} & \omega \to \infty \end{cases} \tag{3.14}$$

$$\mathrm{Var}\left\{Im\left[\widehat{\Phi}(\omega)\right]\right\} = \begin{cases} 0 & \omega \to 0 \\ \frac{1}{2N} & \omega \to \infty \end{cases} \tag{3.15}$$

$$\mathrm{Cov}\left\{Re\left[\widehat{\Phi}(\omega)\right], Im\left[\widehat{\Phi}(\omega)\right]\right\} = \begin{cases} 0 & \omega \to 0 \\ 0 & \omega \to \infty. \end{cases} \tag{3.16}$$

These asymptotic characteristics indicate that the real and imaginary parts of the SCF behave like uncorrelated random variables with a variance inversely proportional to twice the sample size N as the frequency ω approaches infinity. As expected, the greater the sample size, the better the estimate of the characteristic function. On the other hand, the SCF becomes nearly deterministic (zero variance) and equal to the characteristic function ((3.12) and (3.13)) as $\omega \to 0$, indicating that the ideal interval lies at lower frequencies where the SCF most accurately estimates the characteristic function. However, the characteristic function holds no information about the various parameters at $\omega = 0$ where it is always equal to one, independent of the parameters. In addition, the quantity used for the LS regression in (3.6), $\log\left\{-Re\left[\widehat{\Psi}(\omega)\right]\right\}$, is undefined at zero frequency. Since the SCF has unity value at zero frequency ($\widehat{\Phi}(0) = 1$), the real portion of the log-SCF (2.9) has a value of zero ($\widehat{\Psi}(0) = 0$) and, therefore, the quantity for the LS regression in (3.6), which requires the logarithm of the log-SCF ($\log\left\{-Re\left[\widehat{\Psi}(\omega)\right]\right\}$), is undefined at zero frequency and constitutes a singularity. Therefore, a compromise is necessary in which the parameter estimation is performed at lower frequencies, where the SCF best approximates the characteristic function, but sufficiently removed from the singularity at zero frequency so as to ensure that the estimator is well-behaved.

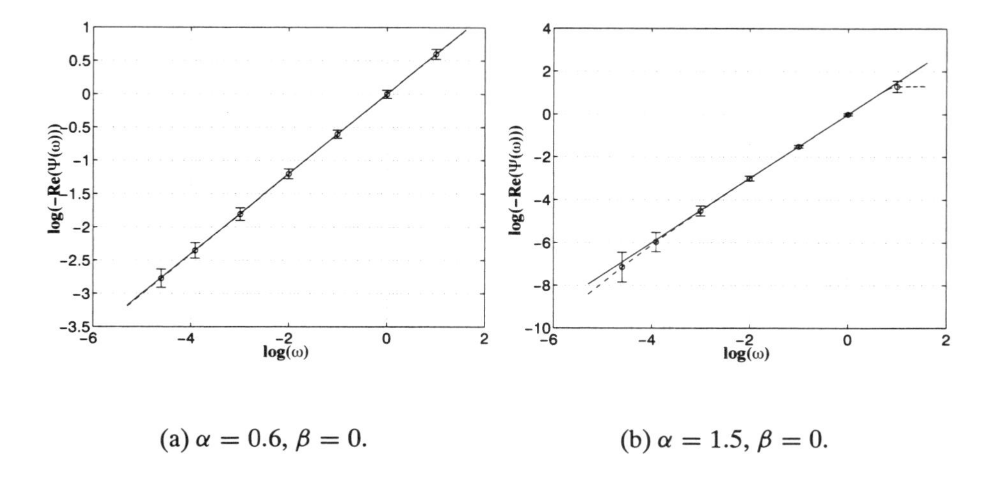

(a) $\alpha = 0.6$, $\beta = 0$.

(b) $\alpha = 1.5$, $\beta = 0$.

Figure 1: Mean and standard deviation of $\log\left\{-Re\left[\widehat{\Psi}\,(\omega)\right]\right\}$ versus $\log(\omega)$ (dashed line). The solid line is the value computed from the true characteristic function.

To illustrate the performance of the SCF as an estimator of the characteristic function, we compute the SCF from a set of data samples. A total of 1000 i.i.d. samples from a stable distribution were generated using the method due to Chambers, Mallows, and Stuck (CMS) [CMS76, ST94] with the spread and the location parameters set to $c = 1$ and $\delta = 0$. The quantities studied are those computed for the LS regression equations in (3.6) and (3.7). Figure 1 shows the values of $\log\{-Re[\Psi(\omega)]\}$ derived from the true characteristic function (solid line) and the mean of $\log\left\{-Re\left[\widehat{\Psi}\,(\omega)\right]\right\}$ (dashed line) along with its standard deviation for 1000 independent realizations of the 1000 data samples, versus $\log|\omega|$ for values of $\alpha = 0.6$ and $\alpha = 1.5$ ($\beta = 0$ in both cases). Clearly, the best estimates are at the smaller frequencies ($\omega < 1$, $\log|\omega| < 0$) that are also sufficiently removed from the singularity at zero frequency. Similarly, a plot of the mean value of the imaginary part of the log-SCF (dashed line) with standard deviations and the values of the imaginary part of the true log-characteristic function (solid line) versus frequency are shown in Figure 2 for $\beta = 0$ and $\beta = 0.5$ ($\alpha = 1.5$ in both cases). Again, the ability of the SCF to accurately estimate the true characteristic function degrades severely at larger frequencies. Both of these figures agree with the asymptotic statistics in (3.12-3.16) and reinforce the argument for an estimation interval on the SCF at low frequencies, sufficiently removed from zero frequency. Similar results were experienced as both α and β were varied across the parameter space. Based on these findings, we recommend the use of the interval $0.1 \leq \omega \leq 1.0$ on the SCF for the estimation of stable distribution parameters.

The last thing to consider is the number, K, of equally spaced points ($\omega = 0.1, 0.1+$

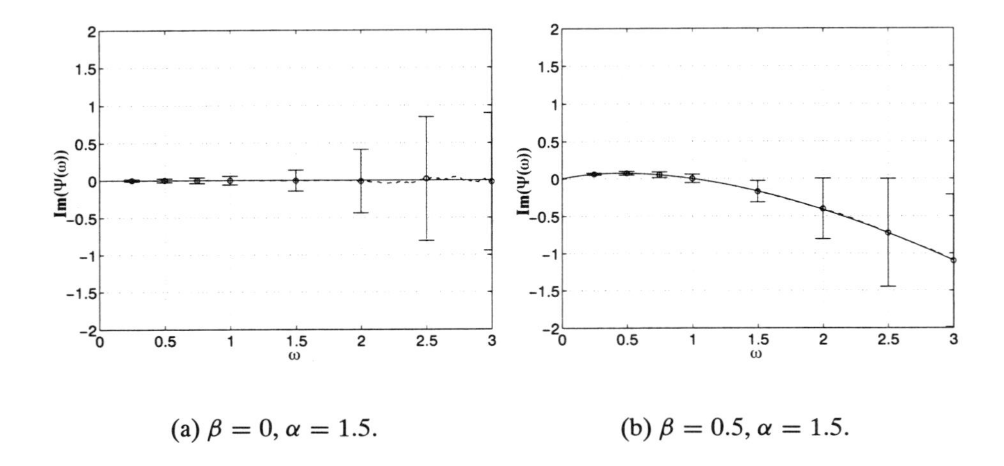

(a) $\beta = 0, \alpha = 1.5$.

(b) $\beta = 0.5, \alpha = 1.5$.

Figure 2: Mean and standard deviation of $Im\left[\widehat{\Psi}(\omega)\right]$ versus frequency ω (dashed line). The solid line is the value computed from the true characteristic function.

$\frac{0.9}{K-1}, \ldots, 1.0 - \frac{0.9}{K-1}, 1.0)$ on the SCF to be used in the regression for the parameter estimation. The estimation is performed ideally with as few points as possible in order to minimize the amount of computation. A simulation was performed in which data samples from stable distributions were generated, followed by the estimation of the parameters using the proposed method. The number of equally spaced points on the interval $\omega \in [0.1, 1.0]$ was varied in order to determine the fewest number of points required to maintain an acceptable level of performance. The values of α and β were varied across the parameter space for a range of sample sizes ($200 \leq N \leq 5000$), and the numbers of points used were 50 points (solid line), 20 points (dashed line), 10 points (starred line), and 5 points (dash-dot line). The results are shown in Figure 3(a) for various values of α and in Figure 3(b) for the various values of β in terms of root mean-square error (RMSE) versus sample size. The performance of the estimator does not vary a great deal for the different numbers of points, although there is a slight drop-off in performance when only 5 points are used for the estimation of α. Therefore, it is recommended that 10 equally spaced points on the interval $\omega \in [0.1, 1.0]$ be used.

In summary, the fixed-interval SCF (FI-SCF) stable distribution parameter estimator is implemented by first normalizing the data samples with initial estimates of c and ζ, followed by the OLS regression of (3.6) and (3.7) over the interval $\omega \in [0.1, 1.0]$ with 10 points. This method has the following key differences with Koutrouvelis' method:

- A modified location parameter ζ is estimated rather than directly estimating δ due to a modification to the regression equation to remove the discontinuity at $\alpha = 1$,

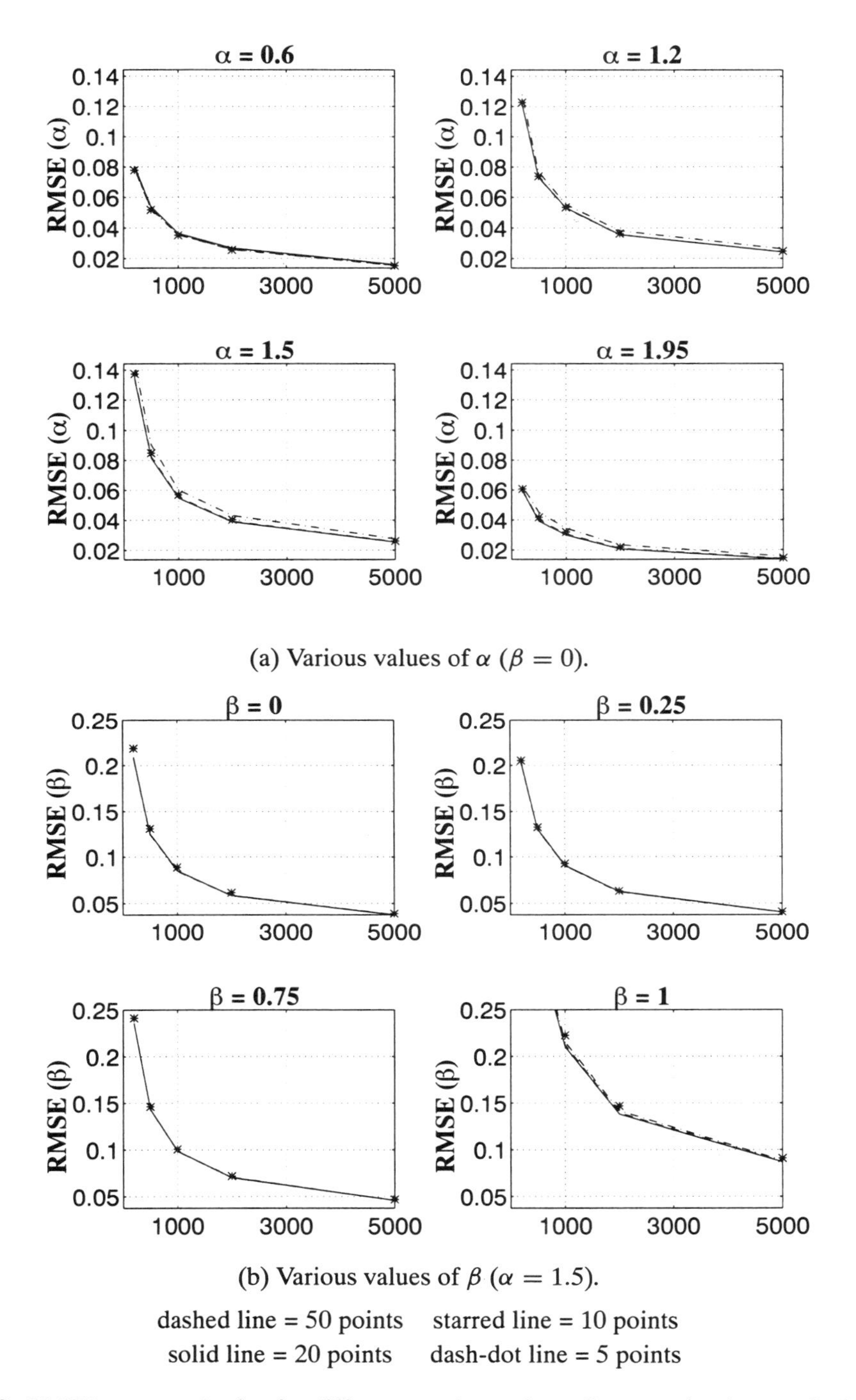

(a) Various values of α ($\beta = 0$).

(b) Various values of β ($\alpha = 1.5$).

dashed line = 50 points starred line = 10 points
solid line = 20 points dash-dot line = 5 points

Figure 3: RMSE vs. sample size for different numbers of equally spaced points on the interval $\omega \in [0.1, 1.0]$.

$\beta \neq 0$.

- The LS regressions are performed only once, as opposed to the numerous iterations for Koutrouvelis' method (typically between 5 and 10). Also, the regression is performed using OLS rather than GLS, thus avoiding the computation of the covariances and the inverse of the covariance matrix.
- The frequency interval is independent of sample size and parameter values which eliminates the lookup table required by the K-SCF method. In addition, the number of points on the SCF used for the FI-SCF method is 10, whereas for the K-SCF method the number of points is variable (between 9 and 118), again depending on sample size and parameter estimates from the previous iteration.
- The initial estimates of c and ζ are found using McCulloch's quantile method. On the other hand, Koutrouvelis uses the Fama and Roll quantile method to find $\hat{c}_0$ and a 25% truncated mean for $\hat{\delta}_0$.

4. Parameter Estimation Results

Monte Carlo simulations were performed in order to demonstrate the performance of the proposed fixed-interval SCF (FI-SCF) parameter estimator relative to other estimation methods. In addition to the proposed method (solid line in Figures 4-7), the techniques considered were Koutrouvelis' SCF (K-SCF) GLS regression method [Kou81] (dashed line), McCulloch's quantile method [McC86] (dash-dot line), and an extreme-value theory (EVT) method [TN96]. The number of non-overlapping intervals used in the EVT technique was set to $L = 50$. Note that the EVT method is intended for symmetric distributions, so comparisons with this method are possible only for the symmetric cases. For each parameter value and sample size, 1000 independent realizations of i.i.d. data samples were generated using the stable random number generator proposed by Chambers et al. [CMS76]. The sample sizes used were $N = 200$, 500, 1000, 2000, and 5000 points. The spread and location parameters were set to $c = 1$ and $\delta = 0$ ($\zeta = \beta \tan \frac{\pi\alpha}{2}$ for $\alpha \neq 1$) for all simulations. All results are reported in terms of the root mean-square error (RMSE) of the parameter estimate

$$\mathrm{RMSE}(\hat{\theta}) = \sqrt{\frac{1}{M} \sum_{m=1}^{M} \left| \theta - \hat{\theta}(m) \right|^2}, \tag{4.1}$$

where $M = 1000$ is the number of realizations, θ is the true parameter value and $\hat{\theta}(m)$ are the estimated parameter values. The results are plotted as RMSE versus sample size. Tabulated results for the proposed FI-SCF estimator are also given in Appendix A.

4.1 Characteristic Exponent Estimation

The first parameter considered is the characteristic exponent α. We begin with simulations for symmetric stable distributions ($\beta = 0$). Stable random variables are generated with different values of the characteristic exponent α ($\alpha = 1.0, 1.2, 1.5, 1.95$) selected to cover most of the parameter space, while emphasizing those values most often encountered in practice [SK74]. The parameter estimation results are shown in Figure 4(a). Note that the performance of the FI-SCF method is slightly worse than the K-SCF method at the lower values of α where it is closer in performance to the quantile method. However, both SCF based estimators outperform the quantile method as α increases. For the greater values of α, the FI-SCF method has almost the same performance as the K-SCF method and actually is slightly better for $\alpha = 1.95$. In comparison to the EVT method, both SCF based methods and the quantile method clearly have much better performance for all values of α for all sample sizes. In fact the RMSE for the EVT method was greater than 0.25 for $\alpha = 1.5$ and $\alpha = 1.95$ so that it could not fit on the plots. Contained in the RMSE of the EVT method is what appears to be a severe bias in the estimator.

Next we consider the estimation of α when the stable distribution is asymmetric with $\beta = 0.5$. The same values of the characteristic exponent were chosen as in the symmetric case ($\alpha = 1.0, 1.2, 1.5, 1.95$) and the performance plots are shown in Figure 4(b). For the moment, only considering the greater values of α (i.e., $\alpha > 1.0$), the performance of the two SCF based estimators is almost identical to the symmetric case, as is to be expected because the estimators are based on the real part of the log-characteristic function ($Re\,[\Psi\,(\omega)] = -\,|c\omega|^{\alpha}$) which is independent of the symmetry parameter β. On the other hand, the performance of the quantile method has dropped off somewhat with a noticable difference with respect to the FI-SCF method even in the case of $\alpha = 1.2$. In the case of $\alpha = 1.0$, the performance of the proposed FI-SCF method is similar to what we expect given the other values of α, while the accuracy of the other two methods has degraded significantly. The drop-off in performance for the K-SCF method can be attributed to the initial estimates of c and δ which assumed a symmetric distribution. Though it did not severely impact performance for the other values of α in the asymmetric case ($\beta = 0.5$), the effect becomes amplified in the vicinity of the discontinuity in the model at $\alpha = 1$ when $\beta \neq 0$. However, the FI-SCF estimator, based on the alternate formulation (3.1), uses McCulloch's quantile method to find and normalize by c and ζ and is able to reasonably preserve the performance of the estimator in the vicinity of the discontinuity.

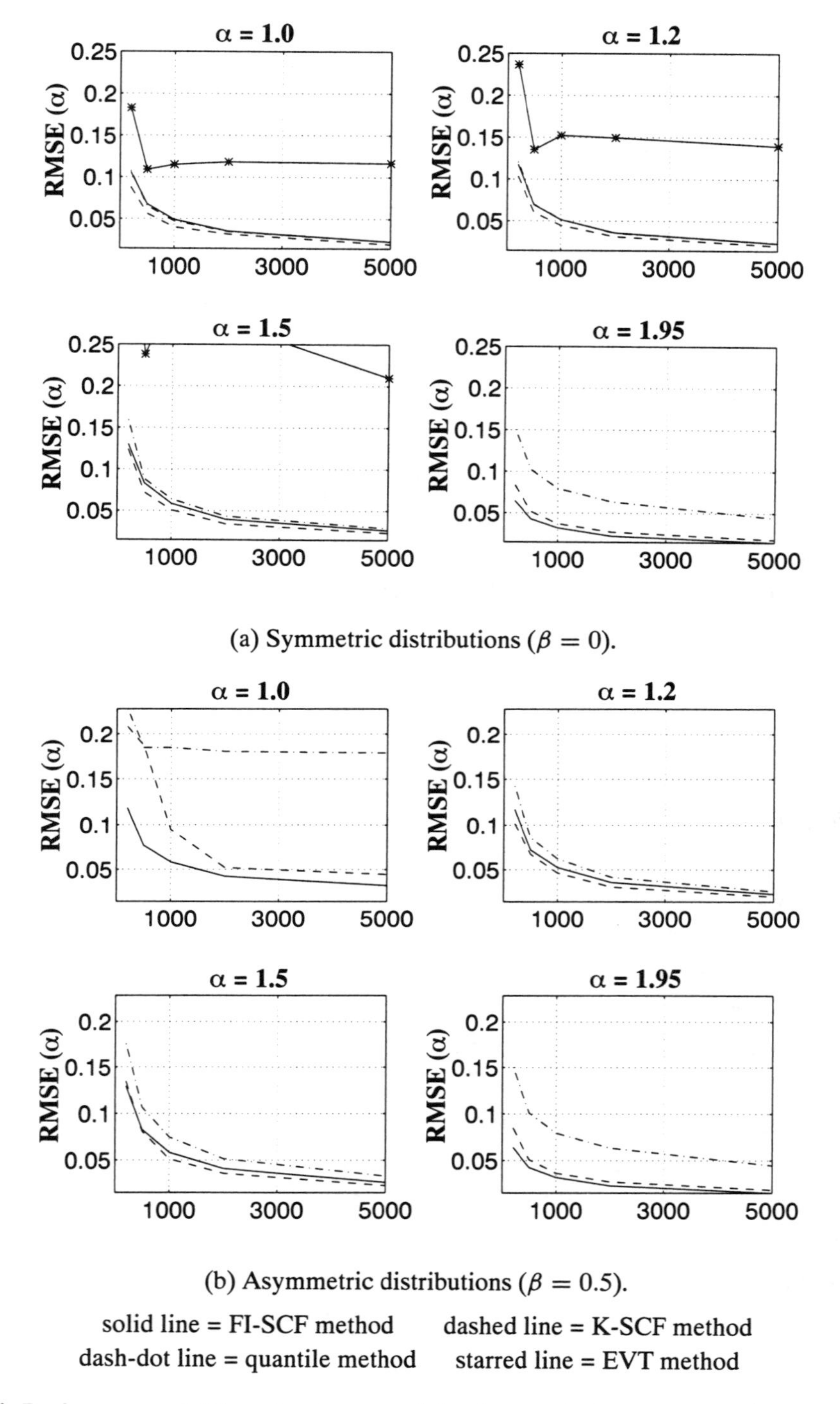

(a) Symmetric distributions ($\beta = 0$).

(b) Asymmetric distributions ($\beta = 0.5$).

solid line = FI-SCF method dashed line = K-SCF method
dash-dot line = quantile method starred line = EVT method

Figure 4: Performance of different estimators of the characteristic exponent α plotted as RMSE vs. sample size.

4.2 Symmetry Index Estimation

The different methods, except for the EVT method, are considered for the estimation of the symmetry index β. The first set of simulations considers the estimation for different values of α with the symmetry index set to $\beta = 0.5$ with the results shown in Figure 5(a). The most drastic difference in performance is at $\alpha = 1$ where the K-SCF estimator has a RMSE greater than 0.5 so it could not be plotted. The large errors are a direct result of the discontinuity in the characteristic function formulation in (2.1) that was used to derive the K-SCF estimator of β. The quantile method also does not perform as well for $\alpha = 1$. On the other hand, the proposed FI-SCF is able to maintain an acceptable level of performance since the estimator is continuous in both α and β. For the other values of α, the K-SCF estimator is better than both the FI-SCF and the quantile estimators which have similar performance. Note that all three estimators drop off significantly for $\alpha = 1.8$. Since the Gaussian distribution ($\alpha = 2.0$) is constrained to be symmetric, the effect of the symmetry index becomes less significant as $\alpha \to 2$. In fact, the symmetry index is arbitrary for Gaussian distributions since the $\tan \frac{\pi\alpha}{2}$ term in (2.2) and (3.2) is zero for $\alpha = 2$. Therefore, any estimator will have a difficult time resolving different values of β for values of α approaching this degenerative case.

Next, the symmetry index β was varied ($\beta = 0.2, 0.4, 0.6, 0.8$), while holding the characteristic exponent constant at $\alpha = 1.5$ with the results shown in Figure 5(b). Only positive values of β were considered since the performance is identical for the corresponding negative values. The proposed FI-SCF method is not as accurate as the K-SCF method, yet better than the quantile method for larger values of β. For smaller values of β, the FI-SCF method is slightly worse than the quantile method.

4.3 Spread and Location Parameter Estimation

The same simulations used for examining estimates of the characteristic exponent α in Section 4.1 were used to evaluate the performance of the spread parameter estimators. The results are shown in Figures 6(a) and (b) for the symmetric and asymmetric cases, respectively. In the symmetric case, the FI-SCF and K-SCF estimators have similar performance with the FI-SCF estimator being narrowly outperformed for smaller values of α. The quantile method has very similar performance to both SCF based methods, except at $\alpha = 1.95$, where it is clearly worse. In all cases, the EVT method performs significantly worse than the other methods with a RMSE greater than 0.2 for $\alpha = 1.5$ and $\alpha = 1.95$ so that it could not be plotted with the other methods. In the asymmetric case, as expected, we find the performance of the SCF based methods to be very similar to the symmetric case for $\alpha \neq 1$. A slight drop-off in performance of the quantile method is noted with respect to the symmetric case. However, as in the estimation of α in section 4.1, the discontinuity at $\alpha = 1$ for $\beta \neq 0$ causes the K-SCF method to have

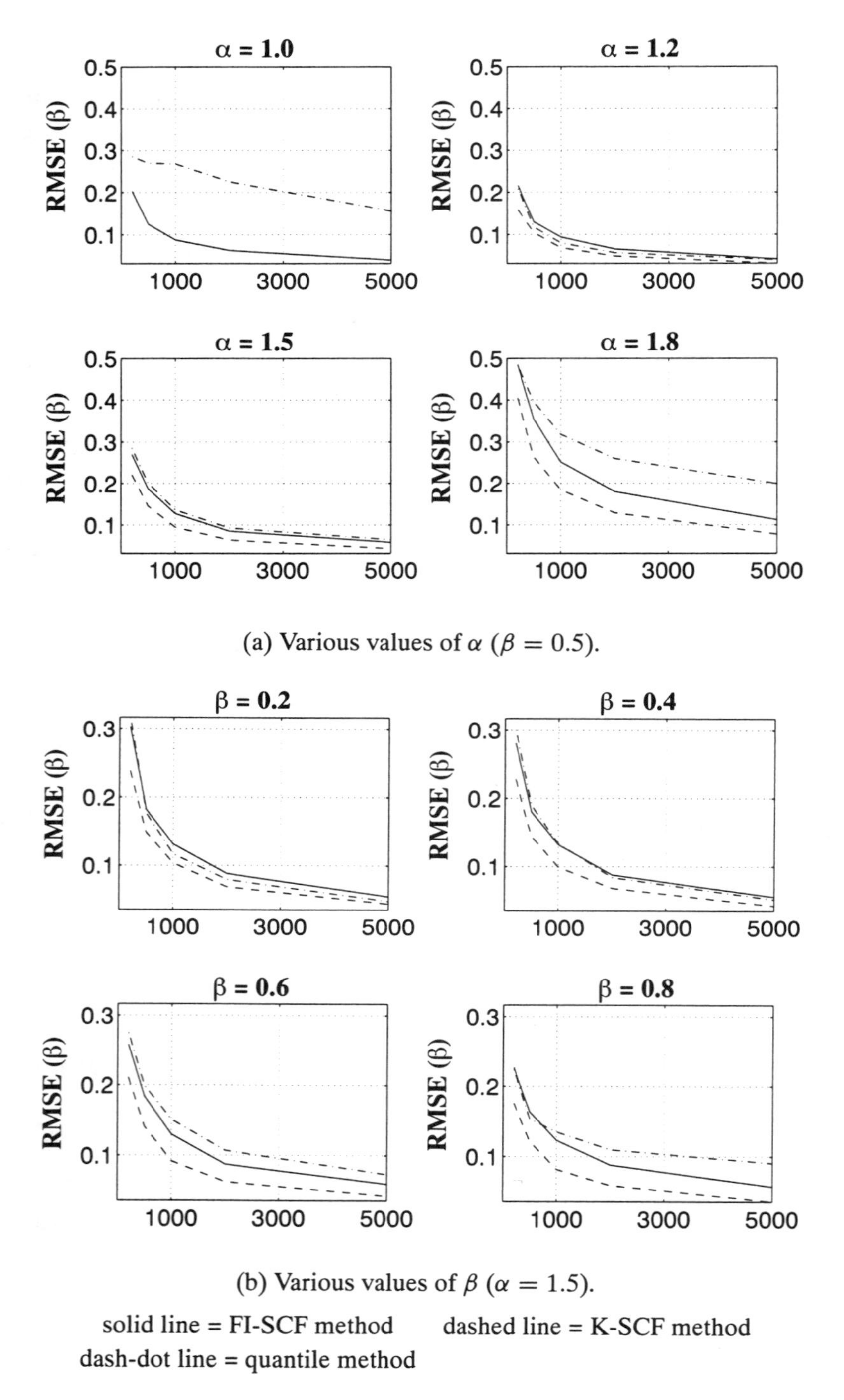

(a) Various values of α ($\beta = 0.5$).

(b) Various values of β ($\alpha = 1.5$).

Figure 5: Performance of different estimators of the symmetry index β plotted as RMSE vs. sample size.

very poor performance. Again, the FI-SCF method did not experience this drop-off in performance since its formulation eliminated this discontinuity. The quantile method had RMSE greater than 0.2 for $\alpha = 1$ so that it could not be plotted with the SCF based methods.

The location parameter estimation results are shown in Figure 7, first for various values of α (α = 1.2, 1.5, 1.8, 1.95) with $\beta = 0.5$, and then for various values of β (β = 0.2, 0.4, 0.6, 0.8) with $\alpha = 1.5$. Note that $\alpha = 1.0$ was not used since it corresponds to a case where the location parameter has no meaning. By examining (3.3), the relationship between δ and ζ becomes discontinuous as β moves away from zero. Even for $\alpha = 1.2$, it is apparent that the performance of all of the estimators (except the EVT method) is significantly worse as $\alpha \to 1.0$. In all of the other cases, the FI-SCF method, the K-SCF method, and the quantile method all have very similar performance. At the lower values of α, the EVT method, which uses the median as an estimate of the location parameter, has significantly better performance than the other methods but this might not be a fair comparison since it assumes $\beta = 0$. The other methods could also realize this significant improvement in performance if they made the same assumption. Since the median is a biased estimate of the location parameter for asymmetric distributions, it is not recommended for general purpose use. The parameter estimation results for various values of β shown in Figure 7(b) show the K-SCF method to have slightly better performance for all values of β except at small sample sizes where the FI-SCF method is significantly better. The FI-SCF and quantile methods have very similar performance except for $\beta = 0.8$ where the quantile method has a small drop-off in performance.

5. Conclusions

In this chapter, we have outlined a procedure for the estimation of the parameters of a stable distribution with an OLS regression of the SCF onto the characteristic function based on modifications to Koutrouvelis' method. The proposed method is derived using an alternate formulation in order to eliminate the discontinuity at $\alpha = 1$ for $\beta \neq 0$ found in the model used by Koutrouvelis. We have attempted to address some of the vital issues concerning such an estimation scheme, namely the selection of the interval on the SCF and the number of points to be used in the OLS regression. Initial estimates of the spread and modified location parameters are computed and the signal is normalized. The parameters are then estimated using the regression of the SCF onto the stable distribution model on a fixed set of frequency points. The spread and modified location parameters are updated with the regression estimates and the location parameter is computed from the modified location parameter estimate.

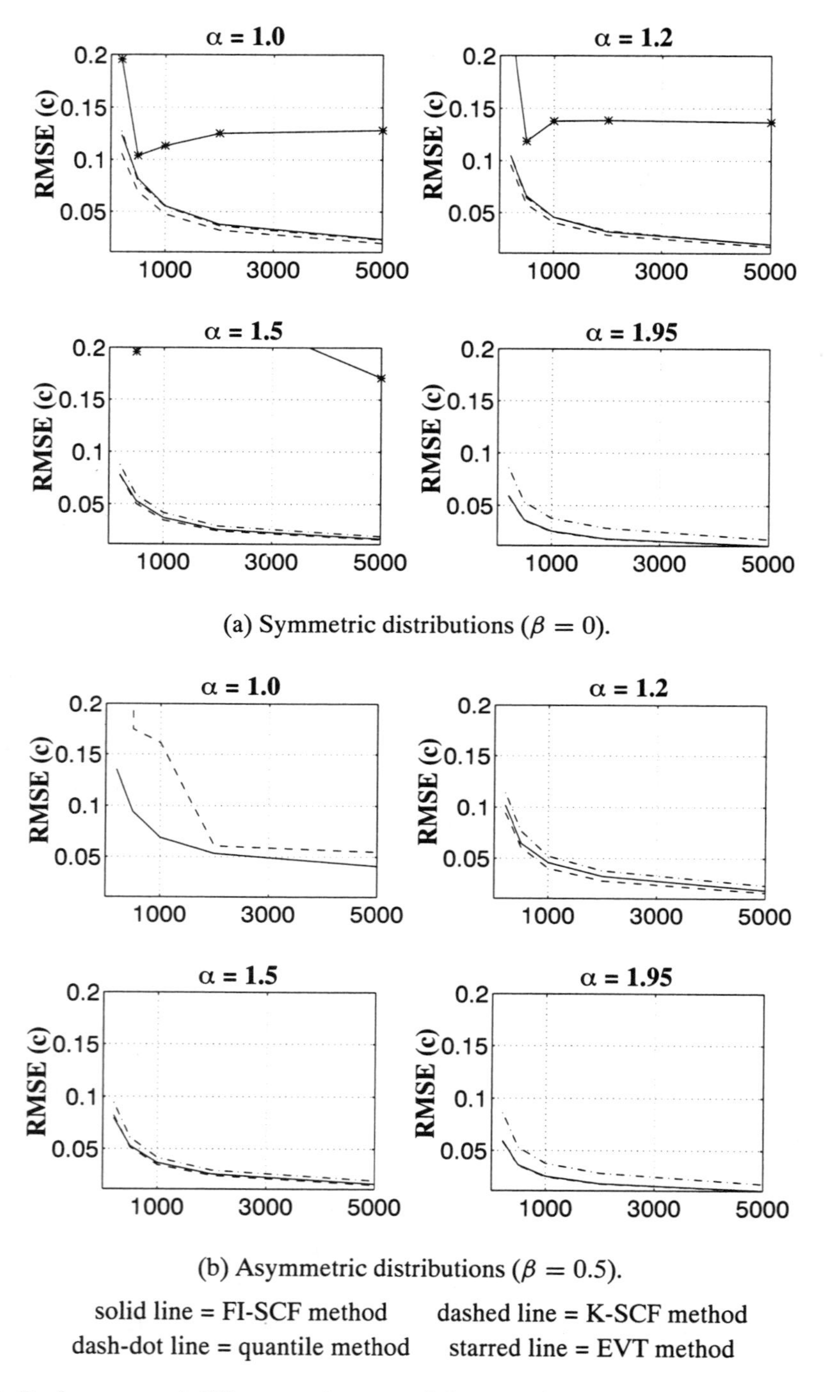

Figure 6: Performance of different estimators of the spread parameter c plotted as RMSE vs. sample size.

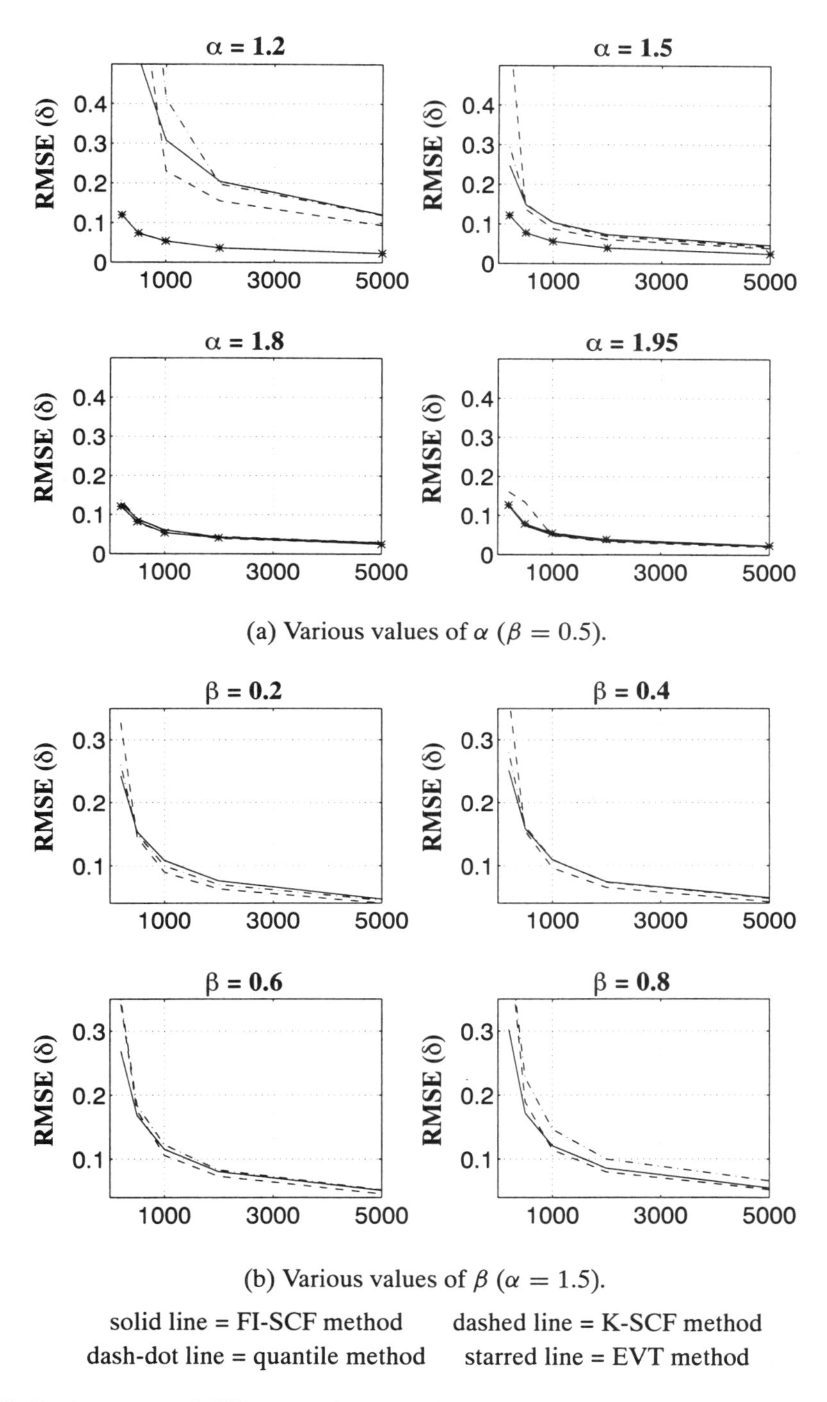

Figure 7: Performance of different estimators of the location parameter δ plotted as RMSE vs. sample size.

The proposed FI-SCF estimator was compared to other existing estimation techniques as the parameters α and β were varied across the parameter space. The results show that the FI-SCF method compares favorably to the K-SCF method with significantly better performance near $\alpha = 1$ for $\beta \neq 0$ due to the elimination of the discontinuity in the model used by Koutrouvelis. Only for the estimation of β did the K-SCF method outperform the FI-SCF method by a noticable margin. However, this gap in performance can be narrowed by incorporating GLS regressions, as done by Koutrouvelis [Kou81], at a mild computational cost. One of the goals in studying characteristic function based estimators was to improve on the K-SCF method, namely to eliminate the need for numerous iterations and lookup tables to find the "best" interval on the SCF for estimation. Also, the new estimator is implemented using OLS instead of GLS for the K-SCF method, eliminating the need for computing covariances and the inversion of the covariance matrix. In addition, the number of points on the SCF to be computed was reduced to 10, compared to as many as 118 depending on the lookup tables.

A. Appendix: Performance Tables

The results of the Monte Carlo simulations with the proposed FI-SCF method using OLS regression are tabulated in this appendix for the various parameter values and sample sizes. The results are given in terms of the mean of the parameter estimate and plus or minus the RMSE.

α	$N = 200$	$N = 500$	$N = 1000$	$N = 2000$	$N = 5000$
1.0	1.02 ± 0.10	1.01 ± 0.07	1.01 ± 0.05	1.01 ± 0.04	1.01 ± 0.02
1.2	1.21 ± 0.12	1.20 ± 0.07	1.20 ± 0.05	1.20 ± 0.04	1.20 ± 0.02
1.5	1.51 ± 0.13	1.50 ± 0.08	1.50 ± 0.06	1.50 ± 0.04	1.50 ± 0.03
1.8	1.81 ± 0.12	1.80 ± 0.07	1.80 ± 0.05	1.80 ± 0.04	1.80 ± 0.02
1.95	1.95 ± 0.06	1.95 ± 0.04	1.95 ± 0.03	1.95 ± 0.02	1.95 ± 0.02

Table 1: Estimation of α for various α ($\beta = 0$).

α	$N = 200$	$N = 500$	$N = 1000$	$N = 2000$	$N = 5000$
1.0	1.04 ± 0.12	1.03 ± 0.08	1.03 ± 0.06	1.03 ± 0.04	1.02 ± 0.03
1.2	1.21 ± 0.12	1.20 ± 0.07	1.20 ± 0.05	1.20 ± 0.04	1.20 ± 0.02
1.5	1.51 ± 0.13	1.50 ± 0.08	1.50 ± 0.06	1.50 ± 0.04	1.50 ± 0.03
1.8	1.81 ± 0.12	1.80 ± 0.07	1.80 ± 0.05	1.80 ± 0.04	1.80 ± 0.02
1.95	1.95 ± 0.06	1.95 ± 0.04	1.95 ± 0.03	1.95 ± 0.02	1.95 ± 0.02

Table 2: Estimation of α for various α ($\beta = 0.5$).

α	$N = 200$	$N = 500$	$N = 1000$	$N = 2000$	$N = 5000$
1.0	0.47 ± 0.20	0.49 ± 0.12	0.49 ± 0.09	0.49 ± 0.06	0.49 ± 0.04
1.2	0.50 ± 0.21	0.49 ± 0.13	0.50 ± 0.09	0.50 ± 0.07	0.50 ± 0.04
1.5	0.52 ± 0.27	0.50 ± 0.19	0.50 ± 0.13	0.50 ± 0.09	0.50 ± 0.06
1.8	0.49 ± 0.48	0.51 ± 0.36	0.52 ± 0.25	0.51 ± 0.18	0.51 ± 0.11
1.95	0.21 ± 0.90	0.27 ± 0.79	0.39 ± 0.63	0.45 ± 0.52	0.49 ± 0.35

Table 3: Estimation of β for various values of α ($\beta = 0.5$).

α	$N = 200$	$N = 500$	$N = 1000$	$N = 2000$	$N = 5000$
1.0	1.00 ± 0.12	1.00 ± 0.08	1.00 ± 0.06	1.00 ± 0.04	1.00 ± 0.02
1.2	0.99 ± 0.11	1.00 ± 0.07	1.00 ± 0.05	1.00 ± 0.03	1.00 ± 0.02
1.5	0.99 ± 0.08	1.00 ± 0.05	1.00 ± 0.04	1.00 ± 0.03	1.00 ± 0.02
1.8	0.99 ± 0.07	1.00 ± 0.04	1.00 ± 0.03	1.00 ± 0.02	1.00 ± 0.01
1.95	0.99 ± 0.06	1.00 ± 0.04	1.00 ± 0.03	1.00 ± 0.02	1.00 ± 0.01

Table 4: Estimation of c for various values of α ($\beta = 0$).

Acknowledgements.

The authors would like to thank the anonymous reviewer for performing a thorough review and providing many useful comments, particularly on the estimation of the location parameter.

α	$N = 200$	$N = 500$	$N = 1000$	$N = 2000$	$N = 5000$
1.0	1.04 ± 0.14	1.04 ± 0.09	1.03 ± 0.07	1.03 ± 0.05	1.03 ± 0.04
1.2	0.99 ± 0.10	1.00 ± 0.06	1.00 ± 0.05	1.00 ± 0.03	1.00 ± 0.02
1.5	0.99 ± 0.08	1.00 ± 0.05	1.00 ± 0.04	1.00 ± 0.03	1.00 ± 0.02
1.8	0.99 ± 0.07	1.00 ± 0.04	1.00 ± 0.03	1.00 ± 0.02	1.00 ± 0.01
1.95	0.99 ± 0.06	1.00 ± 0.04	1.00 ± 0.03	1.00 ± 0.02	1.00 ± 0.01

Table 5: Estimation of c for various values of α ($\beta = 0.5$).

β	$N = 200$	$N = 500$	$N = 1000$	$N = 2000$	$N = 5000$
0.2	0.21 ± 0.30	0.20 ± 0.18	0.21 ± 0.13	0.21 ± 0.09	0.20 ± 0.05
0.4	0.43 ± 0.28	0.41 ± 0.18	0.40 ± 0.13	0.40 ± 0.09	0.40 ± 0.06
0.6	0.61 ± 0.26	0.60 ± 0.18	0.60 ± 0.13	0.60 ± 0.09	0.60 ± 0.06
0.8	0.76 ± 0.23	0.79 ± 0.16	0.80 ± 0.12	0.80 ± 0.09	0.80 ± 0.06

Table 6: Estimation of β for various values of β ($\alpha = 1.5$).

α	$N = 200$	$N = 500$	$N = 1000$	$N = 2000$	$N = 5000$
1.2	0.93 ± 12.44	0.17 ± 1.09	0.10 ± 0.59	0.03 ± 0.33	0.02 ± 0.20
1.5	0.00 ± 0.26	0.00 ± 0.17	0.00 ± 0.11	0.00 ± 0.08	0.00 ± 0.05
1.8	-0.04 ± 0.14	-0.01 ± 0.09	0.00 ± 0.06	0.00 ± 0.04	0.00 ± 0.03
1.95	-0.03 ± 0.12	-0.02 ± 0.08	-0.01 ± 0.05	-0.01 ± 0.04	0.00 ± 0.02

Table 7: Estimation of δ for various values of α ($\beta = 0.5$).

β	$N = 200$	$N = 500$	$N = 1000$	$N = 2000$	$N = 5000$
0.2	0.00 ± 0.24	0.01 ± 0.15	0.00 ± 0.11	0.00 ± 0.08	0.00 ± 0.05
0.4	0.01 ± 0.25	0.00 ± 0.16	0.00 ± 0.11	0.00 ± 0.07	0.00 ± 0.05
0.6	0.00 ± 0.27	0.00 ± 0.17	0.00 ± 0.12	0.00 ± 0.08	0.00 ± 0.05
0.8	-0.04 ± 0.30	-0.02 ± 0.17	0.00 ± 0.12	0.00 ± 0.09	0.00 ± 0.05

Table 8: Estimation of δ for various values of β ($\alpha = 1.5$).

References

[AL89] V. Akgiray and C. G. Lamoureux. Estimation of stable-law parameters: A comparative study. *J. Business & Statistics*, 7(1):85–93, Jan. 1989.

[Ara80] R. W. Arad. Parameter estimation for symmetric stable distribution. *International Economic Review*, 21(1):209–220, Jan. 1980.

[BB81] P. J. Brockwell and B. M. Brown. High-efficiency estimation for the positive stable laws. *J. American Statistical Association*, 76(375):626–631, 1981.

[CMS76] J. M. Chambers, C. L. Mallows, and B. W. Stuck. A method for simulating stable random variables. *J. American Statistical Association*, 71(354):340–344, June 1976.

[DuM73] W. H. DuMouchel. On the asymptotic normality of the maximum-likelihood estimate when sampling from a stable distribution. *The Annals of Statistics*, 3:948–957, 1973.

[Fel71] W. Feller. *An Introduction to Probability Theory and its Applications*, volume 2. John Wiley & Sons, New York, NY, 3rd edition, 1971.

[FM81a] A. Feuerverger and P. McDunnough. On the efficiency of empirical characteristic function procedures. *J. Royal Statistical Society, Series B*, 43(1):20–27, 1981.

[FM81b] A. Feuerverger and P. McDunnough. Chapter in *Statistics and Related Topics*, On Efficient Inference in Symmetric Stable Laws and Processes. Ed. M. Csörgö *et al.*, North-Holland Publishing Co., New York, NY, 1981.

[FR71] E. Fama and R. Roll. Parameter estimates for symmetric stable distributions. *Journal of the American Statistical Association*, 66(334):331–338, June 1971.

[Gol89] G. H. Golub and C. F. Van Loan *Matrix Computations*. The Johns Hopkins University Press, Baltimore, MD, 1989.

[JP76] E. Jakeman and P. N. Pusey. A model for non-Rayleigh sea echos. *IEEE Trans. on Antennas and Propagation*, 24(6):806–814, Nov. 1976.

[KM96] S. M. Kogon and D. G. Manolakis. Signal modeling with self-similar α-stable processes: the fractional Lévy stable motion model. *IEEE Trans. on Signal Processing*, 44(4):1006–1010, April 1996.

[Kou80] I. A. Koutrouvelis. Regression-type estimation of the parameters of stable laws. *J. American Statistical Association*, 75(372):919–928, 1980.

[Kou81] I. A. Koutrouvelis. An iterative procedure for the estimation of the parameters of stable laws. *Comm. in Statistics-Simulation and Computation*, 10(1):17–28, 1981.

[Man63] B. B. Mandelbrot. The variation of certain speculative prices. *J. Business*, 36:394–419, 1963.

[McC] J. H. McCulloch. On the parameterization of the afocal stable distributions. To appear in *Bulletin of London Mathematical Society*.

[McC86] J. H. McCulloch. Simple consistent estimators of stable distribution parameters. *Comm. in Statistics-Simulation and Computation*, 15(4):1109–1138, 1986.

[NS95] C. L. Nikias and M. Shao. *Signal Processing with Alpha-Stable Distributions and Applications*. John Wiley & Sons, New York, NY, 1995.

[Pai96] S. Painter. Evidence of non-Gaussian scaling behavior in heterogeneous sedimentary formations. *Water Resources Research*, 32(5):1183–1195, May 1996.

[Pap91] A. Papoulis. *Probability, Random Variables, and Stochastic Processes*. McGraw-Hill, New York, NY, 3rd edition, 1991.

[PD85] A. S. Paulson and T. A. Delehanty. Modified weighted squared error estimation procedures with special emphasis on the stable laws. *Comm. in Statistics-Simulation and Computation*, 14(4):927–972, 1985.

[PHL75] A. S. Paulson, E. W. Holcomb, and R. A. Leitch. The estimation of the parameters of the stable laws. *Biometrika*, 62:163–170, 1975.

[Pre72] S. J. Press. Estimation of univariate and multivariate stable distributions. *J. American Statistical Association*, 67(340):842–846, 1972.

[Shu88] R. H. Shumway. *Applied Statistical Time Series Analysis*. Prentice-Hall, Englewood Cliffs, NJ, 1988.

[SK74] B. W. Stuck and B. Kleiner. A statistical analysis of telephone noise. *Bell Systems Technical Journal*, 53:1262–1320, 1974.

[ST94] G. Samorodnitsky and M. S. Taqqu. *Stable Non-Gaussian Random Processes: Stochastic Models with Infinite Variance*. Chapman & Hall, New York, NY, 1994.

[TN96] G. A. Tsihrintzis and C. L. Nikias. Fast estimation of the parameters of alpha-stable impulsive interference. *IEEE Trans. on Signal Processing*, 44(6):1492–1503, June 1996.

[Zol86] V. M. Zolotarev. *One-dimensional Stable Distributions*. American Mathematical Society, Providence, RI, 1986.

Stephen M. Kogon
MIT Lincoln Laboratory
244 Wood Street
Lexington, MA 02173-9108

and

Douglas B. Williams
Center for Signal and Image Processing
School of Electrical and Computer Engineering
Georgia Institute of Technology
Atlanta, GA 30332-0250

IV. Regression

Bootstrapping Signs and Permutations for Regression with Heavy-Tailed Errors: a Robust Resampling

Raoul LePage [1], *Krzysztof Podgórski* [2], *Michal Ryznar and Alex White*

Abstract

For multiple linear regression with heavy-tailed errors, we describe and discuss in detail two resampling methods: one based on resampling permutations of residuals from the least squares fit, the other, to some extent a complementary one, exploiting random sign flips. Both succeed, without moment assumptions, in recovering conditional distributions of the least squares estimators. We outline consequences of the results for effective statistical inference for regression models with outsized errors.

1. Introduction

Introduced in 1979 by Stanford statistician Dr. Bradley Efron, bootstrap advances the notion that by sampling from our data we may learn the sampling variations of statistical estimates (see [Efr]). This idea is a powerful one capable of achieving, with the assistance of even a desktop computer, truly remarkable results heretofore available only through the application of deep theoretical approximations and tables.

Bootstrap requires modification to work with time series or other dependent data and may be sensitive to measurement populations that generate outsized measurements with appreciable frequency (whether or not the population mean square is technically finite). In the context of multiple linear regression we will describe an alternative bootstrap, permutation of residuals, which largely circumvents the second limitation. We call this *robust resampling* since it represents a type of resampling that is robust against outsized errors. To achieve this, we renounce the usual bootstrap objective of estimating the sampling error distribution, which for heavy-tailed error distributions is only achievable through such devices as bootstrapping with reduced sample size ([ArGi], [HLeP]). Instead, we consistently estimate *conditional* sampling errors conditional on the order statistics actually present (unobserved) in the data. A corollary benefit is that our confidence intervals for regression coefficients are conditional (on order statistics of errors) and with high probability are narrower in the presence of heavy-tailed error distributions. Careful statements of these results are contained within the text below.

[1] Research partially supported by ONR Grant N00014-91-J-1087

[2] Research partially supported by ONR Grant N00014-93-1-0043 when on the leave at Center for Stochastic Processes, University of North Carolina, Chapel Hill

Let us start with a classical statistical technique based on the Central Limit Theorem, the 95% confidence interval for the population mean μ. The solution given for the case of n independent samples $Y_1, \ldots, Y_n$ is the interval range $\overline{Y} \pm 1.96 * s_Y/\sqrt{n}$, where s_Y^2 is the $(n-1)$-divisor sample variance. Amazingly, another solution, the bootstrap, will give about the same quality results but without any explicit reference to 1.96, s_Y, or $\sqrt{n}$. In practice we:

1. *perform thousands of experiments, each experiment consisting of averaging a sample $Y_1^*, \ldots, Y_n^*$ selected with-replacement from our original sample $Y_1, \ldots, Y_n$,*
2. *find the symmetrically placed interval around our original sample mean $\overline{Y}$ that will contain approximately 95% of these thousands of averages.*

In step 2, we are attempting to find an interval containing exactly 95% of all possible averages from samples of n selected with-replacement from our data $Y_1, \ldots, Y_n$. But there exist n^n such samples! By taking only thousands of such samples at random we are relying on the law of averages. Where possible we should take tens of thousands.

In Figure 1 below, the normal distribution, the theoretical base for the classical technique, is compared with the bootstrap distribution obtained by 10,000 repetitions of resampling from the following data:

```
144 143 142 143 146 141 146 141 148 143 143 141 143 143 146
143 144 144 143 145 142 143 145 144 143 140 143 141 144 141
```

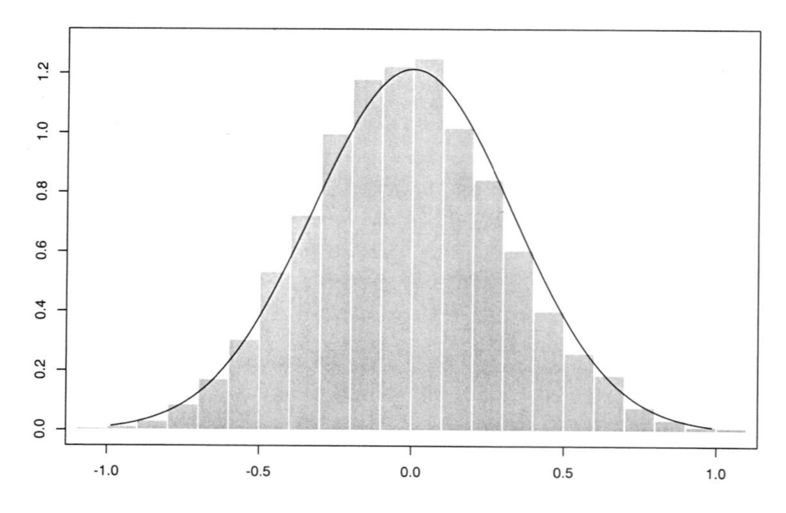

Figure 1: The normal density approximating the distribution of $\overline{Y} - \mu$ vs. a histogram of 10,000 bootstrap samples $\overline{Y^*} - \overline{Y}$.

Remarkably, the 95% confidence interval [142.633, 143.900] produced by the bootstrap is about the same as [142.623, 143.910] yielded by the $1.96 * s_n/\sqrt{n}$ method. This is no accident for the bootstrap method just described (known as the naive bootstrap to distinguish it from more subtle methods of bootstrap) has theoretical properties almost identical to the classical method in this problem and will generally agree pretty closely with the classical solution. Its rather like two archers who both hit the target 95% of the time and whose arrows nearly always hit close together as well. However with naive bootstrap we do not see 1.96, s_n, $\sqrt{n}$. The "theory operates completely in the background, its analytical tools kept completely out of sight in applications."

There are caveats with both 95% methods of course:

- *sample measurements are assumed to be independent,*
- *the population must have a finite mean square,*
- *the number of sample measurements n must be large enough for the limiting 0.95 probability approximation to take hold,*
- *the number of bootstrap replicate samples must be large enough for the law of averages to ensure a good approximation of the bootstrap confidence interval, which is in principle defined for the population of all n^n possible bootstrap replicate samples.*

For the sake of illustration, let us consider the population of 40 containing some "outsized" values:

1 1 1 1 1 1 1 1 1 1 1 1 1 1 1 1 1 1 1 100
2 2 2 2 2 2 2 2 2 2 2 2 2 2 2 2 2 2 2 1000

To estimate the actual coverage of the population mean by the bootstrap "95%" confidence intervals we selected 1000 samples of n=30 from this population. From every one of these we resampled 1000 bootstrap samples of n=30 to produce a bootstrap confidence interval. One thousand of these confidence intervals covered the population mean only approximately 53.3% of the time, much below 95% level. Clearly, this effect is due to "outsized" measurements which greatly affect the population mean but rarely fall in a sample.

Finite population mean square is really a matter of degree since, as we have seen, outsized measurements in the population can increase the minimum n required for the bootstrap approximation or the classical approximation $1.96 * s_n / \sqrt{n}$ to become accurate. In some cases robust solutions do exist. In the sequel we consider two such methods in connection with fitting multiple linear regression by least squares. The first one is based on resampling permutations of residuals of the least squares fit. This solution is robust for all coefficient estimates except the constant term. A primary application would be the familiar one of fitting a straight line through (x, y) data points. It can serve as a model for our discussion. In this case the method about to be proposed is robust for the slope but not for the intercept. Estimation of the latter, in many cases, can be reduced to estimation of the population mean for which, if errors in the model have symmetric distribution, we propose yet another bootstrapping by random coin flipping (resampling signs) which is discussed in Section 3. Before turning to the formal description of the proposed methods of resampling, let us briefly survey the traditional bootstrap for straight line regression.

The classical model for straight line regression is

$$y = a + bx + \epsilon. \tag{1.1}$$

It assumes that whenever we select an x (input) for observation, the y (response) is generated by the above relation in which a, b are unknown constants and ϵ is a random error. A typical assumption is that the errors are independently sampled from a population having a mean of zero and a finite population mean squares. A finite set of n data points (x, y) is obtained under this model and a straight line is fitted to the plot of these n points by the method of least squares. Having fit a line to our noisy data, how much statistical variation is there in the fitted line? Put another way, how would the fitted line vary for different samples of n points which we might have gotten? In particular, can one give a 95% confidence interval for the unknown slope b? As it happens, the slope we fit to a plot is obtained via the scalar dot product $\widehat{b} = v \cdot y$ of some vector v with the y observations. It is worth noting that for the slope (and more generally for every parameter other than the constant term in a multiple linear regression model with constant term) the associated v is a *contrast*, i.e. its entries sum to zero. Actually, for the slope it can be written explicitly in *contrast form*

$$v = \frac{1}{(n-1)s_x^2}(x_1 - \bar{x}, \ldots, x_n - \bar{x}).$$

We have the important relation $v \cdot y = b + v \cdot \epsilon$, so the error in estimating b is $\widehat{b} - b = v \cdot \epsilon$, which is a contrast applied to the errors. General statistical theory now kicks in and tells us that a 95% confidence interval for slope b is $v \cdot y \pm 1.96 * \sqrt{n} s_v s_{\epsilon^\perp}$ where $\epsilon^\perp = y - \widehat{a} - \widehat{b}x$ is the vector of residuals which are the vertical discrepancies between the data plot and the line we fit. By comparison with the classical solution, the naive bootstrap solution is:

1. *perform thousands of experiments, each experiment consisting of computing* $v \cdot \epsilon^*$ *for a vector* ϵ^* *obtained by with-replacement sampling of n from the n residuals,*
2. *find an interval [-#, #] around zero which contains 95% of these numbers* $v \cdot \epsilon^*$,
3. *the naive bootstrap 95% confidence interval is then* $v \cdot y \pm \#$.

Our robust resampling replaces "with-replacement sampling of n from the n residuals" with "random permutation of residuals" in 1.

To conclude this introduction, let us emphasize that methods which consistently estimate conditional distributions, which is the case for our robust resampling methods, can be more attractive than their unconditional counterparts when they produce on average shorter confidence intervals by reducing some variability in estimation errors. In Figure 2, we illustrate this feature for permuting residuals and resampling signs discussed in this presentation and applied to fitting a cubic polynomial and population mean, respectively (see Sections 2 and 3). Using the Monte Carlo method, we generated in both cases 1000 confidence intervals based on the relevant conditional distributions of estimation errors. The resulting histograms of the logarithms of their widths are presented in Figure 2 and compared to the logarithm of the width of the confidence interval based on the corresponding unconditional distributions. As our numerical experiments, reported in Section 2, indicate, it can even happen that the conditional approach will successfully compete with the *exact* unconditional method for models with gaussian errors. This makes it a very attractive tool of statistical analysis whether or not we deal with heavy-tailed distributions.

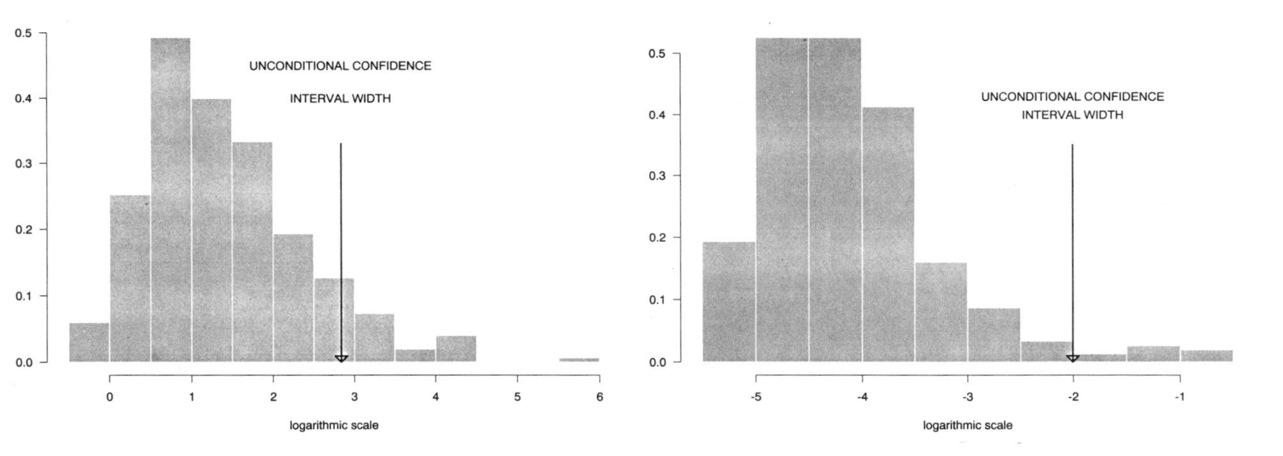

Figure 2: The distribution of widths of conditional confidence intervals vs. the width of the unconditional ones.

2. Permuting residuals

Like the classical solution the naive bootstrap works well for errors having population mean zero and finite population mean square, but can go badly wrong for errors departing from the letter or even the spirit of these assumptions. As mentioned above, our robust solution, introduced originally in [LePP], simply replaces "with-replacement sampling of n from n residuals" with "random shuffle (permutation) of n residuals" in step 1 just above (such a bootstrap shuffle of $\epsilon^{\perp}$ will be denoted by $\pi\epsilon^{\perp}$). It is a method which was proposed years ago as a "descriptive method" (see [FrLa]). We have proved that subject to very minor assumptions this operation has high probability of recovering the sampling variations of $v \cdot \epsilon$ relative to random shuffles of ϵ. In other words, our robust solution targets what is known as the sampling distribution of $v \cdot \epsilon$ *conditional* on the ordered values of the errors actually present (but unobserved) in the data. Theoretical justification for shuffling residuals rests in part on a formula which says that the mean of the squared discrepancy $(v \cdot \pi\epsilon^{\perp} - v \cdot \epsilon)^2$ under shuffling ϵ is only the $(d-1)/(n-1)$ part of the mean of $(v \cdot \epsilon)^2$ under shuffling ϵ, regardless of the contrast vector v, error vector ϵ, number of indėpendent variables d in an arbitrary multiple linear regression model with constant term, or the number n of data points. This implies the following robust version of the bootstrap solution alluded to above:

1. *for a vector of residuals $\epsilon^{\perp}$ and for a contrast v, generate thousands of independent uniformly distributed permutations π, for each of them compute $v \cdot \pi\epsilon^{\perp}$,*

2. *find an interval [-#, #] around zero which contains 95% of these numbers,*

3. *the robust bootstrap 95% confidence interval for b corresponding to v is then $v \cdot y \pm \#$.*

Illustration of resampling permutations. Let us consider a simple three-point regression $y = 1 + x + \epsilon$ with $(x_1, x_2, x_3) = (-0.75, -0.25, 1)$ and $(\epsilon_1, \epsilon_2, \epsilon_3) = (-0.75, 1.5, 1.78)$. Figure 3 shows the least square fit to the data and five other fits resulting from permuting the three residuals. The six-point distribution $v \cdot \pi\epsilon^{\perp} \big| Y$, where v is the contrast corresponding to slope, resides on

$$(-0.82, -0.44, 0.00, 0.08, 0.82, 0.94)$$

while $v \cdot \epsilon \big| \mathcal{F}$, which could be obtained by permuting actual errors, occupies

$$(-1.51, -1, 43, 0.22, 0.52, 1.00, 1.21).$$

Of course, this simple example illustrates only the implementation of the method. In practice, the sample size usually is much larger than 3. Typically, we do not consider all $n!$ permutations but only a Monte Carlo sample.

Let us set this up in the usual matrix notation for general linear regression

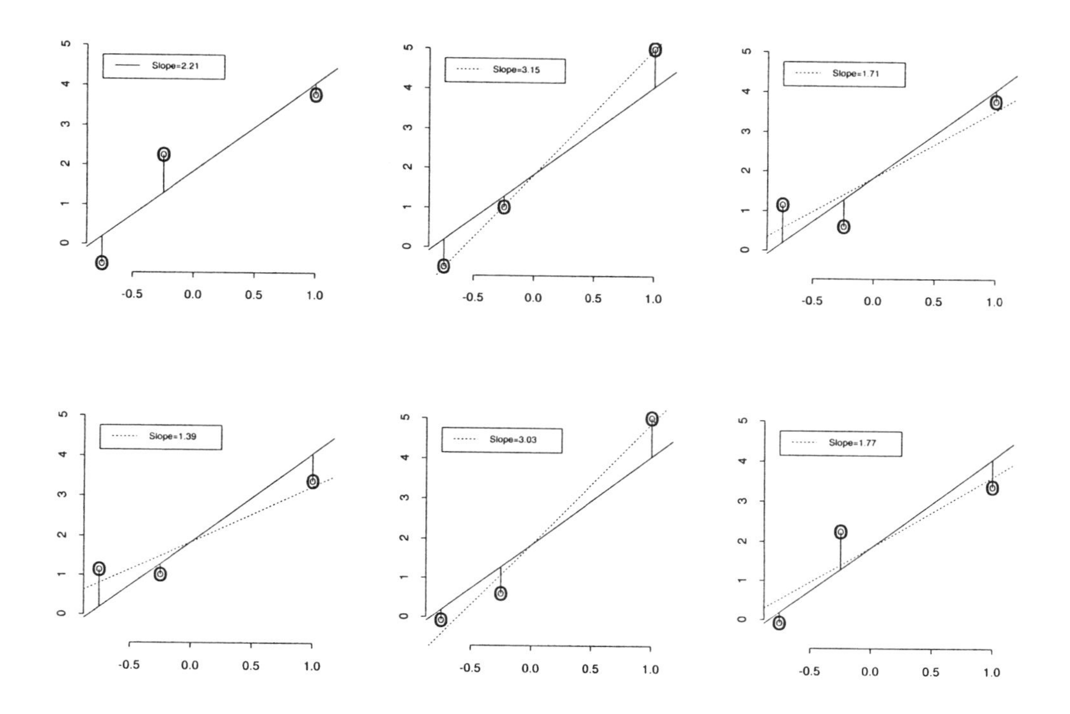

Figure 3: Permuting residuals in three-point regression.

$$Y = \mathbf{X}\beta + \epsilon \tag{2.1}$$

in which $\mathbf{X}$ is an $(n \times d)$ matrix of input values and β is a length-d vector of unknown regression coefficients. For this model, having a concrete set of data, we produce thousands of fits, or more precisely thousands of d-dimensional vectors, each one a vector of coefficients fitted by feeding shuffled (permuted) residuals into the model. There is a way to view all of these thousands of vectors. A parallel plot views coefficient vectors in many dimensions by treating them as functions. For example, the point in four dimensional Euclidean space with coordinates (2,7,3,5) is graphed as a broken line path joining 2 on the first vertical with 7 on the second vertical, etc. When applied to the least squares coefficient estimators obtained by shuffling residuals the plot gives us a picture of the statistical variation in the estimates.

Figure 4 uses parallel plots to give a convincing visual demonstration of the performance of robust resampling in the case of a polynomial regression

$$y = \beta_0 + \beta_1 x + \beta_2 x^2 + \beta_3 x^3 + \epsilon \tag{2.2}$$

for "bad" errors ϵ. For our "bad" errors we chose samples from the Cauchy distribution, which has an infinite population mean. Inputs (i.e. the x values under observation) are

101 equally spaced points from $x = 0$ to $x = 1$. The left hand side plot shows the actual joint sampling variation of the least squares estimators of $\beta_1, \beta_2, \beta_3$. This plot is obtained by permuting the actual errors present in the data and is thus unavailable to the experimenter who sees only the noisy observations (x, y). However, by using only the data and permuting residuals (i.e. by robust resampling) we obtain the right side plot which surprisingly recovers even delicate features of the left side plot in spite of the "bad" Cauchy error distribution.

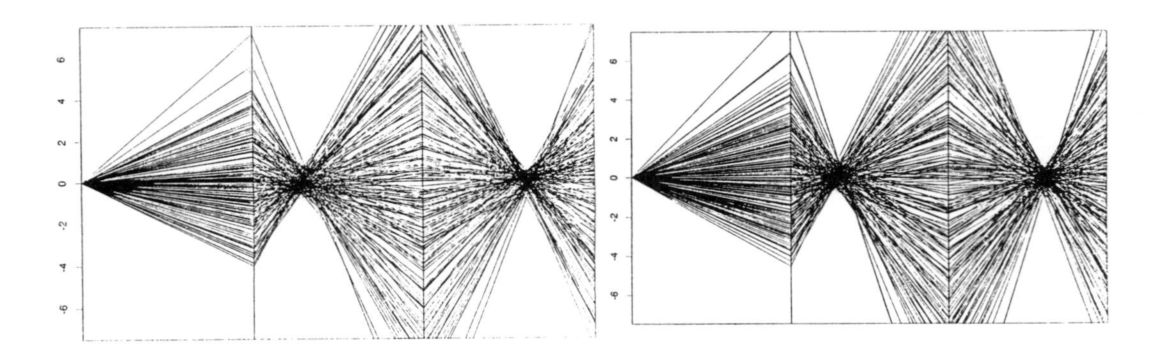

Figure 4: Parallel plots of the conditional distribution of the estimation error (left) vs. the distribution obtained by robust resampling residuals (right).

Another illustration is provided in Figure 5 through density estimates of the sampling distributions corresponding to estimation of β_2. Here, the density estimates were based on 10,000 data points.

Our methods were intended to cope with heavy-tailed distributions. But they can also be applied successfully if the coordinates of ϵ are assumed to be independent samples from any distribution with mean zero and finite variance. One can use Hájek's central limit theorem for rank tests to show that $\upsilon \cdot \epsilon$ conditional on the order statistics of the errors is asymptotically normal [Hjk]. Together with (4.4), this result says that the robust bootstrap will approximate the Gaussian limit just as well as the classical methods. To demonstrate this, let us carry out computer studies for several designs and with normal errors in the models, comparing the solution offered by robust resampling to the classical one based on the normal tables. These show our method to be very competitive.

More specifically, we have considered three regression models with the designs $\mathbf{X}_1$, $\mathbf{X}_2$, $\mathbf{X}_3$ all having $\mathbf{1}$ as the first column and where: $\mathbf{X}_1$ is the 100×2 matrix with the second column being the contrast having the first half of coordinates equal to one and the second one equal to minus one; $\mathbf{X}_2$ is the 101×2 matrix with the second column: $(-50, -49, \ldots, 0, \ldots, 49, 50)$; $\mathbf{X}_3$ is the 35×2 matrix with the second column: $(-2^8, \ldots, -2^0, \ldots, -2^{-8}, 0, 2^{-8}, \ldots, 2^0, \ldots, 2^8)$. For all these models we have considered errors possessing the standard normal distribution and estimation of the parameters corresponding to the non-constant columns of the designs. Using the Monte Carlo method, for each of the models, 1000 vectors of errors have been generated and for each such sample two 95% confidence intervals have been computed: the first one based on the exact, unconditional normal distribution, and the second one based on 1000 permutations of residuals as described in our method. Their half-widths were recorded together with coverage of the true parameter. Since observed coverage probability varied slightly between the two methods and among the models, to properly compare the half-widths the normal distribution confidence intervals were scaled so that their coverage probabilities were equal to that of our method.

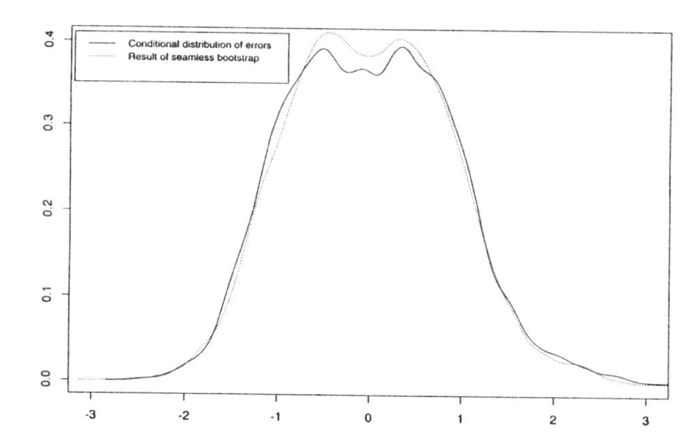

Figure 5: Density estimates of the conditional sampling distributions and its robust recovery for the third parameter in the example dealing with fitting a cubic polynomial.

The results of these numerical studies, summarized in Table 1, indicate that on average, confidence intervals based on our method are very competitive with those based on the exact unconditional normal distribution.

In addition, we can compare the naive and robust bootstraps. To do this we use a special construction linking the two bootstrap algorithms together (see Section 4). For this construction Equation (3.11) in [Hjk] yields

$$\frac{E^{\epsilon}\left((v \cdot \pi\epsilon^{\perp}) - (v \cdot \epsilon^{*})\right)^2}{E^{\epsilon}\,(v \cdot \epsilon^{*})^2} < \frac{2\sqrt{2}n}{n-1} \frac{\max_{1 \le i \le n} \mid \epsilon_i^{\perp} \mid}{\parallel \epsilon^{\perp} \parallel}. \tag{2.3}$$

Model	Method	Mean	Median	St. Dev.
1	C	0.1940	0.1938	
	P	0.1936	0.1934	0.000177
2	C	0.006550	0.006530	
	P	0.006570	0.006560	0.0000587
3	C	0.00448	0.00447	
	P	0.00450	0.00450	0.000148

TABLE 1: The results of numerical studies of the classical (**C**) and resampling permutation (**P**) methods for models with normal errors; Mean and Median refer to the mean and the median of the half-widths of the obtained confidence intervals, St. dev. refers to the standard deviation of the differences of these half-widths.

Under the assumption of independent and identically distributed errors with finite second moment, the right hand side tends to zero as $n \to \infty$ a.s. These details are also discussed in Section 4. Hence for a fixed set of errors we expect the naive and robust bootstrap to perform similarly when viewed as approximations of the normal limit. They do not perform the same however when viewed as approximations to the "exact permutation interval" (see below).

Let us investigate Model 3 more closely. The other models behave comparably. In addition to computations shown above, for each error vector we also compute the confidence interval for the naive bootstrap and an interval for the exact permutation $v \cdot \pi\epsilon$. The interval for $v \cdot \pi\epsilon$ is computed using the same algorithm as for the robust bootstrap but with $\epsilon^{\perp}$ replaced by ϵ.

Arguments given above show that the classical method, and the naive and robust bootstraps, will produce half-widths which approach the asymptotic value $1.96 * \sqrt{n}s_v$ (we have assumed $\sigma = 1$ and our confidence level is .95). Remarkably, our simulations for the Gaussian case suggest that half-widths obtained from the classical method, the naive bootstrap, and the robust bootstrap, besides all being close to $1.96 * \sqrt{n}s_v$, also have *correlations very close to unity.* Closer inspection of the simulations indicates that the cause for this correlation is the effect of $s_{\epsilon^{\perp}}$. If we scale the half-widths by dividing by $s_{\epsilon^{\perp}}$, so that the classical method gives the constant value $1.96 * \sqrt{n}s_v$, then the correlation between the robust and naive bootstrap disappears. This effect is very apparent in Figure 6 which plots the half-widths from Model 3 for the naive vs. robust bootstraps. The line $y = x$ is plotted for visual comparison as well as the point $(1.96 * \sqrt{n}s_v, 1.96 * \sqrt{n}s_v)$ which corresponds to the normal limit.

What explanation remains for the difference in performance of the robust and naive bootstraps? Even when we scale by dividing by $s_{\epsilon^{\perp}}$, there is considerable correlation between the scaled half-widths of the robust bootstrap and the exact permutation, while there is none between the naive bootstrap and the exact permutation, see Figure 6. A

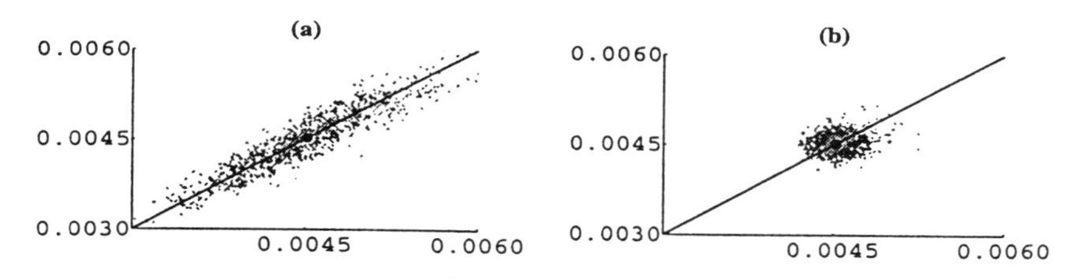

Half-widths for naive (x-axis) vs. robust bootstraps. (a) unscaled, (b) scaled by $s_{\epsilon\perp}$

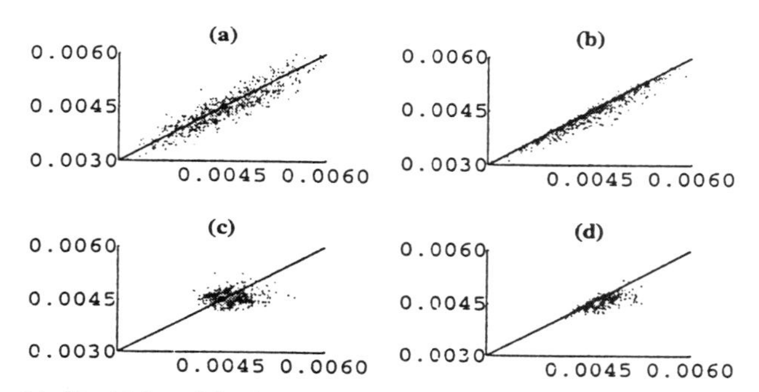

Comparison of half-widths with the exact permutation procedure. (a) and (c) are exact permutation vs. naive bootstrap, (b) and (d) are exact permutation vs. robust bootstrap, (a) and (b) are unscaled, (c) and (d) are scaled by $s_{\epsilon\perp}$.

Figure 6: Simulation results for Model 3.

plausible explanation is that even in the Gaussian case, the robust bootstrap is pursuing its target, namely the distribution of $v \cdot \epsilon$ conditional on the order statistics. The naive bootstrap has *no apparent conditional interpretation.*

3. Resampling signs

As has been mentioned, permutation of residuals is not applicable to estimation of a constant term in regression. The reason is rather obvious: shuffling coordinates of residuals produces only one single value of the dot product with a vector having all entries the same. Thus other methods have to be considered in relation to this problem. Among them are ones based on the traditional bootstrap with replacement such as smaller resample size bootstrapping (see [ArGi], [HLeP]) or bootstrapping M-estimates as in [Lah]. However, if errors are distributed symmetrically around zero, there is a resampling available which is more in the spirit of the previous one as it aims at a conditional law of estimation errors. The method was introduced for this purpose in [LeP].

First, note that (1.1) can be written equivalently as

$$y = \mu + b\tilde{x} + \epsilon, \tag{3.1}$$

where $\mu = a + b\bar{x}$ and $\tilde{x} = x - \bar{x}$. The least squares estimate of the slope in (3.1) is $\widehat{b}$, the same as before, and that of the intercept is $\widehat{\mu} = \bar{y}$. But since $\tilde{x}$ is centered at zero, we have $\bar{y} = \mu + \bar{\epsilon}$ and the estimation error is the same as for the estimation of μ by $\bar{y}$ in the model:

$$y = \mu + \epsilon. \tag{3.2}$$

It can be shown that in many important cases the method described below can be applied for estimation of μ by $\overline{Y}$ both in (3.1) and (3.2) (see Proposition 3.1 below).

Having observations $Y_1 \ldots Y_n$, we resample residuals $(Y_1 - \overline{Y}, \ldots, Y_n - \overline{Y})$ by assigning to them n random signs according to independent flips of a symmetric coin. This method for symmetric but possibly heavy-tailed errors leads to asymptotic approximation of the conditional distribution of the estimation error $\overline{Y} - \mu = \bar{\epsilon}$ conditionally on absolute values of ϵ. Thus, for symmetric errors, we propose the following bootstrap:

1. *compute the residuals* $Y_1 - \overline{Y}, \ldots, Y_n - \overline{Y}$, *assign to them thousands of sequences* $\delta_1, \ldots \delta_n$ *of signs chosen by flipping a symmetric coin, for each assignment computing a sample mean of obtained numbers:* $\overline{\delta(Y - \overline{Y})}$,

2. *find an interval [-#, #] around zero which contains around 95% of these means,*

3. *the bootstrap 95% confidence interval for* μ *is then* $\overline{Y} \pm$ #.

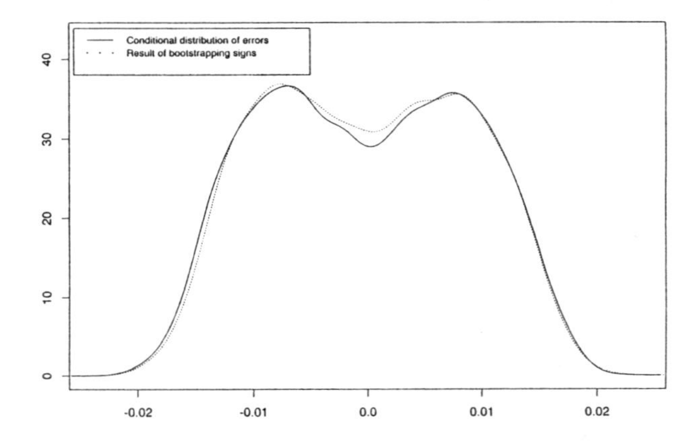

Figure 7: Recovery of the conditional distribution of estimation error by 10,000 random sign flips.

In Figure 7 we see the performance of the method applied to (3.2) with $n = 101$ and errors having symmetric stable distribution with index of stability 1.1, so having the finite moments of order less than 1.1 but not larger or equal than this. For such errors and, more generally, for errors with a symmetric distribution from stable domains of attraction, the method can be justified theoretically through the conditional invariance principle given in [LePPR].

The validation of the method is based on the closeness of the two distributions: $\bar{\epsilon}\,\big|\,|\epsilon|$ and $\overline{\delta(Y-\overline{Y})}\,\big|\,Y$, where the latter for (3.2) is equal to $\overline{\delta(\epsilon-\bar{\epsilon})}\,\big|\,\epsilon$. In [LePPR], using the conditional invariance principle for distributions from stable domains with index $\alpha \in (1,2)$, it was shown that the conditional distributions $a_n\bar{\epsilon}\,\big|\,|\epsilon|$ and $a_n\overline{\delta(\epsilon-\bar{\epsilon})}\,\big|\,\epsilon$, where $a_n = n^{1-1/\alpha}/L(1/n)$ for a slowly varying L, have the same limiting distribution with probability one. For (3.1), however,

$$\overline{\delta(Y-\overline{Y})}\,\big|\,Y = \overline{\delta(\epsilon-\bar{\epsilon})}\,\big|\,\epsilon + b\overline{\delta(x-\bar{x})} \tag{3.3}$$

and thus some restrictions on x are needed.

Proposition 3.1 *For the model (3.1) assume errors from a stable domain with index $\alpha \in (1,2)$ and let $s_n^2 = \sum_{i=1}^n x_i^2$. If $s_n \to \infty$ and for some $\epsilon > 0$*

$$\bar{x} = o(n^{1/\alpha-1/2-\epsilon}/\sqrt{\log\log n}), \quad s_n\sqrt{\log\log s_n} = o(n^{1/\alpha-\epsilon}), \tag{3.4}$$

then we have with probability one

$$\lim_{n\to\infty} a_n\bar{\epsilon}\,\big|\,|\epsilon| \stackrel{d}{=} \lim_{n\to\infty} a_n\overline{\delta(Y-\overline{Y})}\,\big|\,Y.$$

PROOF. By (3.3), it is enough to show that $\lim_{n\to\infty} a_n\overline{\delta(x-\bar{x})} = 0$. Note that

$$\begin{aligned} a_n\overline{\delta(x-\bar{x})} &= \frac{1}{n^{1/\alpha}L(1/n)}\left(\sum_{i=1}^n \delta_i x_i - \bar{x}\sum_{i=1}^n \delta_i\right) \\ &= \frac{n^{-\epsilon}}{L(1/n)}\frac{\sqrt{s_n^2\log\log s_n}}{n^{1/\alpha-\epsilon}}\frac{\sum_{i=1}^n \delta_i x_i}{\sqrt{s_n^2\log\log s_n}} \\ &\quad - \frac{n^{-\epsilon}}{L(1/n)}\frac{\bar{x}\sqrt{\log\log n}}{n^{1/\alpha-\epsilon-1/2}}\frac{\sum_{i=1}^n \delta_i}{\sqrt{n\log\log n}}. \end{aligned}$$

Thus the result is a direct consequence of the properties of slowly varying functions and the Law of Iterated Logarithm. □

Remark 3.2 *The condition (3.4) is naturally satisfied if, for example, both $s_n/\sqrt{n}$ and $\bar{x}$ are bounded.*

4. Theoretical base of permutation bootstrap

We consider the general linear regression model given by (2.1), assuming that $\mathbf{X}$ is of rank d, and the n-vector $\mathbf{1} = (1, \dots, 1)$ belongs to a space spanned by columns of $\mathbf{X}$. Without losing generality, we can assume that $\mathbf{1}$ is the first column of $\mathbf{X}$. About the errors ϵ we assume that they have an exchangeable distribution, i.e. for each permutation σ of $(1, \dots, n)$ we have

$$(\epsilon_1, \dots, \epsilon_n) \stackrel{d}{=} (\epsilon_{\sigma(1)}, \dots, \epsilon_{\sigma(n)}).$$

In particular, if ϵ_i's are i.i.d. then the above condition holds.

Recall that a vector v is called a *contrast* if its entries add to zero and, equivalently, if the Euclidean scalar product $v \cdot \mathbf{1} = 0$. The least squares estimator of β is of the form $\widehat{\beta} = (v_1 \cdot Y, \dots, v_d \cdot Y)$, where $(v_1, \dots, v_d)$ are rows of $\mathbf{M} = (\mathbf{X}^T\mathbf{X})^{-1}\mathbf{X}^T$, for which we have $\mathbf{MX} = \mathbf{I}$. This last equality implies that v_i for $i = 2, \dots, d$ are orthogonal to the first column of $\mathbf{X}$ which is $\mathbf{1}$ and thus they are *contrasts*.

Considering the least square estimator in (2.1), one is interested in the joint distribution of the estimation errors, i.e. the joint distribution of

$$\widehat{\beta}_i - \beta_i = v_i \cdot Y - v_i \cdot \mathbf{X}\beta = v_i \cdot \epsilon, \tag{4.1}$$

for $i = 1, \dots, d$. Instead of approximating or estimating the unconditional joint distribution, we propose, without moment assumptions, to estimate the joint distribution of (4.1), for $i = 2, \dots, d$, *conditionally* on information contained in the order statistics of ϵ. For this purpose we use randomly permuted residuals, i.e. we approximate the joint conditional distribution of

$$v_i \cdot \epsilon \Big| \mathcal{F},\ i = 2, \dots, d, \tag{4.2}$$

where $\mathcal{F}$ is the sigma field generated by the order statistics of ϵ, by the joint distribution of

$$v_i \cdot \pi\epsilon^{\perp} \Big| Y,\ i = 2, \dots, d, \tag{4.3}$$

where π is a random permutation of coordinates in $\mathbb{R}^n$, independent of ϵ, and $\epsilon^{\perp} = Y - \mathbf{X}\widehat{\beta}$ is a vector of residuals.

The theoretical validation of this approach comes from the result of [LePP], where it was proven that in the general case without moment/distribution assumptions other than exchangeability of ϵ, we have

$$E^{\mathcal{F}} \frac{\sum_{k=2}^{d} E^{\epsilon}(v_k \cdot \pi\epsilon - v_k \cdot \pi\epsilon^{\perp})^2}{\sum_{k=2}^{d} E^{\epsilon}(v_k \cdot \pi\epsilon)^2} = \frac{d-1}{n-1}, \tag{4.4}$$

where E^{ϵ} and $E^{\mathcal{F}}$ denote conditional expectations. By this relation and the Markov inequality, provided d/n is small, (4.3) provides *with high probability* a close approximation of (4.2) when both are scaled by $C = \sqrt{\sum_{k=2}^{d} E^{\mathcal{F}}(v_k \cdot \epsilon)^2}$.

When applying this general result, we encounter a problem of ignoring the unobserved scaling C, for example, in the construction of confidence intervals. In [LePP], it has been demonstrated that the scaling can be ignored if $(\sqrt{n}/C)v \cdot \epsilon \big| \mathcal{F}$ converges to infinity with probability one. This happens for i.i.d. errors with an arbitrary, possibly non-symmetric, distribution belonging to a stable domain of attraction. Formal arguments go through [LePPR].

Turning our attention for a moment to the finite second moment case, suppose ϵ_i's are i.i.d. with $\mathrm{E}\epsilon_i = 0$ and $\mathrm{E}\epsilon_i^2 = \sigma^2$. The following theorem in [Hjk] gives the asymptotic normality of $(v \cdot \epsilon)$ conditional on the order statistics.

Theorem 4.1 *Suppose π is a uniformly distributed permutation, $\eta_1, \eta_2, \ldots$ is a sequence of real numbers, $\eta = (\eta_1, \ldots, \eta_n)$ and $v_n = (v_n 1, \ldots, v_n n)$ is a sequence of contrast vectors such that*

$$\lim_{n\to\infty} \frac{\max_{1\le i\le n} (\eta_i - \bar{\eta})^2}{\| \eta - \bar{\eta} \|^2} = 0 \qquad \lim_{n\to\infty} \frac{\max_{1\le i\le n} v_{ni}^2}{\| v_n \|^2} = 0.$$

Then $(v \cdot \pi\eta)$ is asymptotically normal with mean 0 and variance $\| v_n \|^2 \| \eta - \bar{\eta} \|^2 /(n-1)$, if and only if, for any $\tau > 0$

$$\lim_{n\to\infty} 1/n \sum_{|\delta|>\tau} \delta_{ij}^2 = 0$$

where $\delta_{ij} = (v_i (\eta_i - \bar{\eta}))/(n^{-1/2} \| v_n \|^2 \| \eta - \bar{\eta} \|^2)$.

To apply this to our case, note that $s_{\epsilon,n}^2 = \frac{\|\epsilon\|^2}{n}$ converges a.s. to σ^2. Thus

$$\frac{(\epsilon_n)^2}{n} = s_{\epsilon,n}^2 - \frac{n-1}{n} s_{\epsilon,n-1}^2 \to 0 \quad \text{a.s.},$$

which implies that $\max_{1\le i\le n} (\epsilon_i)^2/n$ converges to 0 a.s. Consequently

$$\frac{\max_{1\le i\le n} (\epsilon_i)^2}{\| \epsilon \|^2} = \frac{\max_{1\le i\le n} (\epsilon_i)^2/n}{s_{\epsilon,n}^2} \to 0 \quad \text{a.s.}$$

Assumptions on contrasts appearing in [LePPR] imply that $\lim_{n\to\infty} n \max_{1\le i\le n} v_{ni}^2 / \| v_n \|^2 = 0$. Hájek's theorem applies since

$$\delta_{ij}^2 < \frac{n \max_{1\le i\le n} v_{ni}^2}{\| v_n \|^2} \frac{\max_{1\le i\le n} (\epsilon_i)^2}{\| \epsilon \|^2}.$$

Together with (4.4) this gives the asymptotic normality of $(v \cdot \pi\epsilon^{\perp})$ as well.

To compare the naive bootstrap and robust bootstraps in the finite second moment case let $U_1, \ldots, U_n$ be i.i.d. Uniform(0,1). We can generate a uniformly distributed permutation, π, and a bootstrap sample ϵ^* by defining $\pi_i = \text{Rank}(U_i)$, the rank of U_i among $(U_i)_{i=1}^n$, and $\epsilon_i^* = \epsilon_{\lfloor nU_i \rfloor+1}^{\perp}$, where $\lfloor\cdot\rfloor$ is the greatest integer part. As mentioned in Section 3, this construction leads to (2.3). To show that the right hand side of (2.3) converges to 0 a.s., note that $s^2_{\epsilon^\perp,n} = \| \epsilon^\perp \|^2 /(n-1)$ converges a.s. to σ^2 and use similar arguements to those considered above to check the conditions of Hájek's theorem.

5. Data-reduced models

Although the described method can be applied to a quite general class of exchangeable distributions, sometimes the conditional distribution $v \cdot \epsilon \big| \mathcal{F}$ itself will possess certain unpleasant properties as in the case of heavy-tailed distributions when some coordinates of ϵ can take relatively very large values. By our result the distribution $v \cdot \pi\epsilon^\perp \big| Y$ will inherit this. For the sake of illustration, let us consider the following simple example.

The linear model is defined by $\beta = (0, 1)$ and the 50×2 design $\mathbf{X}$ consisting of two columns: $X_1 = \mathbf{1}$ and X_2 having the first half of coordinates equal to one and the second one equal to minus one. Clearly X_2 is a contrast vector and thus orthogonal to X_1. Assume further that the vector of errors happens to be

150	2	-8	-7	-3	-2	-4	4	-7	2	-7	0	-1	-2	-5	7	6
-3	-9	2	-3	4	0	2	2	3	-4	-1	-4	7	-1	-7	-4	-4
1	1	4	2	1	-5	-5	4	-6	0	4	-3	2	-3	0	3	

Here, the first value dominates all other which typifies what can happen when the distribution of errors has a heavy tail. If we consider the contrast vector $v = X_2/n$ which corresponds to the least square estimator to the second parameter, then the conditional distribution $v \cdot \epsilon \big| \mathcal{F} = v \cdot \pi\epsilon \big| \epsilon$ has the form as in Figure 8 (continuous line). Direct computation gives the vector of residuals:

145.2	-2.8	-12.8	-11.8	-7.8	-6.8	-8.8	-0.8	-11.8	-2.8	-11.8	-4.8	-5.8
-6.8	-9.8	2.2	1.2	-7.8	-13.8	-2.8	-7.8	-0.8	-4.8	-2.8	-2.8	3.6
-3.4	-0.4	-3.4	7.6	-0.4	-6.4	-3.4	-3.4	1.6	1.6	4.6	2.6	1.6
-4.4	-4.4	4.6	-5.4	0.6	4.6	-2.4	2.6	-2.4	0.6	3.6		

Although we observe closeness of the two distributions, the confidence regions based on them should have the form of a union of two widely separate intervals. This effect, when dealing with heavy-tailed distributions, is responsible for unreasonably wide confidence intervals. In such a case, it seems reasonable to "improve" the distribution of interest by excluding observations with outsized disturbance by errors and then apply our method of resampling to the *data-reduced model*. Our results allow for such an improved method.

Data-reduced models come from (2.1) by deleting rows of the design matrix together with corresponding observations and errors. Since it is our intention to remove those rows for which errors are extreme, we consider deletions based on the ranks of errors.

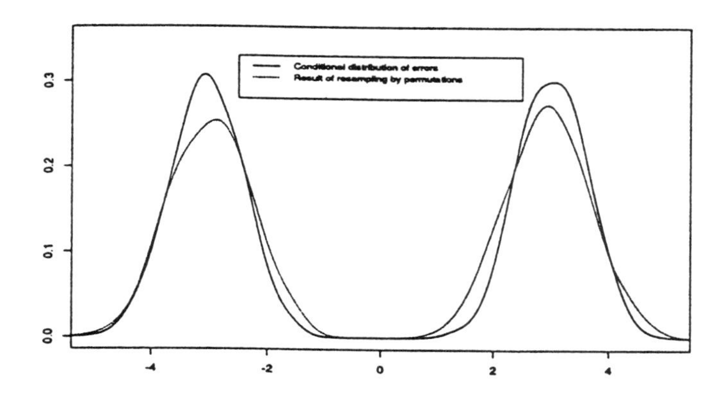

Figure 8: The conditional distribution of estimation errors vs. the permutation bootstrap distribution in the example with an outsized error.

Namely, let $R_i = R_i(\epsilon)$ be the rank of ϵ_i among $(\epsilon_i)_{i=1}^n$ counted from the bottom and defined in such a way that $(R_i)_{i=1}^n$ is a permutation of all $\{1, \ldots, n\}$ regardless of ties in ϵ. The last condition can be preserved, for example, if among tied coordinates of ϵ, a larger rank will be assigned to a coordinate with a larger index. For any subset $M \subseteq \{1, \ldots, n\}$ we define a reduced regression model

$$Y_M = \mathbf{X}_M + \epsilon_M, \tag{5.1}$$

where Y_M, $\mathbf{X}_M$, and ϵ_M all consist of those rows of Y, $\mathbf{X}$, and ϵ indexed by i whose R_i is in M. For example, by taking $M = \{2, \ldots, n-1\}$ we exclude two observations with the smallest and the largest errors. For such reduced models an extension of result (4.4) holds, justifying resampling permutations of residuals $\epsilon_M^{\perp}$ in the reduced model against contrasts $v_{M,i}$, $i = 2, \ldots, d$, corresponding to the design $\mathbf{X}_M$ (cf. [LePP]).

To apply this result in practice, one has to resolve the problem of identifying the extreme ranks of unobserved errors. One possible approach was suggested in [LePP].

Let us consider $n + 1$ models based on (2.1), consisting of the original one and n models coming from "one-at-a-time" deletions of rows (no assessment of ranks is necessary for these deletions). Using our resampling method we find $n + 1$ confidence regions I_0, I_1, I_n at level $(1 - \alpha)$, and take the one with the smallest volume. Denote this region by Γ and note that Γ still consists of a $(1 - \alpha)$ confidence region for $(\beta_2, \ldots, \beta_d)$. We can use this region to determine ranks of errors by computing residuals

$$\epsilon^{\perp}(\gamma) = Y - \overline{Y} \cdot \mathbf{1} - \mathbf{X} \ominus \mathbf{1}\gamma,$$

for $\gamma \in \Gamma$ and applying the following result (see [LePP]).

Proposition 5.1 *Let Γ be a $(1-\alpha)$ confidence region for $(\beta_2, \dots, \beta_d)$ and M be such that for any $\gamma, \gamma' \in \Gamma$.*

$$R_M(\epsilon^{\perp}(\gamma)) = R_M(\epsilon^{\perp}(\gamma')) \tag{5.2}$$

Then with probability $1-\alpha$ for any $\gamma \in \Gamma$ we have

$$R_M(\epsilon^{\perp}(\gamma)) = R_M(\epsilon).$$

In the above result R_M denotes the set of ranks R_i which are in M. Finding a set M for which (5.2) holds makes it random (dependent on the confidence region Γ). Thus a modification of our results for random M is needed. Such a result is also provided in [LePP] together with the proof of Proposition 5.1.

The following resulting method of inference and its extensions are studied in further detail in [Iva]:

1. *compute $n+1$ confidence intervals by resampling permutations of residuals in the $n+1$ regression models consisting of the original one and those obtained by "one-at-a-time" deletions of rows,*
2. *compute the two sets of ranks of residuals against the two end-points of the shortest one among the intervals from above,*
3. *find a group of smallest and largest ranks which is the same in the two sets of ranks computed above,*
4. *remove from the model the rows corresponding to this group of ranks,*
5. *apply least squares estimation to the reduced model and perform resampling permutations to assess the estimation error.*

Let us consider the same example as at the beginning of this section, but to make it more illustrative assume that the last error instead of being 3 is now equal to -95. Using the S-Plus© package, we have computed 51 confidence intervals as described above. These intervals are depicted in Figure 9a). The ranks of residuals against the endpoints of the narrowest interval are computed and shown in Figure 9b). Note that the following group of ranks remains the same: R_1, R_3, R_{19}, R_{50}. In the next step, we have removed rows 1, 3, 19, and 50 from the model.

Results of resampling permutations for this reduced model are shown on Figure 9c) together with the resampling distribution obtained for the unreduced model. Improvement can be clearly seen. The analogous illustration for the same model but with a sample of errors generated from the Cauchy distribution is presented on Figure 9d).

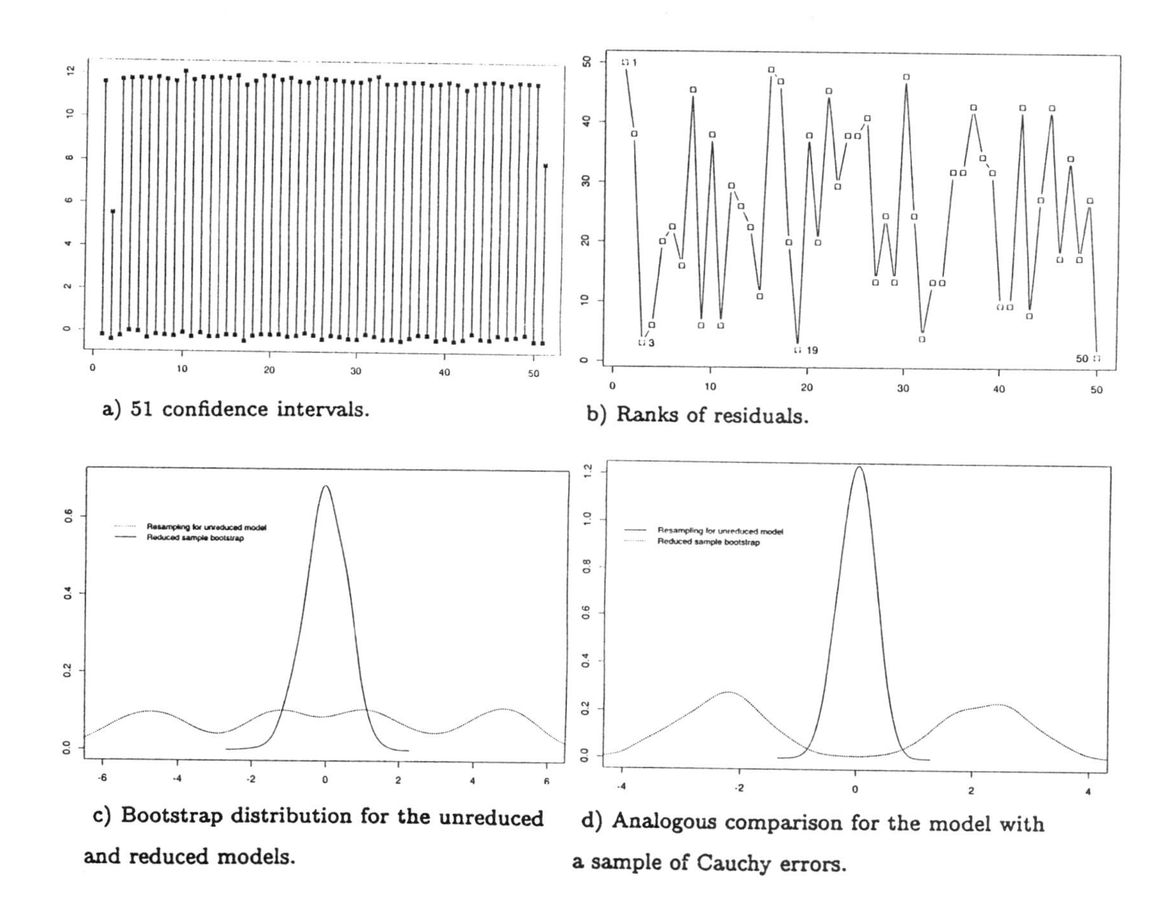

a) 51 confidence intervals.

b) Ranks of residuals.

c) Bootstrap distribution for the unreduced and reduced models.

d) Analogous comparison for the model with a sample of Cauchy errors.

Figure 9: Illustration of inference based on reduced models.

References

[ArGi] M. A. Arcones and E. Giné, The bootstrap of the mean with arbitrary sample size, *Ann. Inst. Henri Poincaré* **25** (1989), 457-481.

[Efr] B. Efron, Bootstrap methods: Another look at the jackknife, *Ann. Statist.* **7** (1979), 1-26.

[FrLa] D. A. Freedman and D. Lane, A nonstochastic interpretation of reported significance levels, *J. Business Econ. Statist.* **1** (1983), 292-298.

[Hjk] J. Hájek, Some Extensions of the Wald-Wolfowitz-Noether Theorem, *Ann. Math. Statist.* **32** (1961), 506-523.

[HLeP] P. Hall and R. LePage, On bootstrap estimation of the distribution of the studentized mean, to appear in *Ann. Inst. Statist. Math.*

[Iva] J. Ivashina, Conditional bootstrap inference for regression with outliers, Master Thesis (1996), Depart. of Math. Sci., Indiana University–Purdue University at Indianapolis.

[Lah] S. N. Lahiri, Bootstrapping M-estimators of a multiple linear regression parameter, *Ann. Statist.* **20** (1992), 1548-1570.

[LeP] R. LePage, Bootstrapping signs, in: *Exploring the limits of bootstrap*, R. LePage and L. Billard Eds., John Wiley and Sons Publ.,(1992).

[LePP] R. LePage and K. Podgórski, Resampling permutations in regression without second moments, *J. Multiv. Analysis* **57** (1996), 119-141.

[LePPR] R. LePage, K. Podgórski, and M. Ryznar, Strong and conditional invariance principles for samples attracted to stable laws, *Probab. Theory Related Fields* to appear.

[Wu] C. F. J. Wu, Jackknife, bootstrap and other resampling methods in regression analysis, *Ann. Statist.* **14** (1986), 1261-1295.

Department of Statistics and Probability
Michigan State University
East Lansing, MI 48824
e-mail: entropy@msu.edu

Department of Mathematical Sciences
Indiana University–Purdue University at Indianapolis
Indianapolis, IN 46202-3216
e-mail: kpodgorski@math.iupui.edu

Institute of Mathematics
Technical University of Wroclaw
50-370 Wroclaw, Poland
e-mail: ryznar@graf.im.pwr.wroc.pl

Department of Statistics and Probability
Michigan State University
East Lansing, MI 48824
e-mail: whiteale@pilot.msu.edu

Linear Regression with Stable Disturbances

J. Huston McCulloch

Abstract

A linear regression with symmetric stable disturbances may be estimated quickly by maximum likelihood (ML), using the symmetric stable density approximation of [MC6]. The resulting estimator is robust, in the sense that it effectively gives less weight to outlier observations than it does to the observations that conform more closely to the regression line or surface. Despite the infinite variance of the stable disturbances, the ML estimators are asymptotically normal, with asymptotic standard errors that may be computed from the Hessian of the log-likelihood function.

The method is illustrated by means of the [HR] Hedonic Housing Price Equation studied by Belsley et al., [BKW], and by the Daniel and Wood, [DW] Pilot-Plant Regression studied by Rousseeuw and Leroy [RL]. The symmetric stable ML estimates compare favorably to results obtained by means of the metric Winsorization advocated by Huber [HPJ] and to those obtained by trimmed least squares as employed by Rousseeuw and Leroy.

Symmetric stable ML linear regression may be implemented most efficiently by means of a quasi-Newton-Raphson method based on the Outer Product of Gradients (OPG) matrix, similar to that used for non-linear least squares by [BHH]. However, the Nelder-Mead polytope method [PTV] also gives satisfactory results.

Strictly speaking, the method does assume that the disturbances are truly symmetric stable, and not merely in the domain of attraction of a stable law. However, if the disturbances merely have a distribution that is approximated by a symmetric stable distribution, the method may to that extent be regarded as approximately valid.

1. Introduction

Many problems of statistical inference fit into the familiar linear regression model

$$y_i = \sum_{j=1}^{k} x_{i,j}\gamma_j + \varepsilon_i \qquad i = 1, \dots, n, \tag{1.1}$$

where y_i is an observed dependent variable, the $x_{i,j}$ are observed independent variables, the γ_j are unknown coefficients to be estimated, and the error terms ε_i are identically and independently distributed (iid) with some probability density function $f(\varepsilon)$ to be specified. An intercept term may be included in (1.1) by taking $x_{i,1} = 1$. The estimation of a simple mean or location parameter is then the special case $k = 1$. Equation (1.1) may be written in matrix form as

$$\mathbf{y} = \mathbf{X}\,\gamma + \varepsilon, \tag{1.2}$$

where

$$\mathbf{y} = \begin{pmatrix} y_1 \\ \vdots \\ y_n \end{pmatrix}, \mathbf{X} = (x_{ij}), \ \gamma = \begin{pmatrix} \gamma_1 \\ \vdots \\ \gamma_k \end{pmatrix}, \ \varepsilon = \begin{pmatrix} \varepsilon_1 \\ \vdots \\ \varepsilon_n \end{pmatrix}. \tag{1.3}$$

The $n \times k$ matrix $\mathbf{X}$ is assumed to be of rank k, with $k << n$. $\mathbf{X}$ is assumed to be either nonstochastic, or at least statistically independent of ε.

The errors ε_i in (1.1) often may be thought of as arising as the sum of a large number of more or less independent unobserved contributions. According to the Generalized Central Limit Theorem, if the sum of a large number of independently and identically distributed (iid) random variables has a limiting distribution after appropriate shifting and scaling, the limiting distribution must be a member of the *stable* class [ZVM]. The Gaussian or normal distribution is the most familiar and tractable stable distribution, and therefore it has, in the past, routinely been postulated to govern error terms in the linear regression model (1.1).

However, as Student [GW] himself once pointed out, many if not most routine analyses contain outliers, i.e. observations reflecting errors that are improbably large to have come from a common normal distribution with the rest of the data. The non-normal stable distributions all have heavier tails than the normal, and therefore are the logical extension of the Gaussian distribution whenever the observed errors are the sum of a large number of unobserved shocks, and such outliers are found to be present. Modern interest in stable distributions for economic and financial models was initiated by Benoit Mandelbrot [MB]. [MC4] surveys more recent developments.

2. Stable Distributions

The full class of stable distributions is characterized by four parameters, often designated α, β, c, and δ. The *characteristic exponent* $\alpha, 0 < \alpha \leq 2$, determines the thickness of the tails. The *skewness parameter* $\beta, -1 \leq \beta \leq 1$, determines the skewness of the distribution. When $\beta = 0$, the distribution is symmetrical about its mode. The *location parameter* δ shifts the distribution left or right, while the *scale parameter* $c, c > 0$, expands or contracts the distribution about δ. The cumulative distribution function (cdf) of a stable random variable X may therefore be written

$$P(X < x) = S(x; \alpha, \beta, c, \delta) = S((x - \delta)/c; \alpha, \beta, 1, 0) \tag{2.1}$$

(See [DM2], [MC3]. The reader is cautioned that not all authors, e.g. [ST], follow this convention for $\alpha = 1$.) We will write the standard stable cdf $S_{\alpha,\beta}(x)$ as

$$S_{\alpha,\beta}(x) = S(x; \alpha, \beta, 1, 0), \tag{2.2}$$

and designate the corresponding probability density functions (pdfs) by $s(x; \alpha, \beta, c, \delta)$ and $s_{\alpha,\beta}(x)$. Equation (2.1) implies

$$s(x; \alpha, \beta, c, \delta) = \frac{1}{c} s_{\alpha,\beta}((x - \delta)/c). \tag{2.3}$$

If a random variable X has distribution $S(x;\alpha,\beta,c,\delta)$, we will write $X \sim S(\alpha,\beta,c,\delta)$.

Figure 1 shows the standard symmetric stable pdfs for $\alpha = 1.0, 1.5$, and 2.00. Like the Gaussian density, the stable densities are bell-shaped and unimodal. They thus extend the concept of a natural "bell curve." When $\alpha = 2$, the stable distribution in fact is normal, with variance $2c^2$. For $\alpha < 2$, the variance is infinite, although the scale parameter remains well defined. In these cases, c is approximately (though not exactly) the old-fashioned "probable error" or semi-interquartile range, i.e. about half the probability lies between $\delta - c$ and $\delta + c$. If $\alpha > 1$, the mean exists and equals δ, but when $\alpha \leq 1$, the mean is undefined. In all symmetric cases, δ is the median. The case $\alpha = 1, \beta = 0$ yields the Cauchy (arctangent) distribution.

Expansions due to [BH] imply that as x becomes large in absolute value, the non-normal stable distributions have power tails that behave like a Pareto distribution:

$$\begin{aligned} \lim_{x\downarrow-\infty} S_{\alpha,\beta}(x)/\mid x\mid^{-\alpha} &= (1-\beta)k_\alpha, \\ \lim_{x\uparrow\infty}(1-S_{\alpha,\beta}(x))/x^{-\alpha} &= (1+\beta)k_\alpha, \end{aligned} \tag{2.4}$$

where $k_\alpha = (1/\pi)\Gamma(\alpha)\sin(\pi\alpha/2)$. The non-normal stable distributions are therefore sometimes said to be "Paretian stable," particularly in the finance literature. Equation (2.4) shows that β is the limiting value of the ratio of the difference of the tail probabilities (lower minus upper) to their sum. As $\alpha \uparrow 2$, $k_\alpha \downarrow 0$, and β loses its effect.

Unfortunately, there is no exact finite arithmetic expression for the stable distribution or density functions that can be computed on a digital computer. This is a problem stable distributions share with other familiar transcendental functions, such as the logarithm, sine, etc. However, [MC6] has developed a fast numerical approximation to the symmetric stable (SS) distribution and density that makes Maximum Likelihood (ML) and other standard statistical techniques practical, even on a personal computer, at least in the special case $\beta = 0$.

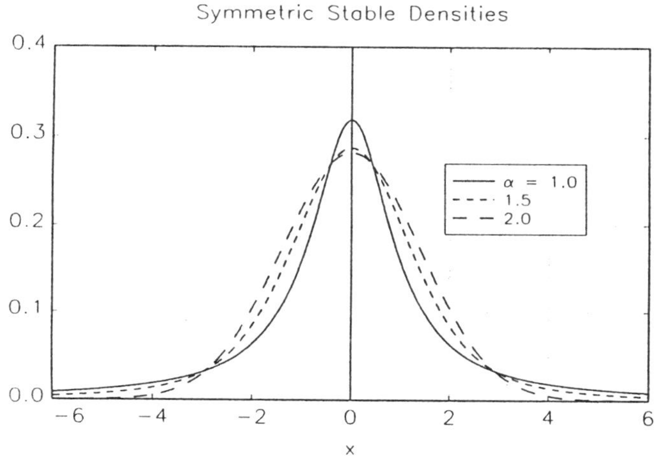

Figure 1: Symmetric stable probability density function for $\alpha = \mathbf{1.00, 1.50, 2.00}$.

3. Symmetric Stable Linear Regression

We assume in this paper that the iid error terms in (1.1) are indeed symmetric stable (SS) with median 0 and scale c, i.e. $\varepsilon_i \sim S(\alpha, 0, c, 0)$. [DM1][DM2] has shown that in such cases, maximum likelihood (ML) estimation is efficient, and gives asymptotically normal estimates of the $k+2$ parameters. Their asymptotic variances are governed by the Fisher information matrix, which he calculates, or, what is equivalent asymptotically, by the Hessian matrix of the log-likelihood function. The ML estimate of a simple location parameter converges on the true parameter value at the rate $n^{-1/2}$, and I would conjecture that this rate is also valid if the rows of $\mathbf{X}$ repeat themselves nonstochastically or are drawn from a finite variance distribution.[1]

When $\alpha < 2$, applying Ordinary Least Squares (OLS) instead of ML to (1.1) will give an estimate of the coefficient vector

$$\hat{\gamma}_{OLS} = (\mathbf{X}'\mathbf{X})^{-1}\mathbf{X}'\mathbf{y} = \gamma + (X'X)^{-1}X'\,\varepsilon \tag{3.1}$$

that is centered on γ, but has an infinite variance stable distribution with the same characteristic exponent α as the underlying error terms. So long as $\alpha > 1$, the OLS estimates will converge on the true parameter values as the sample becomes large, but only at the rate $n^{1/\alpha-1}$, rather than the $n^{-1/2}$ rate attainable with ML. Furthermore, conventional t statistics for the parameter estimates will no longer have the standard Student t distribution, but instead will be concentrated in the vicinity of ± 1 ([LMR]). The OLS estimator is not altogether useless, but it does have a zero asymptotic efficiency relative to ML.

OLS is inefficient when the disturbances are heavy-tailed, because it gives too much *influence* to outliers. Let $f(\varepsilon)$ be a general probability density function for the errors in (1.1), so that in the SS case $f(\varepsilon) = s_{\alpha,0}(\varepsilon/c)/c$ as in (2.3) above. Let $\phi(\varepsilon) = \log(f(\varepsilon))$. Then maximizing the likelihood for (1.1) with respect to γ and any parameters such as c and/or α that may enter the definition of $f(\varepsilon)$ is equivalent to maximizing the log-likelihood function

$$\mathcal{L} = \log L = \sum_{i=1}^{n} \phi(\hat{\varepsilon}_i), \text{ where } \hat{\varepsilon}_i = y_i - \hat{y}_i, \hat{\mathbf{y}} = \mathbf{X}\,\gamma. \tag{3.2}$$

It may easily be shown that the first order conditions for maximizing $\mathcal{L}$ with respect to γ imply

$$\hat{\gamma}_{ML} = (\mathbf{X}'\mathbf{W}\mathbf{X})^{-1}\mathbf{X}'\mathbf{W}\mathbf{y}, \tag{3.3}$$

where

$$\mathbf{W} = \text{ diag } (w(\hat{\varepsilon}_i)), \tag{3.4}$$

[1] Drawing the rows of $\mathbf{X}$ from a thicker tailed distribution or even from a non-stationary process should increase the rate of convergence, by generating high-leverage observations. These issues deserve rigorous treatment, but go beyond the scope of the present paper.

$$w(\varepsilon) = -\phi\prime(\varepsilon)/\varepsilon.$$

In other words, the first order conditions for ML are equivalent to those for Weighted Least Squares, where the weight w_i on the i-th observation is equal to $w(\hat{\varepsilon}_i)$. In the Gaussian case, the weighting function $w(\varepsilon)$ is independent of ε, while in the non-Gaussian stable cases, it is a declining function of the absolute value of ε, as shown in Figure 2 (drawn for $c = 1$). If the true value of α is less than 2, OLS thus gives too much weight to outliers with large positive or negative values of $\hat{\varepsilon}_i$, and too little weight to observations with small values of $\hat{\varepsilon}_i$.

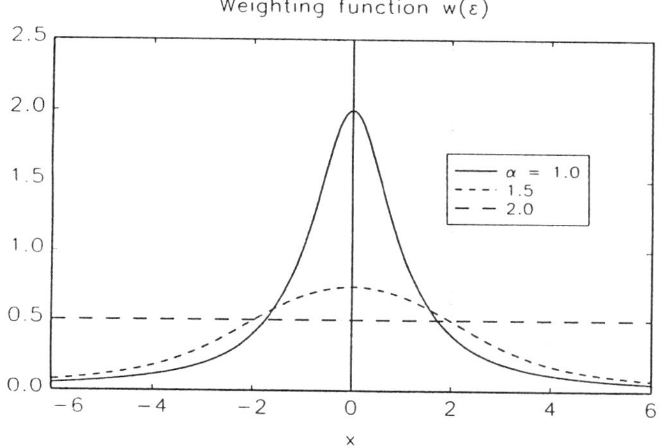

Figure 2: The weighting function $w(\varepsilon)$, for standard symmetric stable distributions.

Huber's [HPJ] function $\psi(\varepsilon) = \varepsilon w(\varepsilon) = -\phi\prime(\varepsilon)$ indicates the proportionate impact on $\hat{\gamma}_{ML}$ of the addition of a new n-th observation to an already large data set, as a function of the error it contains. When estimating a simple location parameter, $\psi(\varepsilon)$ is called the *influence* function.[2] It is shown in Figure 3 for SS distributions, with $c = 1$. For the Gaussian distribution, this function is directly proportional to ε and grows without bound, while in the non-Gaussian cases, the influence function is bounded; it grows with small values of ε, but then peaks out and eventually declines toward 0 for very large values of ε. Stable ML is therefore a *robust* estimator, in that it is less sensitive to large outliers than is OLS. However, it has the advantages over other robust estimators that it is based on ML, and that it uses a criterion function which, like least squares, is naturally motivated by the Central Limit Theorem. By estimating α simultaneously with the other parameters, it is possible to let the data dictate an optimal divergence from OLS, unlike other robust regression procedures as customarily implemented. Because it is solidly based on ML, hypotheses may be tested using likelihood ratio (LR) statistics, or by means of Wald statistics computed from asymptotic standard errors.

[2]In a regression context, the full "influence" of an observation is customarily $\psi(\varepsilon_i)$ times a factor that measures the leverage the observation has on the regression coefficients. See [RL], p. 13. This leverage factor goes beyond the scope of the present paper.

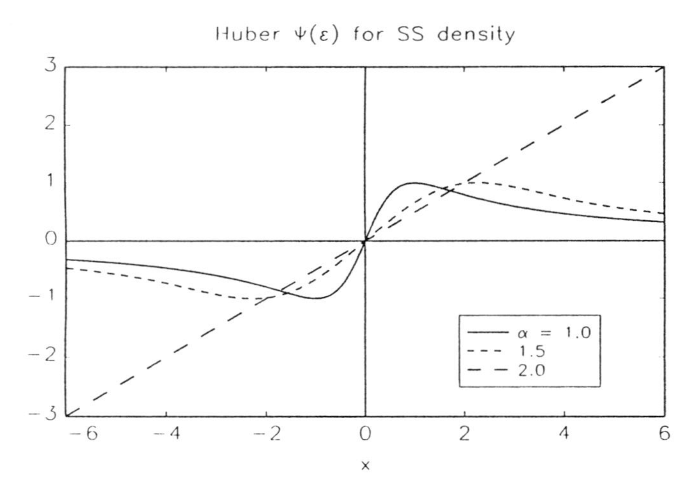

Figure 3: Huber influence function for standard symmetric stable distributions.

We illustrate symmetric stable ML regression with two examples. The first is a classic study already examined by [BKW], p. 229-261, namely the Harrison-Rubinfeld [HR] Hedonic Housing Price Equation that utilizes data from each of $n = 506$ Boston SMSA census tracts on housing prices and 13 explanatory variables, drawn from the 1970 census. The dependent variable, LMV, is the logarithm of the median value of owner-occupied homes. The independent variables are listed in Table 1.[3]

Table 2 shows OLS estimates of the $k = 14$ coefficients (counting the intercept term), along with SS ML estimates. The asymptotic standard errors for the SS ML estimates are based on the Hessian matrix of the maximized log-likelihood function. The SS ML Wald statistics are the ratios of the estimated coefficients to their asymptotic standard errors. These are asymptotically $N(0, 1)$ if the true coefficients are zero.

The estimate of the characteristic exponent, $\hat{\alpha}_{ML}$, is 1.422, indicating considerable leptokurtosis. SS ML increases the log-likelihood by 59.47 over OLS, for an LR statistic (twice the difference in log L) of 118.95. Because the null hypothesis $\alpha = 2.00$ is on the boundary of the parameter space, this LR statistic does not have its customary $\chi^2(1)$ distribution. Nevertheless, [MC5] has tabulated the distribution of this statistic by Monte Carlo methods for $n = 30, 100, 300$, and 1000, and found the $\chi^2(1)$ critical values to be uniformly too conservative. We may therefore overwhelmingly reject normality in favor of the alternative of non-normal symmetric stability.[4]

[3]The data were optically scanned from Appendix 4A of [BKW], and manually corrected. Our OLS estimates agree with those of [BKW] et al. to the precision shown, but not always in the next decimal place, indicating that while not all scanner errors were caught by proofreading, the net effect of the remaining ones is trivial. As this goes to press, we have learned that [GP] have identified eight miscodings in the original [HR] data. However, they show that these have negligible impact on the OLS results.

[4]The critical values in [MC3] are for the case in which only a location parameter is being estimated. However, the addition of non-trivial regressors should only bias the test toward acceptance of normality, since they allow OLS to change the shape of the residuals so as to fit normality more closely.

TABLE 1
Definition of Harrison-Rubinfeld Variables

j	Symbol	Definition
1	Intercept	
2	CRIM	per capita crime rate by town
3	ZN	proportion of town's residential land zoned for lots greater than 25,000 square feet
4	INDUS	proportion of nonretail business acres per town
5	CHAS	dummy variable with value 1 if tract bounds on the Charles River
6	NOXSQ	nitrogen oxide concentration (parts per 10^8) squared
7	RM	average number of rooms squared
8	AGE	proportion of owner-occupied units built prior to 1940
9	DIS	logarithm of the weighted distances to five employment centers in the Boston region
10	RAD	logarithm of index of accessibility to radial highways
11	TAX	full-value property-tax rate (per $10,000)
12	PTRATIO	pupil-teacher ratio by town
13	B	$(Bk - .63)^2$, where Bk is the proportion of blacks in the population
14	LSTAT	logarithm of the proportion of the population that is lower status

The estimated standard errors are uniformly (with the sole exception of the crime variable CRIM) smaller than for OLS, reflecting the greater efficiency of SS ML when $\alpha < 2$. The most important change in coefficient estimates is that for the older housing variable AGE, whose coefficient changes from slightly positive and insignificant to negative and strongly significant. Apparently older housing typically reduces median house prices, but there are a few exceptional neighborhoods with an "antique" premium that masks this tendency in the OLS estimates. SS ML treats these unusual census tracts as outliers, and clearly shows the more typical relationship. Perhaps interacting with AGE is the average number of rooms per unit squared variable RM, which effectively measures the presence of mansions. Its SS ML coefficient is positive and more than double its OLS coefficient.

Also noteworthy is the effect of SS ML on RAD, B, LSTAT, and NOXSQ: RAD, which measures access to radial highways, remains significant but is considerably smaller in size. The coefficient on B, an asymmetric measure of racial homogeneity, almost doubles in size. LSTAT, an index of lower status, becomes considerably smaller in absolute value, but remains highly significant.

TABLE 2
Harrison-Rubinfeld Housing Price Equation

Variable	OLS (s.e.)	OLS t stat.	SS ML (s.e.)	SS ML Wald stat.	Huber
Intercept	9.76 (0.15)	65.21	9.57 (0.11)	89.56	9.63
CRIM	-0.0119 (0.0012)	-9.53	-0.0097 (0.0014)	-6.98	-0.011
ZN	0.00008 (0.00051)	0.16	0.00010 (0.00032)	0.33	0.00004
INDUS	0.0002 (0.0024)	0.10	0.0017 (0.0015)	1.15	0.0012
CHAS	0.091 (0.033)	2.75	0.062000 (0.022)	2.78	0.077
NOXSQ	-0.0064 (0.0011)	-5.64	-0.00419 (0.00085)	-4.92	-0.00505
RM	0.0063 (0.0013)	4.83	0.0157 (0.0011)	13.71	0.0115
AGE	0.00009 (0.00053)	0.17	-0.00125 (0.00038)	-3.28	-0.00066
DIS	-0.191 (0.033)	-5.73	-0.149 (0.024)	-6.21	-0.164
RAD	0.096 (0.019)	5.00	0.054 (0.013)	4.18	0.070
TAX	-0.00042 (0.00012)	-3.43	-0.000357 (0.000082)	-4.33	0.000361
PTRATIO	-0.0311 (0.0050)	-6.21	-0.0274 (0.0032)	-8.63	-0.0290
B	0.36 (0.10)	3.53	0.668 (0.093)	7.17	0.551
LSTAT	-0.371 (0.025)	-14.84	-0.207 (0.022)	-9.24	-0.281
α	2.00 (constrained)		1.422 (0.072)		
log c			-2.519 (0.054)		
c			0.0806		
log L	149.899		209.372		

And finally, the coefficient on the air pollution variable NOXSQ, whose determination was the main objective of the original Harrison-Rubinfeld study, falls by more than 50%. However, it again remains highly significant, thanks to an almost equiproportionate reduction in its standard error.[5]

[5]Two caveats are in order concerning this regression. The first is that the exponential of the dependent variable, LMV, is often \$50, 000, but never any higher. When it is lower, it is typically in odd multiples of \$100. This strongly suggests that it was censored at \$50, 000. This censoring is not obvious from the table in Appendix 4A of [BKW], which

Our second example is the Daniel-Wood [DW] Pilot-Plant data studied by [RL], p. 21-38. A simple regression of Titration on Extraction with the true data from their Table 1 yields the OLS results in the first column of Table 3. SS ML converges on $\alpha = 2$, and so gives the same coefficient estimates, with a tight linear fit and a highly significant positive slope coefficient. [RL] then add a "keypunch" error to one of the 20 points, by changing its "x" value from 37 to 370. This contaminated data gives very different OLS estimates, as shown in the second column. The high leverage of the contaminated point reduces the slope coefficient nearly to zero. The standard errors of the coefficients go up by almost tenfold, so that the reduced slope coefficient does not appear to be significantly different from 0 at the 95% confidence level. The contaminated regression appears to reject the uncontaminated regression coefficient at the .01 level of significance, both individually with t-tests, and jointly with an F test.

The third column of Table 3 shows SS ML coefficients using the contaminated data. The estimate of α is 1.40. The LR statistic for $\alpha = 2$ is 74.35, which overwhelmingly rejects normality and therefore the validity of the OLS t- and F-tests. The SS ML regression coefficients are only trivially different from the uncontaminated data coefficients in the first column. They do, however, have moderately higher standard errors, so that the contamination is not entirely costless. Despite this, the slope is overwhelmingly significantly different from 0. SS ML thus sees right through the keypunch error.

The contaminated point stands out much more clearly when computed using the SS ML coefficients than using the contaminated OLS coefficients, allowing it to be double checked and corrected if at all possible.[6] If the keypunch error cannot be located, however, the results indicate that the distribution of the full data generating process is not significantly different from a Cauchy ($\alpha = 1$) process, and therefore that its mean may not even be defined. This should ordinarily be a cause for concern.

4. Comparison to other Robust Estimators

The last column of Table 2 shows Huber robust estimates of the Harrison-Rubinfeld housing price equation, as computed by Belsley et al ([BKW]). The Huber estimator of location $\hat{\gamma}_H$ is based on a criterion function $\phi(\varepsilon)$ that is quadratic in the center and linear in the tails, but continuously differentiable.[7] This criterion function thus implies an influence function $\psi(\varepsilon)$ that passes through the origin, is linear in the center and flat in the tails, and is continuous. The Huber estimator, which amounts to metric Winsorization, thus has bounded influence, but still gives far more influence to the largest outliers than does SS ML.

---

shows the variable only in logarithmic form. The second problem is that a routinely computed Durbin-Watson statistic on the OLS regression was only 1.10, strongly indicating non-independent residuals. Although this is not a time series, sequentially numbered census tracts tend to be adjacent, so there is evidently a proximity-dependent correlation that should be appropriately taken into account. Both these problems go beyond the scope of the present paper. [PG] address the latter problem in the [HR] data with a spatial autoregression, but assume normality.

[6] The estimated error on the contaminated point is 45 times as large in absolute value as the second largest estimated error using SS ML, versus 2.9 times as large using OLS.

[7] Huber actually minimizes the sum over the sample of a criterion function $\rho(\varepsilon)$, which is simply the negative of our $\phi(\varepsilon)$.

TABLE 3 Daniel-Wood Pilot Plant Regression (Titration on Extraction; n = 20)				
	True Data	– Contaminated Data –		
	OLS (s.e.)	OLS (s.e.)	SS ML (s.e.)	Trimmed LS (s.e.)
Intercept	35.46 (0.64)	58.9 (6.6)	35.45 (0.72)	35.32 (0.70)
Slope	0.3216 (0.0056)	0.081 (0.047)	0.3204 (0.0064)	0.3226 (0.0060)
α		2.00 (constrained)	1.40 (0.34)	
log c			-0.10 (0.22)	
c			0.904	
log L		-82.269	-45.095	

When the splice point between the quadratic and linear portions is at infinity, it is equivalent to OLS. When the splice point is at the origin, it is equivalent to Least Absolute Deviations (LAD). It may be seen from Table 2 that in the critical cases discussed above, the Huber estimates are generally intermediate between the OLS and SS ML estimates, as would be expected.

It is quite feasible to treat the Huber criterion function as a log density function, simply by adding a constant term that makes its exponential (or the exponential of its negative if it is taken as a loss function) integrate to unity. Its scale and even its splice point may then be estimated by ML and asymptotic standard errors computed from the Hessian as in the SS case. However, it is not customary to do this. Instead, [BKW] set the splice point at an intermediate value based on experience with unrelated data sets, and then estimate the scale by incorporating it into the objective function in an ad hoc manner that is not equivalent to ML. They admit that this estimate of scale has very little theoretical basis, and do not even attempt to compute standard errors for their Huber coefficient estimates.

Rousseeuw and Leroy [RL] propose identifying outliers by minimizing the least median squared residual (LMS). This is equivalent to minimizing the interquartile range of the estimated errors. This in turn is equivalent to ML using a density function that is uniform in the center with probability 1/2, exponential in the tails with probability 1/4 each, and continuous. As with the Huber estimator, the scale could be estimated by ML and the regression coefficients given asymptotic standard errors, but Rousseeuw and Leroy do not do this in practice.

Having thus identified the outliers, Rousseeuw and Leroy [RL], p. 15, 38, perform trimmed Least Squares on the "non-outliers", i.e. they omit outliers greater than some threshold, and then perform OLS on the remaining sample to obtain standard errors for the coefficients (cf. also [BKW], exhibit 4.32, p. 242). This implies a weighting function $w(\varepsilon)$ which is constant out to the threshold, and then drops discontinuously to 0. This in turn is equivalent to ML using a density $f(\varepsilon)$ which is proportional to the Gaussian inside the threshold, but which is zero outside. However, if such a density is taken literally and applied to the full sample, the log-likelihood becomes minus infinity, since the outliers have zero likelihood. If the appropriate threshold is then estimated by ML, it must be placed just outside the largest outlier in order to make the log-likelihood finite, so that the entire sample is effectively included and we are back to OLS. Or if the threshold is fixed, but the scale estimated by ML, the scale must be just large enough that the scaled threshold is just outside the largest outlier, so that the full sample is again included. Although trimmed LS can be useful for initializing ML search routines (see below), there can therefore never be an acceptable basis, from an ML point of view, for completely discarding outliers as a final estimator. SS ML, on the other hand, is a statistically sound method of down-weighting such outliers to whatever extent is dictated by the data.

Rousseeuw and Leroy's trimmed LS estimates for the Pilot-Plant regression are shown in the last column of Table 3. They identify these as "Reweighted LS", but they are equivalent to OLS with the contaminated point eliminated. The coefficient estimates are again only trivially different from SS ML or from OLS using the uncontaminated data. The reported standard errors are slightly higher than those from uncontaminated OLS, primarily because of the reduced effective sample size ($n = 19$). However, there is no statistical basis for using these standard errors, since they are proportional to a scale that was not estimated by ML. The SS ML standard errors are a little bit larger, even despite the fact that ML makes no correction for degrees of freedom. They are larger because SS ML correctly recognizes that the data is generated by a distribution that occasionally produces huge outliers with non-zero density, and therefore has an element of uncertainty swept under the rug by trimmed OLS.

Another approach that would give results even more similar to SS ML than the Huber or Rousseeuw and Leroy estimators would be to perform ML, using the Student t distribution with ν degrees of freedom for the error density function $f(\varepsilon)$. Like the SS class, the Student t distributions include the Gaussian ($\nu = \infty$) and Cauchy ($\nu = 1$) as special cases. Because the pdf has power tails and is symmetrical, the influence and weighting functions are qualitatively similar to those of the SS distributions. Nevertheless, this class lacks the CLT properties that motivate the stable distributions and make the latter the more natural extension of the Gaussian distribution when heavy tails are present. McCulloch ([MC5], [MC4]) discusses recent empirical and financial literature on power tails and the Student t versus stable distributions at greater length.

5. Implementing SS ML

The log-likelihood function (3.2) may be maximized with respect to α, c, and the k components of γ by any of a number of standard constrained numerical optimization procedures. It is of course critical that α be constrained above by 2.00. It is also necessary that α be bounded below by some $\alpha_{\min} > 0$, since as α falls toward 0, the likelihood becomes infinite when γ fits any of the observations exactly (see [DM1]). Even aside from this consideration, if the numerical density approximation of [MC6] is employed, α should be bounded below by 0.84, since the precision of the approximation is greatly reduced below this value. There is no particular reason to impose a finite mean, i.e. $\alpha > 1$. Regardless of the method used, it is expedient to replace c by $c^* = \log(c)$ so that the scale variable can be estimated without further restrictions.

One method that works well is the following quasi-Newton-Raphson procedure. Let

$$\boldsymbol{\theta} = (\alpha, c^*, \gamma')' \tag{5.1}$$

be the $(k+2) \times 1$ vector of parameters, $\boldsymbol{\theta}_0$ be an initial guess, and $\boldsymbol{\theta}_h$ be the h-th iteration. Let $\mathbf{g}$ be the gradient of $\mathcal{L}$ with respect to $\boldsymbol{\theta}$ at $\boldsymbol{\theta}_h$, and define $\mathbf{H}^*$ to be the Outer Product of Gradients (OPG) proxy for the Hessian, computed as

$$\mathbf{H}^* = -\sum_{\mathbf{i}=1}^{\mathbf{n}} \mathbf{g_i g_i}' \tag{5.2}$$

where $\mathbf{g_i}$ is the gradient of $\phi(\hat{\varepsilon}_i)$ with respect to $\boldsymbol{\theta}$ at $\boldsymbol{\theta}_h$. A normal iteration is then computed as

$$\boldsymbol{\theta}_{h+1} = \boldsymbol{\theta}_h - \lambda \mathbf{H}^{*-1}\mathbf{g} \tag{5.3}$$

with $\lambda = 1$.

If the Hessian $\mathbf{H}$ of $\mathcal{L}$ with respect to $\boldsymbol{\theta}$ at $\boldsymbol{\theta}_h$ were negative definite and constant, (5.3) with $\lambda = 1$ and $\mathbf{H}$ in place of $\mathbf{H}^*$ would take us to the global maximum in one step (ignoring for the moment the necessary restrictions on α). However, in our case, $\mathbf{H}$ is not necessarily negative definite away from the maximum, let alone constant, so that it may actually lead us in the wrong direction. $\mathbf{H}^*$, on the other hand, is always negative definite, and, by the information identity, differs from $\mathbf{H}$ at the true parameter vector only by a term whose mean is 0 and whose relative effect vanishes in large samples. By using $\mathbf{H}^*$ instead of $\mathbf{H}$, we therefore obtain an iteration that is in a good direction, and usually of a reasonable size.[8]

It does sometimes happen that applying (5.3) with $\lambda = 1$ so overshoots the maximum that it leads to a candidate for $\boldsymbol{\theta}_{h+1}$ that actually has lower log-likelihood than $\boldsymbol{\theta}_h$. In this case, λ may simply be split in half, repeatedly, until a $\boldsymbol{\theta}_{h+1}$ is found at which the log-likelihood is indeed higher. A sufficiently small move in the direction of the gradient, or in the direction of the gradient times any positive definite matrix, always can be found that results in an increase in $\mathcal{L}$.

The initial guess $\boldsymbol{\theta}_0$ may ordinarily be found by .5 trimmed OLS, as follows: First calculate $\hat{\gamma}_{OLS}$ and sort by the OLS residuals. Then run OLS again on the middle 50% of the sorted sample. Compute the full-sample residuals using these .5 truncated OLS coefficients, and then apply the quantile method of [MC1] to these (constraining $\beta = 0$) to obtain initial estimates of α and c. Compare the likelihood using these parameters to that obtained with full-sample OLS, and start with whichever gives the higher value of $\mathcal{L}$.

The required restrictions on α may be imposed, when these restrictions are not binding at $\boldsymbol{\theta}_h$, by scaling down λ, as necessary, so as just to obey the pertinent restriction. When $\alpha = 2(\alpha = \alpha_{\min})$ at $\boldsymbol{\theta}_h$ and the first component of $\mathbf{g}$ is positive (negative), the iteration is performed only with respect to the remaining $k + 1$ parameters.

The illustrative regression of Table 2 was performed using the above quasi-Newton-Raphson method, with a convergence criterion of $.0001\lambda$ for the change in log-likelihood. This method required 25 iterations for the regression in question, and took 17 seconds on a (corrected) Intel P5-100 processor using GAUSS 3.2.12. The Pilot-Plant regression of Table 3 took 2.6 seconds (37 iterations) by this method.

[8] Compare [BHH]. The computational problems of linear regression with non-Gaussian residuals are quite similar to those of non-linear regression with Gaussian residuals.

Another satisfactory method is the Nelder and Mead polytope (a.k.a. downhill simplex) method , as described by [PTV], p. 402-406.[9] The starting values were taken as $\alpha = 2, c^* = \log(\hat{\sigma}_{OLS}/\sqrt{2})$, and $\gamma = \hat{\gamma}_{OLS}$. The initial polytope was constructed by incrementing these values by -.5, log(.5), and one respective OLS standard error, respectively. Iterations continued until the elements of the polytope varied in absolute value by no more than .001 times the respective initial increments, giving substantially the same results as the OPG Newton-Raphson method for both our illustrative regressions.

The Nelder-Mead polytope method is relentless at finding local extrema, but is relatively slow for large k, since the number of iterations is roughly proportional to the number of parameters being estimated. The Housing regression of Table 2 (16 parameters) thus required 3440 iterations by this method. Nevertheless, this entire calculation took less than 150 seconds on the same personal computer. The Pilot-Plant regression of Table 3 took 4.7 seconds (153 iterations) using the Nelder-Mead polytope method. In the quasi-Newton-Raphson method, the number of iterations is roughly independent of the number of parameters being estimated, and if anything declines with the number of observations. However, each iteration requires considerably more calculations than with the Nelder-Mead polytope method, since $\mathbf{g}$ and $\mathbf{H}^*$ are required. Either way, speed is not a big problem for moderate sized problems with modern personal computers.

Once the maximum is found, by any method, an asymptotic covariance matrix for the estimates may be computed as $-\mathbf{H}^{-1}$, as $-\mathbf{H}^{*-1}$, or even by using the Fisher information matrix (see [DM2]). However, Hessian-based standard errors appear to contain less sampling error than OPG-based standard errors, and are generally to be preferred to them, particularly in smaller samples.[10]

The gradients and/or Hessian required by the maximization procedure and/or the asymptotic covariance matrix may be computed either by analytically differentiating the numerical approximation used for the SS density, or numerically.[11] The standard errors in Table 2 were based on a numerical Hessian computed by moving symmetrically .01 times the initial polytope increment above and below $\hat{\gamma}_{ML}$. The resulting approximately symmetrical matrix was made exactly symmetrical by averaging its upper and lower halves. If either restriction on α is binding, α must be treated as fixed and the Hessian computed with respect to only the unconstrained parameters.[12]

If α is known, ML estimates of the regression coefficients and scale parameter could, in principle, be found by iterating on (3.3). However, such an iteration proves to be very slow to converge in similar problems, because the weights change only gradually (cp. [MC2], which deals with a similar problem). Furthermore, it does not help us to estimate α. The latter could be estimated by means of a univariate search, but then the slow iteration would have to be repeated at each value of α. This method is not recommended as a practical method of performing SS ML.

The intialization procedures described above are adequate for most problems that arise in

[9]The two constraints $\alpha \leq 2$ and $\alpha \geq .84$ may be imposed by proportionately truncating reflections and expansions that would have otherwise violated these restrictions, and by not even attempting reflections if the centroid of the points being reflected through lies on either boundary. The latter restriction is required to preserve the dimensionality of the polytope.

[10]See [MC2].

[11]If the gradients for the quasi-Newton-Raphson search are computed numerically, it is important that they be based on equal steps above and below the current estimate. Otherwise, the estimator will be biased by half the numerical step size.

[12]The author is grateful to Jerry Thursby for pointing this out.

practice. However, even more extreme examples than that of Table 3 can be constructed, in which the outliers have such high degree of leverage that the likelihood becomes bimodal, with one mode at OLS, and another, higher one near the true coefficients, with a very low α. In these cases, the offending points may so control the OLS regression that they do not even appear to be outliers. Such cases may respond well to a regression coefficient initialization based on the median of OLS estimates found by deleting each observation, one at a time, with α and c then initialized from the resulting residuals by a method such as that of [MC1].

6. Advanced estimation problems

Signal extraction: Simple state-space problems of the type

$$\begin{aligned} y_t &= x_t + \varepsilon_t, \quad t = 1, \ldots, T, \\ x_t &= a + bx_{t-1} + \eta_t, \quad t = 2, \ldots, T, \end{aligned} \tag{6.1}$$

where y_t is an observed time series and x_t is an unobserved state variable, may be estimated under the assumptions that $\varepsilon_t \sim$ iid $S(\alpha, 0, c_\varepsilon, 0)$, $\eta_t \sim$ iid $S(\alpha, 0, c_\eta, 0)$, and that the $\varepsilon's$ and $\eta's$ are mutually independent, by means of the Sorenson-Alspach filtering algorithm [SA], (see also [KG]). This algorithm for the filter density requires computing a series of numerical integrals, but so long as there is only one state variable as in (6.1), the required integrals are all univariate, and are computationally tractable. The hyperparameters a, b, α, c_ε and c_η may then be estimated by ML, and full sample posteriors for x_t computed by means of the smoothing algorithm of [KG].

This stable signal extraction procedure extends the Kalman filter to the non-Gaussian stable cases. The results are qualitatively similar, but with the major difference that the updating procedure is non-linear: Large outliers are at first largely ignored as being "noise", but if they persist, the posterior distribution eventually detects a large "signal" shock, and its mean then adjusts rapidly.

[OH] has used (6.1) to estimate the time-variant term premium implicit in bond returns, finding $\hat{\alpha}_{ML}$ ranging from 1.61 to 1.80 even after removing substantial Generalized Autoregressive Conditional Heteroskedasticity (GARCH) effects. [BM] use the special case of (6.1) with $a = 0$ and $b = 1$ to model U.S. inflation, finding $\hat{\alpha}_{ML} = 1.81$ and significantly different from 2 by means of the LR test of [MP1].

Multivariate stable distributions: Multivariate stable distributions have a much richer set of structures than multivariate Gaussian distributions, since they are not, except in the special elliptical cases, completely described by a vector of locations and a matrix of correlations (See [BNR]; [ST]; [MC4]). While the estimation of these distributions is still in its infancy, [NPM] find that their so-called spectral representation is amenable to estimation by a method that depends on a series of univariate ML estimates as described above using different projections of the data, coupled with the solution of a quadratic program by standard methods.

Skew-stable distributions: At present, methods such as the above that are based on stable ML are for most practical purposes restricted to the symmetric case $\beta = 0$, simply because no fast numerical approximation is currently available for the general density function. Brorsen and

Preckel [BP] do perform general stable linear regression, but laboriously compute the required stable densities at each iteration with Zolotarev's [ZVM] integral representation.

It is to be hoped that in the near future a general stable density approximation will be developed. Once this is available, general stable ML (subject to the additional restrictions $-1 \le \beta \le 1$) will be basically no more difficult than SS ML. The maximally skew-stable tabulation of [MP1] takes the first step toward developing such an approximation. In the meantime, the quantile method of [MC1] provides easily computed and finitely efficient estimates of the four stable parameters when only a simple location parameter is required.

It should be noted that relaxing the restriction $\beta = 0$ introduces considerable uncertainty about the location parameter (or intercept term in a regression context), particularly when α is far from 2. If we know the distribution is symmetrical, the location parameter is simply the median, which is also the mode and is relatively well identified. But if the distribution might be skewed, uncertainty about the skewness parameter could place the mean well to the right or the left of the mode, even if $\hat{\beta} = 0$. When α is near 1.00, this problem is particularly severe. In these cases, the alternative location parameter

$$\zeta = \begin{cases} \delta + \beta c \tan(\pi\alpha/2), & \alpha \neq 1, \\ \delta, & \alpha = 1 \end{cases}, \tag{6.2}$$

suggested by [MC1] is more easily estimated, but unfortunately lacks any practical interpretation.

Stable random number generation: Even without a numerical approximation to the stable cdf, it has long been relatively simple to compute stable quasi-random variables, by the method of [CMS]. Let U and V be two iid uniform (0, 1) quasi-random variables, computed by standard methods. Let $W = \log(1/U)$ and $\Phi = \pi(V - .5)$. Then for $\alpha \neq 1$,

$$X = \frac{\sin(\alpha\Phi) + \beta\tan(\pi\alpha/2)\cos(\alpha\Phi)}{\cos\Phi} z^{(1-\alpha)/\alpha} \sim S(\alpha, \beta, 1, 0), \tag{6.3}$$

where

$$z = \frac{\cos((1-\alpha)\Phi) + \beta\tan(\pi\alpha/2)\sin((1-\alpha)\Phi)}{W\cos\Phi}.$$

For $\alpha = 1$,

$$X = \frac{2}{\pi}\left((\pi/2 + \beta\Phi)\tan\Phi - \beta\log\frac{(\pi/2)W\cos\Phi}{\pi/2 + \beta\Phi}\right) \sim S(1, \beta, 1, 0). \tag{6.4}$$

Unless α is very near 1 (e.g. within 10^{-8}), (6.3) and (6.4) may be computed directly, provided 8 byte precision (approximately 16 digits) is used. If α is very small, the probability that the answer will be an overflow becomes non-trivial, but so long as $\alpha \ge 0.1$ this is not a problem.[13]

[13]The Chambers [CMS] equation (4.1) yields a stable random variable with ζ (as defined above) $= 0$, rather than with $\delta = 0$. To recover the standard parameterization, we have deleted their intitial $\tan(\alpha\Phi_0)$ (i.e. added $\beta\tan(\pi\alpha/2)$ back on).

7. Summary

A linear regression with symmetric stable disturbances may be estimated quickly by maximum likelihood (ML), using the symmetric stable density approximation of [MC6]. The resulting estimator is robust, in the sense that it effectively gives less weight to outlier observations than it does to the observations that conform more closely to the regression line or surface. Despite the infinite variance of the stable disturbances, the ML estimators are asymptotically normal, with asymptotic standard errors that may be computed from the Hessian of the log-likelihood function.

The method is illustrated by means of the Harrison-Rubinfeld [HR] Hedonic Housing Price Equation studied by [BKW], and by the Daniel-Wood [DW] Pilot-Plant Regression studied by [RL]. The symmetric stable ML estimates compare favorably to results obtained by means of the metric Winsorization advocated by [HPJ] and to those obtained by trimmed least squares as employed by [RL].

Symmetric stable ML linear regression may be implemented most efficiently by means of a quasi-Newton-Raphson optimization method based on the Outer Product of Gradients (OPG) matrix, similar to that used for non-linear least squares by [BHH]. However, the Nelder-Mead polytope method [PTV] also gives satisfactory results.

Strictly speaking, the method does assume that the disturbances are truly symmetric stable, and not merely in the domain of attraction of a stable law. However, if the disturbances merely have a distribution that is approximated by a symmetric stable distribution, the method may to that extent be regarded as approximately valid.

Acknowledgments. The author is grateful to Jerry Thursby, William DuMouchel, Christopher Baum, David Belsley, and an anonymous referee for helpful suggestions, and to James Breece for computational assistance. Early work on this research was supported by NSF grant SOC 78-13780.

References

[BKW] Belsley, D.A., E. Kuh, and R.E. Welsch (1980). *Regression diagnostics: Identifying Influential Data*. John Wiley & Sons, New York.

[BHH] Berndt, E.K., B.H. Hall, R.E. Hall, and J.A. Hausman (1974). Estimation and inference in nonlinear structural models. *Annals of Economic and Social Measurement* **3/4**: 653-65.

[BH] Bergström, H. (1952). On some expansions of stable distribution functions. *Arkiv für Mathematik* **2**: 375-8.

[BM] Bidarkota, P.V., and J.H. McCulloch (1996). State-space modeling with symmetric stable shocks: The case of U.S. inflation. Ohio State Univ. Economics Dept. Working Paper #96-02.

[BP] Brorsen, B.W., and P.V. Preckel (1993). Linear Regression with stably distributed residuals. *Comm. Statist. Thy. Meth.* **22**: 659-67.

[BNR] Byczkowski, T., J.P. Nolan, and B. Rajput (1993). Approximation of multidimensional stable densities. *J. Multivariate Anal.* **46**: 13-31.

[CMS] Chambers, J.M., C.L. Mallows, and B.W. Stuck (1976). A method for simulating stable random variables. *J. Amer. Statist. Assoc.* **71**: 340-4. Corrections **82** (1987): 704, **83** (1988): 581. See also Weron (c. 1975).

[DW] Daniel, C., and F.S. Wood (1971). *Fitting Equations to Data.* John Wiley & Sons, New York.

[DM1] DuMouchel, W.H. (1973). On the Asymptotic Normality of the Maximum-Likelihood Estimate when Sampling from a Stable Distribution. *Ann. Statist.* **1**: 948-57.

[DM2] DuMouchel, W.H. (1975). Stable distributions in statistical inference: 2. Information from stably distributed samples. *J. Amer. Statist. Assoc.* **70**: 386-93.

[GP] Gilley, O.W., and Pace, R. K. (1996). On the Harrison and Rubinfeld Data. *J. Environmental Economics and Management*, **31:** 403-405.

[HR] Harrison, D., and Rubinfeld, D. L. (1978). Hedonic housing prices and the demand for clean air. *J. Environmental Economics and Management* **5**: 81-102.

[HPJ] Huber, P.J. (1981). *Robust Statistics.* John Wiley & Sons, New York.

[KG] Kitagawa, G. (1987). Non-Gaussian state-space modeling of nonstationary time series. *J. Amer. Statist. Assoc.* **82** : 1032-63.

[LMR] Logan, B.F., C.L. Mallows, S.O. Rice, and L.A. Shepp (1973). Limit distributions of self-normalized sums. *Annals of Probability* **1:** 788-809.

[MB] Mandelbrot, B. (1963). New methods in statistical economics. *J. Political Econ.* **71**: 421-40.

[MC1] McCulloch, J.H. (1986). Simple consistent estimators of stable distribution parameters. *Comm. Statist. Sim. & Comput.* **15**: 1109-1136.

[MC2] McCulloch, J.H. (1994). Time series analysis of state-space models with symmetric stable errors by posterior mode estimation. Ohio State Univ. Econ. Dept. W.P. 94-01.

[MC3] McCulloch, J.H.(1996a). On the parameterization of the afocal stable distributions. *Bull. London Math. Soc.*, **28:** 651-655.

[MC4] McCulloch, J.H. (1996b). Financial applications of stable distributions. In G.S. Maddala and C.R. Rao, eds., *Statistical Methods in Finance (Handbook of Statistics* **14**: 393-425). Elsevier Science, Amsterdam.

[MC5] McCulloch, J.H.(1997). Measuring tail thickness in order to estimate the stable index A critique. *J. Business Econ. Statist.*, in press.

[MC6] McCulloch, J.H. (this volume) Numerical approximation of the symmetric stable distribution and density.

[MP1] McCulloch, J.H. and D.B. Panton (1996). Precise fractiles and fractile densities of the maximally-skewed stable distributions. *Computational Statistics and Data Analysis*, in press.

[MP2] McCulloch, J. W., and Panton, D. B. (this volume). Tables of the Maximally-Skewed Stable Distributions and Densities.

[NPM] Nolan, J.P., A.K. Panorska, and J.H. McCulloch (1996). Estimation of stable spectral measures. The American Univ., Univ. of Tennessee at Chattanooga, and the Ohio State Univ.

[OH] Oh, C.S. (1994). *Estimation of Time Varying Term Premia of U.S. Treasury Securities: Using a STARCH Model with Stable Distributions.* Ph.D. dissertation, Ohio State Univ.

[PG] Pace, R. K., and Gilley, O.W. (1997). Using spatial configuration of the data to improve estimation. *J. Real Estate and Financial Economics,* in press.

[PTV] Press, W.H., S.A. Teukolsky, W.T. Vetterling, and B.P. Flannery (1992). *Numerical Recipes in FORTRAN: The Art of Scientific Computing*, 2nd ed. Cambridge Univ. Press, Cambridge.

[RL] Rousseeuw, P.J., and A.M. Leroy (1987). *Robust Regression with Outlier Detection.* John Wiley & Sons, New York.

[ST] Samorodnitsky, G., and M.S. Taqqu (1994). *Stable Non-Gaussian Random Processes.* Chapman and Hill, New York.

[SA] Sorenson, H.W., and D.L. Alspach (1971). Recursive Bayesian estimation using Gaussian sums. *Automatica* **7**: 465-79.

[GW] Student [W.S. Gosset] (1927). Errors of routine analysis. *Biometrika* **19**: 151-64.

[ZVM] Zolotarev, V. M. (1986). *One-Dimensional Stable Laws.* Amer. Math. Soc., 1986. (Translation of *Odnomernye Ustoichivye Raspredeleniya,* Nauka, Moscow, 1983.)

Economics Dept.
Ohio State University
1945 N. High St.
Columbus, OH 43210
(614) 292-0382
(614) 292-3906 (FAX)
mcculloch.2@osu.edu

V. Signal Processing

Deviation from Normality in Statistical Signal Processing: Parameter Estimation with Alpha-Stable Distributions

Panagiotis Tsakalides and Chrysostomos L. Nikias [1]

Abstract

The importance of extending the statistical signal processing methodology to the alpha-stable framework is apparent. First, signal processing engineers have started to appreciate the elegant stability, scaling, and self-similarity properties of stable distributions. Additionally, real life applications exist in which impulsive channels tend to produce large-amplitude, short-duration interferences more frequently than Gaussian channels do. The stable law has been shown to successfully model noise over certain impulsive channels. With this motivation, we describe new robust techniques for parameter estimation in the presence of signals and/or noise modeled as complex isotropic stable processes. In particular, we address the solution of the signal parameter estimation problem through the use of sensor array data retrieved in the presence of impulsive interference. First, we present optimal, maximum likelihood-based approaches to the direction-of-arrival estimation problem and we introduce the Cauchy Beamformer. Then, we develop subspace methods based on fractional lower-order moments and covariations, and introduce the Robust Covariation-Based Multiple Signal Classification (ROC-MUSIC) algorithm. Simulation experiments demonstrate the performance of the proposed methods.

Key Words: Stable processes, statistical signal processing, radar sensor array, direction of arrival, parameter estimation, maximum likelihood, Cramér-Rao bound.

1 Introduction

Most of the theoretical work in statistical signal processing has focused on the case where signals and/or noise are assumed to follow a Gaussian or second-order model. The Gaussian assumption is frequently motivated by the physics of the problem and, most importantly, it often ensures analytical tractability. However, in many practical instances, experimental results have been reported where data series are impulsive in nature and cannot be appropriately modeled by means of the Gaussian or other expo-

[1] Research supported by the Office of Naval Research, grant N00014-92-J-1034 and by Rome Laboratory, grant F30602-95-1-0001.

nential distributions with finite second- and higher-order moments. As an example, analysis of measured monostatic clutter return data from the US Air Force Mountaintop Database provides strong evidence that a heavy-tailed modeling is appropriate for this data [RNP96]. Other types of data from a diverse field of applications, such as active sonar returns [TN96a], CPU time to complete a job, and network traffic related measures such as packet inter-arrival and call holding times, appear to be generated by heavy tailed distributions [Res95].

Theoretical advances in detection and parameter estimation theory have been playing a very important role in the development of state of the art radar and sonar systems. Future advanced airborne radar systems must be able to detect, identify, and estimate the parameters of a target in severe interference backgrounds. As a result, the problem of clutter suppression has been the focus of considerable research in the radar engineering community. It is recognized that effective clutter suppression can be achieved only on the basis of appropriate statistical modeling. As a result, a number of distributions, based on empirical as well as theoretical grounds, have been proposed for the modeling of non-Gaussian clutter and interference environments [Kas88, Mid84].

Recently, a statistical model for impulsive clutter has been proposed, which is based on the theory of symmetric alpha-stable ($S\alpha S$) random processes [NS95]. The model is of a statistical-physical nature and has been shown to arise under very general assumptions and to describe a broad class of impulsive interference. In particular, it has been shown in [NS95] that the first order distribution of the amplitude of the radar return follows a SαS law, while the first-order joint distribution of the quadrature components of the envelope of the radar return follows an isotropic stable law. In addition, the theory of *multivariate sub-Gaussian* random processes provides an elegant and mathematically tractable framework for the solution of the detection and parameter estimation problems in the presence of impulsive correlated radar clutter.

In this paper, we address the solution of the signal parameter estimation problem through the use of sensor array data retrieved in the presence of impulsive interference, which can be modeled as a complex isotropic stable process. One of the most interesting problems in this area is the estimation of the direction of arrival (DOA) of narrow-band source signals having the same known center frequency. In the past, the problem has been studied extensively under the assumption of Gaussian distributed signals and/or noise, and a variety of methods for its solution have been proposed. As a result of the Gaussian assumption, most methods are based on the second-order statistics of the signals.

Maximum Likelihood (ML) was one of the first methods to be applied in the area of array processing [JD93]. When applying the ML technique to the source localization problem, two different assumptions for the signal waveforms result into two different methods. According to the *Stochastic ML* (SML), the signals are modeled as Gaussian random processes. This is often motivated by the Central Limit Theorem and results in mathematically convenient expressions. On the other hand, in the *Deterministic ML* (DML) the signals are considered as unknown, deterministic quantities that need to be estimated in conjunction with the direction of arrival. This is a natural model for digital communication applications where the signals are far from being normal random variables, and where estimation of the signal is of equal interest. The importance of the ML technique comes from the mathematical property that, under certain regularity conditions, the ML estimator is known to be asymptotically efficient, i.e., it achieves the Cramér-Rao bound (CRB) for the estimation error variance. In this sense, ML has the best possible asymptotic properties.

However, due to the high computational load of the multivariate nonlinear maximization problem involved in the ML estimator, sub-optimal methods have also been developed [Cap69, Red79, RPK86, Sch86]. The better known ones are cited here: Minimum Variance Distortionless method of Capon [Cap69] and the so-called eigenvector-based methods including the MUSIC [Sch86], Minimum Norm [Red79, KT83], and the ESPRIT method [RPK86]. The performance of the aforementioned methods is inferior to that of the ML method, especially for low SNR values or when the number of observation snapshots is small.

The MUSIC method, a generalization of Pisarenko's harmonic retrieval method, has received the most attention and triggered the development of a large number of algorithms referred to as *eigenvector* or *subspace* techniques. Besides offering a new geometric interpretation of the array processing problem, MUSIC uses concepts from complex vector spaces and well-known tools from linear algebra, such as the singular value decomposition (SVD), in order to achieve high resolution while keeping the computational complexity relatively low compared to that of the ML methods.

This paper is devoted to parameter estimation of multiple sources in the presence of additive interference which is modeled as a complex isotropic α-stable process. The paper is organized as follows: In Section 2, the statistical model, based on the class of *bivariate symmetric α-stable (SαS) distributions*, is introduced. This model is well-suited for describing noise processes that are impulsive in nature and contains the Gaussian process as a special case. In Section 3, the direction of arrival (DOA) estimation problem is formulated. In Section 4, the ML estimator is developed and the Cramér-Rao bound is derived for the additive Cauchy noise case. We introduce the Cauchy Beamformer and show that it performs robustly over the whole range of additive stable noise environments corresponding to values of α in $1 \leq \alpha \leq 2$. In Section 5, we discuss the development of subspace techniques in the presence of α-stable distributed signals and noise. Our analysis is based on the formulation of the covariation matrix of the array sensor outputs. Finally, Monte-Carlo simulation results are presented in Section 6, and conclusions are drawn in Section 7.

2 Symmetric Alpha-Stable ($S\alpha S$) Distributions

A complex random variable (r.v.) $X = X_1 + \jmath X_2$ is symmetric α-stable ($S\alpha S$) if X_1 and X_2 are jointly $S\alpha S$ and then its characteristic function is written as

$$\begin{aligned} \varphi(\omega) &= E\{\exp[\jmath \Re(\omega X^*)]\} = E\{\exp[\jmath(\omega_1 X_1 + \omega_2 X_2)]\} \\ &= \exp\left[-\int_{S_2} |\omega_1 x_1 + \omega_2 x_2|^\alpha d\Gamma_{X_1,X_2}(x_1, x_2)\right], \end{aligned} \tag{2.1}$$

where $\omega = \omega_1 + \jmath\omega_2$, $\Re[\cdot]$ is the real part operator, and Γ_{X_1,X_2} is a symmetric measure on the unit sphere S_2, called the *spectral measure* of the random variable X. The *characteristic exponent* α is restricted to the values $0 < \alpha \leq 2$ and it determines the shape of the distribution. The smaller the characteristic exponent α, the heavier the tails of the density.

A complex random variable $X = X_1 + \jmath X_2$ is *isotropic* if and only if (X_1, X_2) has a uniform spectral measure [ST94]. In this case, the characteristic function of X can be written as

$$\varphi(\omega) = E\{\exp(\jmath\Re[\omega X^*])\} = \exp(-\gamma|\omega|^\alpha), \tag{2.2}$$

where γ ($\gamma > 0$) is the *dispersion* of the distribution. The dispersion plays a role analogous to the role that the variance plays for second-order processes. Namely, it determines the spread of the probability density function around the origin.

Unfortunately, when $\alpha \neq 1$ or $\alpha \neq 2$, no closed form expressions exist for the density function of the isotropic complex stable random variable. By using the polar coordinate $\rho = |X| = \sqrt{X_1^2 + X_2^2}$, the density function can be written as $f_{\alpha,\gamma}(X_1, X_2) = \chi_{\alpha,\gamma}(\rho)$, and is expressed in a power series expansion form [NS95]. In particular, for $0 < \alpha < 1$,

$$\chi_{\alpha,\gamma}(\rho) = \frac{1}{\pi^2\gamma^{2/\alpha}} \sum_{k=1}^{\infty} \frac{2^{\alpha k}(-1)^{k-1}}{k!} (\Gamma(\alpha k/2 + 1))^2 \sin(\frac{k\alpha\pi}{2})(\frac{\rho}{\gamma^{1/\alpha}})^{-\alpha k-2} \tag{2.3}$$

while for $1 < \alpha < 2$,

$$\chi_{\alpha,\gamma}(\rho) = \frac{1}{\pi\alpha\gamma^{2/\alpha}} \sum_{k=0}^{\infty} \frac{(-1)^k}{2^{2k+1}(k!)^2} \Gamma(\frac{2k+2}{\alpha})(\frac{\rho}{\gamma^{1/\alpha}})^{2k}. \tag{2.4}$$

For $\alpha = 1$ we obtain the Cauchy density

$$\chi_{1,\gamma}(\rho) = \frac{\gamma}{2\pi(\rho^2 + \gamma^2)^{3/2}} \tag{2.5}$$

while for $\alpha = 2$ we get the Gaussian density

$$\chi_{2,\gamma}(\rho) = \frac{1}{4\pi\gamma} \exp(-\frac{\rho^2}{4\gamma}). \tag{2.6}$$

An important difference between the Gaussian and the other distributions of the $S\alpha S$ family is that only moments of order less than α exist for the non-Gaussian $S\alpha S$ family members. The so called *fractional lower order moments* (FLOM) of a $S\alpha S$ random variable with zero location parameter and dispersion γ are given by:

$$E|X|^p = C(p,\alpha)\gamma^{\frac{p}{\alpha}} \qquad \text{for } 0 < p < \alpha \tag{2.7}$$

where

$$C(p,\alpha) = \frac{2^{p+1}\Gamma(\frac{p+2}{2})\Gamma(-\frac{p}{\alpha})}{\alpha\Gamma(\frac{1}{2})\Gamma(-\frac{p}{2})}, \tag{2.8}$$

and $\Gamma(\cdot)$ is the Gamma function defined by

$$\Gamma(x) = \int_0^\infty t^{x-1}\exp(-t)dt. \tag{2.9}$$

Several complex r.v.'s are jointly $S\alpha S$ if their real and imaginary parts are jointly $S\alpha S$. When $X = X_1 + \jmath X_2$ and $Y = Y_1 + \jmath Y_2$ are jointly $S\alpha S$ with $1 < \alpha \le 2$, the *covariation* of X and Y is defined by

$$[X,Y]_\alpha = \int_{S_4}(x_1 + \jmath x_2)(y_1 + \jmath y_2)^{<\alpha-1>} d\Gamma_{X_1,X_2,Y_1,Y_2}(x_1, x_2, y_1, y_2), \tag{2.10}$$

where we use throughout the convention

$$Y^{<\beta>} = |Y|^{\beta-1}Y^*. \tag{2.11}$$

It can be shown that for every $1 \le p < \alpha$, the covariation can be expressed as a function of moments [CM89]

$$[X,Y]_\alpha = \frac{E\{XY^{<p-1>}\}}{E\{|Y|^p\}}\gamma_Y, \tag{2.12}$$

where γ_Y is the dispersion of the r.v. Y given by

$$\gamma_Y^{p/\alpha} = \frac{E\{|Y|^p\}}{C(p,\alpha)} \qquad \text{for } 0 < p < \alpha. \tag{2.13}$$

Obviously, from (2.12) it holds that

$$[X,X]_\alpha = \gamma_X. \tag{2.14}$$

Also, the *covariation coefficient* of X and Y is defined by

$$\lambda_{X,Y} = \frac{[X,Y]_\alpha}{[Y,Y]_\alpha}, \tag{2.15}$$

and by using (2.12) it can be expressed as

$$\lambda_{X,Y} = \frac{E\{XY^{<p-1>}\}}{E\{|Y|^p\}} \qquad \text{for } 1 \le p < \alpha. \tag{2.16}$$

The covariation of complex jointly $S\alpha S$ r.v.'s is not generally symmetric and has the following properties [Cam83]:

P1 If X_1, X_2 and Y are jointly $S\alpha S$, then

$$[aX_1 + bX_2, Y]_\alpha = a[X_1, Y]_\alpha + b[X_2, Y]_\alpha \tag{2.17}$$

for any complex constants a and b.

P2 If Y_1 and Y_2 are independent and Y_1, Y_2 and X are jointly $S\alpha S$, then

$$[aX, bY_1 + cY_2]_\alpha = ab^{<\alpha-1>}[X, Y_1]_\alpha + ac^{<\alpha-1>}[X, Y_2]_\alpha \tag{2.18}$$

for any complex constants a, b and c.

P3 If X and Y are independent $S\alpha S$, then $[X, Y]_\alpha = 0$.

We used the $S\alpha S$ family of distributions to model a variety of real data series. Consider Figure 1 which shows results on the modeling of the amplitude statistics of real radar clutter by means of $S\alpha S$ distributions. Clutter is a group of unwanted radar returns due to scattering centers such as precipitation, mountains, and ocean waves. The received clutter signal can be represented in terms of its *in-phase* (I) and *quadrature* (Q) components. A typical sample set of the sea clutter data is shown in Figure 1(a). A comparison is made between the $S\alpha S$ amplitude probability density (APD) and the Gaussian APD on how they approximate the empirical APD corresponding to the real radar clutter time series. The estimation of the parameters of the stable distribution from the real clutter data was achieved by methods based on fractional lower-order and negative-order moments, as described in [MN95]. For the particular clutter series shown here, the characteristic exponent of the $S\alpha S$ distribution which best fits the data was calculated to be approximately $\alpha = 1.7$. To fit a Gaussian model, the variance was estimated by calculating the sample variance of the data. The impulsive nature of the clutter data is obvious in Figure 1(a). Figure 1(b) shows that the $S\alpha S$ distribution fits the tails of the real data more accurately than the Gaussian density. These results indicate that the $S\alpha S$ distribution is superior to the Gaussian distribution for modeling the particular radar clutter data under study.

3 Parameter Estimation with a Sensor Array

The field of science which engineers call *estimation theory* and statisticians call *parameter estimation* deals with the problem of collecting measurements and estimating the numerical value of a real or complex vector of parameters which best describes a physical system under study. The estimation processor is essentially an algorithm which optimizes an appropriately chosen cost function with respect to the observation vectors. The cost function is formed by assuming certain statistics about the data based on the physics of the problem and by using an optimization criterion.

In this paper, we address the parameter estimation problem in the framework of array signal processing. In radar and sonar systems, an array of sensors detect incoming energy and observe a combination of intentional energy from a target as well as environmental energy such as noise, multipath signals, and interference. Consider an array of r sensors with arbitrary locations and arbitrary directional characteristics, that receive signals generated by q narrow-band sources with known center frequency ω and locations $\theta_1, \theta_2, \ldots, \theta_q$ (cf. Figure 2). Since the signals are narrow-band, the propagation delay across the array is much smaller than the reciprocal of the signal bandwidth, and it follows that, by using a complex envelope representation, the array output can be expressed as

$$\mathbf{x}(t) = \sum_{k=1}^{q} \mathbf{a}(\theta_k) s_k(t) + \mathbf{n}(t), \tag{3.19}$$

where

- $\mathbf{x}(t) = [x_1(t), \ldots, x_r(t)]^T$ is the vector of the signals received by the array sensors
- $s_k(t)$ is the signal emitted by the kth source as received at the reference sensor 1 of the array
- $\mathbf{a}(\theta_k) = [1, \exp(-\jmath\omega\tau_2(\theta_k)), \ldots, \exp(-\jmath\omega\tau_r(\theta_k))]^T$ is the steering vector of the array toward direction θ_k
- $\tau_i(\theta_k)$ is the propagation delay between the first and the ith sensor for a waveform coming from direction θ_k
- $\mathbf{n}(t) = [n_1(t), \ldots, n_r(t)]^T$ is the noise vector

Equation (3.19) can be expressed in a compact form as

$$\mathbf{x}(t) = \mathbf{A}(\boldsymbol{\theta})\mathbf{s}(t) + \mathbf{n}(t), \tag{3.20}$$

where $\mathbf{A}(\boldsymbol{\theta})$ is the $r \times q$ matrix of the array steering vectors

$$\mathbf{A}(\boldsymbol{\theta}) = [\mathbf{a}(\theta_1), \ldots, \mathbf{a}(\theta_q)], \tag{3.21}$$

and $\mathbf{s}(t)$ is the $q \times 1$ vector of the signals

$$\mathbf{s}(t) = [s_1(t), \ldots, s_q(t)]^T. \tag{3.22}$$

Assuming that M snapshots are taken at time instants $t_1, \ldots, t_M$, the data can be expressed as

$$\mathbf{X} = \mathbf{A}(\boldsymbol{\theta})\mathbf{S} + \mathbf{N}, \tag{3.23}$$

where $\mathbf{X}$ and $\mathbf{N}$ are the $r \times M$ matrices

$$\mathbf{X} = [\mathbf{x}(t_1), \ldots, \mathbf{x}(t_M)], \tag{3.24}$$

$$\mathbf{N} = [\mathbf{n}(t_1), \ldots, \mathbf{n}(t_M)], \tag{3.25}$$

and $\mathbf{S}$ is the $q \times M$ matrix

$$\mathbf{S} = [\mathbf{s}(t_1), \ldots, \mathbf{s}(t_M)]. \tag{3.26}$$

Our objective is to estimate the directions of arrival $\theta_1, \ldots, \theta_q$ of the sources from the M snapshots of the array $\mathbf{x}(t_1), \ldots, \mathbf{x}(t_M)$. Although for simplicity of discussion, we only consider the DOA's, other parameters associated with the targets, such as range and Doppler shifts (relative target velocity) may be embedded in vector $\boldsymbol{\theta}$.

For the problem to be well posed, we are going to make the following assumptions regarding the array, the signals, and the noise:

A.1 The number of signals is known and is smaller than the number of sensors, i.e., $q < r$.

A.2 The set of any q steering vectors is linearly independent.

A.3 The noise samples $n_i(t_j)$; $i = 1, \ldots, r$; $j = 1, \ldots, M$, come from a complex (bivariate) *isotropic stable distribution.*

A.4 The noise samples $n_i(t_j)$ are statistically independent from one another both along the array sensors, namely, along index i, and along time, namely, along index j.

Assumptions A.1 and A.2 guarantee the uniqueness of the solution. Assumption A.3 draws a new element into our analysis as we deviate from the conventional assumption that the noise in sensor arrays is a complex-valued Gaussian process. Our assumption incorporates a wide range of noise environments which are impulsive in nature and includes the Gaussian noise as a special case.

4 Maximum Likelihood Estimation in Alpha-Stable Noise

In this section, we develop the Maximum Likelihood (ML) estimator of the source locations in the presence of noise modeled as a *complex isotropic Cauchy process* with dispersion γ. There are two reasons for choosing the Cauchy density to model the additive noise: First, the Cauchy distribution has a closed-form expression for its density function. This results in a fairly straight-forward implementation of the maximum likelihood estimation, with closed form expressions for the Cramér-Rao bound. Secondly, it is shown through simulations that the Cauchy Beamformer is very robust in different impulsive noise environments, i.e., its performance does not change significantly when the parameter α of the $S\alpha S$ noise varies in the interval $1 \leq \alpha \leq 2$.

4.1 ML Bearing Estimation of Sources in Cauchy Noise

In this section, we assume that the noise present at the array sensors is modeled as a *complex isotropic Cauchy process* with pdf given by (2.5, $\alpha = 1$). We are most interested in estimating the bearings of the sources, and therefore we consider the signals themselves, as well as the noise dispersion, γ, as nuisance parameters in the estimation problem.

Under assumption A.4, it follows from (3.19) and (2.5) that the joint density function of the sampled data is given by

$$f(\mathbf{X}) = \prod_{t=1}^{M}\prod_{i=1}^{r} \chi_{1,\gamma}\left(\left|x_i(t) - \sum_{k=1}^{q} a_i(\theta_k)s_k(t)\right|\right) \tag{4.27}$$

or

$$f(\mathbf{X}) = \prod_{t=1}^{M}\prod_{i=1}^{r} \frac{1}{2\pi} \frac{\gamma}{\left(\gamma^2 + \left|x_i(t) - \sum_{k=1}^{q} a_i(\theta_k)s_k(t)\right|^2\right)^{3/2}}, \tag{4.28}$$

where $a_1(\theta_k) = 1$ and $a_i(\theta_k) = \exp(-\jmath\omega\tau_i(\theta_k))$; $i = 2, \ldots, r$. Hence the log-likelihood function $L(\mathbf{X}; \gamma, \mathbf{S}, \boldsymbol{\theta})$, ignoring constant terms, is expressed as:

$$L(\mathbf{X}; \gamma, \mathbf{S}, \boldsymbol{\theta}) = Mr\log(\gamma) - \frac{3}{2}\sum_{t=1}^{M}\sum_{i=1}^{r} \log\left(\gamma^2 + \left|x_i(t) - \sum_{k=1}^{q} a_i(\theta_k)s_k(t)\right|^2\right). \tag{4.29}$$

The ML estimator is obtained by maximizing $L(\mathbf{X}; \gamma, \mathbf{S}, \boldsymbol{\theta})$ with respect to γ, $\mathbf{S}$, and $\boldsymbol{\theta}$, i.e.,

$$\max_{\gamma,\mathbf{S},\boldsymbol{\theta}} L(\mathbf{X}; \gamma, \mathbf{S}, \boldsymbol{\theta}). \tag{4.30}$$

To reduce the dimension of this optimization problem, we first fix γ and $\boldsymbol{\theta}$, and minimize $L(\mathbf{X}; \gamma, \mathbf{S}, \boldsymbol{\theta})$ with respect to the signal $\mathbf{S}$. For fixed t we take the derivative of $L(\mathbf{X}; \gamma, \mathbf{S}, \boldsymbol{\theta})$ with respect to $s_k(t)$:

$$\frac{\partial L}{\partial s_k(t)} = -3\sum_{i=1}^{r} \frac{a_i(\theta_k)\left[x_i(t) - a_i(\theta_k)s_k(t)\right]^*}{\gamma^2 + |x_i(t) - a_i(\theta_k)s_k(t)|^2}. \tag{4.31}$$

Unfortunately, no explicit solution of (4.31) is possible. In order to be able to obtain closed form expressions for the signals, we resort to the application of the pseudo maximum likelihood (PML) estimation. PML estimation is an important method in applications where probability models abound for which the analytical derivation of the maximum likelihood estimate for all the parameters is virtually impossible. The problem formulated in [GS81] can be stated as follows:

Let $X_1, \ldots, X_n$ be i.i.d. random variables with probability distribution $f(X; \boldsymbol{\theta}, \mathbf{S})$ indexed by two sets of parameters. Let $\hat{\mathbf{S}} = \hat{\mathbf{S}}(X_1, \ldots, X_n)$

be an estimate of $\mathbf{S}$ *other than the maximum likelihood estimate, and let* $\hat{\boldsymbol{\theta}}$ *be the solution of the likelihood equation* $\partial/\partial\boldsymbol{\theta} \ \log L(\mathbf{X}; \boldsymbol{\theta}, \hat{\mathbf{S}}) = 0$ *which maximizes the likelihood. Then,* $\hat{\boldsymbol{\theta}}$ *is called a pseudo maximum likelihood estimate of* $\boldsymbol{\theta}$*, and under certain conditions it is consistent and asymptotically normal.*

The PML estimator $\hat{\boldsymbol{\theta}}$ has good large sample properties when $\hat{\mathbf{S}}$ does. In general, the asymptotic analysis for $\hat{\boldsymbol{\theta}}$ will depend on the asymptotic characteristics of $\hat{\mathbf{S}}$.

Returning to the optimization problem described in (4.30), we observe that maximizing $L(\mathbf{X}; \gamma, \mathbf{S}, \boldsymbol{\theta})$ with respect to the signal $\mathbf{S}$ is equivalent to the following minimization problem:

$$\min_{\mathbf{S}} \mathcal{L}(\mathbf{X}; \gamma, \mathbf{S}, \boldsymbol{\theta}) = \min_{\mathbf{S}} \left\{ \sum_{t=1}^{M} \sum_{i=1}^{r} \log(1 + \frac{|x_i(t) - \sum_{k=1}^{q} a_i(\theta_k) s_k(t)|^2}{\gamma^2}) \right\}. \quad (4.32)$$

As we can see, (4.32) involves minimizing a double sum expression of logarithmic functions of the form $\log(1+z)$. In the unit disc $B_1(0) = \{z \in \mathcal{C} : |z| < 1\}$, the function $\log(1+z)$ can be expressed as an infinite series:

$$\log(1+z) = \sum_{n=1}^{\infty} \frac{(-1)^{n-1}}{n} z^n = z - \frac{z^2}{2} + - \cdots; \quad |z| < 1 \quad (4.33)$$

Hence for $|x_i(t) - \sum_{k=1}^{q} a_i(\theta_k) s_k(t)| < \gamma$ the functional $\mathcal{L}(\mathbf{X}; \gamma, \mathbf{S}, \boldsymbol{\theta})$ can be written in the form

$$\mathcal{L}(\mathbf{X}; \gamma, \mathbf{S}, \boldsymbol{\theta}) = \sum_{t=1}^{M} \sum_{i=1}^{r} \sum_{n=1}^{\infty} \frac{(-1)^{n-1}}{n\gamma^{2n}} |x_i(t) - \sum_{k=1}^{q} a_i(\theta_k) s_k(t)|^{2n}. \quad (4.34)$$

A first order approximation of the above expression results in the following $\mathcal{L}^{(1)}(\mathbf{X}; \gamma, \mathbf{S}, \boldsymbol{\theta})$ functional:

$$\mathcal{L}^{(1)}(\mathbf{X}; \gamma, \mathbf{S}, \boldsymbol{\theta}) = \frac{1}{\gamma^2} \sum_{t=1}^{M} \sum_{i=1}^{r} |x_i(t) - \sum_{k=1}^{q} a_i(\theta_k) s_k(t)|^2 \quad (4.35)$$

which, by using (3.20), can be written in a more compact form as:

$$\mathcal{L}^{(1)}(\mathbf{X}; \gamma, \mathbf{S}, \boldsymbol{\theta}) = \frac{1}{\gamma^2} \sum_{t=1}^{M} |\mathbf{x}(t) - \mathbf{A}(\boldsymbol{\theta}) \mathbf{s}(t)|^2. \quad (4.36)$$

Hence the minimization of $\mathcal{L}^{(1)}(\mathbf{X}; \gamma, \mathbf{S}, \boldsymbol{\theta})$ with respect to $\mathbf{S}$ is equivalent to the Least Squares (LS) estimation of $\mathbf{S}$. This problem has a well-known solution:

$$\hat{\mathbf{s}}(t) = (\mathbf{A}^H(\boldsymbol{\theta}) \mathbf{A}(\boldsymbol{\theta}))^{-1} \mathbf{A}^H(\boldsymbol{\theta}) \mathbf{x}(t). \quad (4.37)$$

The dispersion γ can be estimated by using the method of moments. Namely, in the expression for the FLOM of the noise given in (2.7), $E|X|^p$ can be approximated by an average sum:

$$\hat{\gamma} = \frac{\left[\frac{1}{M}\sum_{t=1}^{M}\sum_{i=1}^{r}|x_i(t) - \sum_{k=1}^{q} a_i(\theta_k)\hat{s}_k(t)|^p\right]^{\frac{1}{p}}}{[C(p,1)]^{\frac{1}{p}}}, \tag{4.38}$$

where $p < 1$, and $C(p,1)$ is given by:

$$C(p,1) = p2^p \frac{\Gamma(\frac{p}{2})\Gamma(-p)}{\Gamma(-\frac{p}{2})}. \tag{4.39}$$

By using the above estimates for the signal $\mathbf{S}$ and the noise dispersion γ, we obtain the following reduced optimization problem:

$$\max_{\boldsymbol{\theta}} L(\mathbf{X};\hat{\gamma},\hat{\mathbf{S}},\boldsymbol{\theta}) = \max_{\boldsymbol{\theta}}\{Mr\log(\hat{\gamma}) - \frac{3}{2}\sum_{t=1}^{M}\sum_{i=1}^{r}\log(\hat{\gamma}^2 + |x_i(t) - \sum_{k=1}^{q} a_i(\theta_k)\hat{s}_k(t)|^2)\}. \tag{4.40}$$

An iterative procedure based on the gradient descent principle can be applied in order to solve for $\boldsymbol{\theta}$. In general, the cost function described in (4.40) is non-convex and the optimization procedure has to be initialized sufficiently close to the global extremum. Suboptimal DOA estimators such as the MUSIC algorithm or the ROC-MUSIC method [TN96b] can be used to obtain initial bearing estimates.

4.2 The Cramér-Rao Bound for Cauchy Noise

Under the assumptions stated in Section 3 and for the case of complex isotropic Cauchy noise, the following theorem was proven in [TN95]:

Theorem 1 *The CRB for $\boldsymbol{\theta}$ and γ is given by*

$$CRB(\boldsymbol{\theta}) = \frac{5\gamma^2}{3}\left\{\sum_{t=1}^{M}\Re\left\{\mathbf{S}^H(t)\mathbf{D}^H\left[I - \mathbf{A}\left(\mathbf{A}^H\mathbf{A}\right)^{-1}\mathbf{A}^H\right]\mathbf{D}\mathbf{S}(t)\right\}\right\}^{-1} \tag{4.41}$$

and

$$CRB(\gamma) = \frac{5}{4}\frac{\gamma^2}{Mr}, \tag{4.42}$$

where

$$\mathbf{S}(t) = \begin{bmatrix} s_1(t) & 0 & \cdots & 0 & 0 \\ 0 & s_2(t) & \cdots & 0 & 0 \\ \vdots & \vdots & & \vdots & \vdots \\ 0 & 0 & \cdots & s_{q-1}(t) & 0 \\ 0 & 0 & \cdots & 0 & s_q(t) \end{bmatrix}; \quad t = 1,\ldots,M, \tag{4.43}$$

$$\mathbf{D} = [\mathbf{d}(\theta_1), \ldots, \mathbf{d}(\theta_q)], \tag{4.44}$$

$\mathbf{d}(\theta_i) = \partial \mathbf{a}(\theta_i)/\partial\theta_i;\quad i = 1, \ldots, q$, *and* $\Re\{\cdot\}$ *is the real part operator.*

The above expression for the CRB is very similar to the one derived in [SN89] for the case of additive Gaussian noise. We should note that the above bound can be achieved only when there exist unbiased estimators of all the model parameters γ, $\mathbf{S}$, and $\boldsymbol{\theta}$. A useful insight on the CRB can be gained if we consider the case of a single source ($q = 1$) impinging from direction θ in a linear array whose sensors are spaced a half-wavelength apart. In this case,

$$\mathbf{A} = [1, \exp(-\jmath\pi \sin(\theta)), \ldots, \exp(-\jmath(r-1)\pi \sin(\theta))]^T, \tag{4.45}$$

and

$$\mathbf{D} = [0, -\jmath\pi \cos(\theta) \exp(-\jmath\pi \sin(\theta)), \ldots, -\jmath(r-1)\pi \cos(\theta) \exp(-\jmath(r-1)\pi \sin(\theta))]^T. \tag{4.46}$$

So, it holds that

$$\mathbf{A}^H\mathbf{A} = r, \tag{4.47}$$

$$\mathbf{D}^H\mathbf{D} = \pi^2 \frac{r(r-1)(2r-1)}{6} \cos^2(\theta), \tag{4.48}$$

and

$$\mathbf{D}^H\mathbf{A} = \jmath\pi \frac{r(r-1)}{2} \cos(\theta). \tag{4.49}$$

Then,

$$CRB(\theta) = \frac{20}{\pi^2} \cdot \frac{1}{r(r^2-1)\cos^2(\theta)} \cdot \frac{\gamma^2}{\sum_{t=1}^{M} |s(t)|^2}. \tag{4.50}$$

The term $\frac{\gamma^2}{\sum_{t=1}^{M} |s(t)|^2}$ in the above expression for the CRB can be viewed as the inverse of a quantity analogous to the signal-to-noise ratio (SNR) for the Gaussian case, i.e., a generalized SNR, so to speak. The larger the dispersion γ of the noise, the higher the CRB.

5 Subspace Techniques in the α-Stable Framework

The ML techniques employed in Section 4 are often regarded as exceedingly complex due to the high computational load of the multivariate nonlinear optimization problem involved with these techniques. Hence, sub-optimal methods need to be developed for the solution of the DOA problem in the presence of impulsive noise, when reduced computational cost is a crucial design requirement.

In this section, we assume that the q signal waveforms are non-coherent, statistically independent, complex isotropic $S\alpha S$ ($1 < \alpha \leq 2$) random processes with zero location parameter and covariation matrix $\Gamma_S = \text{diag}(\gamma_{s_1}, \ldots, \gamma_{s_q})$. Also, the noise vector $\mathbf{n}(t)$ is a complex isotropic $S\alpha S$ random process with the same characteristic exponent α as the signals. The noise is assumed to be independent of the signals with covariation matrix $\Gamma_N = \gamma_n \mathbf{I}$.

Equation (3.20) can be written as

$$\mathbf{x}(t) = \mathbf{w}(t) + \mathbf{n}(t), \tag{5.51}$$

where $\mathbf{w}(t) = \mathbf{A}(\boldsymbol{\theta})\mathbf{s}(t)$. By the stability property, it follows that $\mathbf{w}(t)$ is also a complex isotropic $S\alpha S$ random vector with components

$$w_i(t) = \mathbf{A}_i(\boldsymbol{\theta})\mathbf{s}(t) = a_i(\theta_1)s_1(t) + \cdots + a_i(\theta_q)s_q(t) \quad i = 1, \ldots, r. \tag{5.52}$$

Also, it holds that $\mathbf{w}(t)$ is independent of $\mathbf{n}(t)$.

Now, we define the *covariation matrix*, Γ_X, of the observation vector process $\mathbf{x}(t)$ as the matrix whose elements are the covariations $[x_i(t), x_j(t)]_\alpha$ of the components of $\mathbf{x}(t)$. We have that

$$\begin{aligned}
[x_i(t), x_j(t)]_\alpha &= [w_i(t) + n_i(t), w_j(t) + n_j(t)]_\alpha \\
&= [w_i(t), w_j(t)]_\alpha + [w_i(t), n_j(t)]_\alpha \\
&\quad +[n_i(t), w_j(t)]_\alpha + [n_i(t), n_j(t)]_\alpha.
\end{aligned} \tag{5.53}$$

By the independence assumption of $\mathbf{w}(t)$ and $\mathbf{n}(t)$ and by property P3 we have that

$$[w_i(t), n_j(t)]_\alpha = 0, \quad \text{and} \quad [n_i(t), w_j(t)]_\alpha = 0. \tag{5.54}$$

In addition, by using (5.52) and properties P1 and P2 it follows that

$$\begin{aligned}
[w_i(t), w_j(t)]_\alpha &= [\sum_{k=1}^{q} a_i(\theta_k)s_k(t), w_j(t)]_\alpha = \sum_{k=1}^{q} a_i(\theta_k)[s_k(t), w_j(t)]_\alpha \\
&= \sum_{k=1}^{q} a_i(\theta_k)[s_k(t), \sum_{l=1}^{q} a_j(\theta_l)s_l(t)]_\alpha \\
&= \sum_{k=1}^{q} a_i(\theta_k)a_j^{<\alpha-1>}(\theta_k)\gamma_{s_k},
\end{aligned} \tag{5.55}$$

where $\gamma_{s_k} = [s_k, s_k]_\alpha$. Finally, due to the noise assumption made earlier, it holds that

$$[n_i(t), n_j(t)]_\alpha = \gamma_n \delta_{i,j}, \tag{5.56}$$

where $\delta_{i,j}$ is the Kronecker delta function. Combining (5.53)-(5.56) we obtain the following expression for the covariations of the sensor measurements:

$$[x_i(t), x_j(t)]_\alpha = \sum_{k=1}^{q} a_i(\theta_k) a_j^{<\alpha-1>}(\theta_k)\gamma_{s_k} + \gamma_n \delta_{i,j} \quad i, j = 1, \ldots, r. \tag{5.57}$$

Also, the dispersion and covariation coefficients of the array sensor measurements are given respectively by

$$\gamma_{x_j(t)} = \sum_{k=1}^{q} |a_j(\theta_k)|^\alpha \gamma_{s_k} + \gamma_n \quad j = 1, \ldots, r, \tag{5.58}$$

and

$$\lambda_{x_i(t), x_j(t)} = \frac{\sum_{k=1}^{q} a_i(\theta_k) a_j^{<\alpha-1>}(\theta_k)\gamma_{s_k} + \gamma_n \delta_{i,j}}{\sum_{k=1}^{q} |a_j(\theta_k)|^\alpha \gamma_{s_k} + \gamma_n} \quad i, j = 1, \ldots, r. \tag{5.59}$$

In matrix form, (5.57) gives the following expression for the covariation matrix of the observation vector:

$$\Gamma_X \triangleq [\mathbf{x}(t), \mathbf{x}(t)]_\alpha = \mathbf{A}(\boldsymbol{\theta})\Gamma_S \mathbf{A}^{<\alpha-1>}(\boldsymbol{\theta}) + \gamma_n \mathbf{I}, \tag{5.60}$$

where the (i, j)th element of matrix $\mathbf{A}^{<\alpha-1>}(\boldsymbol{\theta})$ results from the (j, i)th element of $\mathbf{A}(\boldsymbol{\theta})$ according to the operation

$$[\mathbf{A}^{<\alpha-1>}(\boldsymbol{\theta})]_{i,j} = [\mathbf{A}(\boldsymbol{\theta})]_{j,i}^{<\alpha-1>} = |\,[\mathbf{A}(\boldsymbol{\theta})]_{j,i}\,|^{\alpha-2}\,[\mathbf{A}(\boldsymbol{\theta})]_{j,i}^{*}\,. \tag{5.61}$$

Clearly, when $\alpha = 2$, i.e., for Gaussian distributed signals and noise, the expression for the covariation matrix is identical to the well-known expression for the covariance matrix:

$$\mathbf{R}_X = \mathbf{A}(\boldsymbol{\theta})\Sigma\mathbf{A}^H(\boldsymbol{\theta}) + \sigma^2\mathbf{I}, \tag{5.62}$$

where Σ is the signal covariance matrix.

When the amplitude response of the sensors equals unity, i.e., for steering vectors of the form $\mathbf{a}(\theta_k) = [1, \exp(-\jmath\omega\tau_2(\theta_k)), \ldots, \exp(-\jmath\omega\tau_r(\theta_k))]^T$, it follows that

$$[\mathbf{A}^{<\alpha-1>}(\boldsymbol{\theta})]_{i,j} = |\exp(-\jmath\omega\tau_j(\theta_i))|^{\alpha-2}\exp(\jmath\omega\tau_j(\theta_i)) = [\mathbf{A}(\boldsymbol{\theta})]_{j,i}^{*}\,, \tag{5.63}$$

and thus the covariation matrix can be written as

$$\Gamma_X = \mathbf{A}(\boldsymbol{\theta})\Gamma_S\mathbf{A}^H(\boldsymbol{\theta}) + \gamma_n\mathbf{I}. \tag{5.64}$$

Also, from (5.58) and (5.59) the dispersion and covariation coefficients of the array sensor measurements can be written as

$$\gamma_{x_j(t)} = \sum_{k=1}^{q} \gamma_{s_k} + \gamma_n, \quad j = 1, \ldots, r, \tag{5.65}$$

and

$$\lambda_{x_i(t),x_j(t)} = \frac{\sum_{k=1}^{q} a_i(\theta_k)a_j^*(\theta_k)\gamma_{s_k} + \gamma_n\delta_{i,j}}{\sum_{k=1}^{q}\gamma_{s_k} + \gamma_n} \quad i, j = 1, \ldots, r. \tag{5.66}$$

Observing (5.64), we conclude that standard subspace techniques can be applied to the covariation or the covariation coefficient matrices of the observation vector to extract the bearing information. More specifically, it follows that the rank of matrix $\mathbf{A}(\boldsymbol{\theta})\Gamma_S\mathbf{A}^H(\boldsymbol{\theta})$ is q, with the smallest $(r-q)$ of its eigenvalues equal to zero. In other words, if we let $\rho_1 \geq \rho_2 \geq \cdots \geq \rho_r$ denote the eigenvalues of matrix Γ_X, then

$$\rho_{q+1} = \cdots = \rho_r = \gamma_n. \tag{5.67}$$

We will refer to the new algorithm resulting from the eigendecomposition of the array covariation coefficient matrix as the **Robust Covariation-Based MUSIC** or **ROC-MUSIC**. By denoting the corresponding eigenvectors of Γ_X by $\{\mathbf{v}_i\}_{i=1}^{r}$, the ROC-MUSIC spectrum can be expressed as

$$S_{ROC-MUSIC}(\theta) = \frac{1}{\sum_{i=q+1}^{r} |\mathbf{a}^H(\theta)\mathbf{v}_i|^2}. \tag{5.68}$$

The locations of the source signals are determined by the values of θ for which the spectrum given by (5.68) peaks. In practice, we have to estimate the covariation matrix from a finite number of array sensor measurements. One such estimator for the covariation coefficient $\lambda_{X,Y}$, is called the *fractional lower order (FLOM) estimator*. The FLOM estimator was proposed in [SN93] and is based on fractional lower order moments of the stable process. Expanding the p-support range of the FLOM estimator we define the *modified FLOM (MFLOM)* estimator

$$\hat{\lambda}_{X,Y}(p) = \frac{\sum_{i=1}^{n} X_i Y_i^{<p-1>}}{\sum_{i=1}^{n} |Y_i|^p} \tag{5.69}$$

for independent observations $(X_1, Y_1), \ldots, (X_n, Y_n)$. In the above expression, $1/2 < p < \alpha$ if X and Y are real $S\alpha S$ random variables, and $0 < p < \alpha$ if X and Y are complex isotropic $S\alpha S$ random variables. The estimator given in (5.69) is a moments-based estimator of the *modified covariation coefficient function* defined as

$$\lambda_{X,Y}(p) = \frac{E\{XY^{<p-1>}\}}{E\{|Y|^p\}}, \quad \begin{array}{ll} 1/2 < p < \alpha & \text{if } X \text{ and } Y \text{ are real} \\ 0 < p < \alpha & \text{if } X \text{ and } Y \text{ are complex.} \end{array} \tag{5.70}$$

The modified covariation coefficient function is well defined (finite) for the aforementioned values of the parameter p as shown in [Tsa95]. The theoretical performance of the MFLOM estimator is studied also in [Tsa95]. Clearly, when $1 \leq p < \alpha$ the function $\lambda_{X,Y}(p)$ equals the covariation coefficient $\lambda_{X,Y}$ as defined in (2.15)-(2.16).

Figure 3 illustrates the influence of the parameter p to the performance of the MFLOM estimator of the covariation coefficient. As we can see, for the case of non-Gaussian stable signals ($1 < \alpha < 2$), the values of p in the range $(1/2, \alpha/2)$ result in the smallest standard deviations. For Gaussian signals, the optimal value of p is 2 and the resulting MFLOM estimator is simply the least squares estimator, as expected.

6 A Performance Study

To demonstrate the performance of the proposed methods for the solution of the parameter estimation problem, we conducted several simulation experiments where we compared the ML estimator based on the Cauchy noise assumption (Cauchy Beamformer, MLC) with the ML estimator based on the Gaussian noise assumption (Gaussian Beamformer, MLG). In addition, we studied the resolution capabilities of the introduced ROC-MUSIC algorithm versus MUSIC.

In all the experiments the array is linear with five sensors spaced a half-wavelength apart. The noise is assumed to follow the bivariate isotropic stable distribution. We chose p to be $p = 0.5$ when we estimated the noise dispersion γ via expression (4.38). In theory, expression (2.7) holds for any value of p as long as $0 < p < \alpha$. In practice, the variance of the estimator is a function of p. As shown in Figure 3, values of p in the range $[1/2, \alpha/2]$ give estimators with the smallest variance.

In every experiment we perform 500 Monte-Carlo runs and compute the mean square error (MSE) of the DOA estimates. The optimization for the MLC and MLG methods is performed by a steepest-descent algorithm with variable stepsize selected by means of Armijo's rule [Lue84]. Since the alpha-stable family for $\alpha < 2$ determines processes with infinite variance, we define two alternative signal-to-noise ratios (SNR's). Namely, we define the *Generalized SNR (GSNR)* to be the ratio of the signal power over the noise dispersion γ:

$$GSNR = 10\log(\frac{1}{\gamma M}\sum_{t=1}^{M}|s(t)|^2). \tag{6.71}$$

Also, for finite sample realizations, we define the *Pseudo-SNR (PSNR)* as:

$$PSNR = 10\log(\frac{\sum_{t=1}^{M}|s(t)|^2}{\sum_{t=1}^{M}|n(t)|^2}). \tag{6.72}$$

In the following, we present a comparative simulation study for the estimation accuracy and the resolution capabilities of the aforementioned algorithms.

Table 1: GSNR and average PSNR for different values of α.

	Noise Characteristic Exponent, α					
	$\alpha = 1.0$	$\alpha = 1.2$	$\alpha = 1.4$	$\alpha = 1.6$	$\alpha = 1.8$	$\alpha = 2.0$
GSNR [dB]	22.7416 ($\gamma = 1$)					
PSNR [dB]	0.5371	6.2171	10.1869	13.1035	15.3269	20.0383

6.1 Estimation Accuracy of the Cauchy Beamformer

In this example we study the estimation accuracy of MLC and MLG as a function of two parameters, namely the noise characteristic exponent α and the noise dispersion γ.

Characteristic exponent α. The importance of this experiment rests in its study of the robustness of the algorithms in different noise environments. Of course, by design, the MLG estimator is optimal for additive Gaussian noise ($\alpha = 2$), and the introduced MLC estimator is optimal for additive Cauchy noise ($\alpha = 1$). In real world applications, an important property of any processor is to be able to perform reasonably well when the assumptions made during its design are not longer true. In this context the beamformer should localize the source signals in a wide range of noise environments ($1 < \alpha < 2$). Here, we test the performance of the estimators when the characteristic exponent, α, of the noise stable law is changing.

Figure 4 shows the resulting MSE curves as functions of the characteristic exponent α. The number of snapshots available to the algorithms is $M = 20$. The GSNR is 22.7416 dB ($\gamma = 1$) and is shown together with the average PSNR, on Table 1.

The CRB for values of α other than 1 and 2 was calculated in [TN95] by using a linear approximation of the Fisher information matrix, and by means of the characteristic function given by (2.2). As we can clearly see, the Cauchy beamformer is practically insensitive to the changes of α, and for exact signal knowledge it almost achieves the CRB for the whole range of values α. On the other hand, both the MLG and the MUSIC algorithms exhibit very large mean-square estimation errors for non-Gaussian noise environments. Note that when $\alpha = 2$, i.e., for the Gaussian noise case, the MLG method has the least MSE, as expected.

The experiment demonstrates that for values of α in $1 \leq \alpha \leq 2$, the ML method based on the Cauchy noise assumption exhibits performance very close to the optimum. This observation, combined with the fact that the ML method based on the Cauchy assumption has computational complexity similar to the ML method based on the Gaussian assumption, justify the importance of the Cauchy Beamformer for the parameter estimation problem in practice.

Noise dispersion γ. In the second experiment we study the influence of the noise dispersion γ, i.e., the influence of the GSNR on the performance of the methods. Here, the noise follows the bivariate isotropic Cauchy distribution with dispersion γ. The number of snapshots available to the algorithms is $M = 20$. The GSNR and average PSNR for this experiment are shown on Table 2.

Table 2: GSNR and average PSNR for different values of γ.

	Noise Dispersion, γ						
	$\gamma = 0.5$	$\gamma = 1$	$\gamma = 2$	$\gamma = 4$	$\gamma = 6$	$\gamma = 8$	$\gamma = 10$
GSNR [dB]	25.7519	22.7416	19.7313	16.7210	14.9601	13.7107	12.7416
PSNR [dB]	6.5577	0.5371	-5.4835	-11.5041	-15.0259	-17.5247	-19.4629

Figure 5 shows the resulting MSE of the estimated DOA as a function of the GSNR. Again the MLC estimate has the best performance. As evident in Figure 5(b), the performance of the MLC using a LS estimate for the signal degrades more rapidly for large values of the noise dispersion, γ (low GSNR values). The reason is that for large values of the noise dispersion, the MLC estimate suffers from suboptimal signal estimation. It is noted however that the MLC estimate still has the best performance.

6.2 Resolution Capability of ROC-MUSIC versus MUSIC

In this experiment we compare the performance of the proposed ROC-MUSIC algorithm to MUSIC. The sample covariation coefficient matrix (SCCM), as estimated by (5.69), is not symmetric and hence it has complex eigenvalues in general. The more snapshots are available at the array sensors, the more nearly symmetric SCCM becomes. We come around this problem by performing the eigenvalue decomposition to the sum of the sample covariation coefficient matrix and its Hermitian transpose.

In every experiment we compute the resolution event probability, and the mean-square error (MSE) of the direction-of-arrival estimates averaged for the two sources. The MSE of the DOA estimates was calculated by taking into consideration only the Monte-Carlo runs for which the two algorithms resolved the two sources.

The resolution analysis of the two algorithms was studied by using a popular resolution criterion defined by the following threshold equation [KB86],[Zha95]:

$$\Lambda(\theta_1, \theta_2) \triangleq P(\theta_m) - \frac{1}{2}\{P(\theta_1) + P(\theta_2)\} > 0, \tag{6.73}$$

where θ_1 and θ_2 are the angles of arrival of the two signals, $\theta_m = (\theta_1 + \theta_2)/2$ is the mid-range between them, and the *null spectrum* $P(\theta) \triangleq 1/S(\theta)$ is defined as the reciprocal of the spatial spectrum $S(\theta)$ given in (5.68). The two signals are said to be resolvable if inequality (6.73) holds. The inequality implies that the null spectrum magnitude at the mid-angle should lie above the line segment linking the two signal valleys, in order for the two sources to be resolvable. In our experiments we estimated the null spectrum from a finite number of array sensor measurements, and the probability of resolution was determined by averaging the resolution events over the independent Monte Carlo runs.

In the following simulation experiments, we study the resolution capability and estimation accuracy of ROC-MUSIC versus MUSIC as a function of the noise characteristic exponent α and the angular separation of the two sources.

Characteristic exponent α. Figure 6 illustrates the performance of the two algorithms in a wide range of noise environments, from the more impulsive (α in the neighborhood of 1) to the Gaussian ones ($\alpha = 2$). The angles of arrival for the two signals are $\theta_1 = -5^o$ and $\theta_2 = 5^o$. The number of snapshots available to the algorithms is $M = 1000$. The GSNR is 22.3 dB ($\gamma = 1$). The characteristic exponent α of the additive noise is unknown to the ROC-MUSIC algorithm. We use two values of the parameter p in the estimation of the covariation matrix (cf. (5.69)): $p = 0.8$ and $p = 0.4$. Clearly, MUSIC can be thought as a special case of ROC-MUSIC with $p = 2$.

Figure 6 depicts the improved performance of ROC-MUSIC over that of MUSIC both in terms of resolution probability and MSE, for values of α in the range $(1, 2)$. Note that for $\alpha < 1.2$. MUSIC does not resolve the two sources in any of the 500 Monte Carlo runs. The results suggest that in impulsive noise environments modeled under the stable law, it is beneficial to use the covariation matrix (lower-order moments) instead of the covariance matrix (second-order moments). Of course, for Gaussian additive noise ($\alpha = 2$) the use of second-order moments ($p = 2$) gives better results. Figure 6(b) shows that the choice of $p = 0.8$ in ROC-MUSIC gives better results than the ones obtained when using $p = 0.4$, especially when $\alpha \geq 1.6$ in which case $p < \alpha/2$. It is not clear however from this result how the parameter p affects the performance of ROC-MUSIC. An analytical study on the way the choice of p affects the localization and resolution capabilities of the algorithm is currently under way.

Angular separation. Figure 7 illustrates the variation of the algorithmic performance with respect to the angle separation of the two incoming signals, for $M = 1000$, GSNR= 22.3 dB, PSNR= -1.56 dB. The additive noise is modeled as complex $S\alpha S$ with $\alpha = 1.5$. As expected, the resolution capability of both algorithms improves with increased angle separation between the two sources. But for a given probability of resolution, the ROC-MUSIC algorithm requires a lower angle separation threshold than MUSIC.

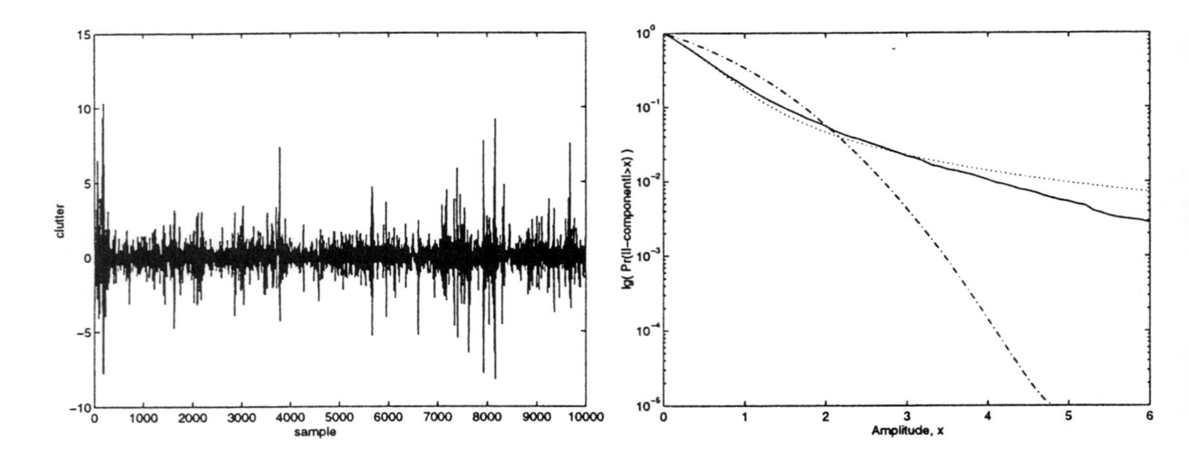

Figure 1: The in-phase component of the clutter time series (left) and the corresponding amplitude probability density curves (right) (Empirical: solid, $S\alpha S$: dotted, Gaussian: dash-dotted).

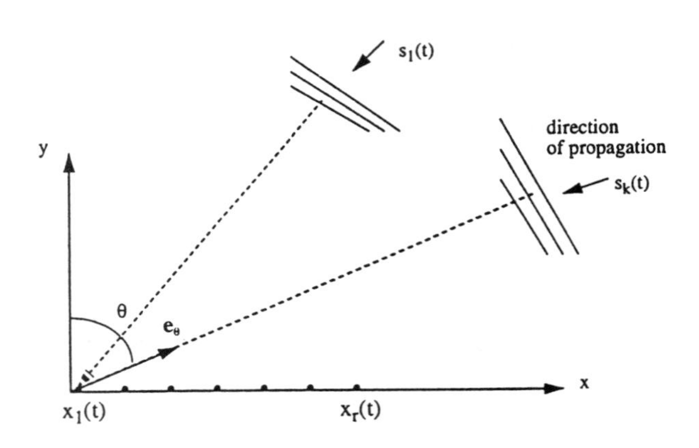

Figure 2: Array processing setup.

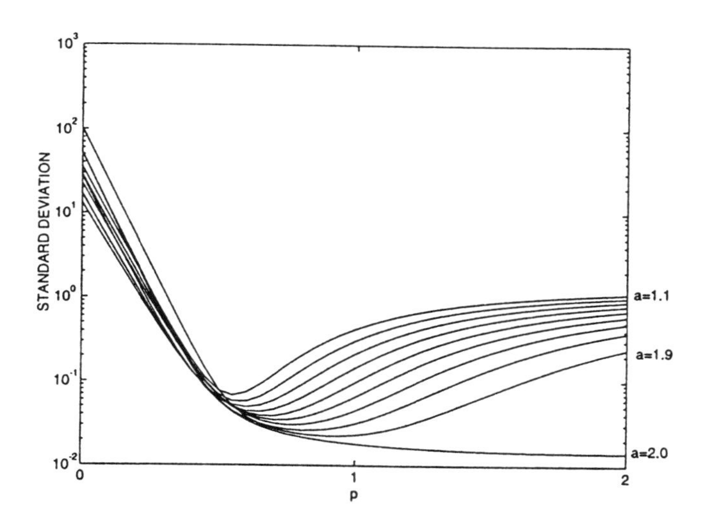

Figure 3: Standard deviation of the MFLOM estimates of the modified covariation coefficient as a function of the parameter p.

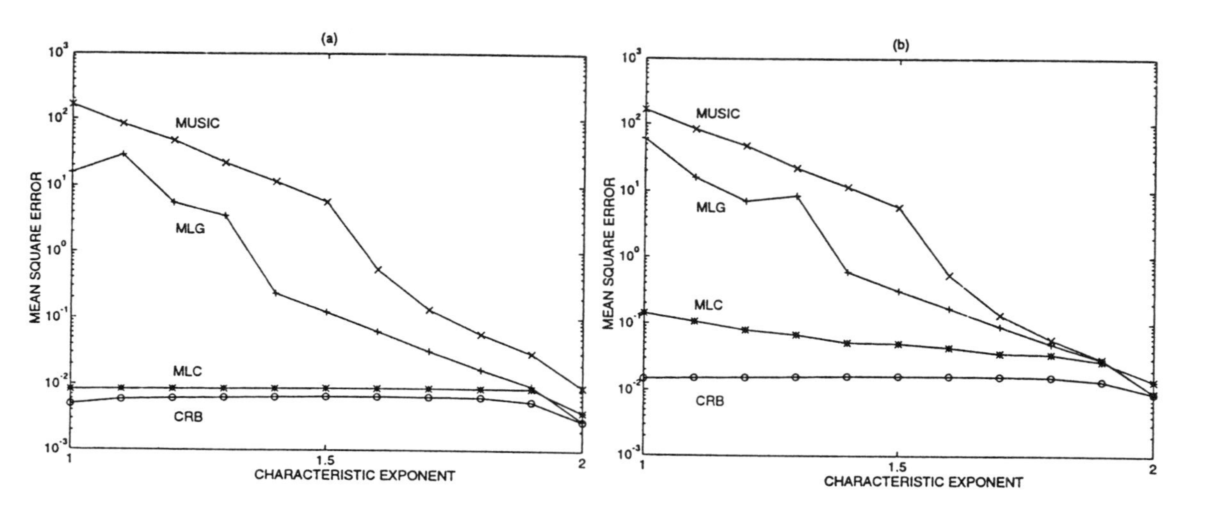

Figure 4: MSE of the estimated DOA and CRB as functions of the characteristic exponent α. (a) Exact signal knowledge, (b) Least-squares estimate of the signal.

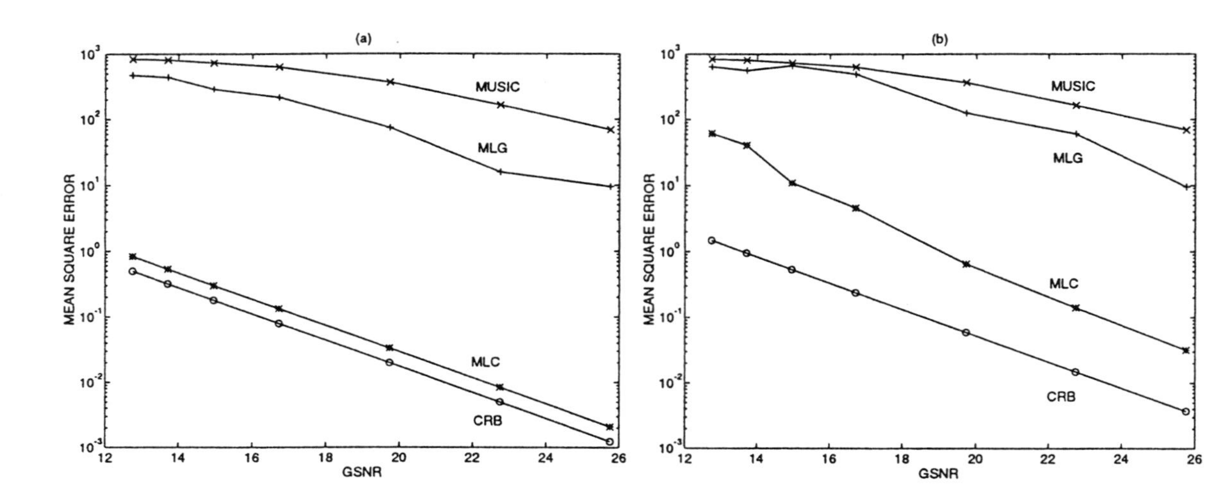

Figure 5: MSE of the estimated DOA and CRB as functions of the GSNR. (a) Exact signal knowledge, (b) Least-squares estimate of the signal.

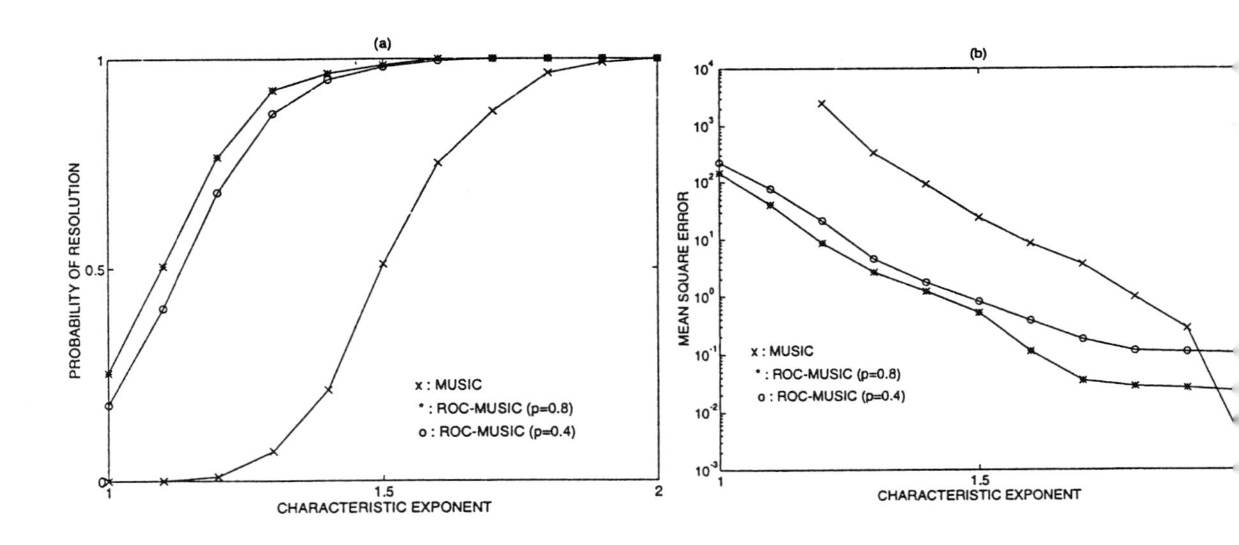

Figure 6: Probability of resolution (a) and mean square error (b) as a function of the characteristic exponent α.

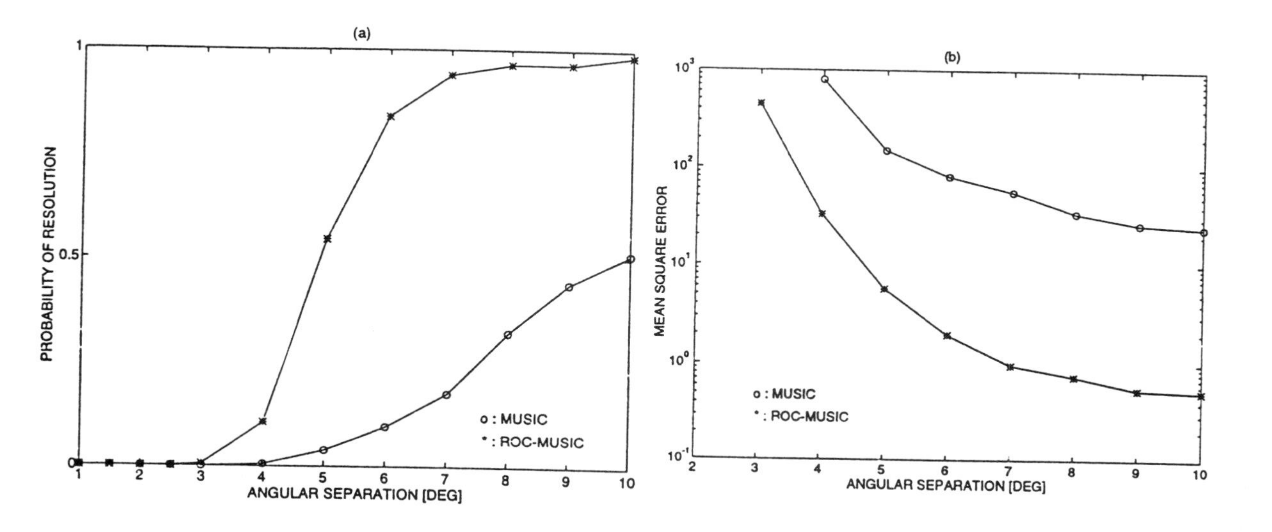

Figure 7: Probability of resolution (a) and mean square error (b) as a function of the source angular separation ($\alpha = 1.5$).

7 Concluding Remarks

In this paper, we have demonstrated the importance of extending the statistical array signal processing methodology to the so-called alpha-stable framework. We developed statistical signal processing methods for a larger class of random processes which include the Gaussian processes as special elements. Our proposed methods can be applied in environments which, while sharing many characteristics, also differ from Gaussian environments in significant ways. In particular, we developed techniques such as the Cauchy Beamformer and the ROC-MUSIC algorithm, which perform very robustly in detecting and localizing sources buried in a wide range of impulsive noise environments.

References

[Cam83] S Cambanis. Complex symmetric stable variables and processes. In P Sen, editor, *Contributions to Statistics: Essays in Honor of Norman L. Johnson*, pages 63–79. North-Holland, New York, 1983.

[Cap69] J Capon. High-resolution frequency-wavenumber spectrum analysis. *Proc. IEEE*, 57:1408–1418, 1969.

[CM89] S Cambanis and A G Miamee. On prediction of harmonizable stable processes. *Sankhyā: The Indian Journal of Statistics*, 51:269–294, 1989.

[GS81] G Gong and F J Samaniego. Pseudo maximum likelihood estimation:Theory and applications. *Ann. Statist.*, 9:861–869, 1981.

[JD93] D Johnson and D Dudgeon. *Array Signal Processing:Concepts and Techniques*. Prentice Hall, Englewood Cliffs, 1993.

[Kas88] S A Kassam. *Signal Detection in Non-Gaussian Noise*. Springer-Verlang, Berlin, 1988.

[KB86] M Kaveh and A J Barabell. The statistical performance of the MUSIC and the minimum-norm algorithms in resolving plane waves in noise. *IEEE Trans. Acoust., Speech, and Signal Process.*, 34:331–341, 1986.

[KT83] R Kumaresan and D W Tufts. Estimating the angles of arrival of multiple plane waves. *IEEE Trans. Aerosp. Electron. Syst.*, 19:134–139, 1983.

[Lue84] D G Luenberger. *Linear and Nonlinear Programming*. Addison Wesley, Menlo Park, second edition, 1984.

[Mid84] D Middleton. Threshold detection in non-Gaussian interference environments: Exposition and interpretation of new results for EMC applications. *IEEE Trans. Electrom. Compat.*, 26:19–28, Feb. 1984.

[MN95] X Ma and C L Nikias. Parameter estimation and blind channel identification for impulsive signal environments. *IEEE Trans. Signal Processing*, 43:2884–2897, December 1995.

[NS95] C L Nikias and M Shao. *Signal Processing with Alpha-Stable Distributions and Applications*. John Wiley and Sons, New York, 1995.

[Red79] S S Reddi. Multiple source location–A digital approach. *IEEE Trans. Aerosp. Electron. Syst.*, 15:95–105, 1979.

[Res95] S Resnick. Heavy tail modeling and teletraffic data. *Preprint*, 1995.

[RNP96] I S Reed, C L Nikias, and V Prasanna. Multidisciplinary research on advanced high-speed, adaptive signal processing for radar sensors. Technical report, University of Southern California, Jan. 1996.

[RPK86] R Roy, A Paulraj, and T Kailath. ESPRIT–A subspace rotation approach to estimation of parameters of cisoids in noise. *IEEE Trans. Acoust., Speech, and Signal Process.*, 34:1340–1342, 1986.

[Sch86] R O Schmidt. Multiple emitter location and signal parameter estimation. *IEEE Trans. Antennas Prop.*, 34:276–280, 1986.

[SN89] P Stoica and A Nehorai. MUSIC, maximum likelihood, and Cramer-Rao bound. *IEEE Trans. Acoust., Speech, and Signal Process.*, 37:720–741, 1989.

[SN93] M Shao and C L Nikias. Signal processing with fractional lower order moments: Stable processes and their applications. *Proc. IEEE*, 81:986–1010, 1993.

[ST94] G Samorodnitsky and M S Taqqu. *Stable Non-Gaussian Random Processes: Stochastic Models with Infinite Variance*. Chapman and Hall, New York, 1994.

[TN95] P Tsakalides and C L Nikias. Maximum likelihood localization of sources in noise modeled as a stable process. *IEEE Trans. Signal Processing*, 43:2700–2713, Nov. 1995.

[TN96a] P Tsakalides and C L Nikias. Advanced classification methods with lower-order statistics for shallow water sonar environments. Technical report, University of Southern California, June 1996.

[TN96b] P Tsakalides and C L Nikias. The robust covariation-based MUSIC (ROC-MUSIC) algorithm for bearing estimation in impulsive noise environments. *IEEE Trans. Signal Processing*, 44:1623–1633, July 1996.

[Tsa95] P Tsakalides. *Array Signal Processing with Alpha-Stable Distributions.* PhD thesis, University of Southern California, Los Angeles, California, December 1995.

[Zha95] Q T Zhang. Probability of resolution of the MUSIC algorithm. *IEEE Trans. Signal Processing*, 43:978–987, 1995.

Panagiotis Tsakalides and Chrysostomos (Max) Nikias
Signal and Image Processing Institute
Department of Electrical Engineering – Systems
University of Southern California
Los Angeles, California 90089-2564

tsakalid@sipi.usc.edu
nikias@sipi.usc.edu

Statistical Modeling and Receiver Design for Multi-User Communication Networks

George A. Tsihrintzis

Abstract

In this article, alpha-stable statistical models for impulsive noise are applied to the modeling of interference arising from multiple users in communication networks. It is shown that the traditionally assumed Gaussian model is a special case of the more general alpha-stable model. Next, the performance is evaluated of asymptotically optimum multichannel structures for incoherent detection of amplitude-fluctuating bandpass signals in impulsive interference modeled as a bivariate, isotropic, symmetric, alpha-stable process. A test statistic is proposed that takes into account the infinite variance in the noise model and approaches the performance of the strictly optimum algorithm for fixed (non-zero) error probabilities as the sample size becomes infinite and the input signal-to-noise ratio tends to zero. It is shown that, for sufficiently low probability of false alarm, the detector derived on the basis of a Cauchy assumption performs better than the detector derived on the basis of a Gaussianity assumption in non-Gaussian bivariate, isotropic, symmetric, alpha-stable interference.

Key words: Multi-User Communication, Incoherent Receiver, Fluctuating Target, Impulsive Noise, Stable Distribution, Robust Detection

1 Introduction

Traditional communication theory and techniques have been dominated by the Gaussianity assumption. This tradition is often justified by the observation that the ideal Gaussian model arises from the Central Limit Theorem and, therefore, is valid, at least within a good approximation. In addition, the Gaussian algorithms rely only on second-order statistics (SOS) of the observed data, generally have a simple linear structure, and do not require prohibitively fast or sophisticated computer hardware or software for their real-time implementation.

Despite their simplicity, the Gaussian algorithms may, however, be severely suboptimal when applied on data that are clearly non-Gaussian. A well-known example is the matched filter for coherent reception of completely known deterministic signals in additive white Gaussian noise [34]. If the noise distribution deviates from Gaussianity, then the performance of the matched filter degrades significantly and, as a result, the false alarm rate or the error probability increase unacceptably. On the other hand, more sophisticated, usually modestly nonlinear, signal processing restores the performance to levels comparable to those of the matched filter in ideal Gaussian noise (see, for example, [31, 29, 32] for a treatment and a performance analysis of problems in detection of signals in impulsive noise).

The problem of designing signal processing algorithms for non-Gaussian environments has been addressed in the past and a lot of literature has been produced in which statistics of the data of order higher than the second have been utilized. Collectively, this field has been known as *Nonlinear, Higher-Order Statistics-Based Signal Processing* [19] and has enjoyed significant applications in Radar, Communications, Image Processing, and Control. Higher-order statistics (HOS) provided the solution to many problems that second-order statistics (SOS) alone could not solve. In particular, the problem of detection of random signals in colored additive Gaussian noise of arbitrary, unknown correlation was addressed and solved with the use of the HOS of the output of a matched filter in [5]. The justification was that the Gaussian noise would be suppressed, since the HOS of arbitrary Gaussian processes are identically zero. Therefore, for long enough observations, a HOS domain would also be a high signal-to-noise ratio domain for signal detection.

However, there exist a broad class of non-Gaussian phenomena encountered in practice which are characterized by sharp spikes, occasional bursts, and heavy outliers. These phenomena are collectively referred to as "impulsive" and are described by probabilistic models with tails significantly heavier than those of the Gaussian distribution. The sources that generate such processes come in abundance and include lightning in the atmosphere, switching transients in power lines, ice cracking in underwater acoustics, accidental hits in telephone lines, and reflections from moving vehicles, high buildings, and sea waves in radar clutter [11, 26, 3, 36, 20, 30]. With data collected from this type of phenomena, both the Gaussian (SOS-based) and the HOS-based algorithms perform poorly, especially when the observation length is short. Therefore, a need arises for new algorithms specifically designed to mitigate the effects of impulsiveness in the data, which at the same time perform well with limited observations. Significant gains in signal processing algorithm performance can be obtained if their design is based on more appropriate statistical-physical models for the impulsive interference [22, 18, 31, 29].

In the past, serious attempts to quantify and predict the clearly non-Gaussian behavior of these processes have resulted in a large volume of statistical models, the most common of which include the statistically-physically derived Middleton mixture models [12, 13, 14, 15, 17, 16], empirical Gaussian mixtures, and a number of other heavy-tailed distributions such as the Weibull, the K, and the log-normal. Particularly accurate are Middleton's statistical-physical models, which are based on the filtered-impulse mechanism and can be classified in one of three classes, namely A, B, and C. Interference in class A is "coherent" in narrowband receivers, causing a negligible amount of transients.

Interference in class B, however, is "impulsive," consisting of a large number of overlapping transients. Finally, interference in class C is the sum of the other two interferences. The Middleton model has been shown to describe real impulsive interferences with high fidelity; however, it is mathematically involved for signal processing applications. This is particularly true of the class B model, which contains seven parameters, one of which is purely empirical and in no way relates to the underlying physical model. Moreover, mathematical approximations must be used in the derivation of the Middleton model, which [1] are equivalent to changes in the assumed physics of the noise and lead to ambiguities in the relation between the mathematical formulae and the physical scenario. Very recently, an alternative to the Middleton model was proposed [20], that is based on the theory of symmetric, α-stable (SαS) distributions [23].

In particular, it was shown by Nikias and Shao [20] that, under general assumptions, the first-order distribution of impulsive interference follows a SαS law. The stable model was then tested with a variety of real data and was found in all cases examined to match the data with excellent fidelity [20]. The performance of optimum and suboptimum receivers in the presence of SαS impulsive interference was examined by Tsihrintzis and Nikias [31], both theoretically and via Monte-Carlo simulation, and a method was presented for the real time implementation of the optimum nonlinearities. From this study, it was found that the corresponding optimum receivers perform quite well in the presence of SαS impulsive interference, while the performance of Gaussian and other suboptimum receivers is unacceptably low. It was also shown that a receiver designed on a Cauchy assumption for the first-order distribution of the impulsive interference performed only slightly below the corresponding optimum receiver, provided that a reasonable estimate of the noise dispersion was available, as could, for example, be obtained from the algorithms in [28].

In the present article, the alpha-stable model is proposed for the statistical description of the interference arising in modern multi-user communication networks. It is shown that the alpha-stable class provides a valid asymptotic distribution for a large class of interferences arising in multi-user communication networks and includes the traditional Gaussian model as a special case. On the basis of this model, asymptotically optimum, spatially-diverse detectors are derived for fading signals in alpha-stable impulsive interference. More specifically, the article is organized as follows: Section 2 is devoted to the derivation of the alpha-stable model for multi-user interference in communication networks and a brief summary of the properties of alpha-stable distributions. Section 3 is devoted to a study of the performance of spatially diverse receivers in bivariate, isotropic, symmetric, alpha-stable interference. Finally, Section 4 summarizes the article and suggests avenues for possible future research.

2 Statistical Modeling of Multi-User Communication Networks

2.1 Multi-User Communication Networks

The radio channel is fundamentally a broadcast communication medium. As a result, signals transmitted by one user can potentially be received by all other users within range of the transmitter. Although this high connectivity is very useful in some applications, like broadcast radio or television, it requires stringent access control in certain multi-user communication networks, such as mobile cellular systems, personal communication systems, and cordless phones, to avoid, or at least to limit, interference between transmissions. The necessary insulation between signals in different connections is achieved by assigning to each transmission different components of the domains that contain the signals, such as the *spatial*, *frequency*, *time*, or *code* domains.

In most code-division multiple-access systems, for example, different users employ signals that have very small cross-correlation. Thus, correlators can be used to extract individual signals from a mixture, even though they are simultaneously transmitted and in the same frequency band. If the transmitted signals are perfectly orthogonal and the channel only adds white Gaussian noise, matched-filter receivers are optimal for extracting a signal from the superposition of waveforms. If, however, the channel is dispersive because of multipath, the signals arriving at the receiver will no longer be orthogonal and will introduce some multi-user interference, i.e., signal components from other signals that are not rejected by the matched filter.

Multi-user detection is still an emerging technique. The fundamental idea is to model multi-user interference and devise receivers that reject or cancel the undesired signals. A variety of techniques have been proposed, ranging from optimum maximum-likelihood sequence estimation via multistage schemes, reminiscent to decision feedback algorithms, to linear decorrelating receivers. A survey of the theory and practice of multi-user detection is given in [35].

2.2 Statistical-Physical Models for Multi-User Interference[1]

Problem Formulation and Physical Assumptions

Let us assume, without loss of generality, that the origin of the spatial coordinate system is at the point of observation. The time axis is taken in the direction of the past with its origin at the time of observation, i.e., t is the duration from the time of pulse emission to the time of observation. Consider a region Ω in $\mathcal{R}^n$, where $\mathcal{R}^n$ may be a plane

[1]This section has been excerpted from: M. Shao, *Symmetric, Alpha-Stable Distributions: Signal Porcessing with Fractional, Lower-Order Statistics*, Ph.D. Dissertation, University of Southern California, Los Angeles, California, 1993. The interested reader is also referred to [20, 33] for further details.

($n = 2$) or the entire three-dimensional space ($n = 3$). For simplicity, we assume that Ω is a semi-cone with vertex at the point of observation. Inside this region, there is a collection of interfering sources (e.g., interfering users) which randomly generate transient pulses. It is assumed that all sources share a common random mechanism so that these elementary pulses have the same type of waveform, $aD(t; \underline{\theta})$, where the symbol $\underline{\theta}$ represents a collection of time-invariant random parameters that determine the scale, duration, and other characteristics of the interference, and a is a random amplitude. We shall further assume that only a countable number of such sources exist inside the region Ω, distributed at random positions $\mathbf{x}_1, \mathbf{x}_2, \cdots$. These sources independently emit pulses $a_i D(t; \underline{\theta}_i)$, $i = 1, 2, \cdots$, at random times $t_1, t_2, \cdots$, respectively. This implies that the random amplitudes $\{a_1, a_2, \cdots\}$ and the random parameters $\{\underline{\theta}_1, \underline{\theta}_2, \cdots\}$ are both i.i.d. sequences, with the prespecified probability densities $p_a(a)$ and $p_{\underline{\theta}}(\underline{\theta})$, respectively. The location $\mathbf{x}_i$ and emission time t_i of the ith source, its random parameter $\underline{\theta}_i$ and amplitude a_i are assumed to be independent for $i = 1, 2, \cdots$. The distribution $p_a(a)$ of the random amplitude a is assumed to be symmetric, implying that the location parameter of the interference is zero.

When an elementary transient pulse $aD(t; \underline{\theta})$ passes through the medium and the receiver, it is distorted and attenuated. The exact nature of the distortion and the attenuation can be determined from knowledge of the beam patterns of the source and the antenna, the source locations, the impulse response of the receiver, and other related parameters [14]. For simplicity, we will assume that the effect of the transmission medium and the receiver on the transient pulses may be separated into two multiplicative factors, namely filtering and attenuation. Without attenuation, the medium and the receiver together may be treated as a deterministic linear, time-invariant filter. In this case, the received transient pulse is the convolution of the impulse response of the equivalent filter and the original pulse waveform $aD(t; \underline{\theta})$. The result is designated by $aE(t; \underline{\theta})$. The attenuation factor is generally a function of the source location relative to the receiver. For simplicity, we shall assume that the sources within the region of consideration have the same isotropic radiation pattern and the receiver has an omnidirectional antenna. Then the attenuation factor is simply a decreasing function of the distance from the source to the receiver. A good approximation is that the attenuation factor varies inversely with a power of the distance [6, 14], i.e.,

$$g(\mathbf{x}) = c_1 / r^p, \tag{2.1}$$

where $c_1, p > 0$ are constants and $r = |\mathbf{x}|$. Typically, the attenuation rate exponent p lies between $\frac{1}{2}$ and 2.

Combining the filtering and attenuation factors, one finds that the waveform of a pulse originating from a source located at $\mathbf{x}$ is $aU(t; \mathbf{x}, \underline{\theta})$, where

$$U(t; \mathbf{x}, \underline{\theta}) = \frac{c_1}{r^p} E(t; \underline{\theta}). \tag{2.2}$$

Further assuming that the receiver linearly superimposes the interference, pulses, the observed instantaneous interference amplitude at the output of the receiver and at the time of observation is

$$X = \sum_{i=1}^{N} a_i U(t_i; \mathbf{x}_i, \underline{\theta}_i), \tag{2.3}$$

where N is the total number of interference pulses arriving at the receiver at the time of observation.

In our model, we maintain the usual basic assumption for the interference generating processes that the number N of arriving pulses is a Poisson point process in both space and time, the intensity function of which is denoted by $\rho(\mathbf{x}, t)$ [4, 6, 14]. The intensity function $\rho(\mathbf{x}, t)$ represents approximately the probability that an interference pulse originating from a unit area or volume and emitted during a unit time interval will arrive at the receiver at the time of observation. Thus, it may be considered as the spatial and temporal density of the interference sources. In this article, we shall restrict our consideration to the common case of time-invariant source distribution, i.e., we set $\rho(\mathbf{x}, t) = \rho(\mathbf{x})$. In most applications, $\rho(\mathbf{x})$ is a non-increasing function of the range $r = |\mathbf{x}|$, implying that the number of sources that occur close to the receiver is usually larger than the number of sources that occur farther away. This is certainly the case, for example, for the tropical atmospheric noise where most lightning discharges occur locally, and relatively few discharges occur at great distances [6]. If the source distribution is isotropic about the point of observation, i.e., if there is no preferred direction from which the pulses arrive, then it is reasonable to assume that $\rho(\mathbf{x})$ varies inverse-proportionately with a certain power of the distance r [14, 6]:

$$\rho(\mathbf{x}, t) = \frac{\rho_0}{r^\mu}, \tag{2.4}$$

where μ and $\rho_0 > 0$ are constants.

The Asymptotic Statistical Model

Our method for calculating the characteristic function $\varphi(\omega)$ of the interference amplitude X is similar to the one used in [37] for the model of point sources of influence. We first restrict our attention to interference pulses emitted from sources inside the region $\Omega(R_1, R_2)$ and within the time interval $[0, T)$, where $\Omega(R_1, R_2) = \Omega \cap \{\mathbf{x} : R_1 < |\mathbf{x}| < R_2\}$. The amplitude of the truncated interference is then given by

$$X_{T,R_1,R_2} = \sum_{i=1}^{N_{T,R_1,R_2}} a_i U(t_i; \mathbf{x}_i, \underline{\theta}_i), \tag{2.5}$$

where $N_{T,R1,R2}$ is the number of pulses emitted from the space-time region $\Omega(R_1, R_2) \times [0, T)$. The observed interference amplitude X is understood to be the limit of X_{T,R_1,R_2} as T, $R_2 \to \infty$ and $R_1 \to 0$ in some suitable sense.

Note that N_{T,R_1,R_2} is a Poisson random variable with parameter

$$\lambda_{T,R_1,R_2} = \int_{\Omega(R_1,R_2)} \int_0^T \rho(\mathbf{x}, t) d\mathbf{x} dt \tag{2.6}$$

and its *factorial* moment-generating function is given by

$$\mathcal{E}\{t^{N_{T,R_1,R_2}}\} = \exp[\lambda_{T,R_1,R_2}(t-1)]. \tag{2.7}$$

Let the actual source locations and their emission times be $(\mathbf{x}_i, t_i)$, $i = 1, \cdots, N_{T,R_1,R_2}$. Then, the random pairs $(\mathbf{x}_i, t_i)$, $i = 1, \cdots, N_{T,R_1,R_2}$, are i.i.d., with a common joint density function given by

$$f_{T,R_1,R_2}(\mathbf{x}, t) = \frac{\rho(\mathbf{x}, t)}{\lambda_{T,R_1,R_2}}, \qquad \mathbf{x} \in \Omega(R_1, R_2), \quad t \in [0, T). \tag{2.8}$$

In addition, N_{T,R_1,R_2} is independent of the locations and emission times of all the sources. All of the above results are the consequences of the basic Poisson assumption [21].

By our previous assumptions, $\{(a_i, t_i, \mathbf{x}_i, \underline{\theta}_i)\}_{i=1}^{\infty}$ is an i.i.d. sequence. Hence, X_{T,R_1,R_2} is a sum of i.i.d. random variables with a random number of terms. Its characteristic function can be calculated as follows:

$$\begin{aligned} \varphi_{T,R_1,R_2}(\omega) &= \mathcal{E}\{\exp(i\omega X_{T,R_1,R_2})\} \\ &= \mathcal{E}\{[\psi_{T,R_1,R_2}(\omega)]^{N_{T,R_1,R_2}}\}, \end{aligned} \tag{2.9}$$

where

$$\psi_{T,R_1,R_2}(\omega) = \mathbf{E}\{\exp(i\omega a_1 U(t_1; \mathbf{x}_1, \underline{\theta}_1)) \mid T, \Omega(R_1, R_2)\}. \tag{2.10}$$

By Eq.(2.7),

$$\varphi_{T,R_1,R_2}(\omega) = \exp(\lambda_{T,R_1,R_2}(\psi_{T,R_1,R_2}(\omega) - 1)). \tag{2.11}$$

Since $a_1, \underline{\theta}_1$ and $(\mathbf{x}_1, t_1)$ are independent, with pdfs $p_a(a)$, $p_{\underline{\theta}}(\underline{\theta})$ and $f_{T,R_1,R_2}(\mathbf{x}, t)$, respectively, one obtains

$$\begin{aligned} \psi_{T,R_1,R_2}(\omega) &= \int_{-\infty}^{\infty} p_a(a) da \int_{\Theta} p_{\underline{\theta}}(\underline{\theta}) d\underline{\theta} \int_0^T dt \\ & \int_{\Omega(R_1,R_2)} \frac{\rho(\mathbf{x}, t)}{\lambda_{T,R_1,R_2}} \exp(i\omega a U(t; \mathbf{x}, \underline{\theta})) d\mathbf{x}. \end{aligned} \tag{2.12}$$

Combining (3-29), (3-31), (3-39) and (3-38), one can easily show that the logarithm of the characteristic function of X_{T,R_1,R_2} is

$$\begin{aligned} \log \varphi_{T,R_1,R_2}(\omega) &= \rho_0 \int_{-\infty}^{\infty} p_a(a) da \int_{\Theta} p_{\underline{\theta}}(\underline{\theta}) d\underline{\theta} \int_0^T dt \\ & \int_{\Omega(R_1,R_2)} \frac{\exp(i\omega a c_1 r^{-p} E(t; \underline{\theta})) - 1}{r^{-\mu}} d\mathbf{x}, \end{aligned} \tag{2.13}$$

where $r = |\mathbf{x}|$.

After some tedious algebraic manipulations [20], one can finally show that, in the limit of $T, R_2 \to \infty$, $R_1 \to 0$, the characteristic function of the instantaneous interference amplitude attains the limiting form

$$\varphi(\omega) = \exp(-\gamma|\omega|^{\alpha}), \tag{2.14}$$

where

$$0 < \alpha = \frac{n - \mu}{p} \leq 2 \tag{2.15}$$

with n the dimension of the space occupied by the interference sources. As defined above, α is an effective measure of an average source density with range [14] and determines the degree of impulsiveness of the interference. Hence, we have shown that *under a set of very mild and reasonable conditions, impulsive interference follows, indeed, a stable law.*

Similarly, in the case of narrowband reception, one can show [20] that the joint characteristic function of the quadrature components of the interference attains the form

$$\varphi(\omega_1, \omega_2)) = \exp(-\gamma|\omega_1^2 + \omega_2^2|^{\frac{\alpha}{2}}), \tag{2.16}$$

where

$$\gamma > 0, \quad 0 < \alpha \leq 2.$$

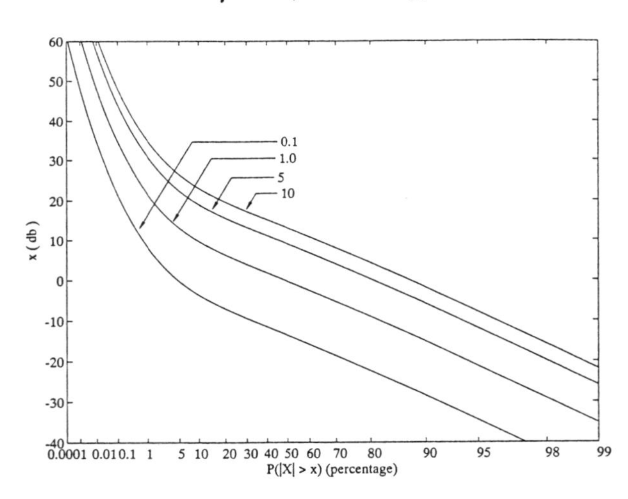

Figure 1. The APD of the instantaneous amplitude of $S\alpha S$ interference for $\alpha = 1.5$ and various values of γ.

Hence, *the joint distribution of the quadrature components is bivariate isotropically stable.* From this result, one can derive the first-order statistics of the envelope and phase [20] of impulsive interference. Specifically, it can be shown that the random phase is *uniformly distributed* in $[0, 2\pi]$ and is *independent* of the envelope. The density of the envelope, on other hand, is given by

$$f(a) = a \int_0^\infty s \exp(-\gamma s^\alpha) J_0(as) ds, \quad a \geq 0, \tag{2.17}$$

where $J_o(\cdot)$ is the Bessel function of the first kind of zeroth order. By integrating Eq.(2.17), one obtains the envelope distribution function

$$F(a) = a \int_0^\infty \exp(-\gamma t^\alpha) J_1(at) dt, \quad a \geq 0., \tag{2.18}$$

where $J_1(\cdot)$ is the Bessel function of the first kind of first order. We note that when $\alpha = 2$, the well-known Rayleigh distribution is obtained for the envelope.

The amplitude probability distribution (APD) of the envelope can now be computed as

$$P(A > a) = 1 - a \int_0^\infty \exp(-\gamma t^\alpha) J_1(at) dt, \quad a \geq 0. \tag{2.19}$$

From [38], it follows that *the envelope distribution and density functions are again heavy-tailed.*

Figs. 1 and 2 plot the APD of SαS interference for various values of α and γ. To fully represent the large range of the exceedance probability $P(|X| > x)$, the coordinate grid used in these two figures employs a highly folded abscissa. Specifically, the axis for $P(|X| > x)$ is scaled according to $-\log(-\log P(|X| > x))$. As clearly shown in Fig. 2, SαS impulsive interference is clearly Gaussian (Rayleigh) for low amplitudes.

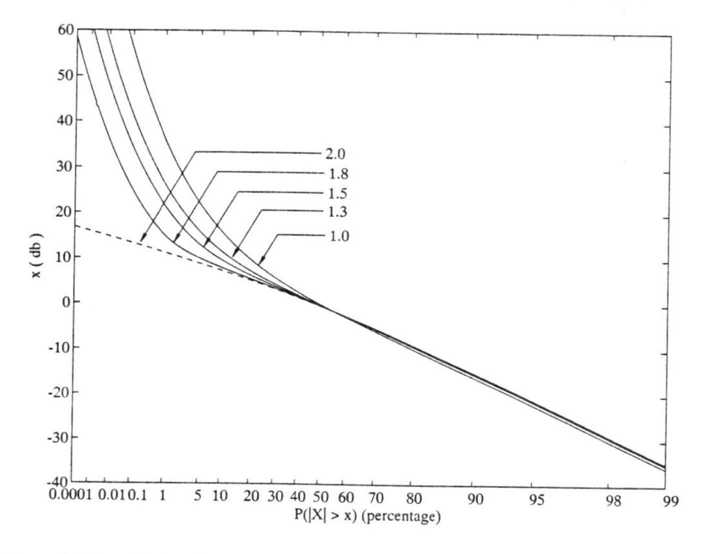

Figure 2. The APD of the instantaneous amplitude of $S\alpha S$ interference for $\gamma = 1$ and various values of α

2.3 Univariate SαS Distributions

A univariate symmetric, α-stable (SαS) pdf $f_\alpha(\gamma, \delta; \cdot)$ is best defined via the inverse Fourier transform integral [10, 25]

$$f_\alpha(\gamma, \delta; x) = \frac{1}{2\pi} \int_{-\infty}^{\infty} \exp(i\delta\omega - \gamma|\omega|^\alpha) e^{-i\omega x}\, d\omega \tag{2.20}$$

and is completely characterized by the three parameters α (*characteristic exponent*, $0 < \alpha \leq 2$), γ (*dispersion*, $\gamma > 0$), and δ (*location parameter*, $-\infty < \delta < \infty$).

The characteristic exponent α relates directly to the heaviness of the tails of the SαS pdf: for large $|x - \delta|$, $f_\alpha(\gamma, \delta; x) \sim 1/|x - \delta|^{\alpha+1}$, and, thus, the smaller its value, the heavier the pdf tails. The value $\alpha = 2$ corresponds to a Gaussian pdf, while the value $\alpha = 1$ corresponds to a Cauchy pdf. For these two pdfs, closed-form expressions exist, namely

$$f_2(\gamma, \delta; \xi) = \frac{1}{\sqrt{4\pi\gamma}} \exp[-\frac{(\xi - \delta)^2}{4\gamma}] \tag{2.21}$$

$$f_1(\gamma, \delta; \xi) = \frac{1}{\pi} \frac{\gamma}{\gamma^2 + (\xi - \delta)^2}. \tag{2.22}$$

For other values of the characteristic exponent, no closed-form expressions are known. All the SαS pdfs can be computed, however, at arbitrary argument with the real time method developed in [31]. The dispersion γ is a measure of the spread of the SαS pdf, in many ways similar to the variance of a Gaussian pdf and equal to half the variance of the pdf in the Gaussian case ($\alpha = 2$). Finally, the location parameter δ is the point of symmetry of the SαS pdf.

The non-Gaussian ($\alpha \neq 2$) SαS distributions maintain many similarities to the Gaussian distribution, but at the same time differ from it in some significant ways. For example, a non-Gaussian SαS pdf maintains the usual bell shape and, more importantly, non-Gaussian SαS random variables satisfy the linear stability property [10]. However, non-Gaussian SαS pdfs have much sharper peaks and much heavier tails than the Gaussian pdf. As a result, only their moments of order $p < \alpha$ are finite, in contrast with the Gaussian pdf which has finite moments of arbitrary order. These and other similarities and differences between Gaussian and non-Gaussian SαS pdfs and their implications on the design of signal processing algorithms are presented in detail in the tutorial paper [25] or, more comprehensively, in the monograph [20]. For illustration purposes, we show in Fig. 3 plots of the SαS pdfs for location parameter $\delta = 0$, dispersion $\gamma = 1$, and for characteristic exponents $\alpha = 0.5, 1, 1.5, 1.99$, and 2. The curves in Fig. 3 have been produced by direct calculation of the inverse Fourier transform integral in Eq.(2.20).

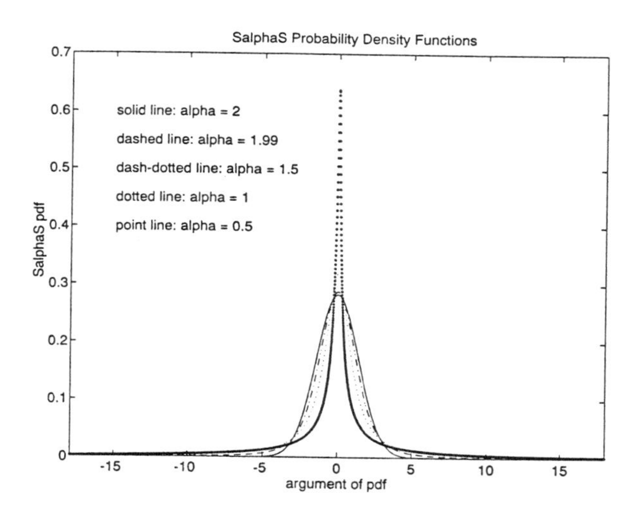

Figure 3. Symmetric, alpha-stable pdfs of zero location parameter, unit dispersion, and various characteristic exponents

2.4 Bivariate Isotropic Symmetric Alpha-Stable (BISαS) Distributions

The bivariate isotropic symmetric alpha-stable (BISαS) probability density function (pdf) $f_{\alpha,\gamma,\delta_1,\delta_2}(x_1, x_2)$ is given by the inverse Fourier transform

$$f_{\alpha,\gamma,\delta_1,\delta_2}(x_1, x_2) = \frac{1}{(2\pi)^2}\int_{-\infty}^{\infty}\int_{-\infty}^{\infty} e^{i(\delta_1\omega_1+\delta_2\omega_2)-\gamma(\omega_1^2+\omega_2^2)^{\alpha/2}} e^{-i(x_1\omega_1+x_2\omega_2)}, \quad (2.23)$$

where the parameters α and γ are termed the *characteristic exponent* and the *dispersion*, respectively, and δ_1 and δ_2 are *location parameters.* The characteristic exponent generally ranges in the interval $0 < \alpha \leq 2$ and relates to the heaviness of the tails, with a smaller exponent indicating heavier tails. The dispersion γ is a positive constant relating to the spread of the pdf. The two marginal distributions obtained from the bivariate distribution in Eq.(2.23) are univariate SαS with characteristic exponent α, dispersion γ, and location parameters δ_1 and δ_2, respectively [25, 20]. In the rest of the article, we are going to assume $\delta_1 = \delta_2 = 0$, without loss of generality, and drop the corresponding subscripts from all our expressions.

Unfortunately, no closed-form expressions exist for the general BISαS, pdf except for the special cases of $\alpha = 1$ (Cauchy) and $\alpha = 2$ (Gaussian):

$$f_{\alpha,\gamma}(x_1, x_2) = \begin{cases} \frac{\gamma}{2\pi(\rho^2+\gamma^2)^{3/2}} & \text{for } \alpha = 1 \\ \frac{1}{4\pi\gamma}\exp(-\frac{\rho^2}{4\gamma}) & \text{for } \alpha = 2, \end{cases} \quad (2.24)$$

where $\rho^2 = x_1^2 + x_2^2$. For the remaining (non-Gaussian, non-Cauchy) BISαS distributions, power series exist [25, 20], but are not of interest to this article and, therefore, are not given here.

2.5 BISαS Models for Impulsive Noise

Consider a narrowband receiver operating in SαS impulsive interference $n(t)$ of dispersion γ. Let

$$n(t) = n_c(t)\cos(2\pi f_0 t) - n_s(t)\sin(2\pi f_0 t) = \Re\{(n_c(t) + i n_s(t))\exp(i2\pi f_0 t)\}, \quad (2.25)$$

where n_c and n_s are the "in-phase" and "quadrature" components of the interference and $n_c + i n_s$ is its corresponding complex amplitude. As pointed out in Section 2.2, the joint pdf of the two components can be shown [20] to be the BISαS pdf of dispersion γ, i.e.

$$f(n_c(t), n_s(t)) = f_{\alpha,\gamma}(n_c(t), n_s(t)). \quad (2.26)$$

We can immediately see that the in-phase and quadrature components of narrowband SαS interference are *not* independent, except in the Gaussian case ($\alpha = 2$) [20]. Moreover, the joint distribution of the two components is heavy-tailed when compared to the Gaussian case. Finally, transformation of the Fourier integral in Eq.(2.23) in radial coordinates ($\rho = \sqrt{n_c^2 + n_s^2}$, $\theta = \arctan(n_s/n_c)$), gives the following joint pdf of the envelope ρ and the phase θ of narrowband SαS interference:

$$f(\rho, \theta) = \frac{1}{2\pi}\rho \int_0^\infty u \exp(-\gamma u^\alpha) J_0(u\rho)\, du \equiv \frac{1}{2\pi} f_{e,\alpha,\gamma}(\rho); \quad \rho > 0, \quad \theta \in [0, 2\pi) \quad (2.27)$$

Eq.(2.27) clearly shows that the phase θ of narrowband SαS interference is uniformly distributed in the interval $[0, 2\pi)$ and independent of the corresponding interference envelope. These facts are in agreement with the corresponding results for the Gaussian case [20]; however, the envelope distribution is heavy-tailed when compared with the Gaussian model. Indeed, it can be shown [20] that the envelope distribution $f_{e,\alpha,\gamma}(\rho)$ has algebraic tails

$$\lim_{\rho \to \infty} \rho^{\alpha+1} f_{e,\alpha,\gamma}(\rho) = 2\pi B(\alpha, \gamma), \quad (2.28)$$

where $B(\alpha, \gamma)$ is a positive constant, independent of ρ. In Fig. 4, we show the APD, $\Pr\{A > a\}$ of the amplitude A of BISαS random variables of unit dispersion and various characteristic exponents and compare them to the Gaussian distribution (dashed line).

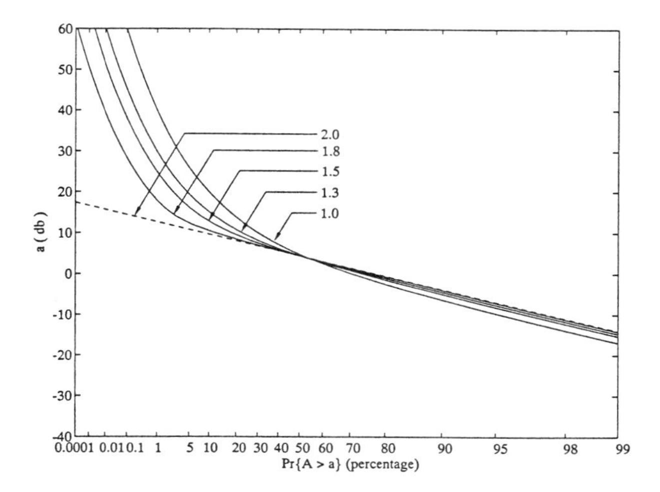

Figure 4. The APD of bivariate, isotropic, symmetric, alpha-stable narrowband interference unit dispersion and various characteristic exponents

3 Diversity Techniques in Communication Networks

3.1 Introduction

Diversity is a technique commonly used in mobile radio systems to combat signal fading. The basic principle of diversity lies in the intuitive concept that if several replicas of the same information carrying signal are transmitted with comparable strengths over multiple channels that exhibit independent fading, there is a good likelihood that at least one of them will not be in a fade at any instant in time. Without diversity techniques, in noise limited conditions, the transmitter will have to deliver a much higher power level to protect the link during the short intervals during which the channel is severely faded. In mobile radio, the power available on the reverse link is severely limited by the battery capacity in handheld subscriber units. Diversity methods play a crucial role in reducing transmit power needs. Also, cellular communication networks are mostly interference limited and once again mitigation of channel fading through use of diversity can translate into improved interference tolerance, which in turn means greater ability to support additional users and therefore higher system capacity.

The basic principles of diversity have been known since 1927, when the first experiments in space diversity were reported. There are many techniques for obtaining independently fading branches and these can be divided into two main classes. The first, called "explicit techniques," utilize explicit redundant signal transmission. In the second, called "implicit techniques," the signal is transmitted only once, but the decorrelating effects of the propagation medium, such as multipaths, are exploited to receive signals over multiple diverse channels. In this article, the performance of spatially-diverse receivers based on Gaussianity assumptions is assessed in an environment of heavy-tailed distributed interference and compared to the performance of proposed spatially-diverse receivers in which the infinite variance of the interference statistical model is accounted for.

3.2 Diversity in SαS Interference

In this Section, the asymptotically optimum detector of Izzo and Paura [8] is modified to take into account the infinite variance of non-Gaussian, BISαS interference models. It is noted that the Izzo-Paura detector and its proposed modification are asymptotically optimum in the sense that their performance approaches that of the strictly optimum algorithm for fixed (non-zero) probabilities of error as the sample size becomes infinite and the input SNR tends to zero [9].

The detection problem examined in this article can be formulated as the situation where one needs to decide between the two hypotheses

$$\begin{aligned} H_0 &: \tilde{r}_{kj} = \tilde{n}_{kj} \\ & \qquad\qquad\qquad\qquad k = 1, 2, \ldots, K; \quad j = 1, 2, \ldots, J \\ H_1 &: \tilde{r}_{kj} = A_k \frac{\xi}{\sqrt{J}} \tilde{S}_j e^{i\theta_k} + \tilde{n}_{kj}. \end{aligned} \tag{3.1}$$

In this problem, $k = 1, 2, \ldots, K$ indexes K independent channels, along which a fading signal is observed, and $j = 1, 2, \ldots, J$ indexes J time samples along observation channels. It is assumed that the noise samples $\tilde{n}_{kj}$ and the signal amplitudes $A_k \geq 0$ are statistically independent across and along channels, while the fading is slow enough to allow for A_k to be constant over the observation interval. Moreover, the random phases θ_k are independent across paths[2] and uniformly distributed in $[0, 2\pi)$ and ξ is descriptive of the transmitted signal strength. Finally, a tilde '$\tilde{\ }$' is used to denote a complex envelope, while an overbar indicates the complex conjugate.

The proposed detector computes the generalized likelihood ratio

$$\Lambda(\tilde{R}_1, \tilde{R}_2, \ldots, \tilde{R}_K) = \frac{\mathcal{E}_{A,\theta}\{f(\tilde{R}_1, \tilde{R}_2, \ldots, \tilde{R}_K | H_1, A_1, A_2, \ldots, A_K, \theta_1, \theta_2, \ldots, \theta_K)\}}{f(\tilde{R}_1, \tilde{R}_2, \ldots, \tilde{R}_K | H_0)}, \tag{3.2}$$

where $\tilde{R}_k = [\tilde{r}_{k1}, \tilde{r}_{k2}, \ldots, \tilde{r}_{kJ}]$ and the expectation is calculated over the random amplitudes A_k and the random phases θ_k, $k = 1, 2, \ldots, K$.

---

[2]The assumption of independent random phases across paths may be violated in practice. In those cases, the phases θ_k are either deterministic constants that can be incorporated in the amplitudes A_k or they are random, the correlation between which needs to be accounted for in the receiver design. However, both the general methodologies and the key findings of this Section still hold.

From the independence assumption for the noise samples, it is immediately concluded that the following ratio provides an equivalent test

$$\log \Lambda(\tilde{R}_1, \tilde{R}_2, \ldots, \tilde{R}_K) = \sum_{k=1}^{K} \log \mathcal{E}_{A_k,\theta_k}\{$$

$$\prod_{j=1}^{J} \frac{f_{\alpha,\gamma}(r_{I,kj} - A_k \frac{\xi}{\sqrt{J}} \Re\{\tilde{S}_j e^{i\theta_k}\}, r_{Q,kj} - A_k \frac{\xi}{\sqrt{J}} \Im\{\tilde{\bar{S}}_j e^{-i\theta_k}\}}{f_{\alpha,\gamma}(r_{I,kj}, r_{Q,kj})}\}, \tag{3.3}$$

where $\tilde{r}_{kj} = r_{I,kj} - i r_{Q,kj}$ is the complex envelope of the jth sample received from the kth channel.

It can be shown [8] that the above test statistic is asymptotically equivalent[3] to the statistic

$$t(\tilde{R}_1, \tilde{R}_2, \ldots, \tilde{R}_K) = \sum_{k=1}^{K} \log \mathcal{E}_{A_k}\{\exp(-A_k^2 \xi^2 P F_k/2) I_0(A_k \xi \sqrt{I_k^2 + Q_k^2})\}, \tag{3.4}$$

where

$$P = \lim_{J\to\infty} \frac{1}{J} \sum_{j=1}^{J} |\tilde{S}_j|^2, \tag{3.5}$$

$I_0(\cdot)$ is the zeroth order, modified Bessel function of the first kind, $\mathcal{E}_{A_k}$ denotes expectation with respect to A_k and

$$I_k = \frac{1}{\sqrt{J}} \sum_{j=1}^{J} g_k(R_{kj}) \Re\{\tilde{R}_{kj} \tilde{\bar{S}}_j\} \tag{3.6}$$

$$Q_k = \frac{1}{\sqrt{J}} \sum_{j=1}^{J} g_k(R_{kj}) \Im\{\tilde{R}_{kj} \tilde{\bar{S}}_j\} \tag{3.7}$$

$$g_k(R_{kj}) = -\frac{1}{R_{kj}} \frac{d}{dR_{kj}} \log f_k(R_{kj}) \tag{3.8}$$

$$F_k = \pi \int_0^\infty u^3 g_k^2(u) f_k(u)\, du, \tag{3.9}$$

and $R_{kj} = |\tilde{r}_{kj}| = \sqrt{r_{I,kj}^2 + r_{Q,kj}^2}$.

Assuming the fading variables A_k to be independent, Rayleigh-distributed and after carrying out the computations in Eq.(3.4), one gets the asymptotically optimum test statistic [8]

$$t = \sum_{k=1}^{K} \frac{w_k}{2 + w_k} (I_k'^2 + Q_k'^2), \tag{3.10}$$

[3] In fact, $\log \Lambda \to t$ in probability as $J \to \infty$ and, therefore, t is an Asymptotically Optimum Detector (AOD) statistic.

where

$$w_k = \mathcal{E}\{A_k^2\}\xi^2 P\pi \int_0^\infty u^3 g_k^2(u) f_k(u)\, du. \tag{3.11}$$

In Eq.(3.10)

$$I_k' = \frac{\frac{1}{\sqrt{J}} \sum_{j=1}^{J} g_k(R_{kj}) \Re\{\tilde{R}_{kj} \tilde{\bar{S_j}}\}}{\sqrt{2PF_k}} \tag{3.12}$$

$$Q_k' = \frac{\frac{1}{\sqrt{J}} \sum_{j=1}^{J} g_k(R_{kj}) \Im\{\tilde{R}_{kj} \tilde{\bar{S_j}}\}}{\sqrt{2PF_k}}. \tag{3.13}$$

In the rest of this Section, the test statistic in Eq.(3.10) is computed for Gaussian (i.e., BIS($\alpha = 2$)S) and Cauchy (i.e., BIS($\alpha = 1$)S) noise. The detectors corresponding to other BISαS noises cannot be computed in closed-form due to lack of a closed-form expression for the corresponding distributions. However, it is shown in Section 4 that the Cauchy detector is robust in the entire class of BISαS interferences.

a. Gaussian Detector

From Eq.(2.24),

$$f_k(u) = f_{2,\gamma}(u) = \frac{1}{4\pi\gamma} e^{-\frac{u^2}{4\gamma}}. \tag{3.14}$$

Therefore

$$g_k(u) = \frac{1}{2\gamma}, \tag{3.15}$$

which will give

$$\begin{aligned} \int_0^\infty u^3 g_k^2(u) f_k(u)\, du &= \frac{1}{16\pi\gamma^3} \int_0^\infty u^3 e^{-\frac{u^2}{4\gamma}}\, du \\ &= \frac{1}{2\pi\gamma} \end{aligned} \tag{3.16}$$

and

$$w_k = \frac{\mathcal{E}\{A_k^2\}\xi^2 P}{2\gamma}. \tag{3.17}$$

To write Eq.(3.16), we made use of the results in [24, p. 342].

Therefore, we can now write the Gaussian detector in the form

$$t_G = \sum_{k=1}^{K} \frac{\mathcal{E}\{A_k^2\}\xi^2 P}{4\gamma + \mathcal{E}\{A_k^2\}\xi^2 P} (I_k'^2 + Q_k'^2), \tag{3.18}$$

which simplifies to[4]

$$t_G = \sum_{k=1}^{K} W_k^G | \sum_{j=1}^{J} \tilde{R}_{kj} \tilde{\bar{S}}_j |^2, \tag{3.19}$$

where

$$W_k^G = \frac{\mathcal{E}\{A_k^2\}\xi^2 P}{4\gamma + \mathcal{E}\{A_k^2\}\xi^2 P} \frac{1}{4\gamma^2 J}.$$

b. Cauchy Detector

From Eq.(2.24),

$$f_k(u) = f_{1,\gamma}(u) = \frac{\gamma}{2\pi(u^2+\gamma^2)^{3/2}}. \tag{3.20}$$

Therefore

$$g_k(u) = \frac{3}{u^2+\gamma^2}, \tag{3.21}$$

which will give

$$\begin{aligned} \int_0^\infty u^3 g_k^2(u) f_k(u)\, du &= 9\gamma \int_0^\infty \frac{u^3\, du}{(u^2+\gamma^2)^{7/2}} \\ &= \frac{9}{2} \frac{\Gamma(2)\Gamma(3/2)}{\Gamma(7/2)} \frac{1}{\gamma^2} \\ &\equiv \frac{C}{\gamma^2} \end{aligned} \tag{3.22}$$

and

$$w_k = \frac{\mathcal{E}\{A_k^2\} C \xi^2 P \pi}{\gamma^2}. \tag{3.23}$$

Therefore, the Cauchy detector becomes

$$t_C = \sum_{k=1}^{K} \frac{\mathcal{E}\{A_k^2\} C \xi^2 P \pi}{2\gamma^2 + \mathcal{E}\{A_k^2\} C \xi^2 P \pi} (I_k'^2 + Q_k'^2) \tag{3.24}$$

and simplifies to

$$t_C = \sum_{k=1}^{K} W_k^C | \sum_{j=1}^{J} \frac{\tilde{R}_{kj} \tilde{\bar{S}}_j}{R_{kj}^2 + \gamma^2} |^2, \tag{3.25}$$

where

$$W_k^C = \frac{9\gamma^2 \mathcal{E}\{A_k^2\}\xi^2}{2J[2\gamma^2 + \mathcal{E}\{A_k^2\} C \xi^2 P \pi]}.$$

From the above, it is clear that *the Gaussian and the Cauchy detectors have similar complexities and there is no computational advantage in preferring one over the other.*

[4]This Gaussian detector is not only asymptotically optimum, but it is also strictly optimum under Rayleigh fading, as can be shown by an easy modification of the derivation of Eq.(4.18) in [7, p. 143].

3.3 Asymptotic Theoretical Performance Analysis

In this Section, we outline the steps in obtaining an asymptotic theoretical analysis of the performance of both the Gaussian and the Cauchy receivers. However, we do not compute the corresponding formulae down to all the details since a Monte-Carlo illustration is a more reliable performance evaluator.

a. Gaussian Detector

The usual asymptotic analysis of a detector performance that is based on the Central Limit Theorem is not applicable in the case of the Gaussian detector operating in non-Gaussian BISαS interference because the test statistic t_G can be shown to have infinite variance.[5] The asymptotic performance analysis can be carried out following steps similar to those in [31], but this approach will not be followed here since a Monte-Carlo simulation is expected to be more accurate and illustrative of our findings.

b. Cauchy Detector

The mean of the test statistic t_C is

$$\begin{aligned}\mu_C &= \sum_{k=1}^{K} \frac{9\gamma^2 \mathcal{E}\{A_k^2\}\xi^2}{2J[2\gamma^2 + \mathcal{E}\{A_k^2\}C\xi^2 P\pi]} \sum_{j=1}^{J} \mathcal{E}\{|\frac{\tilde{R}_{kj}\tilde{\bar{S}}_j}{R_{kj}^2+\gamma^2}|^2\} \\ &= \sum_{k=1}^{K} \frac{9\gamma^2 \mathcal{E}\{A_k^2\}\xi^2}{2J[2\gamma^2 + \mathcal{E}\{A_k^2\}C\xi^2 P\pi]} \sum_{j=1}^{J} \mathcal{E}\{\frac{R_{kj}^2|\tilde{\bar{S}}_j|^2}{(R_{kj}^2+\gamma^2)^2}\} \\ &= \sum_{k=1}^{K} \frac{9\gamma^2 \mathcal{E}\{A_k^2\}\xi^2}{2J[2\gamma^2 + \mathcal{E}\{A_k^2\}C\xi^2 P\pi]} \sum_{j=1}^{J} \int_0^\infty \frac{R_{kj}^2|\tilde{\bar{S}}_j|^2}{(R_{kj}^2+\gamma^2)^2} f_e(R_{kj})\, dR_{kj}, \quad (3.26)\end{aligned}$$

where $f_e(R_{kj})$ is the distribution of the envelope of the signal received through the kth channel in which both the random noise and the channel fading have been taken into account. Under either hypothesis H_0 (noise only) or H_1 (signal plus noise), $f_e \sim \frac{1}{R_{kj}^{\alpha+1}}$ and, therefore, $\mu_C < \infty$, for all $0 < \alpha \leq 2$.

Similarly, the variance of the test statistic is:

$$\sigma_C^2 = \sum_{k=1}^{K} (\frac{9\gamma^2\xi^2\mathcal{E}\{A_k^2\}}{2J[2\gamma^2 + \mathcal{E}\{A_k^2\}C\xi^2 P\pi]})^2 \operatorname{var}\{|\sum_{j=1}^{J} \frac{\tilde{R}_{kj}\tilde{\bar{S}}_j}{R_{kj}^2+\gamma^2}|^2\} \quad (3.27)$$

and $\sigma_C^2 < \infty$, for all $0 < \alpha \leq 2$, under either hypothesis H_0 (noise only) or H_1 (signal plus noise).

Therefore, we can invoke the Central Limit Theorem for the test statistic t_C and claim

$$P_d = \frac{1}{2} \operatorname{erfc}[\frac{\eta - \mu_{C,H_1}}{\sqrt{2\sigma_{C,H_1}^2}}] \quad (3.28)$$

[5] In fact [2], the test statistic t_G will be asymptotically $\frac{\alpha}{2}$-stable and, therefore, will possess neither finite mean, nor finite variance.

$$P_{fa} = \frac{1}{2}\,\mathrm{erfc}[\frac{\eta - \mu_{C,H_0}}{\sqrt{2\sigma^2_{C,H_0}}}], \qquad (3.29)$$

where P_d (P_{fa}) is the probability of detection (false alarm), η is the detector threshold to which the test statistic t_C will be compared, and erfc $(\cdot)$ denotes the complementary error function. Eqs.(3.28) and (3.29) can next be combined into

$$P_d = \frac{1}{2}\,\mathrm{erfc}\,[\frac{\sqrt{2\sigma^2_{C,H_0}}\,\mathrm{erfc}^{-1}(2P_{fa}) - \mu_{C,H_1} + \mu_{C,H_0}}{\sqrt{2\sigma^2_{C,H_1}}}] \qquad (3.30)$$

to obtain the asymptotic operating characteristic of the Cauchy receiver.

c. Conclusions from the Asymptotic Theoretical Analysis of the Gaussian and Cauchy Detectors

From the asymptotic distributions of the Gaussian and Cauchy detectors, the following conclusions can be drawn:

1. The Gaussian will perform below the Cauchy detector in non-Gaussian ($\alpha < 2$) BISαS noise for sufficiently low probability of false alarm. This is expected since, in this case, the test statistic of the Gaussian detector has very heavy tails. Thus, for this detector to attain a reasonably low false alarm probability, its threshold must be set to a very high value, effectively wiping out all but the very strong signals. The Cauchy detector on the other hand suppresses very large inputs, as can be seen from Eq.(3.25) and will, therefore, perform above the Gaussian detector for $\alpha < 2$ and sufficiently low probability of false alarm.

2. The performance of the Gaussian detector in non-Gaussian ($\alpha < 2$) BISαS noise will *not* improve significantly as the number of receiver diversity increases, since the test statistic employed by this detector has infinite mean and variance. The performance of the Cauchy detector, on the other hand, will improve with increasing diversity, as can be seen from Eqs.(3.26), (3.27), and (3.30).

3.4 Monte-Carlo Evaluated Performance

In this Section, the performance is evaluated of the Gaussian and the Cauchy detectors of the previous Sections via an extensive Monte-Carlo simulation. In all of the simulations, the signal of interest was a square pulse of unit height, corresponding to $P = 1$. All the performance curves were derived from 25,000 independent Monte-Carlo runs, in which $\xi = 1$ and the number of signal samples received at each channel was set to $J = 16$.

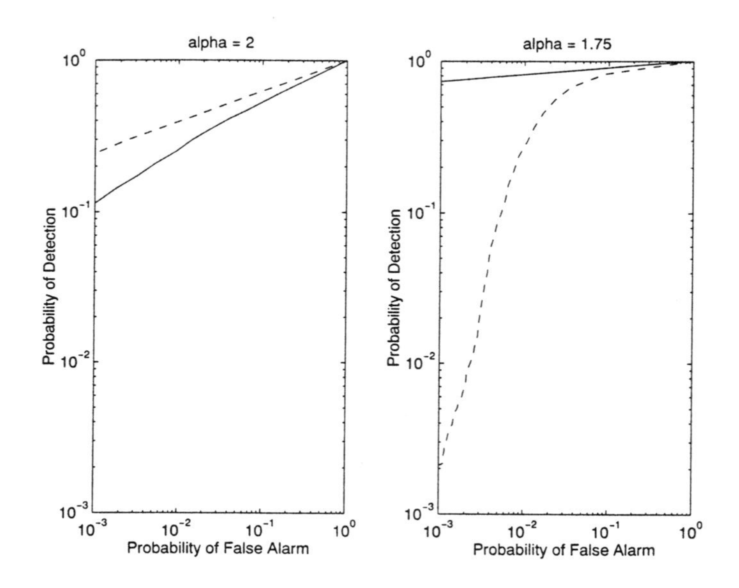

Figure 5 (a). Performance comparison between the Gaussian and the Cauchy detector for Gaussian ($\alpha = 2$) and BIS($\alpha = 1.75$)S noise

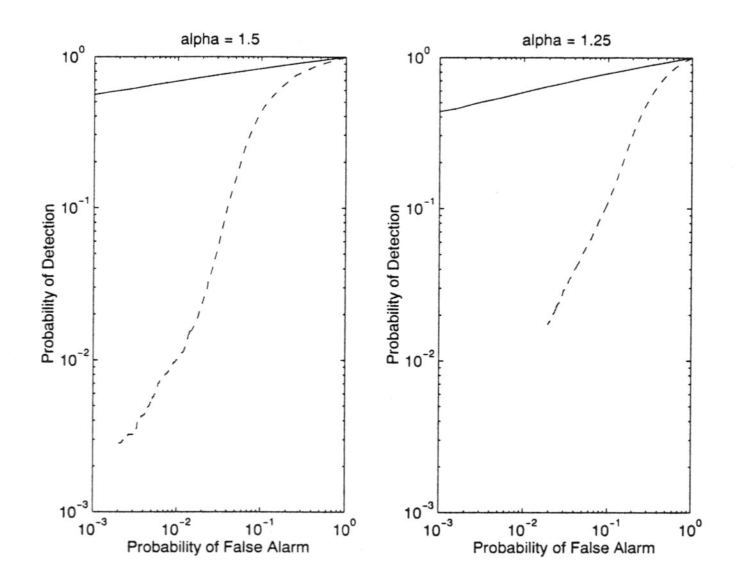

Figure 5 (b). Performance comparison between the Gaussian and the Cauchy detector for BIS($\alpha = 1.5$)S and BIS($\alpha = 1.25$)S noise

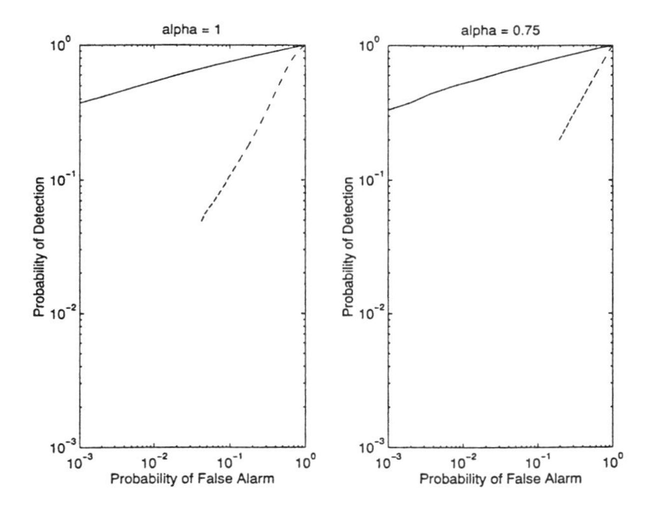

Figure 5 (c). Performance comparison between the Gaussian and the Cauchy detector for BIS($\alpha = 1$)S and BIS($\alpha = 0.75$)S noise

First, the case was simulated of a signal detector consisting of $K = 3$ channels, with $\mathcal{E}\{A_1^2\} = 0.5$, $\mathcal{E}\{A_2^2\} = 0.1$, and $\mathcal{E}\{A_3^2\} = 0.01$. For the noise, we set $\gamma = 1$ and examined the cases $\alpha = 2$, 1.75, 1.5, 1.25, 1, and 0.75. Figs. 5 show comparative plots of the corresponding operating characteristics of the Gaussian (dotted line) and the Cauchy (solid line) receiver. Figs. 6 show the performance of the Gaussian (Fig. 6a) and the Cauchy (Fig. 6b) receiver for different values of α.

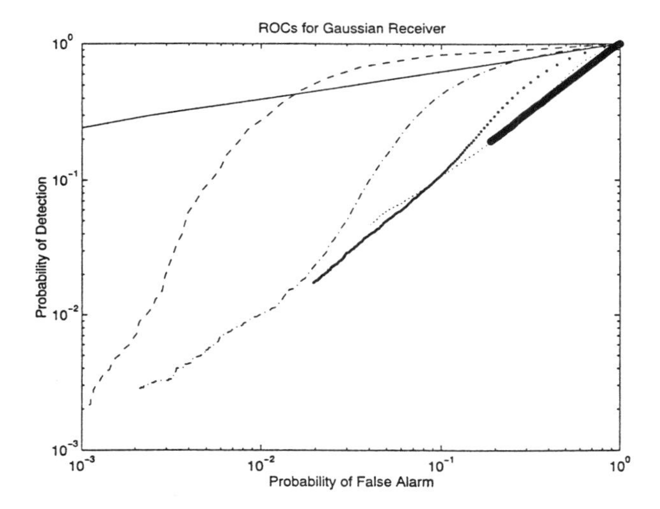

Figure 6(a). Performance of the Gaussian detector in BISαS noise for $\alpha = 2$ (solid line), 1.75 (dashed line), 1.5 (dash-dotted line), 1.25 (point line), 1 (dotted line), and 0.75 (circle line)

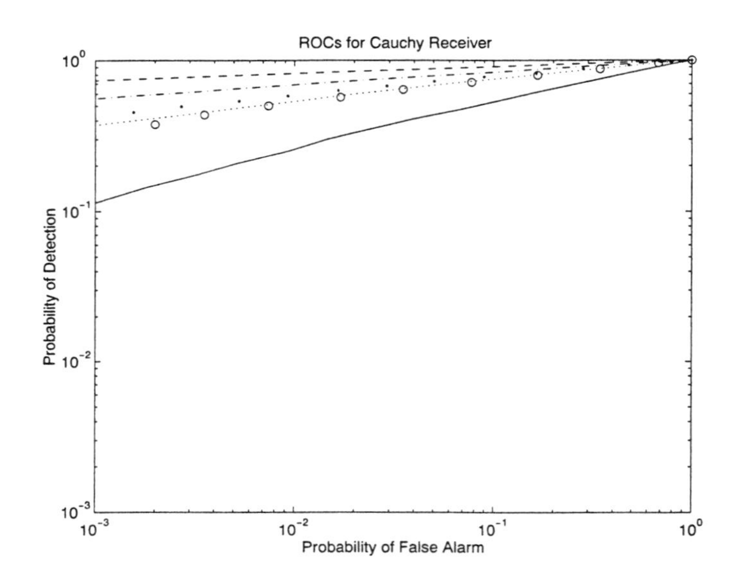

Figure 6(b). Performance of the Cauchy detector in BISαS noise for $\alpha = 2$ (solid line), 1.75 (dashed line), 1.5 (dash-dotted line), 1.25 (point line), 1 (dotted line), and 0.75 (circle line)

The conclusion drawn from this study is that *the Cauchy receiver is robust in the entire class of BISαS noises, while the performance of the Gaussian receiver degrades very quickly even if the Gaussianity assumption is only slightly violated, as for example is the case that corresponds to $\alpha = 1.75$. Moreover, this performance degradation is more serious at the low probabilities of false alarm at which any realistic detector would operate.* As a result, the Cauchy receiver should be preferred over the Gaussian one as a safeguard against impulsiveness in the interference.

Next, the performance was evaluated of the Gaussian and the Cauchy receiver in BIS($\alpha = 1.5$)S noise of unit dispersion as a function of their diversity, i.e., as a function of the number of channels through which the signals are observed. We chose $\mathcal{E}\{A_k^2\} = 0.1$ for $k = 1, 2, \ldots, K$, i.e., we assumed that the K channels were identical. We simulated the cases of $K = 1$, 2, 3, 5, and 10. The resulting receiver operating characteristics are plotted in Figs. 7. In particular, Fig. 7a shows the performance of the Gaussian receiver for $K = 10$ (solid line), $K = 5$ (dashed line), $K = 3$ (dash-dotted line), $K = 2$ (point line), and $K = 1$ (dotted line). The performance of the Cauchy receiver is shown in the corresponding lines of Fig. 7b. The key observation is that, *in non-Gaussian, alpha-stable interference, the performance of the Gaussian receiver, unlike that of the Cauchy receiver, does not improve significantly as the number of channels increases.*

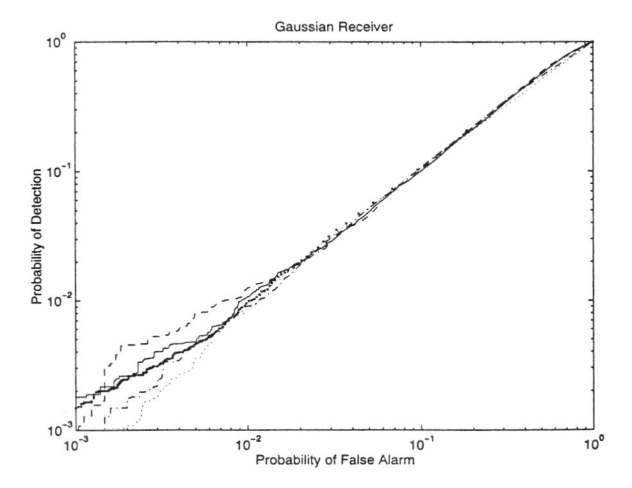

Figure 7 (a). Performance of the Gaussian detector in BIS($\alpha = 1.5$)S noise for $K = 1$ (dotted line), 2 (point line), 3 (dash-dotted line), 5 (dashed line), and 10 (solid line)

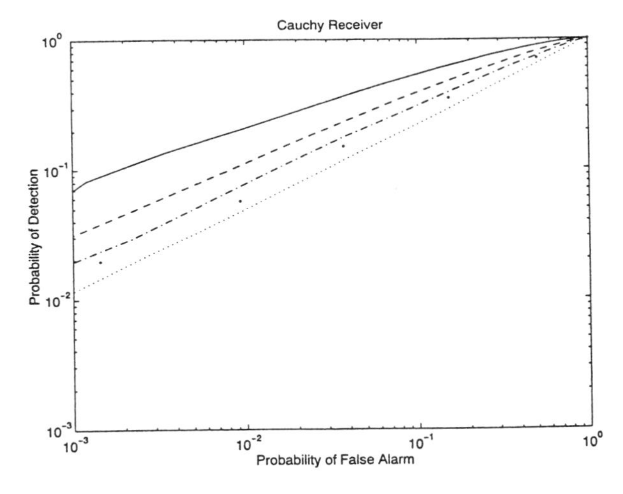

Figure 7(b). Performance of the Cauchy detector in BIS($\alpha = 1.5$)S noise for $K = 1$ (dotted line), 2 (point line), 3 (dash-dotted line), 5 (dashed line), and 10 (solid line)

4 Summary, Conclusions, and Future Research

In this article, alpha-stable statistical models for impulsive noise were applied to the modeling of interference arising from multiple users in communication networks. It was shown that the traditionally assumed Gaussian model is a special case of the more general alpha-stable model. Next, the spatially diverse, Izzo-Paura detector of amplitude-fluctuating signals was modified to account for impulsive interference modeled as a bivariate, symmetric, alpha-stable process. The proper test statistic was derived for incoherent detection of frequency-nonselective, slowly Rayleigh-fading signals in both Gaussian and Cauchy noise and their performance was tested in the entire class of BISαS noises.

From this research, it was found that, for sufficiently low probability of false alarm, the detector derived on the basis of a Cauchy assumption performs better than the detector derived on the basis of a Gaussianity assumption in non-Gaussian bivariate, isotropic, symmetric, alpha-stable interference. A second very important finding was that, unlike the Cauchy detector, the performance of the Gaussian detector did *not* improve signficantly as the number of channels increased, which implies that receiver diversity does *not* improve the performance of the Gaussian detector. Finally, it should be underlined that the complexities of the two detectors are similar, so no implementation gains are expected from use of a Gaussian detector. It is, thus, a safe conclusion that Cauchy detectors ought to be preferred over the traditional Gaussian ones.

Natural future avenues of the work reported in here seem to lead to an extensive study of the form and the performance of Cauchy receivers for detection of various types of fading signals and proper tabulation of the findings. Very important in this task is also the study of the performance of specific communication systems (signals) in impulsive interference. Another avenue that also seems very important is the derivation of diverse receiver structures that rely on moments of the observed signals rather than on likelihood ratios, similar to [27]. This approach would be very useful in the case of highly dependent impulsive noise of unknown structure for which the performance of the present detector would degrade. These ideas are currently under exploration and the findings will be reported in the near future.

References

[1] L A Berry. Understanding Middleton's canonical formula for class A noise. *IEEE Transactions on Electromagnetic Compatibility*, EMC-23:337–344, 1981.

[2] R Davis and S Resnick. Limit theory for moving averages of random variables with regularly varying tail probabilities. *Annals Prob.*, 13:179–195, 1985.

[3] E C Field and M Lewinstein. Amplitude-probability distribution model for VLF/ELF atmospheric noise. *IEEE Trans. Comm.*, COM-26:83–87, 1978.

[4] K. Furutsu and T. Ishida. On the theory of amplitude distribution of impulsive random noise. *J. of Applied Physics*, 32(7), 1961.

[5] G B Giannakis and M K Tsatsanis. Signal detection and classification using matched filtering and higher-order statistics. *IEEE Trans. Acoust. Speech, Sign. Proc.*, ASSP-38:1284–1296, 1990.

[6] A.A. Giordano and F. Haber. Modeling of atmospheric noise. *Radio Science*, 7:1101–1123, 1972.

[7] C W Helstrom. *Elements of Signal Detection and Estimation*. Prentice Hall, Englewood Cliffs, NJ, 1995.

[8] L Izzo and L Paura. Asymptotically optimum space-diversity detection in non-gaussian noise. *IEEE Trans. Comm.*, COM-34:97–103, 1986.

[9] L Izzo and L Paura. Authors' reply to comments on "asymptotically optimum space-diversity detection in non-gaussian noise". *IEEE Trans. Comm.*, COM-43:19, 1995.

[10] P Lévy. *Calcul des Probabilités*, volume II. Gauthier-Villards, Paris, 1925. chapter 6.

[11] P Mertz. Model of impulsive noise for data transmission. *IRE Trans. Comm. Systems*, CS-9:130–137, 1961.

[12] D Middleton. First-order probability models of the instantaneous amplitude, Part I. Report OT 74-36, Office of Telecommunications, 1974.

[13] D Middleton. Statistical-physical models of man-made and natural radio noise, Part II: First-order probability models of the envelope and phase. Report OT 76-86, Office of Telecommunications, 1976.

[14] D Middleton. Statistical-physical models of electromagnetic interference. *IEEE Trans. Electromagnetic Compatibility*, EMC-19(3):106–127, 1977.

[15] D Middleton. Statistical-physical models of man-made and natural radio noise, Part III: First-order probability models of the instantaneous amplitude of Class B interference. Report NTIA-CR-78-1, Office of Telecommunications, 1978.

[16] D Middleton. Canonical non-Gaussian noise models: Their implications for measurement and for prediction of receiver performance. *IEEE Transactions on Electromagnetic Compatibility*, EMC-21(3), 1979.

[17] D Middleton. Procedures for determining the parameters of the first-order canonical models of class A and class B electromagnetic interference. *IEEE Trans. Electromagnetic Compatibility*, EMC-21(3):190–208, 1979.

[18] J H Miller and J B Thomas. Detectors for discrete-time signals in non-Gaussian noise. *IEEE Trans. Inform. Theory*, IT-18:241–250, 1972.

[19] C L Nikias and A Petropulu. *Higher-Order Spectra Analysis: A Nonlinear Signal Processing Framework*. Prentice-Hall, Englewood Cliffs, NJ, 1993.

[20] C L Nikias and M Shao. *Signal Processing with Alpha-Stable Distributions and Applications*. John Wiley & Sons, Inc., New York, NY, 1995.

[21] E. Parzen. *Stochastic Process*. Holden-Day, San Francisco, CA, 1962.

[22] S S Rappaport and L Kurz. An optimal nonlinear detector for digital data transmission through non-Gaussian channels. *IEEE Trans. Comm. Techn.*, COM-14:266–274, 1966.

[23] G Samorodnitsky and M S Taqqu. *Stable, Non-Gaussian Random Processes: Stochastic Models with Infinite Variance*. Chapman & Hall, New York, NY, 1994.

[24] S M Selby, editor. *Standard Mathematical Tables*. CRC Press, Cleveland, OH, 22 edition, 1974.

[25] M Shao and C L Nikias. Signal processing with fractional lower-order moments: Stable processes and their applications. *Proc. IEEE*, 81:986–1010, 1993.

[26] B W Stuck and B Kleiner. A statistical analysis of telephone noise. *The Bell System Technical Journal*, 53:1262–1320, 1974.

[27] G A Tsihrintzis and C L Nikias. Random signal detection using fractional, lower-order statistics. *Signal Processing*. (submitted on June 25, 1995, pp. 18).

[28] G A Tsihrintzis and C L Nikias. Fast estimation of the parameters of alpha-stable impulsive interference. *IEEE Trans. Signal Processing*, SP-44:1492–1503, 1996.

[29] G A Tsihrintzis and C L Nikias. On the detection of stochastic impulsive transients over background noise. *Signal Processing*, 41:175–190, January 1995.

[30] G A Tsihrintzis and C L Nikias. Modeling, parameter estimation, and signal detection in radar clutter with alpha-stable distributions. In *1995 IEEE Workshop on Nonlinear Signal and Image Processing*, Neos Marmaras, Halkidiki, Greece, June 1995.

[31] G A Tsihrintzis and C L Nikias. Performance of optimum and suboptimum receivers in the presence of impulsive noise modeled as an α–stable process. *IEEE Trans. Comm.*, COM-43:904–914, part II of 3 parts, February/March/April 1995.

[32] G A Tsihrintzis and C L Nikias. Incoherent receivers in alpha-stable impulsive noise. *IEEE Trans. Signal Processing*, SP-43:2225–2229, September 1995.

[33] G A Tsihrintzis, M Shao, and C L Nikias. Recent results in applications and processing of alpha-stable-distributed time series. *J. Franklin Institute*, 333B:467–497, 1996.

[34] H L Van Trees. *Detection, Estimation, and Modulation Theory, Part I*. Wiley, New York, 1968.

[35] S Verdu. Multi-user detection. In *Advances in Statistical Signal Processing - Vol. 2: Signal Detection*. JAI Press, Greenwich, CT, 1992.

[36] E J Wegman, S G Schwartz, and J B Thomas, editors. *Topics in Non-Gaussian Signal Processing*. Academic Press, New York, 1989.

[37] V Zolotarev. *One-Dimensional Stable Distributions*. American Mathematical Society, Providence, RI, 1986.

[38] V. M. Zolotarev. Integral transformations of distributions and estimates of parameters of multidimensional spherically symmetric stable laws. In J. Gani and V. K. Rohatgi, editors, *Contribution to Probability: A Collection of Papers Dedicated to Eugene Lukacs*, pages 283–305. Academic Press, 1981.

Center for Electromagnetics Research
Department of Electrical and Computer Engineering
Northeastern University
Boston, MA 02115
Partial support for this research was provided by the Woodrow W. Everett, Jr. SCEEE Development Fund in cooperation with the Southeastern Association of Electrical Engineering Department Heads

VI. Model Structures

Subexponential Distributions

Charles M. Goldie and Claudia Klüppelberg

Abstract

We survey the properties and uses of the class of subexponential probability distributions, paying particular attention to their use in modelling heavy-tailed data such as occurs in insurance and queueing applications. We give a detailed summary of the core theory and discuss subexponentiality in various contexts including extremes, random walks and Lévy processes with negative drift, and sums of random variables, the latter extended to cover random sums, weighted sums and moving averages.

1. Definition and first properties

Subexponential distributions are a special class of heavy–tailed distributions. The name arises from one of their properties, that their tails decrease more slowly than any exponential tail; see (1.4). This implies that large values can occur in a sample with non–negligible probability, and makes the subexponential distributions candidates for modelling situations where some extremely large values occur in a sample compared to the mean size of the data. Such a pattern is often seen in insurance data, for instance in fire, wind–storm or flood insurance (collectively known as catastrophe insurance). Subexponential claims can account for large fluctuations in the surplus process of a company, increasing the risk involved in such portfolios. This situation is treated in Section 2.

Subexponentials play a similar role in queueing models. Situations with extreme service times, modelled by a subexponential distribution, result in huge waiting times in the system (see Example 2.7). The workload process also shows large fluctuations (see Example 6.4).

Linear models are widely used as simple models for (or first order approximations to) dependent data. Extremely large values in the innovations, modelled by subexponential distributions, have immediate consequences for the single observation. Moreover, they cause effects in larger parts of the sample, determined by the linear filter.

In all these models a few large values may determine the long–term behaviour of a system. This can be made very precise by describing the sample path behaviour of resulting stochastic processes as the surplus process in insurance or the workload process of a queue, since the latter models have been the most fully investigated. This is reviewed in Section 6.

Heavy tails are just one of the consequences of the defining property of subexponential distributions, which is designed specially to work well with the probabilistic models commonly employed in the above–mentioned areas of application. The subexponential concept has just the right level of generality to be usable in these models while including as wide a range of distributions as possible. It includes all distributions with regularly varying tails (domains of attraction of sum– or max–stable laws) but is considerably wider (see Table 3.7). Hence it encompasses many more types of behaviour in the extremes (see Section 4).

Subexponential distributions were first studied in 1964 by Chistyakov. Research during the seventies was centred around applications in insurance, queueing and branching processes, based on the Pollaczek–Khinchin formula (2.2), linking a subexponential input df and an output df of interest. In a simple insurance model this output df may be the ruin probability, while in a simple queueing model it may be the df of the stationary waiting time. Methods were rather more analytic than probabilistic at that time. Properties of subexponential moment generating functions, necessary and sufficient conditions for subexponentiality, and closure properties were investigated.

Extensions to more general models followed: renewal arrival streams replaced Poisson arrivals. Modelling in that generality required the tracing of subexponential input distributions through a Wiener–Hopf factorisation. Use of random Markov environments required tracing different input distributions (light– and heavy–tailed), by means of matrix algebra.

Recently, more probabilistic methods have entered the field. Questions like "how does ruin happen?" or "when is ruin most likely to happen?" given it happens at all, or "what does the workload process at a high level look like?" were asked and answered. They necessitated novel methods to investigate path properties using the regenerative structure of models, as well as excursion theory for Markov processes and extreme value theory.

Against this background we present two defining properties of subexponential distributions. The first, more analytic one, is motivated by the Pollaczek–Khinchin formula (2.2) below, while the second probabilistic one provides a more intuitive interpretation of subexponentiality.

Definition 1.1. (Subexponential distribution function)
Let $(X_i)_{i\in\mathbb{N}}$ be iid positive rvs with df F such that $F(x) < 1$ for all $x > 0$. Denote

$$\overline{F}(x) = 1 - F(x)\,, \quad x \geq 0\,,$$

the tail of F and

$$\overline{F}^{n*} = 1 - F^{n*}(x) = P(X_1 + \cdots + X_n > x)$$

the tail of the n–fold convolution of F. F is a subexponential df ($F \in \mathcal{S}$) *if one of the following equivalent conditions holds:*

$$(a)\ \lim_{x\to\infty} \frac{\overline{F}^{n*}(x)}{\overline{F}(x)} = n \text{ for some (all) } n \geq 2\,,$$

$$(b)\ \lim_{x\to\infty} \frac{P(X_1 + \cdots + X_n > x)}{P(\max(X_1, \ldots, X_n) > x)} = 1 \text{ for some (all) } n \geq 2. \qquad \square$$

Remarks 1) Definition (a) goes back to Chistyakov (1964). He proved that the limit (a) holds for all $n \geq 2$ if and only if it holds for $n = 2$. It was shown in Embrechts and Goldie (1982) that (a) holds for $n = 2$ if it holds for some $n \geq 2$.

2) The equivalence of (a) and (b) was shown in Embrechts and Goldie (1980). A proof goes as follows:

$$P(\max(X_1, \dots, X_n) > x) = 1 - F^n(x) = \overline{F}(x) \sum_{k=0}^{n-1} F^k(x) \sim n\overline{F}(x), \quad x \to \infty,$$

($\sim$ means that the quotient of lhs and rhs tends to 1). Hence

$$\frac{P(X_1 + \cdots + X_n > x)}{P(\max(X_1, \dots, X_n) > x)} \sim \frac{\overline{F^{n*}}(x)}{n\overline{F}(x)} \to 1 \quad \Longleftrightarrow \quad F \in \mathcal{S}.$$

3) Definition (b) provides a physical interpretation of subexponentiality: the sum of n iid subexponential rvs is likely to be large if and only if their maximum is likely to be large. This accounts for extremely large values in a subexponential sample.

4) From Definition (a) and the fact that $\mathcal{S}$ is closed with respect to tail–equivalence (see Definition 3.3) we conclude that

$$F \in \mathcal{S} \quad \Longrightarrow \quad F^{n*} \in \mathcal{S}, \quad n \in \mathbb{N}. \tag{1.1}$$

Furthermore, from Definition (b) and the fact that F^n is the df of the maximum of n iid rvs with df F, we conclude that

$$F \in \mathcal{S} \quad \Longrightarrow \quad F^n \in \mathcal{S}, \quad n \in \mathbb{N}.$$

Hence $\mathcal{S}$ is closed with respect to taking sums and maxima of iid rvs. The relationship of subexponentials and maxima will be further investigated in Section 4. Various generalisations of (1.1) will be considered in Section 5.

5) Definition (b) demonstrates the heavy–tailedness of subexponential dfs. It is further substantiated by the implications (first proved by Chistyakov (1964))

$$F \in \mathcal{S} \quad \Longrightarrow \quad \lim_{x\to\infty} \frac{\overline{F}(x-y)}{\overline{F}(x)} = 1 \quad \forall\, y \in \mathbb{R} \tag{1.2}$$

$$\Longrightarrow \quad \int_0^\infty e^{\varepsilon x}\, dF(x) = \infty \quad \forall\, \varepsilon > 0 \tag{1.3}$$

$$\Longrightarrow \quad \overline{F}(x)/e^{-\varepsilon x} \to \infty \quad \forall\, \varepsilon > 0. \tag{1.4}$$

Property (1.4) accounts for the name subexponential df: the tail of F decreases more slowly than any exponential tail. Property (1.3) shows that subexponential dfs have no exponential moments. This prevents any method being applicable that requires the existence of exponential moments. □

2. The supremum of a random walk with negative drift

Subexponential dfs traditionally play an important role in continuous time models with a random walk skeleton. We choose a class of insurance risk models for demonstrating the general method.

The *classical insurance risk process* is defined as

$$R(t) = u + ct - \sum_{i=1}^{N(t)} X_i\,, \quad t \geq 0\,,$$

where $u \geq 0$ is the *initial capital* (or *risk reserve*), and $c > 0$ is the *premium rate*, i.e. premiums are linear in time. $(N(t))_{t\geq 0}$ is a *homogeneous Poisson process with intensity* $\lambda > 0$, counting the number of claims up to time t. $(X_i)_{i\in\mathbb{N}}$ are iid positive claims, independent of $(N(t))$, with df F, finite mean μ and *integrated tail df*

$$F_I(x) = \frac{1}{\mu}\int_0^x \overline{F}(t)\,dt\,, \quad x \geq 0\,. \tag{2.1}$$

Denote by $\psi(u)$ the *ruin probability*, given a risk reserve u, i.e.

$$\psi(u) = P\big(R(t) < 0 \quad \text{for some} \quad t > 0\big).$$

The risk process $(R(t))$ has two important features: the *inter–arrival times* are iid exponential rvs $(E_i)_{i\in\mathbb{N}}$ with mean $1/\lambda$, and ruin can occur only at claim times. Hence if we define

$$S_0 = 0\,, \quad S_n = \sum_{i=1}^{n}(X_i - cE_i)\,, \quad n \in \mathbb{N}\,,$$

then

$$\begin{aligned}\psi(u) &= P\big(S(t) > u \quad \text{for some} \quad t > 0\big)\\ &= P\big(S_n > u \quad \text{for some} \quad n \in \mathbb{N}\big)\\ &= P\Big(\max_{n\geq 1} S_n > u\Big)\,.\end{aligned}$$

Under the *net–profit condition* $\rho = \lambda\mu/c < 1$, the random walk (S_n) has negative drift and $\psi(u) \to 0$ as $u \to \infty$. If we denote by

$$\tau(u) = \inf\{n \geq 0 : S_n > u\}$$

the *ruin time*, then

$$P\Big(\max_{n\geq 1} S_n > u\Big) = P\big(\tau(u) < \infty\big)\,,$$

representing the ruin problem as a problem of first hitting times. The ruin problem can be handled by an analysis of the ladder heights or by solving a renewal equation (see

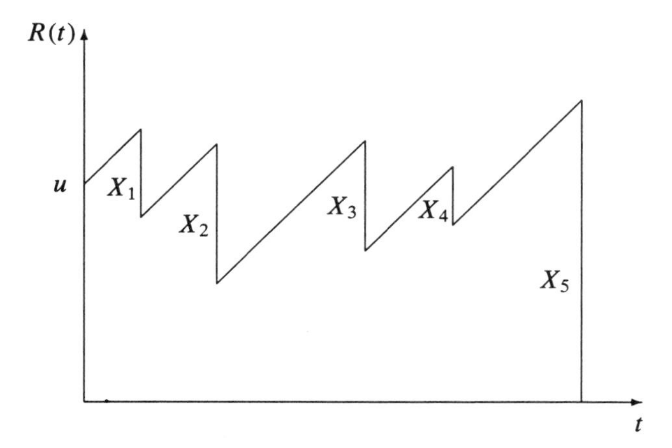

Figure 2.1. *Idealised sample path of the risk process.*

Asmussen(1996), Embrechts, Klüppelberg and Mikosch (1997), Feller (1971), Grandell (1991)), representing the non–ruin probability in terms of the *Pollaczek–Khinchin formula:*

$$1 - \psi(u) = (1-\rho)\sum_{n=0}^{\infty} \rho^n F_I^{n*}(u)\,, \qquad u \geq 0\,, \tag{2.2}$$

where F_I is the integrated tail df (2.1) and $F_I^{0*} = I_{[0,\infty)}$ is the df of Dirac (unit) measure at 0. In this representation ρF_I is the *ladder height df.* The infinite series on the rhs of (2.2) defines a defective renewal measure ($\rho F_I(x) \to \rho < 1$ as $x \to \infty$), and the corresponding renewal process is transient: the sequence of renewals eventually stops, and at each renewal $1-\rho$ is the probability of termination then and there.

If *Cramér's condition* holds, i.e. if there exists some $\gamma > 0$ such that

$$\int_0^{\infty} e^{\gamma x}\overline{F}(x)\,dx = \frac{c}{\lambda}\,, \tag{2.3}$$

the defect can be removed and, under the usual conditions, Smith's key renewal theorem implies that

$$\psi(u)e^{\gamma u} \to C\,, \qquad u \to \infty\,, \tag{2.4}$$

where C is a non–negative constant; thus $\psi(u)$ decreases exponentially fast to 0. It is clear from (1.3) that for $F_I \in \mathcal{S}$ Cramér's condition (2.3) does not hold. But a different approach, as we now describe, shows that subexponentials form the class of *heavy–tailed* distributions that allows for ruin estimates.

We rewrite formula (2.2) in terms of the tails,

$$\psi(u) = (1-\rho)\sum_{n=1}^{\infty} \rho^n \overline{F}_I^{n*}(u)\,, \qquad u \geq 0\,.$$

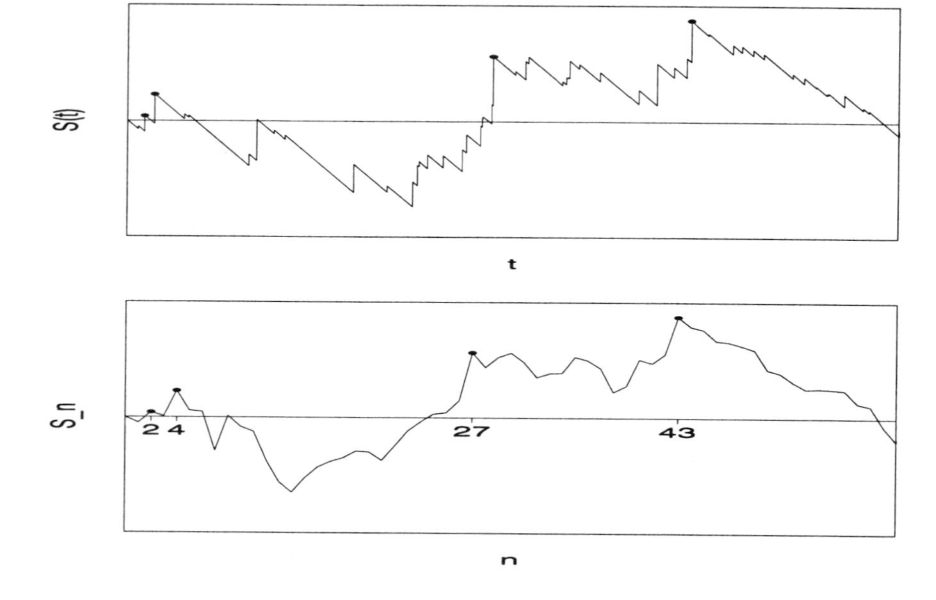

Figure 2.2. *Sample path of the process $(S(t))$ and its embedded random walk (S_n). The ladder points are indicated by dots.*

Dividing both sides by $\overline{F}_I(u)$, we see that Definition 1.1(a) yields an asymptotic estimate for $\psi(u)$ provided that one can safely interchange the limit and the infinite sum. This is ensured by the following lemma due to Kesten (for a proof see Athreya and Ney (1972)), and Lebesgue's dominated convergence theorem.

Lemma 2.3. *If $F \in \mathcal{S}$, then for every $\varepsilon > 0$ there exists some positive constant $K(\varepsilon)$ such that for all $n \in \mathbb{N}$ and $x > 0$,*

$$\frac{\overline{F^{n*}}(x)}{\overline{F}(x)} \le K(\varepsilon)(1+\varepsilon)^n . \qquad \square$$

As remarked, these considerations lead to an asymptotic evaluation of ψ. It turns out that this is not just a consequence of subexponentiality, but is characterised by it, as follows.

Theorem 2.4. (The ruin probability in the classical risk model)

$$F_I \in \mathcal{S} \iff 1-\psi \in \mathcal{S} \iff \lim_{u\to\infty} \frac{\psi(u)}{\overline{F}_I(u)} = \frac{\rho}{1-\rho} . \qquad \square$$

This theorem can be generalised by a Wiener–Hopf factorisation to the more general *Sparre Andersen model*, where the claim arrival process is an arbitrary renewal process.

Theorem 2.5. (The ruin probability in the renewal risk model)

$$1-\psi \in \mathcal{S} \iff F_I \in \mathcal{S} \implies \lim_{u\to\infty} \frac{\psi(u)}{\overline{F}_I(u)} = \frac{\rho}{1-\rho}. \qquad \square$$

The result of Theorem 2.4 has been further extended by Asmussen, Fløe Henriksen and Klüppelberg (1994) to a *Markov–modulated risk model*, where the risk process is not time–homogeneous, but evolves in an environment given by a Markov process with finite state space. A state of the Markov process defines the arrival intensity of the Poisson process and the claim–size distribution. Further results in the realm of this model have been obtained by Asmussen and Højgaard (1995) and Jelenkovič and Lazar (1996).

Asymptotic estimates for the ruin probability change when the company receives interest on its reserves. For regularly varying claim–size df F and a positive force of interest δ the corresponding ruin probability satisfies

$$\psi_\delta(u) \sim c_\delta \overline{F}(u), \quad u \to \infty,$$

for some positive constant c_δ, i.e. it is tail–equivalent to the claim–size df itself. This has been proved in Klüppelberg and Stadtmüller (1996). The case of general subexponential claims has been treated in Asmussen (1996).

Remarks 1) The importance of subexponential dfs for insurance risk theory was recognised by Teugels (1975).

2) A textbook treatment of subexponential distributions in the context of risk theory is to be found in Embrechts, Klüppelberg and Mikosch (1997).

3) Theorem 2.4 is due to Embrechts and Veraverbeke (1982) based on work by Embrechts, Goldie and Veraverbeke (1979). Theorem 2.5 can be found in Embrechts and Veraverbeke (1982); see also Veraverbeke (1977) and Bertoin and Doney (1996). A density version of Theorem 2.4 can be found in Klüppelberg (1989a), Theorem 4.1 (since F_I has a density, so does $1-\psi$).

4) Theorem 2.5 can be further generalised to a general discrete time or continuous time random walk or Lévy process with negative drift and increment variable S_1 with df B such that the right tail of B satisfies $\overline{B}(x) = P(S_1 > x) \sim \overline{F}(x)$ for a subexponential df F. Notice that this is in accordance with the situation for the classical risk process, where $S_1 = X_1 - cE_1$ and

$$\frac{P(X_1 - cE_1 > x)}{P(X_1 > x)} = \int_0^\infty \frac{\overline{F}(x+cy)}{\overline{F}(x)} \lambda e^{-\lambda y}\, dy \to 1, \quad x \to \infty.$$

($\overline{F}(x+cy) \le \overline{F}(x)$ for all $x > 0$ and the quotient tends to 1 by (1.2), hence Lebesgue dominated convergence applies.) What is needed is that the ladder height df F_I is subexponential. This also shows that in this context it is quite natural to define subexponentiality only for positive rvs. $\square$

Similar results to those for the risk models have been derived in the context of branching processes and queueing theory.

Example 2.6. (Branching processes)
Let $(Z(t))_{t\geq 0}$ denote the population size in the Bellman–Harris model, i.e. the particles produce (independently of each other) at the end of their lifetime a random number of offspring. Let F be the lifetime df of a particle and $m < 1$ the mean number of offspring. A renewal–type argument similar to the argument leading to equation (2.2) yields, for $\mu(t) = EZ(t)$,

$$\mu(t) = \left(\sum_{n=0}^{\infty} m^n F^{n*}\right) * (1-F)(t) = \sum_{n=0}^{\infty} m^n \left(\overline{F}^{(n+1)*}(t) - \overline{F}^{n*}(t)\right).$$

An application of Lemma 2.3 yields together with Definition 1.1 an obvious analogue of Theorem 2.4. Early references in this context are Athreya and Ney (1972), Chistyakov (1964) and Chover, Ney and Wainger (1973). □

Example 2.7. (Queueing models)
Consider a GI/G/1 queue with renewal arrival stream and general service time df F. Let F have finite mean μ and integrated tail distribution (2.1). We consider a stable queue, i.e. with traffic intensity $\rho < 1$. Then the stationary waiting time df can be represented as the df of the maximum of a random walk (Feller (1971), VI.9). Hence analogues of Theorem 2.4 (corresponding to an M/G/1 queue) and Theorem 2.5 are immediate. Early results were derived by Pakes (1975), Smith (1972) and Veraverbeke (1977). □

3. Conditions for subexponentiality

It should be clear from the definition that a characterisation of subexponential dfs or even of dfs whose integrated tail df is subexponential (as needed in the risk and queueing models) will not be possible in terms of simple expressions involving the tail.

Recall that all subexponential dfs have property (1.2), hence the class of such dfs provides potential candidates for subexponentiality. The class is named as follows.

Definition 3.1. (The class $\mathcal{L}$)
Let F be a df on $(0,\infty)$ such that $F(x) < 1$ for all $x > 0$. We say $F \in \mathcal{L}$ if

$$\lim_{x\to\infty} \frac{\overline{F}(x-y)}{\overline{F}(x)} = 1 \quad \forall\, y > 0\,.$$ □

Unfortunately, $\mathcal{S}$ is a proper subset of $\mathcal{L}$. Examples for a df in $\mathcal{L}$ but not in $\mathcal{S}$ can be found in Embrechts and Goldie (1980) and Pitman (1980).

A famous subclass of $\mathcal{S}$ is the class of dfs with regularly varying tail. For a positive measurable function f we write $f \in \mathcal{R}(\alpha)$ for $\alpha \in \mathbb{R}$ (f is *regularly varying with index* α) if

$$\lim_{x\to\infty} \frac{f(tx)}{f(x)} = t^{\alpha} \quad \forall\, t > 0\,.$$

A function $f \in \mathcal{R}(0)$ is called *slowly varying*. For further properties of regularly varying functions we refer to the monograph by Bingham, Goldie and Teugels (1989).

Example 3.2. (Distribution functions with regularly varying tails)
Let $\overline{F} \in \mathcal{R}(-\alpha)$ for $\alpha \geq 0$, then it has the representation

$$\overline{F}(x) = x^{-\alpha}\ell(x), \quad x > 0,$$

for some $\ell \in \mathcal{R}(0)$. Notice first that $F \in \mathcal{L}$, hence it is a candidate for $\mathcal{S}$. We check Definition 1.1(a). Let X_1, X_2 be iid rvs with df F. Now use the decomposition

$$\begin{aligned} P(X_1 + X_2 > x) &= P(X_1 \leq \frac{x}{2}, X_1 + X_2 > x) + P(X_2 \leq \frac{x}{2}, X_1 + X_2 > x) \\ &\quad + P(X_1 > \frac{x}{2}, X_2 > \frac{x}{2}). \end{aligned}$$

Then

$$\frac{\overline{F^{2*}}(x)}{\overline{F}(x)} = 2\int_0^{x/2} \frac{\overline{F}(x-y)}{\overline{F}(x)}\, dF(y) + \frac{\overline{F}^2(x/2)}{\overline{F}(x)}.$$

Immediately, by the definition of $\mathcal{R}(-\alpha)$, the last term tends to 0. The integrand satisfies $\overline{F}(x - y)/\overline{F}(x) \leq \overline{F}(x/2)/\overline{F}(x)$ for $0 \leq y \leq x/2$, hence Lebesgue dominated convergence applies and, since $F \in \mathcal{L}$, the integral on the rhs tends to 1 as $x \to \infty$.

Examples of dfs with regularly varying tail are Pareto, Burr, log–gamma and stable dfs (see Table 3.7). If $\alpha > 1$ then F has finite mean and, by Karamata's theorem, $F_I \in \mathcal{R}(-(\alpha - 1))$, giving $F_I \in \mathcal{S}$ as well. □

In much of the present discussion we are dealing only with the right tail of a df. This notion can be formalised, starting with the following definition.

Definition 3.3. (Tail–equivalence)
Two dfs F and G with support unbounded to the right are called tail–equivalent *if* $\lim_{x\to\infty} \overline{F}(x)/\overline{G}(x) = c \in (0, \infty)$. □

The next representation is a consequence of Theorem 1.3.1 of Bingham, Goldie and Teugels (1989) and the fact that

$$F \in \mathcal{L} \iff \overline{F} \circ \ln \in \mathcal{R}(0).$$

Lemma 3.4. (Representation of dfs in $\mathcal{L}$)
$F \in \mathcal{L}$ if and only if it has representation

$$\overline{F}(x) = c(x) \exp\left\{-\int_z^x q(t)\, dt\right\}, \quad x \geq z \geq 0,$$

where c and q are non–negative measurable functions such that $c(x) \to c \in (0, \infty)$ and $q(x) \to 0$, as $x \to \infty$, and $\int_z^\infty q(t)\, dt = \infty$. □

This implies in particular that each $F \in \mathcal{L}$ is tail–equivalent to an absolutely continuous df with hazard rate q which tends to 0 (for a definition see after Remark 3 below).

Since $\mathcal{S}$ is closed with respect to tail–equivalence (Teugels (1975)) it is of interest to find conditions on the hazard rate such that the corresponding df or/and integrated tail df is subexponential. In order to unify the problem of finding conditions for $F \in \mathcal{S}$ and $F_I \in \mathcal{S}$, the following class was introduced in Klüppelberg (1988).

Definition 3.5. (The class $\mathcal{S}^*$)
Let F be a df on $(0, \infty)$ such that $F(x) < 1$ for all $x > 0$. We say $F \in \mathcal{S}^*$ if F has finite mean μ and

$$\lim_{x\to\infty} \int_0^x \frac{\overline{F}(x-y)}{\overline{F}(x)} \overline{F}(y)\, dy = 2\mu\,. \qquad \square$$

The next result makes the class useful for applications.

Proposition 3.6. *If $F \in \mathcal{S}^*$, then $F \in \mathcal{S}$ and $F_I \in \mathcal{S}$.* $\square$

Name	*Tail $\overline{F}$ or density f*	*Parameters*
Pareto	$\overline{F}(x) = \left(\frac{\kappa}{\kappa + x}\right)^{\alpha}$	$\alpha, \kappa > 0$
Burr	$\overline{F}(x) = \left(\frac{\kappa}{\kappa + x^{\tau}}\right)^{\alpha}$	$\alpha, \kappa, \tau > 0$
Log–gamma	$f(x) = \frac{\alpha^{\beta}}{\Gamma(\beta)} (\ln x)^{\beta-1} x^{-\alpha-1}$	$\alpha > 1, \beta > 0$
Truncated α–stable	$\overline{F}(x) = P(\lvert X\rvert > x)$ where X is an α–stable rv	$0 < \alpha < 2$
Lognormal	$f(x) = \frac{1}{\sqrt{2\pi}\,\sigma x} e^{-(\ln x - \mu)^2/(2\sigma^2)}$	$\mu \in \mathbb{R}, \sigma > 0$
Benktander–type–I	$\overline{F}(x) = c(\alpha + 2\beta \ln x)$ $e^{-(\beta(\ln x)^2 + (\alpha+1)\ln x)}$	$c, \alpha, \beta > 0$
Benktander–type–II	$\overline{F}(x) = c\alpha x^{-(1-\beta)} e^{-\alpha x^{\beta}/\beta}$	$c, \alpha > 0$ $0 < \beta < 1$
Weibull	$\overline{F}(x) = e^{-x^{\tau}}$	$0 < \tau < 1$
"Almost" exponential	$\overline{F}(x) = e^{-x(\ln x)^{-\alpha}}$	$\alpha > 0$

Table 3.7. *Subexponential dfs. All of them are in $\mathcal{S}^*$ provided they have finite mean.*

Remarks 1) The class $\mathcal{S}^*$ is "almost" $\mathcal{S} \cap \{F : \mu(F) < \infty\}$, where $\mu(F)$ is the mean of F. A precise formulation can be found in Klüppelberg (1988).

2) The tails of dfs in $\mathcal{S}^*$ are subexponential densities (Klüppelberg (1989a), Willekens (1986)).

3) The class $\mathcal{S}^*$ is closed with respect to tail–equivalence. □

The task of finding easily verifiable conditions for $F \in \mathcal{S}$ or/and $F_I \in \mathcal{S}$ has now been reduced to the finding of simple conditions for $F \in \mathcal{S}^*$. We formulate some of them in terms of the *hazard function* $Q = -\ln \overline{F}$ and its density q, the *hazard rate* of F. (Recall that $\mathcal{S}^* \subset \mathcal{S} \subset \mathcal{L}$, hence by Lemma 3.4 each $F \in \mathcal{S}^*$ is tail–equivalent to an absolutely continuous df whose hazard rate tends to 0.)

Proposition 3.8. (Conditions for $F \in \mathcal{S}^*$)

(*a*) *If* $\limsup_{x\to\infty} xq(x) < \infty$, *then* $F \in \mathcal{S}^*$.

(*b*) *If there exist* $\delta \in (0, 1)$ *and* $v \geq 1$ *such that* $Q(xy) \leq y^\delta Q(x)$ *for all* $x \geq v$, $y \geq 1$ *and* $\liminf_{x\to\infty} xq(x) \geq (2 - 2^\delta)^{-1}$, *then* $F \in \mathcal{S}^*$.

(*c*) *If* q *is eventually decreasing to* 0, *then*

$$F \in \mathcal{S}^* \quad \Longleftrightarrow \quad \lim_{x\to\infty} \int_0^x e^{yq(x)} \overline{F}(y)\, dy = \mu\,.$$ □

Corollary 3.9. (More conditions for $F \in \mathcal{S}^*$)
Suppose

$$\lim_{x\to\infty} q(x) = 0 \quad \textit{and} \quad \lim_{x\to\infty} xq(x) = \infty\,.$$

If additionally one of the following conditions holds, then $F \in \mathcal{S}^*$.

(*a*) $\limsup_{x\to\infty} xq(x)/Q(x) < 1$.

(*b*) $q \in \mathcal{R}(-\delta)$ *for* $\delta \in (0, 1]$.

(*c*) $Q \in \mathcal{R}(\delta)$ *for* $\delta \in (0, 1)$ *and* q *is eventually decreasing.*

(*d*) $q \in \mathcal{R}(0)$, q *is eventually decreasing, and* $Q(x) - xq(x) \in \mathcal{R}(1)$. □

There are many more conditions for $F \in \mathcal{S}$ or $F_I \in \mathcal{S}$ to be found in the literature. We mention Chistyakov (1964), Cline (1986), Goldie (1978), Klüppelberg (1988), Pitman (1980), Teugels (1975); the selection above is taken from Klüppelberg (1988, 1989b).

4. Subexponentials and maxima

Definition 1.1(b) suggests subexponential dfs as appropriate models for extremal events. This immediately warrants an investigation of their relationship to classical extreme value theory. For an introduction to the latter we refer to Embrechts, Klüppelberg and Mikosch (1997), Chapter 3, or Resnick (1987).

Let $(X_n)_{n\in\mathbb{N}}$ be iid rvs with df $F \in \mathcal{S}$ and assume that there exist constants $a_n > 0$ and $b_n \in \mathbb{R}$ such that

$$a_n^{-1}\left(\max(X_1, \ldots, X_n) - b_n\right) \quad \stackrel{d}{\rightarrow} \quad G\,, \quad n \to \infty\,,$$

where G is some non–degenerate df. In this case we say F is in the maximum domain of attraction of G and write $F \in MDA(G)$. If $F \in \mathcal{S}$ its support is unbounded above, hence G is either the Fréchet df $\Phi_\alpha(x) = \exp\{-x^\alpha\}$ for $x \geq 0$, where $\alpha > 0$, or the Gumbel df $\Lambda(x) = \exp\{-e^{-x}\}$ for $x \in \mathbb{R}$. We write $F \in MDA(\Phi_\alpha)$ or $F \in MDA(\Lambda)$, respectively.

It is well known that $F \in MDA(\Phi_\alpha)$ if and only if $\overline{F} \in \mathcal{R}(-\alpha)$. Thus it remains to investigate $\mathcal{S} \cap MDA(\Lambda)$.

A good indicator for the extremal behaviour of a model is the *mean–excess function* (which exists for dfs with finite mean)

$$a(x) = E(X - x \mid X > x) = \int_x^\infty \overline{F}(y)\,dy/\overline{F}(x)\,, \quad x > 0\,.$$

From Karamata's theorem we know that F has finite mean when $\overline{F} \in \mathcal{R}(-\alpha)$ with $\alpha > 1$. Moreover, $\overline{F} \in \mathcal{R}(-(\alpha+1))$ for $\alpha > 0$ if and only if $a(x) \sim x/\alpha$. For the lognormal df we have $a(x) \sim \sigma^2 x/\ln x$, and for the Weibull df $a(x) \sim x^{1-\tau}/\tau$ in the parametrisation of Table 3.7. (Recall that $a(x)$ is constant for the exponential df and converges to 0 for the normal df.)

Necessary conditions and sufficient conditions for $F \in MDA(\Lambda) \cap \mathcal{S}$ have been derived by Goldie and Resnick (1988). The following condition applies to the examples in Table 3.7.

Lemma 4.1. *Let F be a df with finite mean and assume that $a(x)$ is eventually non–decreasing and there exists some $t > 1$ such that*

$$\liminf_{x\to\infty} \frac{a(tx)}{a(x)} > 1\,.$$

Then $F \in MDA(\Lambda) \cap \mathcal{S}$. □

Remarks 1) Pareto, Burr, log–gamma and stable dfs belong to $MDA(\Phi_\alpha)$ for some $\alpha > 0$, while lognormal, Benktander and Weibull dfs are in $MDA(\Lambda)$.

2) For $F \in MDA(\Lambda)$ with infinite right endpoint we have $\lim_{x\to\infty} a(x)/x = 0$. A generalisation of Karamata's theorem ensures that $\overline{F} \in \mathcal{R}(-\infty)$, i.e.

$$\lim_{x\to\infty} \frac{\overline{F}(tx)}{\overline{F}(x)} = \begin{cases} 0 & t > 1\,, \\ \infty & t < 1\,. \end{cases}$$

3) The fact that subexponential dfs may belong to $MDA(\Phi_\alpha)$ and $MDA(\Lambda)$ has consequences when studying extremal events in various models with subexponential input functions; see Theorems 5.4 and 6.2 for examples. □

5. Subexponentials and sums

From (1.1) we know that $\mathcal{S}$ is closed under the operation of taking sums of iid rvs. It is also closed under convolution roots; that is, the converse to (1.1) is true (Embrechts, Goldie and Veraverbeke (1979)). In this section we investigate further closure and other properties related to sums of subexponential rvs.

Convolution closure

A question naturally emerging from (1.1) is whether $\mathcal{S}$ is in general convolution closed, i.e. if $F, G \in \mathcal{S}$, does it always follow that $F * G \in \mathcal{S}$? The (negative) answer was given by Leslie (1989), who found two subexponential dfs whose convolution is not in $\mathcal{S}$. However, this must be a rather pathological example, as the following result covers most "reasonable" cases.

Theorem 5.1. (Convolution closure properties of $\mathcal{S}$)

(*a*) *Let* $F \in \mathcal{S}$ *and* $\overline{G}_i(x) \sim c_i\overline{F}(x)$, *where* $c_i \in (0,\infty)$ *for* $i = 1, 2$. *Then* $\overline{G_1 * G_2}(x) \sim (c_1 + c_2)\overline{F}(x)$.

(*b*) *Let* $F \in \mathcal{S}$ *and* $\overline{G}(x) \sim c\overline{F}(x)$ *for* $c \in [0,\infty)$. *Then* $\overline{F * G}(x) \sim (1 + c)\overline{F}(x)$.

(*c*) *Let* $F, G \in \mathcal{S}$. *Then* $F * G \in \mathcal{S}$ *if and only if* $pF + (1 - p)G \in \mathcal{S}$ *for some (all)* $p \in (0, 1)$. □

Remarks 1) It has been known for a long time (see Feller (1971)) that the subclass of dfs with regularly varying tails is convolution closed. Indeed, if $\overline{F}(x) = x^{-\alpha}\ell_1(x)$ and $\overline{G}(x) = x^{-\alpha}\ell_2(x)$, then $\overline{F * G}(x) \sim x^{-\alpha}(\ell_1(x) + \ell_2(x))$. Notice that the case of two different indices of regular variation is covered by Theorem 5.1(b) for $c = 0$.

2) For a proof of Theorem 5.1 we refer to Embrechts and Goldie (1982); see also Cline (1986). It is possible to develop a special algebra to handle convolution questions. After all everything happens in the convolution semigroup of measures on $(0,\infty)$ and subexponential dfs can be considered as idempotent elements in the factor–semigroup with respect to tail–equivalence. Cline (1987) and Klüppelberg (1990) follow such an approach. □

Random sums

Theorem 2.4, together with (2.2), can be viewed as a generalisation of (1.1) to random (geometric) sums. The following result is due to Embrechts, Goldie and Veraverbeke (1979), Embrechts and Goldie (1982), and Cline (1987).

Theorem 5.2. (Random sums of iid subexponential rvs)
Suppose (p_n) *defines a probability measure on* $\mathbb{N}_0$ *such that* $\sum_{n=0}^{\infty} p_n(1+\varepsilon)^n < \infty$ *for some*

$\varepsilon > 0$ and $p_k > 0$ for some $k \geq 2$. Let

$$G(x) = \sum_{n=0}^{\infty} p_n F^{n*}(x), \quad x > 0. \tag{5.1}$$

Then

$$F \in \mathcal{S} \iff \lim_{x\to\infty} \frac{\overline{G}(x)}{\overline{F}(x)} = \sum_{n=1}^{\infty} n p_n \iff G \in \mathcal{S} \text{ and } \overline{F}(x) \neq o(\overline{G}(x)). \qquad \square$$

Remarks 3) Let $(X_i)_{i\in\mathbb{N}}$ be iid with df F and let N be a rv taking values in $\mathbb{N}_0$ with distribution (p_n). Then G is the df of the random sum $\sum_{i=1}^{N} X_i$ (with the convention $\sum_{i=1}^{0} X_i = 0$) and the result of Theorem 5.2 translates into

$$P\left(\sum_{i=1}^{N} X_i > x\right) \sim ENP(X_1 > x), \quad x \to \infty.$$

If (p_n) is a Poisson or geometric distribution the condition $\overline{F}(x) \neq o(\overline{G}(x))$ in (c) is unnecessary (Cline (1987)). $\square$

A further generalisation of Theorem 5.2 is towards infinite divisibility. Let F be an infinitely divisible df on $(0, \infty)$. Then its moment generating function $\widehat{f}$ has the representation

$$\widehat{f}(s) = \exp\left\{as - \int_0^{\infty} (1 - e^{sx})\, d\nu(x)\right\}, \quad s \geq 0, \tag{5.2}$$

where $a \geq 0$ is a constant and ν is the Lévy measure of F. The following result was proved by Embrechts, Goldie and Veraverbeke (1979). It is based on the representation of F as $F = F_1 * F_2$, where $F_1(x) = o(e^{-\varepsilon x})$ for all $\varepsilon > 0$ and $F_2(x)$ is compound Poisson with the normalised Lévy measure as compounding df. Then $\overline{F}(x) \sim \overline{F}_2(x)$ by Theorem 5.1(b), and the closure of $\mathcal{S}$ with respect to tail–equivalence ensures $F \in \mathcal{S} \iff F_2 \in \mathcal{S}$. From Theorem 5.2 one obtains:

Corollary 5.3. (Infinitely divisible dfs and Lévy measures)

$$F \in \mathcal{S} \iff \nu(1, x]/\nu(1, \infty) \in \mathcal{S} \iff \overline{F}(x) \sim \nu(x, \infty). \qquad \square$$

Remarks 4) This result has been extended to infinitely divisible processes by Rosinski and Samorodnitsky (1993) who relate subadditive functionals of a sample path to a subexponential Lévy measure.

5) The asymptotic behaviour of high quantiles of an infinitely divisible process with regularly varying Lévy measure has been investigated by Embrechts and Samorodnitsky (1995). $\square$

Large Deviations

A further question immediately arises from Definition 1.1, namely what happens if n varies together with x. Hence *large deviations* theory is called for. Notice that the usual "rough" large deviations machinery based on logarithms cannot be applied. Classical results for $\overline{F} \in \mathcal{R}(-\alpha)$ state that

$$P(S_n - ES_n > x) \sim P(\max(X_1, \ldots, X_n) > x) \sim n\overline{F}(x), \quad n \to \infty, \tag{5.3}$$

which relation holds uniformly for $x > \gamma n$ for every fixed $\gamma > 0$. "Uniformly" here is in a ratio sense:

$$\sup_{x \in (\gamma n, \infty)} \left| \frac{P(S_n - ES_n > n)}{n\overline{F}(x)} - 1 \right| \to 0, \quad n \to \infty;$$

see Heyde (1967a, 1967b, 1968), A.V. Nagaev (1969a, 1969b) and Vinogradov (1994). Large deviations results for so–called semi–exponential tails $\overline{F}(x) = \exp\{-x^\alpha \ell(x)\}$, for $\alpha \in (0, 1)$ and $\ell \in \mathcal{R}(0)$, have been derived by S. V. Nagaev (1979); see also Rozovskii (1993). However, for such tails the x–regions, where (5.3) holds, do not in general include all the region $[\gamma n, \infty)$. A very general treatment of large deviation results for subexponentials is given in Pinelis (1985). For references and extensions of (5.3) towards random sums we refer to Klüppelberg and Mikosch (1997), where also certain applications to insurance and finance are treated. Generalisations to mixing sequences are to be found in Gantert (1996).

Weighted sums of subexponential random variables

Weighted sums are the first objects to study on the way to linear processes; they are the one–dimensional objects. The results given in Theorem 5.5 below were derived by Davis and Resnick (1985, 1988), and they have been used in combination with point–process techniques for studying the extremes of linear processes.

Assume that $(Z_j)_{j \in \mathbb{Z}}$ are iid with subexponential df F, and form the weighted sum

$$X = \sum_{j=-\infty}^{\infty} \psi_j Z_j. \tag{5.4}$$

The real sequence (ψ_j) is assumed to have properties such that X is well–defined as an almost–surely converging series. For this application it is natural to extend the notion of subexponentiality to dfs on the real line. Let F be a df on $\mathbb{R}$ and $F(x) < 1$ for all $x \in \mathbb{R}$. F is called a *subexponential df on* $\mathbb{R}$ if there exists a subexponential df G on $(0, \infty)$ such that $\overline{F}(x) \sim \overline{G}(x)$ as $x \to \infty$. In order to derive the tail behaviour of the df of the weighted sum X given in (5.4) we assume the tail balance condition

$$\overline{F}(x) \sim p\, P(|Z| > x), \quad F(-x) \sim qP(|Z| > x) \tag{5.5}$$

for $p \in (0, 1]$ and $q = 1 - p$.

Proposition 5.4. *Assume that Z is a rv with df F, subexponential on $\mathbb{R}$.*

(a) *Let $\overline{F}(x) \sim px^{-\alpha}\ell(x)$ and $F(-x) \sim qx^{-\alpha}\ell(x)$. Then*

$$P(\psi_j Z > x) = \begin{cases} P\big(Z > x/\psi_j\big) \sim \psi_j^\alpha p x^{-\alpha}\ell(x) & if \quad \psi_j > 0, \\ P\big(Z < -x/|\psi_j|\big) \sim |\psi_j|^\alpha q x^{-\alpha}\ell(x) & if \quad \psi_j < 0, \end{cases}$$
$$= |\psi_j|^\alpha x^{-\alpha}\ell(x)\left(pI_{\{\psi_j>0\}} + qI_{\{\psi_j<0\}}\right).$$

(b) *Let $\overline{F} \in \mathcal{R}(-\infty)$ and assume that (5.5) holds, then*

$$\frac{P(\psi_j Z > x)}{\overline{F}(x)} = \begin{cases} 1 & if \quad \psi_j = 1, \\ q/p & if \quad \psi_j = -1, \\ 0 & if \quad |\psi_j| < 1. \end{cases}$$ □

From Theorem 5.1(b) we derive for independent rvs $(X_i)_{|i|\le m}$ such that $P(X_i > x) \sim a_i\overline{F}(x)$, where $a_i \in [0, \infty)$, that

$$P\Big(\sum_{|i|\le m} X_i > x\Big) \sim \overline{F}(x)\sum_{|i|\le m} a_i\,.$$

This result can immediately be applied to the truncated sum

$$X^{(m)} = \sum_{|j|\le m} \psi_j Z_j\,,$$

where the Z_j are iid with subexponential df F on $\mathbb{R}$ satisfying the tail balance condition (5.5). If the sequence (ψ_j) tends sufficiently fast to 0, then the result for the truncated sum $X^{(m)}$ extends to the infinite sum (5.4).

Theorem 5.5. *Let $(Z_j)_{j\in\mathbb{Z}}$ be iid rvs with df F, subexponential on $\mathbb{R}$, and let X be the random sum given by (5.4).*

(a) *If $\overline{F} \in \mathcal{R}(-\alpha)$ for $\alpha \in (0,\infty)$, i.e. $P(|Z_1| > x) = x^{-\alpha}\ell(x)$, and $\sum_{j=-\infty}^{\infty} |\psi_j|^\delta < \infty$ for some $\delta \in (0, \min(\alpha, 1))$, then*

$$P(X > x) \sim x^{-\alpha}\ell(x)\sum_{j=-\infty}^{\infty} |\psi_j|^\alpha\left(pI_{\{\psi_j>0\}} + qI_{\{\psi_j<0\}}\right).$$

(b) *If $F \in MDA(\Lambda) \cap \mathcal{S}$ and $\sum_{j=-\infty}^{\infty} |\psi_j|^\delta < \infty$ for some $\delta \in (0, 1)$ and without loss of generality $\max_j |\psi_j| = 1$ (or else we normalise X), then*

$$P(X > x) \sim (pk^+ + qk^-)P(|Z_1| > x)\,,$$

where k^+ is the total number of times ψ_j takes the value 1 (there can only be finitely many), and k^- is the total number of times ψ_j takes the value -1. □

6. Rare events of a Lévy process with subexponential increments

Let (S_t) be a Lévy process in continuous time or a random walk in discrete time with increment S_1 having df B. Assume furthermore that (S_t) has negative drift, i.e. $-\beta = ES_1 < 0$. Then $M = \max_{t \geq 0} S_t < \infty$ a.s. and, if we define $\tau(u) = \inf\{t > 0 : S_t > u\}$, then $\{M > u\} = \{\tau(u) < \infty\}$. Furthermore, for u large, this event is rare, i.e.

$$\psi(u) = P(\tau(u) < \infty) = P(M > u)$$

is small. Typical large deviations problems are the asymptotic form of $\psi(u)$ as $u \to \infty$ (derived in Section 2), and properties of a sample path leading to an upcrossing of a high level u. Let

$$P^{(u)} = P(\cdot \mid \tau(u) < \infty),$$

then we are interested in the $P^{(u)}$–distribution of the path

$$S_{[0,\tau(u))} = (S_t)_{0 \leq t \leq \tau(u)}$$

leading to the occurrence of a rare event.

We are in particular interested in the following quantities:

$Y(u) = S_{\tau(u)}$	the level of the process after the upcrossing,
$Z(u) = S_{\tau(u)-}$	the level of the process just before the upcrossing,
$Y(u) - u$	the size of the overshoot,
$W(u) = Y(u) + Z(u)$	the size of the increment leading to the upcrossing.

For the sake of contrast we briefly report on the results under a Cramér condition (see Asmussen (1982) and references therein). Suppose there exists some $\gamma > 0$ such that the moment generating function $\widehat{b}(s) = Ee^{sS_1}$ has $\widehat{b}(\gamma) = 1$. Then the $P^{(u)}$–distribution of $S_{[0,\tau(u))}$ is in an appropriate sense the same as the unconditional distribution with respect to the Lévy process obtained by the exponential change of measure defined by

$$dB_\gamma(x) = e^{\gamma x} dB(x).$$

The Lévy process so defined (with increment df B_γ) has positive drift. In particular, if the cumulant generating function $\kappa(s) = \ln \widehat{b}(s)$ has $\kappa'(\gamma) < \infty$ then, as $u \to \infty$,

$$\frac{\tau(u)}{u} \xrightarrow{P^{(u)}} (\kappa'(\gamma))^{-1} \quad \text{and} \quad \frac{S_{t\tau(u)}}{t\tau(u)} \xrightarrow{P^{(u)}} \kappa'(\gamma).$$

We can summarise the situation under a Cramér condition as follows: the behaviour of the sample path of the Lévy process (or random walk) leading to an upcrossing is as if the increment distribution changed from B to B_γ. The main dramatic feature we see in the sample path is a change of drift causing the upcrossing. The intuitive picture is that rare events occur as a consequence of a build–up of claims over a period where the underlying parameters change by exponential change of measure. The sample path of the risk process leading to ruin exhibits a change of drift.

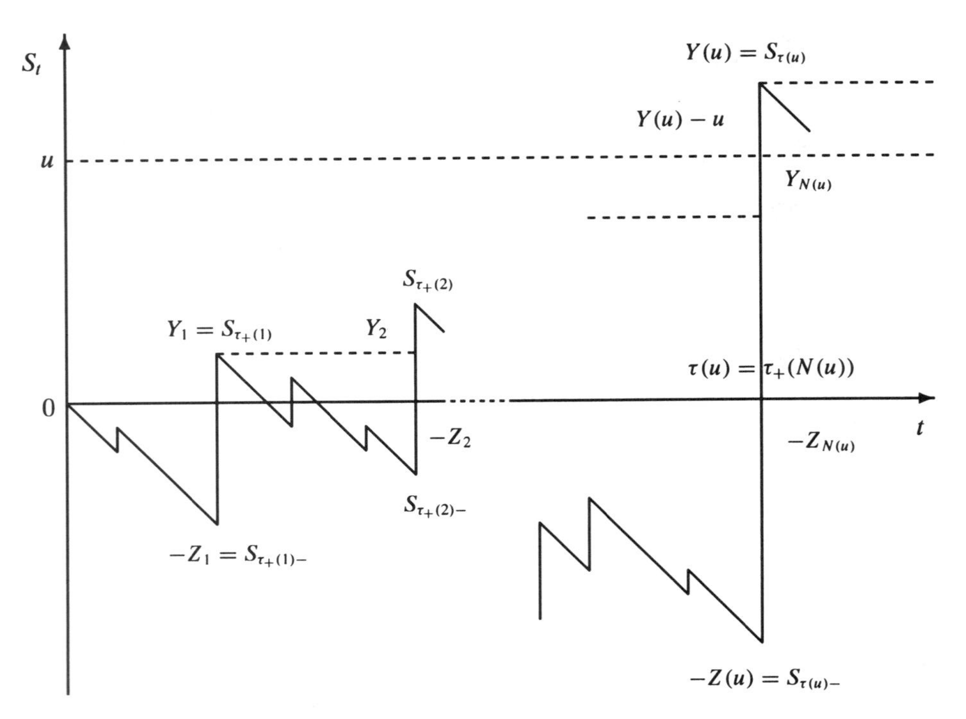

Figure 6.1. *Sample path of a classical ruin process leading to ruin.* $(Y_i)_{i\in\mathbb{N}}$ *are the ladder heights,* $(S_{\tau+(k)})_{k\in\mathbb{N}}$ *are the ladder epochs of* $(S_t)_{t\geq 0}$. *The rv* $N(u)$, *giving the number of ladder segments until ruin, is geometric.*

This picture changes radically for subexponential increment distributions (Asmussen and Klüppelberg (1996); see also Asmussen (1996)). Here an upcrossing happens as a result of one large increment whereas the process behaves in a typical way until the rare event happens. The following result describes the behaviour of the process before an upcrossing, and the upcrossing event itself.

Theorem 6.2. (Sample path leading to ruin)
Assume that the increment S_1 *has df* B *with finite mean* $-\mu < 0$. *Assume furthermore that* $\overline{B}(x) \sim \overline{F}(x)$ *as* $x \to \infty$ *for some* $F \in \mathcal{S}^* \cap MDA(G)$ *for some extreme value distribution* G. *Let* $a(u) = \int_u^\infty \overline{F}(y)\,dy/\overline{F}(u)$. *Then, as* $u \to \infty$,

$$\left(\frac{Z(u)}{a(u)}, \frac{\tau(u)}{a(u)}, \frac{Y(u)-u}{a(u)}, \left(\frac{S_{t\tau(u)}}{\tau(u)}\right)_{0\leq t<1}\right) \longrightarrow \left(V_\alpha, \frac{V_\alpha}{\mu}, T_\alpha, (-\mu t)_{0\leq t<1}\right)$$

in $P^{(u)}$*–distribution in* $\mathbb{R} \times \mathbb{R}_+ \times \mathbb{R}_+ \times \mathbb{D}[0,1)$, *where* $\mathbb{D}[0,1)$ *denotes the space of cadlag* *functions on* $[0, 1)$, *and* V_α *and* T_α *are positive rvs with df satisfying*

$$P(V_\alpha > x, T_\alpha > y) = \overline{G}_\alpha(x+y) = \begin{cases} \left(1 + \dfrac{x+y}{\alpha}\right)^{-\alpha} & if \quad \overline{F} \in \mathcal{R}(-\alpha-1), \\ e^{-(x+y)} & if \quad F \in MDA(\Lambda). \end{cases}$$

(Here α *is a positive parameter, the latter case when* $F \in MDA(\Lambda)$ *being considered as the case* $\alpha = \infty$.*)* □

Remarks 1) The normalising function $a(\cdot)$ is unique only up to asymptotic equivalence. Since

$$a(u) \sim \int_u^\infty \overline{B}(y)\,dy/\overline{B}(u), \quad u \to \infty,$$

the rhs here is also a possible normalising function.
2) Extreme value theory is the foundation of this result: recall first that $\overline{F} \in \mathcal{R}(-(\alpha+1))$ is equivalent to $F \in MDA(\Phi_{\alpha+1})$, and hence to $F_I \in MDA(\Phi_\alpha)$ by Karamata's theorem. Furthermore, $F \in MDA(\Lambda) \cap \mathcal{S}^*$ implies $F_I \in MDA(\Lambda) \cap \mathcal{S}$. Extreme value theory then provides the form of G_α as the only possible limit df for the excess distribution (Balkema and de Haan (1974)). G_α is called a *generalised Pareto distribution*. The normalising function $a(u)$ tends to infinity as $u \to \infty$. For $\overline{F} \in \mathcal{R}(-(\alpha+1))$ Karamata's theorem gives $a(u) \sim u/\alpha$. For $F \in MDA(\Lambda)$ this is Lemma 2.1 in Goldie and Resnick (1988).
3) The limit result for $(S_{t\tau(u)})$ given in Theorem 6.2 substantiates the assertion that the process (S_t) evolves typically up to time $\tau(u)$. □

Theorem 6.2 applies in particular to the models in Section 2. Indeed, stronger results (employing total variation distance) can be obtained for these examples since (S_t) is a downwards skip–free Markov process (with paths having only upwards jumps and deterministic downwards movements). We conclude with two special results which answered questions that were open for some time, but refer to Asmussen and Klüppelberg (1996, 1997) for details.

Example 6.3. (Finite time ruin probability)
Define the ruin probability before time T by

$$\Psi(u, T) = P(\tau(u) \le T).$$

From the limit result on $\tau(u)$ given in Theorem 6.2 one finds the following: if $\overline{F} \in \mathcal{R}(-(\alpha+1))$ for some $\alpha > 0$ then

$$\lim_{u\to\infty} \frac{\psi(u, uT)}{\psi(u)} = 1 - \big(1 + (1-\rho)T\big)^{-\alpha},$$

and if $F \in MDA(\Lambda) \cap \mathcal{S}^*$ then

$$\lim_{u\to\infty} \frac{\psi(u, a(u)T)}{\psi(u)} = 1 - e^{-(1-\rho)T}.$$ □

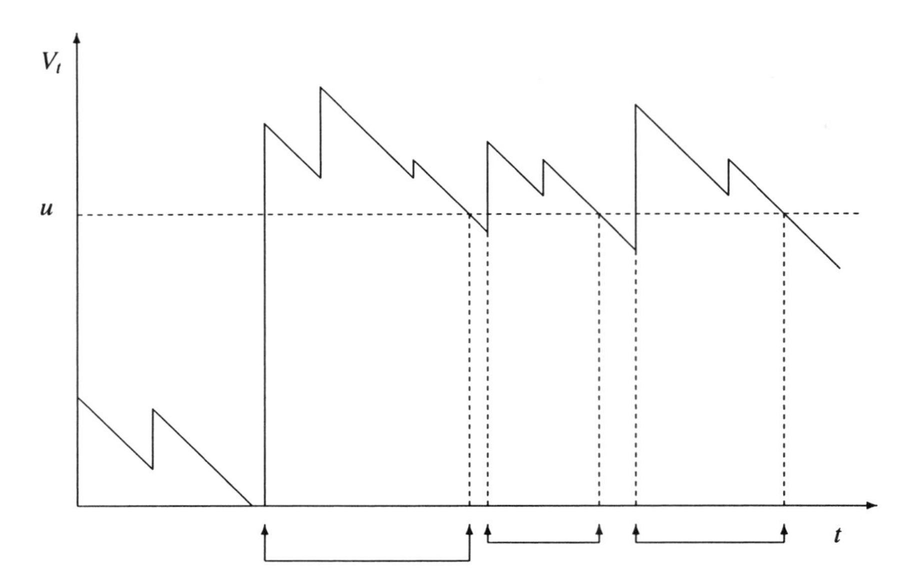

Figure 6.5. *Sample path of the workload process showing three high–level excursions.*

Example 6.4. (Excursions of the workload process of an M/G/1–queue)
Let $P^{(u)}$ denote the distribution of the doubly infinite version $(V_t)_{t\in\mathbb{R}}$ of the workload process, for which a stationary excursion above level u starts at time 0. Assume that $\rho = \lambda\mu < 1$, λ being the arrival rate and μ the mean service time, and let π denote the stationary distribution of (V_t) and F_I the stationary excess distribution. By the Markov property, the existence of a limit law for an excursion is equivalent to $P^{(u)}$–convergence of $V_0 - u$.

In the light–tailed case the excess $V_0 - u$ and hence the whole excursion has a limit as $u \to \infty$. However, if $F_I \in \mathcal{S}$, this limit is defective; more precisely,

$$\lim_{u\to\infty} P^{(u)}(V_0 - u \le y) = \rho F_I(y), \quad y > 0.$$

Furthermore, if $F \in \mathcal{L}$,

$$P^{(u)}(V_0 > u + y \mid V_{0-} = z) = \frac{\overline{F}(u + y - z)}{\overline{F}(u - z)} \to 1, \quad u \to \infty,$$

for all $y, z > 0$. So for $F^* \in \mathcal{S}$ (then $F \in \mathcal{L}$ and $F_I \in \mathcal{S}$) there are two types of excursions.
(1) With probability $1 - \rho$ the excursion starts from $V_{0-} = O(1)$ and the excess is huge. There is one indicated in Figure 6.5, the first one.
(2) With probability ρ the excursion starts from pre–level $u - V_{0-} = O(1)$ and the excess $V_0 - u$ has df F_I. There are two indicated in Figure 6.5, namely the last two.
This can be interpreted as that the process evolves in a typical way, with negative drift, until a very large service time causes an excursion. After the overshoot the drift takes over again, but there may be some smaller excursions on the way down which can be considered as aftershocks caused mainly by the preceding large service time. □

7. Concluding remarks

Recent interest in subexponential distributions concentrates mainly on relations between heavy tails, long range dependence and self–similarity.

If for instance the input stream of a GI/M/1 queue exhibits long range dependence, then the stationary queue size and the stationary waiting time distributions are each heavy–tailed; see Resnick and Samorodnitsky (1997). They describe a special model for a long range dependent arrival stream (the inter–arrival times are stationary with a special long range dependence structure) and derive bounds for the tails of the stationary queue size and the stationary waiting time distributions. Their results are by no means as explicit as the results presented in this paper, but they derive bounds for distribution tails. As stated by the authors, "one simply needs to better understand the behaviour of queues with long range dependent input".

An on/off model for packet transmission has been described in Willinger, Taqqu, Sherman and Wilson (1995). This model explains the slow rate of decay of the covariance function of the data, which is an indicator for long range dependence. An interesting review paper with updated references is Resnick (1996).

Vesilo and Daley (1997) consider long range dependence of point processes with queueing examples. They show for instance that certain regularly varying inter–arrival times or service times lead to a long–range dependent departure process in a queueing model.

On/off models with subexponential on–periods have been considered by Jelenkovič and Lazar (1996), using mainly regular variation arguments. There is also a paper by Heath, Resnick and Samorodnitsky (1996) on this topic.

Acknowledgments

The authors are most grateful to Thomas Mikosch for a careful reading and for suggesting many improvements and clarifications. Any remaining obscurities and infelicities are the responsibility of the authors.

CMG thanks the Department of Mathematics of the Johannes Gutenberg University, Mainz, for hospitality and support.

CK thanks the Center for Applied Probability at Columbia University for the invitation to speak at the Applied Probability Day in April 1996. This review paper is partly based on her seminar at this occasion.

References

[1] Asmussen, S. (1982) Conditioned limit theorems relating a random walk to its associate, with applications to risk reserve processes and the GI/G/1 queue. *Adv. Appl. Prob.* **14**, 143–170.

[2] Asmussen, S. (1996) Rare events in the presence of heavy tails. Preprint. Lund

University.

[3] Asmussen, S. (1996) Subexponential asymptotics for stochastic processes: extremal behaviour, stationary distributions and first passage probabilities. Preprint, University of Lund.

[4] Asmussen, S. (1997) *Ruin Probabilities.* World Scientific, Singapore. To appear.

[5] Asmussen, S., Fløe Henriksen, L. and Klüppelberg, C. (1994) Large claims approximations for risk processes in a Markovian environment. *Stoch. Proc. Appl.* **54**, 29–43.

[6] Asmussen, S. and Højgaard, B. (1995) Ruin probability approximations for Markov–modulated risk processes with heavy tails. Preprint, University of Aalborg.

[7] Asmussen, S. and Klüppelberg, C. (1996) Large deviation results for subexponential tails, with applications to insurance risk. *Stoch. Proc. Appl.* **64**, 103–125.

[8] Asmussen, S. and Klüppelberg, C. (1997) Stationary M/G/1 excursions in the presence of heavy tails. *J. Appl. Probab.* **34**, 208–212.

[9] Athreya, K.B. and Ney, P.E. (1972) *Branching Processes.* Springer, Berlin.

[10] Balkema, A.A. and de Haan, L. (1972) Residual life–time at great age. *Ann. Probab.* **96**, 792–804.

[11] Bertoin, J. and Doney, R.A. (1996) Some asymptotic results for transient random walks. *Adv. Appl. Probab.* **28**, 207–226.

[12] Bingham, N.H., Goldie, C.M. and Teugels, J.L. (1989) *Regular Variation*, revised paperback ed. Cambridge University Press, Cambridge.

[13] Chistyakov, V.P (1964) A theorem on sums of independent positive random variables and its applications to branching random processes. *Theory Probab. Appl.* **9**, 640–648.

[14] Chover, J., Ney, P. and Wainger, S. (1973) Functions of probability measures. *J. Analyse Math.* **26**, 255–302.

[15] Cline, D.B.H. (1986) Convolution tails, product tails and domains of attraction. *Probab. Theory Related Fields* **72**, 529–557.

[16] Cline, D.B.H. (1987) Convolutions of distributions with exponential and subexponential tails. *J. Austral. Math. Soc. Ser. A* **43**, 347–365.

[17] Davis, R.A. and Resnick, S.I. (1985) Limit theory for moving averages of random variables with regularly varying tail probabilities. *Ann. Probab.* **13**, 179–195.

[18] Davis, R.A. and Resnick, S.I. (1988) Extremes of moving averages of random variables from the domain of attraction of the double exponential distribution. *Stoch. Proc. Appl.* **30**, 41–68.

[19] Embrechts, P. and Goldie, C.M. (1980) On closure and factorization theorems for subexponential and related distributions. *J. Austral. Math. Soc. Ser. A* **29**, 243–256.

[20] Embrechts, P. and Goldie, C.M. (1982) On convolution tails. *Stoch. Proc. Appl.* **13**, 263–278.

[21] Embrechts, P., Goldie, C.M. and Veraverbeke, N. (1979) Subexponentiality and infinite divisibility. *Z. Wahrscheinlichkeitstheorie verw. Geb.* **49**, 335–347.

[22] Embrechts, P., Klüppelberg, C. and Mikosch, T. (1997) *Modelling Extremal Events for Insurance and Finance.* Springer, Heidelberg.

[23] Embrechts, P. and Samorodnitsky, G. (1995) Sample quantiles of heavy tailed stochastic processes. *Stoch. Proc. Appl.* **59**, 217–233.

[24] Embrechts, P. and Veraverbeke, N. (1982) Estimates for the probability of ruin with special emphasis on the possibility of large claims. *Insurance: Math. Econom.* **1**, 55–72.

[25] Feller, W. (1971) *An Introduction to Probability Theory and Its Applications II.* Wiley, New York.

[26] Gantert, N. (1996) Large deviations for a heavy–tailed mixing sequence. Preprint, Technical University Berlin.

[27] Goldie, C.M. (1978) Subexponential distributions and dominated–variation tails. *J. Appl. Probab.* **15**, 440–442.

[28] Goldie, C.M. and Resnick, S.I. (1988) Distributions that are both subexponential and in the domain of attraction of an extreme–value distribution. *Adv. Appl. Probab.* **20**, 706–718.

[29] Grandell, J. (1991) *Aspects of Risk Theory.* Springer, Berlin.

[30] Heath, D., Resnick, S., and Samorodnitsky, G. (1996) Heavy tails and long range dependence in on/off processes and associated fluid models. Preprint, Cornell University.

[31] Heyde, C.C. (1967a) A contribution to the theory of large deviations for sums of independent random variables. *Z. Wahrscheinlichkeitstheorie verw. Geb.* **7**, 303–308.

[32] Heyde, C.C. (1967b) On large deviation problems for sums of random variables not attracted to the normal law. *Ann. Math. Statist.* **38**, 1575–1578.

[33] Heyde, C.C. (1968) On large deviation probabilities in the case of attraction to a nonnormal stable law *Sankhya Ser. A* **30**, 253–258.

[34] Jelenkovič, P.R. and Lazar, A.A. (1996) Subexponential asymptotics of a Markov–modulated G/G/1 queue. Preprint, Columbia University, New York.

[35] Jelenkovič, P.R. and Lazar, A.A. (1996) Multiplexing on–off sources with subexponential on periods. CTR Technical Report 457-96-23, Columbia University, New York.

[36] Klüppelberg, C. (1988) Subexponential distributions and integrated tails. *J. Appl. Probab.* **25**, 132–141.

[37] Klüppelberg, C. (1989a) Subexponential distributions and characterisations of related classes. *Probab. Theory Related Fields* **82**, 259–269.

[38] Klüppelberg, C. (1989b) Estimation of ruin probabilities by means of hazard rates. *Insurance: Math. Econom.* **8**, 279–285.

[39] Klüppelberg, C. (1990) Asymptotic ordering of distribution functions on convolution semigroup. *Semigroup Forum* **40**, 77–92.

[40] Klüppelberg, C. and Mikosch, T. (1997) Large deviations of heavy–tailed random sums with applications to insurance and finance. *J. Appl. Probab.* **34**. To appear.

[41] Klüppelberg, C. and Stadtmüller, U. (1996) Ruin probabilities in the presence of heavy–tails and interest rates. *Scand. Actuar. J.* To appear.

[42] Leslie, J.R. (1989) On the non–closure under convolution of the subexponential family. *J. Appl. Probab.* **26**, 58–66.

[43] Mikosch, T. and Nagaev, A.V. (1996) Large deviations for heavy–tailed distributions with applications to insurance. Preprint, University of Groningen.

[44] Nagaev, A.V. (1969a) Limit theorems for large deviations where Cramér's conditions are violated (in Russian). *Izv. Akad. Nauk UzSSR Ser. Fiz.–Mat. Nauk* **6**, 17–22.

[45] Nagaev, A.V. (1969b) Integral limit theorems for large deviations when Cramér's condition is not fulfilled I,II. *Theory Probab. Appl.* **14**, 51–64 and 193–208.

[46] Nagaev, S.V. (1979) Large deviations of sums of independent random variables. *Ann. Probab.* **7**, 745–789.

[47] Pakes, A.G. (1975) On the tails of waiting time distributions. *J. Appl. Probab.* **12**, 555–564.

[48] Pinelis, I.F. (1985) On the asymptotic equivalence of probabilities of large deviations for sums and maxima of independent random variables (in Russian). In: *Limit Theorems in Probability Theory*, pp. 144–173. Trudy Inst. Math. **5**. Nauka, Novosibirsk.

[49] Pitman, E.J.G. (1980) Subexponential distribution functions. *J. Austral. Math. Soc. Ser. A* **29**, 337–347.

[50] Resnick, S.I. (1987) *Extreme Values, Regular Variation, and Point Processes.* Springer, New York.

[51] Resnick, S.I. (1996) Heavy tail modelling and teletraffic data. Preprint. Cornell University.

[52] Resnick, S.I. and Samorodnitsky, G. (1997) Performance decay in a single server exponential queueing model with long range dependence. Operations Research, **45**, 235–243.

[53] Rosinski, J. and Samorodnitsky, G. (1993) Distributions of subadditive functionals of sample paths of infinitely divisible processes. *Ann. Probab.* **21**, 996–1014.

[54] Rozovskii, L.V. (1993) Probabilities of large deviations on the whole axis. *Theory Probab. Appl.* **38**, 53–79.

[55] Smith, W.L. (1972) On the tails of queueing time distributions. Mimeo Series No. 830, Dept. of Statistics, University of North Carolina, Chapel Hill.

[56] Teugels, J.L. (1975) The class of subexponential distributions. *Ann. Probab.* **3**, 1001–1011.

[57] Vesilo, R. and Daley, D.J. (1997) Long range dependence of point processes, with queueing examples. *Stoch. Proc. Appl.* **70**, 265–282.

[58] Veraverbeke, N. (1977) Asymptotic behaviour of Wiener–Hopf factors of a random walk. *Stoch. Proc. Appl.* **5**, 27–37.

[59] Vinogradov, V. (1994) *Refined Large Deviation Limit Theorems.* Longman, Harlow (England).

[60] Willekens, E. (1986) *Hogere Orde Theorie voor Subexponentiële Verdelingen.* Ph.D. Thesis, Katholieke Universiteit Leuven.

[61] Willinger, W., Taqqu, M., Sherman, R. and Wilson, D. (1995) Self–similarity in high–speed packet traffic: analysis and modelling of ethernet traffic measurements. *Statistical Science* **10**, 67–85.

Charles M. Goldie
School of Mathematical Sciences
University of Sussex
Brighton BN1 9QH, England
email: C.M.Goldie@sussex.ac.uk

Claudia Klüppelberg
Center for Mathematical Sciences
Munich University of Technology
D–80290 Munich, Germany
email: cklu@mathematik.tu-muenchen.de

Structure of Stationary Stable Processes

Jan Rosiński

ABSTRACT. This is a survey article on the structure of stationary symmetric α–stable processes, $0 < \alpha < 2$. A decomposition of such processes into three mutually independent parts, a harmonizable process, a mixed moving average, and the third process, is discussed. A comparison with the Gaussian case is made and some fundamental structural differences are displayed.

1. Introduction and Preliminaries

This article is a survey of some recent results on stationary stable processes. We examine this special case of non Gaussian strictly stationary processes because it is now better understood and exemplifies fundamental differences that occur when one departs from the Gaussian case. Various concrete examples of stationary stable processes have been studied in the literature (see [7], [16] and the references therein) but our aim here is to look at a more general picture of the structure of this class. To make the comparison with the Gaussian case explicit, we first recall some basic well–known facts on stationary Gaussian processes. We begin with the *spectral representation* which says that every mean zero measurable stationary Gaussian process $\{X_t\}_{t\in\mathbf{R}}$ is of the form

$$X_t = \int_{\mathbf{R}} e^{its}\, M(ds), \qquad t \in \mathbf{R}, \tag{1.1}$$

where M is Gaussian white noise with finite control measure $\mu(A) = E|M(A)|^2$ (see, e.g., [2]). The spectral representation constitutes a basis for the analysis of stationary Gaussian processes and is crucial in prediction, modeling, and the statistical inference of such processes. The first step in analysis of (1.1) is to decompose the control measure μ, which is also the spectral measure of the covariance function $R(t) = EX_t\overline{X_0}$, into three mutually orthogonal parts

$$\mu = \mu_1 + \mu_2 + \mu_3,$$

1991 *Mathematics Subject Classification.* Primary: 60G10; secondary: 60G07, 60E07, 60G57..

Key words and phrases. Stable processes, stationarity, spectral representation, mixed moving average, harmonizable processes, nonsingular flow, cocycle, ergodic properties.

Research supported in part by NSF Grant DMS-9406294 and the Tennessee Science Alliance.

where μ_1 is the discrete component of μ, μ_2 is the absolutely continuous component, and μ_3 is the continuous singular component. This decomposition induces the corresponding decomposition of the process into three mutually independent stationary Gaussian processes

$$X_t = X_t^{(1)} + X_t^{(2)} + X_t^{(3)}. \tag{1.2}$$

Here $\{X_t^{(1)}\}_{t\in\mathbf{R}}$ is a series of simple harmonics with independent Gaussian amplitudes; its covariance function is uniformly almost periodic. $\{X_t^{(2)}\}_{t\in\mathbf{R}}$ is a *moving average* process, i.e., $\{X_t^{(2)}\}_{t\in\mathbf{R}}$ admits a representation (possibly on an enlarged probability space) of the form

$$X_t^{(2)} = \int_{\mathbf{R}} f(s-t)\, N(ds), \qquad t \in \mathbf{R}, \tag{1.3}$$

where N is Gaussian white noise with control Lebesgue measure λ (informally, $EN^2(ds) = ds$) and $f \in L^2(\mathbf{R}, \lambda)$ (see, e.g., [2], XI.8). The term $X_t^{(3)}$ of (1.2) is of a pathological nature and thus neglected in modeling and theoretical studies. If the spectral measure μ is absolutely continuous (that is, if the spectral density $\mu'(s) = \mu(ds)/ds$ exists), then (1.2) reduces to the middle term, so that $\{X_t\}_{t\in\mathbf{R}}$ has both representations (1.1) and (1.3). This is an important fact allowing alternative views of the process, either as a superposition of harmonics with random amplitude (1.1) or as a filtered noise (1.3).

2. Rigidity of Integral Representations of Stable Processes

We will now consider non-Gaussian Lévy–stable processes. Specifically, we will investigate symmetric α–stable (SαS) processes, $0 < \alpha < 2$, which are defined as follows. A real (complex, resp.) stochastic process $\{X_t\}_{t\in\mathbf{T}}$ is said to be a SαS process if any finite linear combination $\sum a_j X_{t_j}$ has a SαS (isotropic α–stable, resp.) distribution ($\mathbf{T}$ is an arbitrary parameter set). We should mention, at the outset, that not every stationary SαS process is of the form (1.1). In fact, we will show in Section 4 that only a small subclass of such processes admits the representation (1.1). However, SαS processes have (stochastic) integral representations which we will now discuss.

Let $(S, \mathcal{B})$ be a Borel space equipped with a σ–finite measure μ. A countably additive set function $M : \mathcal{B}_0 \mapsto L^0(\Omega, P)$ is said to be a (real) SαS random measure if M is independently scattered and

$$E\exp\{iuM(A)\} = \exp\{-|u|^\alpha \mu(A)\}, \quad u \in \mathbf{R}, \quad A \in \mathcal{B}_0,$$

where $\mathcal{B}_0 = \{A \in \mathcal{B} : \mu(A) < \infty\}$. A complex S$\alpha$S random measure is defined analogously with the symmetry assumption replaced by the isotropy (see [16]). Every (separable in probability) SαS process $\{X_t\}_{t\in\mathbf{T}}$ admits an integral representation

$$\{X_t\}_{t\in\mathbf{T}} \stackrel{d}{=} \left\{\int_S f_t(s)\, M(ds)\right\}_{t\in\mathbf{T}}, \tag{2.1}$$

where M is a SαS random measure on $S = [0, 1]$ with control Lebesgue measure λ, and $\{f_t\}_{t\in\mathbf{T}} \subset L^\alpha(S, \lambda)$ is a family of deterministic functions (see, e.g., [16]). Here "$=^d$" reads "equal in distribution". This representation can be strengthen to the almost sure representation

$$X_t = \int_S f_t(s)\, M_1(ds) \quad a.s., \qquad t \in \mathbf{T}, \tag{2.2}$$

with the same set of functions $\{f_t\}_{t\in\mathbf{T}}$ as in (2.1) but on an enlarged probability space (see [10]). The random measure M_1 has the same distribution (and so the control measure) as M. Finally, we mention that if $\mathbf{T}$ is a separable metric space and $\{X_t\}_{t\in\mathbf{T}}$ is a measurable process, then $\{f_t\}_{t\in\mathbf{T}}$ can be chosen such that the map $\mathbf{T}\times S \ni (t, s) \to f_t(s) \in \mathbf{R}$ (or $\mathbf{C}$) is jointly measurable (see [15]).

The next theorem shows that the representation (2.1) is rigid: the process $\{X_t\}_{t\in\mathbf{T}}$ determines the set of functions $\{f_t\}_{t\in\mathbf{T}}$ up to a change of variable and a multiplier.

Theorem 2.1. ([11], *Th. 3.1)* *Let* $\mathbf{T}$ *be a separable metric space and* $\alpha \in (0, 2)$. *Suppose that*

$$\left\{\int_{S_1} f_t^{(1)}\, dM_1\right\}_{t\in\mathbf{T}} \stackrel{d}{=} \left\{\int_{S_2} f_t^{(2)}\, dM_2\right\}_{t\in\mathbf{T}},$$

where M_i *is a S*α*S random measure on a Borel space* $(S_i, \mathcal{B}_{S_i})$ *with a* σ*–finite control measure* μ_i, *and* $\{f_t^{(i)}\}_{t\in\mathbf{T}} \subset L^\alpha(S_i, \mu_i)$ *is such that the map* $\mathbf{T}\times S \ni (t, s) \to f_t^{(i)}(s) \in \mathbf{R}$ *(or* $\mathbf{C}$*) is Borel measurable* $(i = 1, 2)$. *Then, for every* σ*–finite Borel measure* ν *on* $\mathbf{T}$, *there exist Borel functions* $\phi : S_2 \to S_1$ *and* $h : S_2 \to \mathbf{R}$ *(*$\mathbf{C}$, *resp.) such that*

$$f_t^{(2)}(s) = h(s) f_t^{(1)}(\phi(s)) \qquad \nu \otimes \mu_2 - a.e. \tag{2.3}$$

Note that in the Gaussian case, the integral representation (2.1) determines only the covariance structure of $\{f_t\}_{t\in\mathbf{T}}$, $\int_S f_t \overline{f_{t'}}\, d\mu = EX_t\overline{X_{t'}}$, which gives a large flexibility for possible choices of $\{f_t\}_{t\in\mathbf{T}}$ in (2.1). For instance, a Brownian motion can be represented via a Gaussian white noise M as

$$X_t^{(1)} = \int_{-\infty}^{\infty} (\ln|s - t| - \ln|s|)\, M(ds)$$

and as $X_t^{(2)} = \int_0^\infty 1_{[0,t]}(s)\, M(ds)$, $t \geq 0$. Theorem 2.1 shows that in the SαS case, $\{f_t\}_{t\in\mathbf{T}}$ carries a lot more information about the process than it does in the Gaussian case.

Now we return to stationary processes. A SαS process $\{X_t\}_{t\in\mathbf{R}}$ is said to be *harmonizable* if there exists a complex SαS random measure M on $\mathbf{R}$ with finite control measure μ such that

$$\{X_t\}_{t\in\mathbf{R}} \stackrel{d}{=} \left\{\int_{\mathbf{R}} e^{its}\, M(ds)\right\}_{t\in\mathbf{R}} \tag{2.4}$$

$\{X_t\}_{t\in\mathbf{R}}$ is called a SαS *moving average* process if there exists a function $f \in L^\alpha(\mathbf{R}, \lambda)$ such that

$$\{X_t\}_{t\in\mathbf{R}} \stackrel{d}{=} \left\{\int_{\mathbf{R}} f(s-t)\, M(ds)\right\}_{t\in\mathbf{R}}, \tag{2.5}$$

where M is a SαS random measure on $\mathbf{R}$ with Lebesgue control measure λ. Similarly to (2.1)-(2.2), equalities (2.4) and (2.5) have a.s. versions on an enlarged probability space. Theorem 2.1 gives simple explanation to the following two well–known but surprising facts concerning SαS processes. The first one was proven by Urbanik [18] for non–Gaussian infinitely divisible sequences and later has been extended in several directions (see, e.g., [19], [9], [1], [6]). It says that there can not be any nonzero SαS processes which is both harmonizable and a moving average.

Corollary 2.2. (*Disjointness of harmonizable and moving average processes.*)

Suppose that there is a SαS process which possesses both, harmonizable (2.3) and moving average (2.4) representations. Using Theorem 2.1 for $f_t^{(1)}(s) = e^{its}$, $f_t^{(2)}(s) = f(s-t)$, $t, s \in \mathbf{R}\ (f \in L^\alpha(\mathbf{R}, \lambda))$, we get from (2.3),

$$f(s-t) = h(s)e^{it\phi(s)} \qquad \lambda \otimes \lambda - a.e.$$

Integrating the α-th power of the moduli of both sides of this equation with respect to t over $\mathbf{R}$, we get the left–hand side finite and the right–hand side infinite, unless $h(s) = 0$. Hence $f = 0$, which says that the only SαS process which has both harmonizable and moving average representations is a zero process. $\square$

The second surprising fact, which does not hold in the Gaussian case, has been discovered by Kanter [8]. He proved that the output of a linear filter of a SαS noise identifies the impulse–response function of the filter up to a time shift and the global sign.

Corollary 2.3. (*Identification of impulse-response functions.*)

Suppose that a SαS process possesses two moving average representations (in distribution) with the impulse–response functions f and g. Apply Theorem 2.1 with $f_t^{(1)}(s) = f(s-t)$ and $f_t^{(2)}(s) = g(s-t)$. We get

$$g(s-t) = h(s) f(\phi(s) - t) \qquad \lambda \otimes \lambda - a.e.$$

Fix $s = s_0$ such that this equation holds for λ–almost all t, and put $u = s_0 - t$. Then

$$g(u) = h(s_0) f(u - u_0) \qquad \text{for } \lambda\text{–almost all } u,$$

where $u_0 = s_0 - \phi(s_0)$. Since f and g represent the same process,

$$\begin{aligned}\int_{\mathbf{R}} |g(u)|^\alpha \, du &= |h(s_0)|^\alpha \int_{\mathbf{R}} |f(u-u_0)|^\alpha \, du \\ &= |h(s_0)|^\alpha \int_{\mathbf{R}} |g(u)|^\alpha \, du.\end{aligned}$$

Hence $|h(s_0)| = 1$, i.e., $g(u) = \epsilon f(u - u_0)$, with $|\epsilon| = 1$, which is the conclusion of Kanter's theorem ([8]). $\square$

3. Mixed Moving Averages

Since the classes of harmonizable and moving average processes are disjoint (Corollary 2.2), one may ask how large are these classes as subclasses of all stationary SαS processes. This question will be studied in Section 4. Here we begin with a simpler related question. Namely, it is clear that the class of harmonizable processes is closed under superpositions of independent processes. We ask whether the same is true for SαS moving averages. In the Gaussian case, a superposition of independent moving averages is always a moving average. Indeed, a stationary Gaussian process is a moving average if and only if its spectral measure is absolutely continuous and the sum of absolutely continuous measures is absolutely continuous. The next proposition shows that the situation is very different in the α-stable case.

Proposition 3.1. *Let* $\{X_t^{(i)}\}_{t\in\mathbf{R}}$ *be two independent (non–zero) SαS moving average processes,* $i = 1, 2$, $\alpha \in (0, 2)$. *Then*

$$X_t = X_t^{(1)} + X_t^{(2)}, \qquad t \in \mathbf{R}$$

is a SαS moving average process if and only if there exists a constant k such that

$$\{X_t^{(2)}\}_{t\in\mathbf{R}} \stackrel{d}{=} \{kX_t^{(1)}\}_{t\in\mathbf{R}}$$

Proof. Suppose that f, f_1, and f_2 represent moving averages X, $X^{(1)}$, and $X^{(2)}$, respectively. Consider $S = \{1, 2\} \times \mathbf{R}$ equipped with the product measure $\mu := (\delta_1 + \delta_2) \otimes \lambda$ and let M be a SαS random measure on S with control measure λ. Define a function g on S by

$$g(i, x) = f_i(x), \qquad (i, x) \in S.$$

One can see immediately, by checking the characteristic functions, that

$$\{X_t\}_{t\in\mathbf{R}} \stackrel{d}{=} \left\{ \int_S g(i, s-t)\, M(di, ds) \right\}_{t\in\mathbf{R}}.$$

In view of Theorem 2.1, there exist $\phi : S \mapsto \mathbf{R}$ and $h : S \mapsto \mathbf{R}(\mathbf{C})$ such that

$$g(i, s-t) = h(i, s) f(\phi(i, s) - t) \qquad \lambda \otimes \mu - a.e.$$

This yields

$$f_i(s-t) = h(i, s) f(\phi(i, s) - t) \qquad \lambda \otimes \lambda - a.e.,$$

$i = 1, 2$. Now, as in the proof of Corollary 2.3, one can find $s_i, u_i \in \mathbf{R}$ such that

$$f_i(u) = h(i, s_i) f(u - u_i) \qquad \lambda - a.e.,$$

$i = 1, 2$. Hence

$$f_2(u) = k f_1(u + u_1 - u_2) \qquad \lambda - a.e.$$

where $k = h(2, s_2)/h(1, s_1)$. This proves the proposition. □

It is now natural to consider the smallest class containing SαS moving averages and closed under superpositions and weak limits. Such a class, called the class of *mixed moving averages*, was introduced and studied in [17].

Definition 3.2. *A SαS process is said to be a mixed moving average if there is a Borel space W equipped with a σ–finite measure ν and a function* $g \in L^\alpha(W \times \mathbf{R}, \nu \otimes \lambda)$ *such that*

$$\{X_t\}_{t\in\mathbf{R}} \stackrel{d}{=} \left\{ \int_{W\times\mathbf{R}} g(w, s-t)\, N(dw, ds) \right\}_{t\in\mathbf{R}} \tag{3.1}$$

where N is a $S\alpha S$ random measure on $W \times \mathbf{R}$ with control measure $\nu \otimes \lambda$.

$S\alpha S$ mixed moving averages are still disjoint with harmonizable processes and it is possible to obtain an extension of Kanter's theorem (Corollary 2.3) for mixed moving averages (see [17] and [11]). One of interesting applications of such processes is to construct a reversible reflection positive non Gaussian $S\alpha S$ process (see [17]).

4. Representation and Decomposition of Stationary Stable Processes

Let $\{X_t\}_{t\in\mathbf{R}}$ be a stationary measurable $S\alpha S$ process, $\alpha \in (0, 2)$. As any separable in probability $S\alpha S$ process, it has an integral representation (2.1). The stationarity gives

$$\left\{\int_S f_{u+t}\, dM\right\}_{u\in\mathbf{R}} \stackrel{d}{=} \left\{\int_S f_u\, dM\right\}_{u\in\mathbf{R}},$$

for every $t \in \mathbf{R}$. In view of by Theorem 2.1, for each fixed t,

$$f_{u+t}(s) = h_t(s) f_u(\phi_t(s)) \qquad \lambda \otimes \mu - a.e.$$

Suppose that this equality holds for all s, t, u. Putting $u = 0$ we get

$$f_t(s) = h_t(s) f_0(\phi_t(s)), \qquad t \in \mathbf{R},\ s \in S. \tag{4.1}$$

These are the intuitions leading to the representation given in Theorem 4.1. Before stating it, we will need a few definitions. A *measurable flow* on a Borel space S is a jointly measurable map $\mathbf{R} \times S \ni (t, s) \mapsto \phi_t(s) \in S$ such that $\phi_{t_1+t_2}(s) = \phi_{t_1}(\phi_{t_2}(s))$ and $\phi_0(s) = s$, for every $t_1, t_2 \in \mathbf{R}$ and $s \in S$. A flow is said to be *nonsingular* if $\mu \circ \phi_t^{-1}$ is absolutely continuous with respect to μ, for every t. Let A be a locally compact second countable group. A measurable map $\mathbf{R} \times S \ni (t, s) \mapsto a_t(s) \in A$ is said to be a *cocycle* for the flow $\{\phi_t\}_{t\in\mathbf{R}}$ if

$$a_{t_1+t_2}(s) = a_{t_1}(s) a_{t_2}(\phi_{t_1}(s)) \qquad \text{for all } t_1, t_2 \in \mathbf{R},\ s \in S. \tag{4.2}$$

This equation resembles the chain rule of differentiation; in fact, the Radon-Nikodym derivative $m_t(s) = \frac{d(\mu\circ\phi_t)}{d\mu}(s)$ is a cocycle (taking values in $A = (0, \infty)$) for a nonsingular flow $\{\phi_t\}_{t\in\mathbf{R}}$.

Theorem 4.1. *(Representation of stationary $S\alpha S$ processes;* [12], [13]*).*
Let $\{X_t\}_{t\in\mathbf{R}}$ be as above. Then there exist a nonsingular measurable flow $\{\phi_t\}_{t\in\mathbf{R}}$ on some Borel σ-finite measure space (S, μ), a cocycle $\{a_t\}_{t\in\mathbf{R}}$ for this flow, with values

in $\{-1, 1\}$ *or in* $\{|z| = 1 : z \in \mathbf{C}\}$, *depending whether the process is real or complex valued, and a function* $f \in L^{\alpha}(S, \mu)$ *such that (on a possibly enlarged probability space)*

$$X_t = \int_S a_t \cdot m_t^{1/\alpha} \cdot f \circ \phi_t \, dM \quad a.s., \qquad t \in \mathbf{R}, \tag{4.3}$$

Here M *is a* SαS *random measure on* S *with control measure* μ *and* $m_t = d(\mu \circ \phi_t)/d\mu$. *Conversely, (4.3) defines a measurable stationary* SαS *process for any choice of* $\{\phi_t\}_{t\in\mathbf{R}}$, $\{a_t\}_{t\in\mathbf{R}}$, *and* f *fulfilling above conditions.*

Remarks 4.2.

(a) (4.3) gives a harmonizable process when $S = \mathbf{R}$, $f = 1$, $\phi_t(s) = s$, and $a_t(s) = e^{its}$, for all $t, s \in \mathbf{R}$. To obtain a moving average (2.5), one takes $S = \mathbf{R}$, $\phi_t(s) = s - t$, $a_t(s) = 1$, $\mu = \lambda$. Similarly, a mixed moving average is defined by a flow $\phi_t(w, s) = (w, s - t)$ and $a_t(w, s) = 1$ on $S = W \times \mathbf{R}$.

(b) If instead of the shift on $\mathbf{R}$, one takes a shift modulo 1 on $[0, 1)$, i.e., $\phi_t(s) = s +_1 (-t)$, $s \in [0, 1)$, $t \in \mathbf{R}$, then two different classes of stationary SαS processes can be obtained. The first one is a straightforward analog of moving averages

$$X_t = \int_0^1 f(s +_1 (-t)) \, M(ds), \tag{4.4}$$

and the second one is of the form

$$X_t = \int_0^1 (-1)^{[s-t]} f(s +_1 (-t)) \, M(ds), \qquad t \in \mathbf{R}. \tag{4.5}$$

Here M is a SαS random measure with Lebesgue control measure λ, $f \in L^{\alpha}([0, 1], \lambda)$, and $[x]$ denotes the largest integer not greater than x. The cocycle term (or a "switching sign factor") in (4.5) is of the form $a_t(s) = (-1)^{[s-t]}$, while in (4.4) we have $a_t(s) = 1$. These are the only $\{-1, 1\}$–valued cocycles for the flow $\phi_t(s) = s +_1 (-t)$.

(c) If V is another Borel space with a σ–finite measure ν and $\Phi : V \to S$ is a null-preserving Borel isomorphism, then one can change variable in (4.3) to integrate over V instead of S. This change of variable produces a process in the same form as (4.3),

$$X_t = \int_V c_t \cdot n_t^{1/\alpha} \cdot g \circ \psi_t \, dN \quad a.s., \tag{4.6}$$

where $c_t = a_t \circ \Phi$, $\psi_t = \Phi^{-1} \circ \phi_t \circ \Phi$, $g = b^{-1/\alpha} f \circ \Phi$, and $b = d\nu/d(\mu \circ \Phi)$. Here

$$N(A) = \int_{\Phi A} (b \circ \Phi^{-1})^{1/\alpha} \, dM \tag{4.7}$$

is a SαS random measure on V with control measure ν and $n_t = d(\nu \circ \psi_t)/d\nu$. A change of variable can be used to simplify the representation (4.3) in certain situations. For instance, if the flow $\{\phi_t\}_{t \in \mathbf{R}}$ admits a σ–finite invariant measure ν, then one takes $V = S$ and Φ the identity map. This gives $n_t = 1$ in (4.6), which may be a simpler representation than (4.3). Another simplification occurs when the cocycle term in (4.3) is a coboundary, i.e., $a_t(s) = e(s)/e(\phi_t(s))$, for some Borel function $e : S \mapsto \{-1, 1\}$(or $\{|z| = 1\}$). In this case one may replace in (4.3) f by $e^{-1} f$ and M by N, $N(A) = \int_A e \, dM$; this operation removes the cocycle term from (4.3).

(d) The representation (4.3) is essentially unique up to a change of variable and a removable coboundary term described in the previous remark (see [12]).

(e) Hardin [5] characterized spectral representations of stationary SαS processes in terms of groups of linear isometries on L^α–spaces. Our representation (4.3) gives an explicit form of such groups in terms of flows and cocycles. In this way it is possible to use some results of ergodic theory for further characterizations of stationary SαS processes which we will now consider.

In ergodic theory there are some classical decompositions of measure spaces generated by nonsingular flows. The most trivial one is obtained by considering the set of fixed points

$$S_1 = \{s \in S : \phi_t(s) = s \quad \forall\, t \in \mathbf{R}\}.$$

The flow instantaneously moves the points on $S \setminus S_1$ and is constant on S_1. Both these sets are invariant under the flow. Now, $S \setminus S_1$ admits the Hopf decomposition into the dissipative S_2 and conservative S_3 parts, which are invariant under the flow (see [12] and the references therein). Thus we have a decomposition of S into three invariant sets

$$S = S_1 \cup S_2 \cup S_3.$$

This decomposition yields a decomposition of the stochastic integral in (4.3) into three independent parts, say $X_t^{(1)}$, $X_t^{(2)}$, and $X_t^{(3)}$. This is the idea leading to the following result.

Theorem 4.3. *(Decomposition of stationary $S\alpha S$ processes;* [12], [13].*)*
Every stationary measurable $S\alpha S$ process $\{X_t\}_{t\in\mathbf{R}}$ has a unique decomposition into the sum of three mutually independent stationary $S\alpha S$ processes

$$X_t = X_t^{(1)} + X_t^{(2)} + X_t^{(3)}, \qquad t \in \mathbf{R} \tag{4.8}$$

(on a possibly enlarged probability space) such that $\{X_t^{(1)}\}_{t\in\mathbf{R}}$ is a harmonizable process, $\{X_t^{(2)}\}_{t\in\mathbf{R}}$ is mixed moving average, and $\{X_t^{(3)}\}_{t\in\mathbf{R}}$ does not admit harmonizable or mixed moving average components.

Remarks 4.4

(i) Sometimes the flow and the cocycle may not be given explicitly and one only has a representation (2.2). Then the decomposition (4.8) can obtained directly from $\{f_t\}_{t\in\mathbf{R}}$ as follows

$$X_t^{(i)} = \int_{S_i} f_t \, dM, \quad t \in \mathbf{R}, \quad i = 1, 2,$$

where

$$S_1 = \{s : \; f_{t_1+t_2}(s) f_0(s) = f_{t_1}(s) f_{t_2}(s) \quad \text{for a.a. } t_1, t_2 \in \mathbf{R}\},$$

$$S_2 = \{s : \int_{\mathbf{R}} |f_t(s)|^{\alpha} \, dt < \infty\},$$

and

$$S_3 = S \setminus (S_1 \cup S_2).$$

(see [12], [13]).

(ii) $\{X_t\}_{t\in\mathbf{R}}$ is a mixed moving average if and only if $S = S_2$ (modulo μ). It is the usual moving average process if and only if $S = S_2$ (modulo μ) and $\{\phi_t\}_{t\in\mathbf{R}}$ is ergodic. Indeed, a moving average process is defined by the translation flow $\phi_t(s) = s - t$ which is ergodic and satisfies $S = S_2$. On the other hand, condition $S = S_2$ leads to (3.1) with $\phi_t(w, s) = (w, s - t)$. This flow is ergodic if and only if W is a singleton, i.e., (3.1) defines the usual moving average.

(iii) Despite the obvious similarities, there are big differences between the decompositions (1.2) and (4.8). The third term in (1.2) is usually not present in practical models but this is not so for the third term in (4.8). Many interesting $S\alpha S$ processes can have the form of $X^{(3)}$. This includes sub-Gaussian and sub-stable processes (see [16] for a definition), the periodic processes discussed in Remark 4.2 (b), and the real part of a harmonizable process (see [13]).

(iv) Processes of the form $X^{(3)}$ are represented by (4.3) with a conservative without fixed points flow $\{\phi_t\}_{t\in\mathbf{R}}$. Essentially, any such flow and a related cocycle generates a different process of type $X^{(3)}$. This shows that the class of stationary SαS processes can be difficult to describe in definitive terms. The classes of harmonizable and moving average processes, which are generated by the simplest flows, a constant flow and a shift, constitute relatively small subclasses of all stationary SαS processes.

(v) In the decomposition (4.8), the harmonizable part is never ergodic (see, e.g., [7]), a mixed moving average part is mixing of all orders (and so is ergodic, see [17]). The third part in (4.8) is generated by a infinitely recurrent flow, so that (4.3) yields long–range dependence in $X^{(3)}$. If $\{\phi_t\}_{t\in\mathbf{R}}$ admits a finite invariant measure, then $X^{(3)}$ is not ergodic (make a change of variable (4.6) and use [3] or [7]). On the other hand, if the flow admits a σ–finite invariant measure, then it is possible to construct a mixing process of the type $X^{(3)}$ based on an asymptotically singular conservative flow (see [14]). However, taking into account (4.8), one may say that "typically" a stationary SαS process is not ergodic, unless it is a mixed moving average.

References

[1] CAMBANIS, S. and HOUDRÉ, C. (1993), *Stable processes: moving averages versus Fourier transforms*, Probability Theory and Related Fields **95**, 75–85.

[2] DOOB, J.L. (1953), *Stochastic Processes*, John Wiley.

[3] GROSS, A. (1994), *Some mixing conditions for stationary symmetric stable stochastic processes*, Stochastic Processes and Applications **51**, 277-295.

[4] GROSS, A. and WERON, A. (1994), *On measure-preserving transformations and doubly stationary symmetric stable processes*, Studia Math. **114**, 275-287.

[5] HARDIN, C.D. (1982), *On the spectral representation of symmetric stable processes*, J. Multivariate Analysis **12**, 385–401.

[6] HERNÁNDEZ, M. and HOUDRÉ, C. (1993), *Disjointness results for some classes of stable processes*, Studia Math. **105**, 235–252.

[7] JANICKI, A. and WERON, A. (1994), *Simulation and chaotic behavior of α-stable stochastic processes*, Marcel Dekker.

[8] KANTER, M. (1973), *The L^p norm of sums of translates of a function*, Trans. Am. Math. Soc. **179**, 35–47.

[9] MAKAGON, A. and MANDREKAR, V. (1990), *The spectral representations of stable processes: harmonizability and regularity*, Probability Theory and Related Fields **85**, 1–11.

[10] RAJPUT, B.S. and ROSIŃSKI, J. (1989), *Spectral representations of infinitely divisible processes*, Probability Theory and Related Fields **82**, 451–487.

[11] ROSIŃSKI, J. (1994), *On the uniqueness of the spectral representation of stable processes*, J. Theoret. Probab. **7**, 615-634.

[12] ROSIŃSKI, J. (1995), *On the structure of stationary stable processes*, Ann. Probab. **23**, 1163–1187.

[13] ROSIŃSKI, J. (1996), *Decomposition of stationary α–stable random fields*, Preprint.

[14] ROSIŃSKI, J. and SAMORODNITSKY, G. (1996), *Classes of mixing stable processes*, Bernoulli **2**, 365–377.

[15] ROSIŃSKI, J. and WOYCZYNSKI, W.A. (1986), *On Ito stochastic integration with respect to p–stable motion: inner clock, integrability of sample paths, double and multiple integrals*, Ann. Probab. **14**, 271–286.
[16] SAMORODNITSKY, G. and TAQQU, M.S. (1994), *Non-Gaussian Stable Processes*, Chapman & Hall.
[17] SURGAILIS, D., ROSIŃSKI, J., MANDREKAR, V., and CAMBANIS, S. (1993), *Stable mixed moving averages*, Probab. Th. Rel. Fields **97**, 543–558.
[18] URBANIK, K. (1964), *Prediction of strictly stationary sequences*, Colloq. Math. **12**, 115–129.
[19] URBANIK, K. (1967), *Some prediction problems for strictly stationary processes*, in: Proc. 5th Berkeley Sympos. Math. Statist. Probab., Vol. 2, Part I, Univ. of California Press, 235–258.

Department of Mathematics, University of Tennessee, Knoxville, TN 37996-1300
email: rosinski@math.utk.edu

Tail Behavior of Some Shot Noise Processes

Gennady Samorodnitsky [1] [2]

Abstract

We consider different ways that heavy tails can arise in shot noise models and discuss potential applications of the latter to financial modeling.

1. Introduction

A *shot noise process* is a stochastic process of the form

$$X(t) = \sum_{T_i \le t} R_i(t - T_i), \quad t \in R, \tag{1.1}$$

where T_i's are the points of a point process N on the real line, and $\{R_i(t),\ t \ge 0\}$, $i = 1, 2, \ldots$ is a sequence of i.i.d. measurable stochastic processes, independent of the point process. Regarding $X(t)$ as describing the state of a certain system at time t, there is a very natural interpretation of a model given by (1.1). Think of T_i's being the instances of time a "shock" arrives at the system. The effect of the shock at the system is a dynamic and random process. That is, h units of time later, at time $T_i + h$ this effect is equal to $R_i(h)$. The total state of the system at time t is then the sum of the effects of all shocks occurring up to that time. The stochastic processes $\{R_i(t),\ t \ge 0\}$, $i = 1, 2, \ldots$ are called the *response processes*.

Shot noise stochastic processes are, in certain cases, attractive as stochastic models because many phenomena seem to actually develop as the result of discrete "shocks".

The point process N is parameterized by (Borel) subsets A of the real line, with $N(A)$ equal to the number of shocks located in A. We shall assume that N is *Radon*, i.e. $N(A) < \infty$ with probability 1 for every compact (bounded and closed) set A. It is stationary if $(N(A_1 + s), \ldots, N(A_k + s)) \stackrel{d}{=} (N(A_1), \ldots, N(A_k))$ for all $k \ge 1$, all compact sets $A_1, \ldots, A_k$ and $s \in R$. It is an important feature of shot noise processes that that the shot noise process (1.1) is stationary if the point process N is stationary, quite irrespectively of the structure of the response processes $\{R_i(t),\ t \ge 0\}$, $i = 1, 2, \ldots$. See Resnick [Res87] and Karr [Kar86] for more details on point processes.

The rich literature on shot noise processes typically discusses various particular cases of the process (1.1). Those are obtained by specifying a particular class of response

---

[1] Research supported by the NSF grant DMS-94-00535 and the NSA grant MDA904-95-H-1036.

[2] *Keywords and phrases:* Shot noise processes, heavy tails, financial models.

processes and/or a particular class of point processes $\{T_i,\ i \geq 1\}$. A very common class of response processes is that of the kind

$$R_i(t) = Z_i f(t),\ t \geq 0,$$

where $\{Z_i\}$ is a sequence of i.i.d. random variables, and $\{f(t),\ t \geq 0\}$ is a non-random measurable function. With this choice of response processes one explicitly separates the response amplitude Z_i from the dynamics of the response given by the function f. The resulting shot noise process has then the form

$$X(t) = \sum_{T_i \leq t} Z_i f(t - T_i),\ t \in R. \tag{1.2}$$

The model (1.2) is simpler than the general shot noise model (1.1), but it still forms a class of models that is often adequate. A further simplification of the model is achieved by taking $Z_i \equiv 1,\ i \geq 1$, thus leaving only the shock arrival times T_i's as sources of the randomness in the state of the system.

The intuitive appeal and relative simplicity of shot noise processes have attracted attention of researchers both with a view of further developing the theory of these processes and to discuss their applications in particular situations. General expositions of shot noise processes trace back to Rice [Ric44]. See also Parzen [Par62], Daley [Dal71] and Vervaat [Ver79]. More references are listed in Hsing and Teugels [HT89]. Shot noise processes have been used as models for computer failure times (Lewis [Lew64]) and earthquake aftershocks (Vere-Jones [VJ70]), and have been applied in such diverse fields as acoustics (Kuno and Ikegaya [KI73]), risk theory (Klüppelberg and Mikosch [KM95a] and [KM95b]) and financial processes (Samorodnitsky [Sam95]). The latter work is quoted extensively in the present paper, for it is our purpose presently to discuss in what ways heavy tails can and cannot arise in shot noise models and the implications of this to financial modeling.

The simplest, and a very common, choice of the point process $\{N(A),\ A$ a Borel set$\}$ is that of a time homogeneous Poisson process. A more general setting is that of a renewal process. Here $T_i = Y_1 + \ldots + Y_i$, $T_{-i} = -(Y_{-(i-1)} + \ldots + Y_0)$ for $i \geq 1$, and $\ldots, Y_{-2}, Y_{-1}, Y_0, Y_1, \ldots$ is a positive i.i.d. sequence. Of course, unless the renewal process is, actually, Poisson, the resulting shot noise process will not be stationary. If the step distribution F_Y of Y_i's has a finite mean μ_Y, one can construct a stationary shot noise process by using a stationary renewal process. That is, we define $(T_{-1}, T_1) = (-US, (1-U)S)$, with S and U being independent random variables, such that $P(S \leq x) = (1/\mu_Y) \int_0^x t F_Y(dt)$, $x > 0$, and U uniform in $(0, 1)$, and then $T_i = T_1 + Y_2 + \ldots + Y_i$ and $T_{-i} = T_{-1} - (Y_{-1} + \ldots + Y_{-(i-1)})$ for $i \geq 2$. Here Y_i's are independent of (T_{-1}, T_1).

Taking Poisson or renewal processes as the basis, one often defines a shot noise process with $\{N(A),\ A$ a Borel set$\}$ being a *cluster process*. A cluster point process consists of "secondary" points in clusters centered around "primary" points of the Poisson or renewal process. Intuitively, the clusters of secondary points are thought of as bursts of activity resulting from the event corresponding to a primary point. In this paper we will look only at a cluster Poisson process, which is defined as follows. Let $\{p_k,\ k \geq 1\}$ be a probability law on positive integers, and F_D be a probability law on R. Let $(K_j,\ -\infty < j < \infty)$ be i.i.d. random variables with common law $\{p_k,\ k \geq 1\}$, independent of the array $(D_{ij},\ i \geq 1,\ -\infty < j < \infty)$ of i.i.d. random variables with common law F_D. Finally, let $(C_j,\ -\infty < j < \infty)$ be the points of a time homogeneous Poisson process with intensity λ, independent of $(K_j,\ -\infty < j < \infty)$ and of $(D_{ij},\ i \geq 1,\ -\infty < j < \infty)$. Then for every Borel set A,

$$N(A) = \sum_{j=-\infty}^{\infty} \sum_{i=1}^{K_j} \mathbf{1}(C_j + D_{ij} \in A), \tag{1.3}$$

where $\mathbf{1}$ is the indicator function. That is, C_j's are the cluster centers ("primary points"), K_j's are the cluster sizes, $C_j + D_{ij},\ i = 1, \ldots, K_j$ are the points of the jth cluster, $\{p_k,\ k \geq 1\}$ is the cluster size distribution and F_D is the displacement distribution within a cluster.

A general discussion of the tails of a stationary shot noise process is in the next section. In Section 3 we consider a particular shot noise model, which exhibits features found in many financial processes. Certain technical details are left to the Appendix at the end of the paper.

2. Heavy tails of shot noise processes

Let $\{X(t),\ t \in R\}$ be a shot noise process given by (1.2). Throughout this paper such a process is understood to be defined as

$$X(t) = \lim_{B \to \infty} \sum_{-B \leq T_i \leq t} Z_i f(t - T_i),\ t \in R, \tag{2.1}$$

where the limit exists in probability. Let us assume, for the time being, that T_i's are the points of a time homogeneous Poisson process with intensity $\lambda > 0$. The shot noise processes is then stationary. How can we make the marginal distribution of $X(t)$ heavy-tailed?

It is obvious that the tails of $X(0)$ are at least as heavy as the tails of the amplitude size Z_1. That is, let

$$P(Z_1 > x) = x^{-p} L(x),\ x \to \infty,$$

where $p > 0$ and L is a slowly varying function. If the Lebesgue measure of the set $\{t > 0 :\ f(t) > 0\}$ is non-zero, then

$$\underline{\lim}_{x \to \infty} \frac{P(X(0) > x)}{x^{-p} L(x)} > 0,$$

that is, the marginal distributions of the shot noise process cannot have tails of a smaller order than $x^{-p}L(x)$, which is the tail of Z_1. Can the tails of $X(0)$ be heavy if the tails of Z_1 are not?

Suppose that the function f is bounded. The simple argument in the Appendix shows that, in this case, the tails of $X(0)$ cannot be heavier than those of Z_1. Specifically, if for some $p \geq 1$

$$E|Z_1|^p < \infty, \tag{2.2}$$

then $E|X(0)|^p < \infty$ as well. See Doney and O'Brien [DO91] for a related delicate result in the absence of infinite divisibility.

If the response function f is not bounded, however, then the heavy tails of $X(0)$ can be caused by a special structure of f, even when Z_1 is light tailed. Indeed, the model $X_{\gamma,d,m}$ below uses a (symmetrized) exponentially distributed Z_1. Models in which heavy tails are caused by a special structure of the response function f tend to be more flexible than those in which heavy tails are caused by a heavy-tailed amplitude Z_1, in the following sense. Suppose that we would like to have a stationary model $\{X(t),\ t \in R\}$ in which, for some $n_0 > 1$, the probability tail of $X(0) + X(1) + \ldots + X(n_0 - 1)$ is strictly lighter than that of $X(0)$. This is desirable, for instance, in the case of financial modeling, where daily returns appear to have heavier tails than, say, weekly returns. This "tail cancellation" property is impossible to model by a shot noise process when the heavy tails are caused by a correspondingly heavy-tailed amplitude Z_1, for it remains unaffected by taking linear combinations of the values of the process. On the other hand, forming such linear combinations can affect significantly the structure of the response function f and, through it, the probability tails of the shot noise, if it is the response function that causes heavy tails. We exploit this idea in the sequel.

3. A shot noise model for financial processes

Before describing the particulars of the model, we list some of the more important properties of financial processes, like stock or exchange rate returns.

1. The data has observable periods of high activity and low activity, often measured through volatility.

2. The observations are heavy tailed.

3. As the sampling interval increases, the tails tend to become less heavy. That is, adding up a certain number of consecutive observations may lead to lighter tails.

4. The data has long memory property (long range dependence).

See e.g. Akgiray and Booth [AB88], Guillaume *et al.* [GDD+94] and Müller *et al.* [MDD+95]. We remark that the latter property means that observations far apart in time are highly dependent. The empirical autocorrelation function of the raw returns is often quite small, but the long memory is typically captured by the empirical autocorrelation function of the absolute returns.

Our purpose in this section is to show how we can account for such properties with a shot noise model. We will use a cluster Poisson shot noise process described in the previous section. An extremely important property of shot noise models is its intuitive structure. See Mandelbrot [Man82] for an emphatic argument for intuitive models. For simplicity we construct a symmetric model (even though it is well known that positive excursions of financial processes often look differently from their negative excursions). Non-symmetric models can be constructed in a similar way.

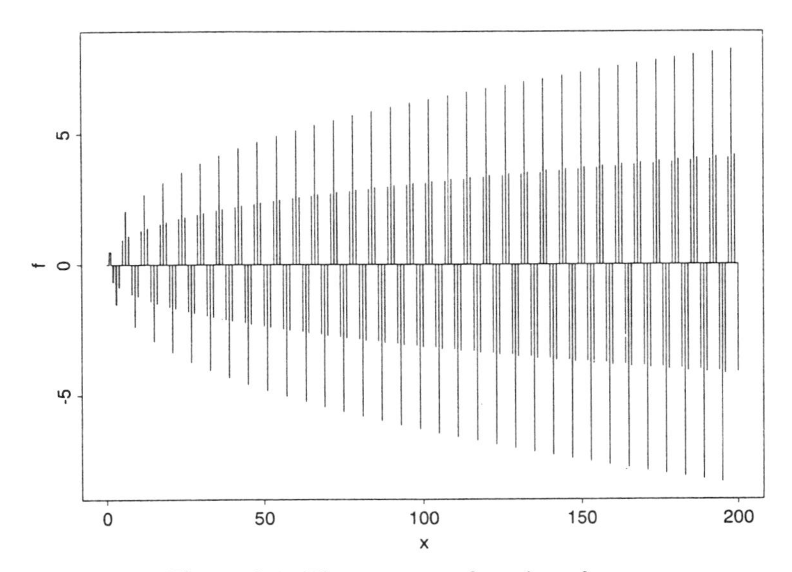

Figure 3.1. The response function $f_{\gamma,d,m}$

Fix a $\gamma > 0$, $d > 0$ and $m \geq 1$, and define

$$f_{\gamma,d,m}(x) = \begin{cases} j^{\gamma} \cos \pi \frac{j}{m} & \text{if } x \in \left(j - (j+1)^{-(d+1)},\, j + (j+1)^{-(d+1)}\right),\ j \geq 1 \\ 0 & \text{otherwise} \end{cases} . \tag{3.1}$$

The function $f_{\gamma,d,m}$ is our response function. It is plotted on Figure 3.1 for $\gamma = .4$, $d = .85$ and $m = 3$. Observe that the "spikes" of the response function tend to grow with time, but they occur on more and more narrow intervals. The spikes grow faster when the parameter γ is big, and they become more narrow when the parameter d is big. We expect, therefore, to see both heavier tails and longer memory when γ is big, and d is small. This becomes transparent from Theorem 3.1 below. The presence of a trigonometric factor in (3.1) will be useful for the purpose of "tail cancellation".

Let $Z_1 \stackrel{d}{=} \epsilon Y$, where ϵ is a Rademacher random variable ($P(\epsilon = 1) = P(\epsilon = -1) = 1/2$), independent of the standard exponential random variable Y, and let $\{X_{\gamma,d,m}(t),\ t \in R\}$ be defined by (1.2) with the response function given by (3.1). This means that Z_1 has a Laplace density $f_Z(z) = .5e^{-|z|}$, all z. We choose a Laplace distribution for Z_1 because it is light tailed, and it makes tail computations more transparent via Tauberian-type arguments. Observe that the "cluster" structure of the shot noise process has built-in periods of high and low activity. The degree of distinction between them can be articulated in the model by varying the cluster size distribution and the displacement within cluster distribution. The following theorem shows that our model has other features listed above as well.

Note that we are discussing here only the equally spaced observations $X_{\gamma,d,m}(i),\ i = 0, 1, 2, \dots$ of $\{X_{\gamma,d,m}(t),\ t \in R\}$. However, the process $\{X_{\gamma,d,m}(t),\ t \in R\}$ is, by its nature, a continuous time process, and, as such, can be naturally used as a continuous-time model.

Theorem 3.1 *(i) Suppose that $EK_1 < \infty$. Then $X_{\gamma,d,m}$ is a well defined stationary infinitely divisible stochastic process. (ii) Assume that*

$$EK_1^2 < \infty \tag{3.2}$$

and that the displacement distribution F_D is absolutely continuous with respect to the Lebesgue measure. Then

$$P(X_{\gamma,d,m}(t) > x) \sim c^{(1)}_{d,\gamma,m} \lambda EK_1 x^{-d/\gamma},\ x \to \infty, \tag{3.3}$$

where

$$c^{(1)}_{d,\gamma,m} = \frac{1}{2\gamma m}\Gamma(d/\gamma) \sum_{i=0}^{2m-1} \left| \cos \pi \frac{i}{m} \right|^{d/\gamma}.$$

(iii) Under the assumptions of (ii), if $\gamma > 1$, then

$$P\Big(\sum_{k=0}^{2m-1} X_{\gamma,d,m}(t+k) > x\Big) \sim \begin{cases} c^{(2)}_{d,\gamma,m}\lambda EK_1 x^{-(d+1)/\gamma} & \text{if } \frac{d+1}{\gamma} < \frac{d}{\gamma-1}, \\ c^{(3)}_{d,\gamma,m}\lambda EK_1 x^{-d/(\gamma-1)} & \text{if } \frac{d+1}{\gamma} > \frac{d}{\gamma-1}, \\ (c^{(2)}_{d,\gamma,m} + c^{(3)}_{d,\gamma})\lambda EK_1 x^{-(d+1)/\gamma} & \text{if } \frac{d+1}{\gamma} = \frac{d}{\gamma-1}, \end{cases} \tag{3.4}$$

and if $\gamma \leq 1$ then

$$P\Big(\sum_{k=0}^{2m-1} X_{\gamma,d,m}(t+k) > x\Big) \sim c^{(2)}_{d,\gamma,m}\lambda EK_1 x^{-(d+1)/\gamma} \tag{3.5}$$

as $x \to \infty$. Here

$$c^{(2)}_{d,\gamma,m} = \frac{1}{m}\Gamma\big((d+\gamma+1)/\gamma\big) \sum_{i=1}^{2m-1} \sum_{j=1}^{2m} \Big| \sum_{k=0}^{i-1} \cos \pi \frac{j+k}{2m} \Big|^{(d+1)/\gamma} > 0$$

and

$$c_{d,\gamma,m}^{(3)} = \frac{1}{\gamma - 1}(\gamma m)^{d/(\gamma-1)}\Gamma\Big(d/(\gamma - 1)\Big).$$

(iv) Assume that (3.2) holds, and that

$$d > 2\gamma. \tag{3.6}$$

Then the process $X_{\gamma,d,m}$ has a finite variance, and its covariance function $R(n)$ satisfies, as $n \to \infty$,

$$R(n) \sim \kappa_{d,\gamma,m}\lambda E K_1 n^{2\gamma-d} \sum_{i=1}^{2m} \cos \pi \frac{i}{m} \cos \pi \frac{i+n}{m} \tag{3.7}$$

in the sense that the product $n^{d-2\gamma}R(n)$ has at most 2m subsequential limits given by (3.7), all of which are finite, and some of which are positive. Here

$$\kappa_{d,\gamma,m} = 2m^{-1} \int_0^\infty x^\gamma (1 + x)^{\gamma-d-1} ds.$$

For the proof see Samorodnitsky [Sam95].

Observe that parts (ii) and (iii) of Theorem 3.1 show that our shot noise process has heavy tails *in spite of light tailed exponential* Z's. The exponent of the tail of a single observation is $-d/\gamma$, but that of the tail of the sum of $2m$ consecutive observations is either $-(d+1)/\gamma$ or $-d/(\gamma-1)$, and, therefore, the latter tail is lighter than that of a single observation. Therefore, as the sample interval has increased, the tail has become less heavy. If one had a reliable estimator of the tail index, it would have been possible to estimate d and γ from (3.3) and (3.4). The trigonometric behavior of the correlation function is due to the choice of the response function $f_{\gamma,d,m}$. However, long memory is obvious. In particular, $\sum_{n=1}^{\infty} |R(n)| = \infty$ if $2\gamma < d \le 2\gamma + 1$.

The plot on Figure 3.2 shows the result of a simulation of the cluster Poisson shot noise model, using a finite number of clusters. We have taken 600 clusters, with cluster distribution being geometric with mean 10. The Poisson arrival rate has been taken to be equal to 1, and the displacement distribution F_D to be uniform in (0, .3). For the purpose of the above simulation we have chosen $m = 2$, $d = 1.2$ and $\gamma = .5$. Note that even though our process is symmetric (and ergodic), the displayed simulated path is mostly negative, which demonstrates the length of dependence in our model.

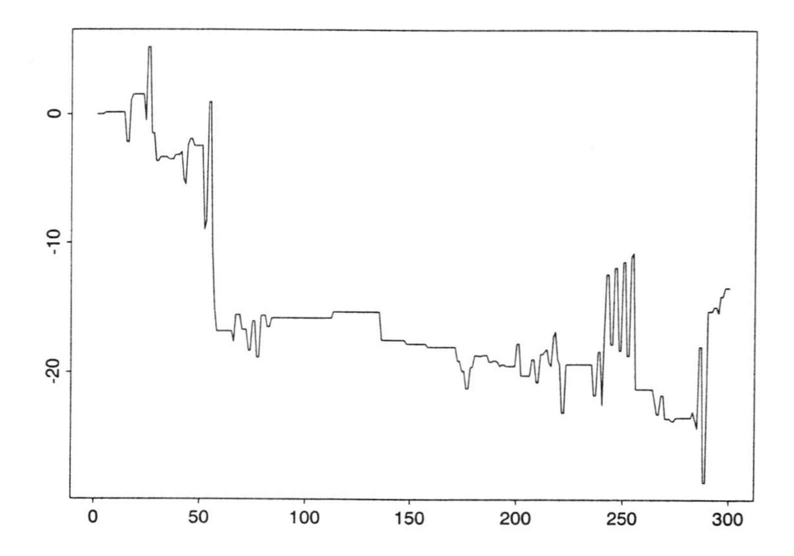

Figure 3.2 A simulated path of a cluster Poisson shot noise process

This is not very characteristic of financial returns processes. However, our goal here is only to demonstrate certain features of cluster Poisson shot noise processes (one gets very different types of behavior with different choices of parameters.) One can easily see on this plot the periods of high and low volatility, and the presence of heavy tails is also clear.

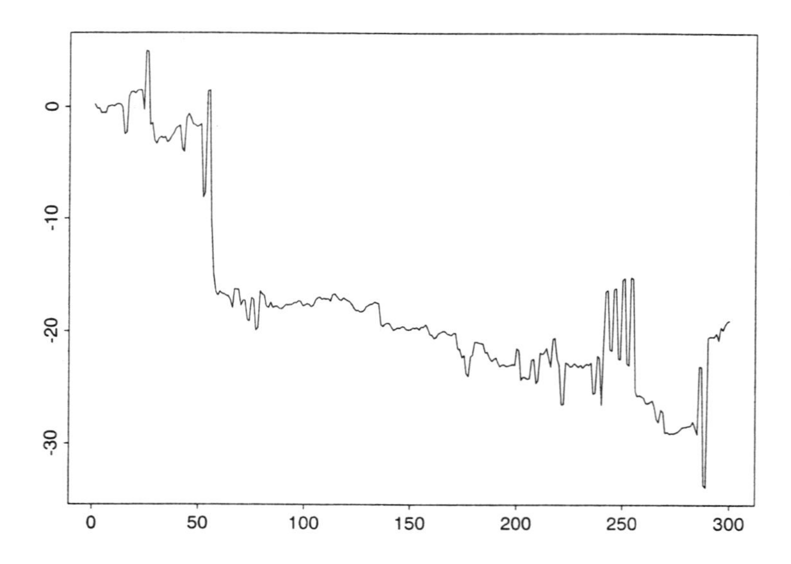

Figure 3.3 A path of a cluster Poisson shot noise process with added Brownian component

It follows from Proposition 4.1 that $X_{\gamma,d,m}(0)$ has a finite Lévy measure. Therefore, so does any increment of the process $\{X_{\gamma,d,m}(t),\ t \in R\}$. Consequently, any increment of the process takes the zero value with positive probability. Since the process is also ergodic, the zero values of the increments will actually be observed.

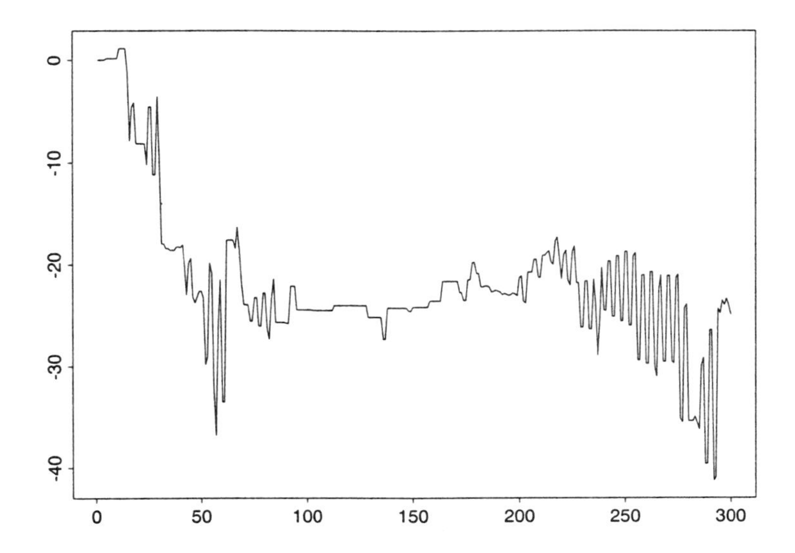

Figure 3.4 A case with longer memory and heavier tails

Unavoidably, therefore, the process will have "flat" spots, observed on the plot. One can regard such a model as accounting only for the "most significant" events on the market. If we model the rest of the activity on the market by an additive Brownian motion, the result is the path shown on Figure 3.3.

Finally, to demonstrate the effect on the length of the memory of the process and the heaviness of the tails, we have reduced in Figure 3.4 the parameter d from 1.2 to 0.8. Figure 3.5 represents the latter path with an additive Brownian motion. The preceding discussion indicates that we should expect to see longer memory and heavier tails. This can be seen from the plot on Figure 3.4. The longer memory can be observed by noticing that the sample path of a symmetric ergodic process plotted on Figure 3.4 spends more time in the negative region than the sample path on Figure 3.2 does. The fact that the sample path on Figure 3.4. oscillates more often and with higher magnitude than does the sample path on Figure 3.2 is an indication of heavier tails in the former case. Note that we have used the same random numbers for the two simulations ($d = 1.2$ and $d = .8$), hence both are skewed towards negative values.

The example of a shot noise process discussed in this section clearly demonstrates the potential of this class of models. Shot noise processes have an intuitively natural structure and offer flexibility of modeling both heavy tails and long memory. It is also important to realize that shot noise processes can be easily modified to model multidimensional processes, e.g. a portfolio of stocks, or of several major currencies. A major piece of work to be done is in identifying most appropriate classes of shot noise models, estimating their parameters, and working out prediction procedures for such models.

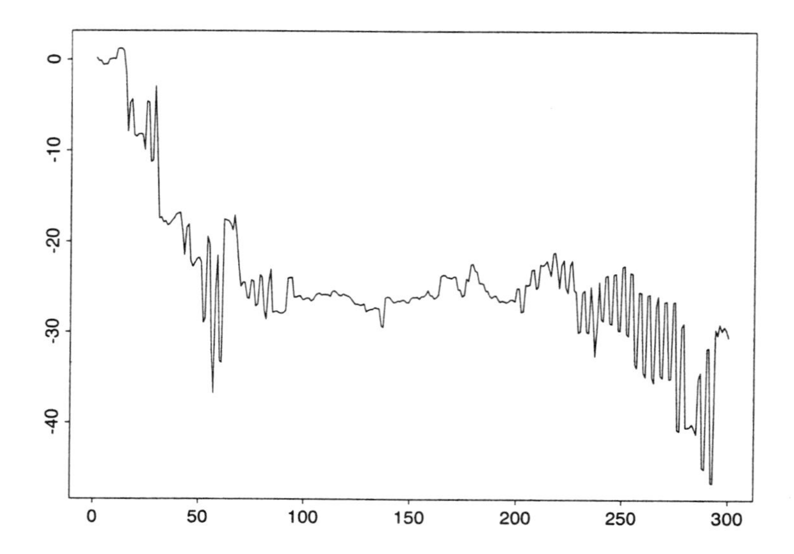

Figure 3.5 A longer memory and heavier tails shot noise process with added Brownian component

4. Appendix: The structure of a cluster Poisson shot noise process

Let $\{X(t),\ t \in R\}$ be a shot noise process (1.2) with the point process $\{N(A),\ A \text{ a Borel set}\}$ being a cluster Poisson process (1.3). We refer to this process as a *cluster Poisson shot noise process*. The following proposition describes some of the basic properties of this process.

Proposition 4.1 *The cluster Poisson shot noise process is well defined by (2.1) if and only if*

$$\int_0^\infty \sum_{k=1}^\infty p_k E\left(1 \wedge \left|\left(\sum_{i=1}^k Z_i f(x + D_{i1})\right)\right|\right) dx < \infty. \tag{4.1}$$

In this case, it is a stationary infinitely divisible process. The marginal Lévy measure μ of $X(0)$ is given by

$$\mu = \lambda \sum_{k=1}^\infty p_k \mu_k, \tag{4.2}$$

where each μ_k is a Lévy measure defined by

$$\mu_k = \left(P \times \text{Leb}\right) \circ H_k^{-1},$$

with $H_k : \Omega \times (0, \infty) \to R$ given by

$$H_k(\omega, x) = \sum_{i=1}^{k} Z_i(\omega) f(x + D_{i1}(\omega)).$$

Moreover, if $EK_1^2 < \infty$, $EZ_1^2 < \infty$ and

$$\int_0^\infty f(y)^2 dy < \infty,$$

then the process $\{X(t),\ t \in R\}$ has a finite second moment, and its covariance function $(R(t),\ t \geq 0)$ is given by

$$R(t) = \lambda E K_1 E Z_1^2 \int_0^\infty f(y) f(y+t) dy + \lambda (E K_1^2 - E K_1)(E Z_1)^2 \delta(t), \tag{4.3}$$

with

$$\delta(t) = \int_0^\infty f(y+t) \Big(\int_R \int_R f(y + x_1 - x_2) F_D(dx_1) F_D(dx_2) \Big) dy \tag{4.4}$$

(where $f(y) = 0$ for $y < 0$).

See Section 2 of Samorodnitsky [Sam95] for the argument. Observe that the effect of the displacement distributions F_D on the covariance function disappears if $EZ_1 = 0$. In the latter case

$$R(t) = \lambda E K_1 E Z_1^2 \int_0^\infty f(y) f(y+t) dy, \tag{4.5}$$

which makes it easy to see how long memory can be modeled with shot noise processes.

We conclude the appendix with the argument that, when the response function f of a shot noise process is bounded, then the assumption (2.2): $E|Z_1|^p < \infty$ for a $p \geq 1$ implies that $E|X(0)|^p < \infty$ as well.

Let M be such that

$$\text{Leb}\{x > 0 : |f(x)| > M\} = 0.$$

The random variable $X(0)$ is infinitely divisible and its Lévy measure is given by

$$\mu = \left(P \times \text{Leb}\right) \circ H^{-1}, \tag{4.6}$$

where $H : \Omega \times (0, \infty) \to R$ is given by

$$H((\omega, x)) = Z_1(\omega) f(x).$$

See also (4.2) below. Moreover,

$$\int_R (|z| \wedge 1)\, \mu(dz) < \infty. \tag{4.7}$$

Now, $E|X(0)|^p < \infty$ will follow once we show that

$$\int_R |z|^p\, \mu(dz) < \infty. \tag{4.8}$$

If Z_1 is essentially bounded, then there is an $0 < a < \infty$ such that $P(|Z_1| \le a) = 1$. Then

$$\begin{aligned}\int_R |z|^p\, \mu(dz) &= \int_R |z|^p \mathbf{1}(|z| \le aM)\, \mu(dz)\\ &\le (aM)^{p-1} \int_R |z| \mathbf{1}(|z| \le aM)\, \mu(dz) < \infty\end{aligned}$$

by (4.7). If Z_1 is not essentially bounded, then there are $0 < a < b < \infty$ such that both $P(a < |Z_1| < b) > 0$ and $P(|Z_1| > b) > 0$. Then by (4.7)

$$\infty > E \int_0^\infty (|Z_1 f(x)| \wedge 1)\, dx$$

$$\begin{aligned}&\ge E\Big(\mathbf{1}(a < |Z_1| < b) \int_0^\infty (|Z_1 f(x)| \wedge 1)\, dx\Big) + E\Big(\mathbf{1}(|Z_1| > b) \int_0^\infty (|Z_1 f(x)| \wedge 1)\, dx\Big)\\ &\ge aP(a < |Z_1| < b) \int_0^\infty |f(x)| \mathbf{1}(|f(x)| \le \frac{1}{b})\, dx\\ &\quad + P(|Z_1| > b) \int_0^\infty \mathbf{1}(|f(x)| > \frac{1}{b})\, dx.\end{aligned}$$

Therefore,

$$\int_0^\infty |f(x)| \mathbf{1}(|f(x)| \le \frac{1}{b})\, dx < \infty \tag{4.9}$$

and

$$\int_0^\infty \mathbf{1}(|f(x)| > \frac{1}{b})\, dx < \infty. \tag{4.10}$$

We therefore have

$$\begin{aligned}\int_R |z|^p\, \mu(dz) &= E \int_0^\infty |Z|^p |f(x)|^p\, dx = E|Z|^p \int_0^\infty |f(x)|^p\, dx\\ &\le E|Z|^p \Big((\frac{1}{b})^{p-1} \int_0^\infty |f(x)| \mathbf{1}(|f(x)| \le \frac{1}{b})\, dx\\ &\quad + M^p \int_0^\infty \mathbf{1}(|f(x)| > \frac{1}{b})\, dx \Big) < \infty,\end{aligned}$$

and so $E|Z_1|^p < \infty$ implies $E|X(0)|^p < \infty$ when f is a bounded function.

References

[AB88] V. Akgirav and G.G. Booth. The stable-law model of stock returns. *J. Bus. Econ. Stat.*, 6:51–57, 1988.

[Dal71] D.J. Daley. The definition of a multidimensional generalization of shot noise. *Journal of Applied Probab.*, 8:128–135, 1971.

[DO91] R.A. Doney and G.L. O'Brien. Loud shot noise. *Ann. Appl. Probab.*, 1:88–103, 1991.

[GDD+94] D.M. Guillaume, M.M. Dacorogna, R.R. Davé, U.A. Müller, R.B. Olsen, and O.V. Pictet. From the bird's eye to the microscope: a survey of new stylized facts of the intra-daily foreign exchange markets. A discussion paper by the O& A Research Group, 1994.

[HT89] T. Hsing and J.L. Teugels. Extremal properties of shot noise processes. *Adv. Applied Probab.*, 21:513–525, 1989.

[Kar86] A.F. Karr. *Point Processes and Their Statistical Inference*. Marcel Dekker, New York and Basel, 1986.

[KI73] A. Kuno and K. Ikegaya. A statistical investigation of acoustic power radiated by a flow of random point sources. *J. Acoustic Soc. Japan*, 29:662–671, 1973.

[KM95a] C. Klüppelberg and T. Mikosch. Explosive Poisson shot noise processes with applications to risk reserves. *Bernoulli*, 1:125–148, 1995.

[KM95b] C. Klüppelberg and T. Mikosch. Modelling delay in claim settlement. *Scand. Actuar. Journal*, pages 154–168, 1995.

[Lew64] P.A. Lewis. A branching Poisson process model for the analysis of computer failure pattern. *Journal Royal Stat. Soc., Ser. B*, 26:398–456, 1964.

[Man82] B.B. Mandelbrot. *The Fractal Geometry of Nature*. W.H. Freeman and Co., San Francisco, 1982.

[MDD+95] U.A. Müller, M.M. Dacorogna, R.D. Davé, R.B. Olsen, O.V. Pictet, and J.E. von Weizsäker. Volatilities of different time resolutions - analyzing the dynamics of market components. Journal of Empirical Finance, to appear, 1995.

[Par62] E. Parzen. *Stochastic Processes*. Holden-Day, San Francisco, 1962.

[Res87] S.I. Resnick. *Extreme Values, Regular Variation and Point Processes.* Springer-Verlag, New York, 1987.

[Ric44] S.O. Rice. Mathematical analysis of random noise. *Bell Syst. Tech. J.*, 23:282–332, 1944.

[Sam95] G. Samorodnitsky. A class of shot noise models for financial applications. In C.C. Heyde, Yu. V. Prohorov, R. Pyke, and S.T. Rachev, editors, *Proceeding of Athens International Conference on Applied Probability and Time Series. Volume 1: Applied Probability*, pages 332–353. Springer, 1995.

[Ver79] W. Vervaat. On a stochastic difference equation and a representation of non-negative infinitely divisible random variables. *Adv. in Appl. Probab.*, 11:750–783, 1979.

[VJ70] D. Vere-Jones. Stochastic models for earthquake occurencies. *Journal Royal Stat. Soc., Ser. B*, 532:1–42, 1970.

Gennady Samorodnitsky
School of Operations Research and Industrial Engineering
Cornell University
Ithaca, NY 14853

VII. Numerical Procedures

Numerical Approximation of the Symmetric Stable Distribution and Density

J. Huston McCulloch

Abstract

This paper develops a numerical approximation to the symmetric stable distribution and density that is accurate to an expected log density precision of 10^{-6} for α in the range [0.84, 2.00], yet requires as little as 34 microseconds per ordinate on a personal computer. This approximation renders accurate maximum likelihood and/or posterior mode estimation with symmetric stable errors computationally tractable. The absolute precision of the distribution is 2.2×10^{-5} for α in the range [.92, 2.00], while that for the density is 6.6×10^{-5} in the same range.

1. Introduction

Let $S_\alpha(x)$ be the cumulative distribution function (c.d.f.) for a random variable X that has a standard symmetric stable distribution with characteristic exponent $\alpha \varepsilon (0, 2]$ and log characteristic function

$$\log E \exp(iXt) = - \mid t \mid^\alpha .$$

Let $s_\alpha(x)$ be the corresponding probability density function. See [ZV] and [ST] for properties of these distributions, [HC] for tables and graphs of the density, [DW] for tables of the distribution, and [MC] for discussion of financial applications.

[MR] note that despite its theoretical attractiveness, "the practical implementation of the stable Paretian distribution is a rather difficult task, and the burden associated with it is considered to be a major drawback for its practical usefulness." The present paper rectifies this situation, at least in the important symmetric case, by providing a quickly computable numerical approximation that is sufficiently accurate for likelihood maximization and most other statistical techniques.

Special cases of the symmetric stable distribution and density for which high precision numerical approximations are readily available are the Cauchy distribution:

$$C(x) = S_1(x) = \frac{1}{\pi} \tan^{-1}(x) + .5 \tag{1.1}$$

and the Gaussian distribution (with variance 2):

$$G(x) = S_2(x) = \frac{1}{2\sqrt{\pi}} \int_{-\infty}^{x} e^{-\xi^2/4} d\xi. \tag{1.2}$$

Our strategy is first to transform the x interval $[0, \infty]$ onto the more tractable interval $[0, 1]$ with a transformation $z = z_\alpha(x)$ which is specified below. Since we are only attempting to fit the symmetric stable distributions, it is sufficient to find an approximation for $x \geq 0$. In order to minimize the relative error in the upper tail, we will fit the complemented c.d.f. $S_\alpha^c(x) = 1 - S_\alpha(x)$ rather than the c.d.f. itself. We then will exploit existing approximations to $C(x)$ and $G(x)$ by interpolating between the complements of these two functions in the transformed space, thus substantially reducing the number of free parameters that will have to be computed. Further reductions in the number of free parameters are obtained by utilizing terms from known series expansions. The residual that remains is fit by a quintic spline across z. The free spline parameters in turn are fit as a quintic polynomial across α. Having fit the c.d.f. as a proper c.d.f., the density may be obtained by analytically differentiating the c.d.f. approximation, with confidence that it will integrate exactly to unity.

2. Series Coefficients

The lead term of the [BH] expansion for $x \to \infty$ implies that stable distributions with $\alpha < 2$ have thick "Paretian" tails behaving asymptotically like the Pareto distribution:

$$\lim_{x \to -\infty} \frac{S_\alpha(x)}{|x|^{-\alpha}} = \lim_{x \to \infty} \frac{S_\alpha^c(x)}{x^{-\alpha}} = \frac{1}{\pi} \Gamma(\alpha) \sin \frac{\pi \alpha}{2}. \tag{2.1}$$

Fitting the distribution for $x \geq 0$ is therefore facilitated by the change of variables

$$z = z_\alpha(x) = 1 - (1 + a_\alpha x)^{-\alpha} \tag{2.2}$$

for some positive constant a_α (which depends upon α and is selected below). This transformation maps $[0, \infty]$ onto the more tractable interval $[0, 1]$, while at the same time straightening out the Paretian tail with a computable slope.

Utilizing this transformation, we may now express the (complemented) symmetric stable distribution for $x \geq 0$ as a linear combination of the (complemented) Cauchy and Gaussian distributions, plus a small residual $R_\alpha(z)$:

$$S_\alpha^c(x) = (2 - \alpha) C^c(x_1) + (\alpha - 1) G^c(x_2) + R_\alpha(z), \tag{2.3}$$

where

$$x_1 = x_1(z),$$

$$x_2 = x_2(z),$$

and

$$x_\alpha(z) = [(1 - z)^{-1/\alpha} - 1]/a_\alpha$$

is the inverse of the transformation $z_\alpha(x)$. We immediately have:

$$R_\alpha(0) = R_\alpha(1) = 0, \quad \alpha \varepsilon (0, 2], \tag{2.4}$$

$$R_1(z) = R_2(z) = 0, \quad z \varepsilon [0, 1]. \tag{2.5}$$

Since *med* $|x| = 1.00 \pm .05$ for $\alpha \geq 0.8$, the choice

$$a_\alpha = 2^{1/\alpha} - 1 \tag{2.6}$$

makes the central value $R_\alpha(.5)$ relatively small in the α range of greatest interest.

The coefficient on the Paretian tail in (2.1) implies

$$R'_\alpha(1) = -\frac{1}{\pi}(a_\alpha)^\alpha \Gamma(\alpha) \sin\frac{\pi\alpha}{2} + \frac{2-\alpha}{\pi}. \tag{2.7}$$

If the approximated residual is forced to obey this equation, the relative error in the tails will be 0 as x becomes infinite, at least for $\alpha < 2$.

A second Bergstrom expansion, for $x \to 0$, implies

$$s_\alpha(0) = \frac{1}{\pi\alpha}\Gamma(\frac{1}{\alpha}), \tag{2.8}$$

$$s'(0) = 0, \tag{2.9}$$

and

$$s''_\alpha(0) = \frac{-1}{\pi\alpha}\Gamma\left(\frac{3}{\alpha}\right). \tag{2.10}$$

These three expressions may be used to further restrict the residual, by determining $R'_\alpha(0)$, $R''_\alpha(0)$, and $R'''_\alpha(0)$:

$$R'(0) = \frac{2-\alpha}{\pi} + \frac{\alpha-1}{4a_2\sqrt{\pi}} - \frac{s_\alpha(0)}{\alpha a_\alpha}, \tag{2.11}$$

$$R''(0) = \frac{2(2-\alpha)}{\pi} + \frac{3(\alpha-1)}{8a_2\sqrt{\pi}} - \frac{s_\alpha(0)(1+\alpha)}{\alpha^2 a_\alpha}, \tag{2.12}$$

$$R'''(0) = \frac{4(\alpha-2)}{\pi} + \frac{\alpha-1}{a_2\sqrt{\pi}}\left(\frac{15}{16} - \frac{1}{32a_2^2}\right) - \frac{1}{\alpha^3 a_\alpha}[(1+\alpha)(1+2\alpha)s_\alpha(0) + \frac{s''_\alpha(0)}{a_\alpha^2}]. \tag{2.13}$$

3. Spline Approximation

The six restrictions imposed on $R_\alpha(z)$ by (2.4), (2.7), and (2.11) - (2.13) are in themselves sufficient to determine a unique quintic polynomial in z for each α. However, these quintics in z are not sufficiently accurate for our purposes. We therefore generalize to model $R_\alpha(z)$ as a *quintic spline* in z : A function $f(x)$ defined on an interval of the real line is said to be a *d-degree spline* iff $f(x)$ is piecewise a d-degree polynomial and has $d-1$ continuous derivatives. The selected points at which $f^{(d)}(x)$ may be discontinuous are called *knotpoints.*

Continuing to impose the above six restrictions, each knotpoint creates one new free parameter for each value of α. It was found that these free parameters could themselves be adequately approximated by quintic polynomials in α. Since these are constrained by the two restrictions (2.5), we have 4 free parameters g_{ij}, $i = 1, 2, 3, 4$, for each knotpoint z_j. As will be seen below, the 19 knotpoints $z_j = .05, .10, \ldots, .95$ were found to be adequate for the precision we sought, making a total of $4 \times 19 = 76$ free parameters in all.

Our specific algorithm for computing the symmetric stable c.d.f. is as follows: Using the g_{ij} given in Table 1, we compute

$$\begin{aligned} r_{\alpha j} &= (2-\alpha)\textstyle\sum_{i=1}^{4} g_{ij}[(2-\alpha)^i - 1], \qquad j = 1, \ldots, 19, \\ z_j &= j/20, \\ q_{\alpha 1} &= R'_\alpha(0), \\ q_{\alpha 2} &= R''_\alpha(0)/2, \\ q_{\alpha 3} &= R'''_\alpha(0)/6, \\ b &= -\textstyle\sum_{i=1}^{3} q_{\alpha i} - \sum_{j=1}^{19} r_{\alpha j}(1-z_j)^5, \\ c &= R'_\alpha(1) - \textstyle\sum_{i=1}^{3} i q_{\alpha i} - 5\sum_{j=1}^{19} r_{\alpha j}(1-z_j)^4, \\ q_{\alpha 4} &= 5b - c, \end{aligned} \tag{3.1}$$

and

$$q_{\alpha 5} = b - q_{\alpha 4}.$$

Then we compute z from (2.2) and approximate $R_\alpha(z)$ by

$$\hat{R}_\alpha(z) = \sum_{i=1}^{5} q_{\alpha i} z^i + \sum_{j=1}^{19} r_{\alpha j} \max(z - z_j, 0)^5. \tag{3.2}$$

Note that (3.1) may be reduced to a simple polynomial

$$\hat{R}(z) = \sum_{i=0}^{5} p_{\alpha i J} z^i, \quad J = 0, \ldots, 19, \tag{3.3}$$

where J is the greatest integer in $20z$, and

$$p_{\alpha i J} = q_{\alpha i} + \binom{5}{i} \sum_{j=1}^{J} r_{\alpha j}(-z_j)^{5-1}, i = 0, \ldots, 5 \tag{3.4}$$

(The above sum is taken as 0 for $J = 0$.) We then substitute (3.2) into (2.3) to obtain our approximation $\hat{S}^c_\alpha(x)$ of the (complemented) stable c.d.f., using received approximations to $C(x)$ in (1.1) and $G(x)$ in (1.2).

The density approximation is then easily obtained by differentiating (2.2), (2.3), and (3.2):

$$\hat{s}(x) = [(2-\alpha)c(x_1)x_1'(z) + (\alpha-1)g(x_2)x_2'(z) - \hat{R}_\alpha'(z)]z_\alpha'(x), \tag{3.5}$$

where

$$c(x) = \frac{1}{\pi(1+x^2)}$$

and

$$g(x) = \frac{1}{2\sqrt{\pi}}e^{-x^2/4}$$

are the Cauchy and Gaussian densities (the latter with variance 2),

$$x_1'(z) = (1-z)^{-2},$$

$$x_2'(z) = \frac{1}{2a_2}(1-z)^{-3/2},$$

$$\hat{R}_\alpha'(z) = \sum_{i=1}^{5} i p_{\alpha i J} z^{i-1},$$

and

$$z_\alpha'(x) = \alpha a_\alpha (1 + a_\alpha x)^{-\alpha-1}.$$

Partial derivatives of the density with respect to x and/or α may also be obtained, as desired.

4. Calibration

In order to obtain the spline coefficients g_{ij} in Table 1, and to determine the precision of the resulting approximation, we carefully evaluated the stable distribution and density at a fine mesh of calibration points, using the proper integral representations for the stable distribution and density developed by [ZV]. His representation for the complemented symmetric stable c.d.f., for positive x, may be written as:

$$S_\alpha^c(x) = \begin{cases} \frac{1}{\pi}\int_0^{\pi/2} e^{-w_\alpha(x)v_\alpha(\phi)}d\phi, & \alpha > 1, \\ \frac{1}{\pi}\int_0^{\pi/2}\left(1 - e^{-w_\alpha(x)v_\alpha(\phi)}\right)d\phi, & \alpha < 1, \end{cases} \tag{4.1}$$

where

$$w_\alpha(x) = x^{\alpha/(\alpha-1)},$$

and

$$v_\alpha(\phi) = [\sin(\alpha\phi)]^{\alpha/(1-\alpha)}\cos[(\alpha-1)\phi](\cos\phi)^{1/(\alpha-1)}.$$

The derived integral representation for the symmetric stable density is

$$s_\alpha(x) = \frac{1}{\pi \mid 1-\alpha \mid}x^{1/(\alpha-1)}\int_0^{\pi/2} v_\alpha(\phi)e^{-w_\alpha(x)v_\alpha(\phi)}d\phi. \tag{4.2}$$

Both integrands are bounded and well-behaved. The integrands in (4.1) both increase monotonically from 0 to 1, while the integrand (4.2) is unimodal and vanishes at both limits of integration. The only difficult cases occur when α is close to 1 or 2 and x is large or near 0. In these cases, the integrand for the density becomes a narrow spike and elsewhere is virtually zero, while that for the distribution moves abruptly from virtually 0 to virtually 1. In these cases it is expedient to isolate the nontrivial region with a binary search before integration.

Having thus narrowed the range of integration as necessary, numerical integration was performed by the following procedure, inspired by [PTVF]: For $n = 1, 2, \ldots$, each integrand was evaluated at $2^n + 1$ equally spaced points (including the end points of the range of integration), and the integral was evaluated by trapezoids to obtain V_n. For $n \geq 3$, V_{n-2}, V_{n-1}, V_n define a geometric series which was continued to $n = \infty$, yielding V_n^*. For $n \geq 5$, the same procedure was applied to V_{n-2}^*, V_{n-1}^*, V_n^* to obtain V_n^{**}. The sequence $\{V_n\}$ converges very slowly, $\{V_n^*\}$ converges rapidly, and $\{V_n^{**}\}$ converges (eventually) even more rapidly. This process was terminated when one of the three series $\{W_n\}$ (with its associated constants) met both of the following absolute and relative convergence criteria:

$$| W_n - W_{n-1} | \leq 10^{-8} \text{ and } | W_n - W_{n-1} | / W_n \leq 10^{-6}.$$

The remaining error was then assumed to be less than the convergence criterion. Convergence was most often achieved with n in the range 6 to 10, though the worst single case (the distribution formula with $\alpha = 1.98$ and $z = .96$, i.e. $x = 9.74$) required $n = 18$, and alone took 35 seconds of CPU time on an IBM 3061, even after the range of integration had been considerably reduced. (See [MP] for alternative numerical approaches to Zolotarev's integral in the asymmetric cases.)

The distribution and density were evaluated as above for the grid $z_k = .02, .04, \ldots, .98$ and $\alpha_h = .50, .52, \ldots, .98, 1.02, 1.04, \ldots, 1.98$, a total of $49 \times 74 = 3626$ points. Our choice of a_α above implies that this equal spacing of z values places equal numbers of calibration points to the left and right of $x = 1$ (the approximate median of $|x|$), so that the inner and outer halves of the distribution are approximately equally represented, for each value of α. The fit of our approximation will of course be exact (to within the precision of the received numerical approximations for the Cauchy and normal distributions) for $\alpha = 1$ and 2, as it is for $z = 0$ and 1.

In order to obtain as small as possible a value for the expected error in log likelihood estimates obtained from our approximation in the particularly important region $1 \leq \alpha \leq 2$, we found (by weighted linear least squares regression) the values of the 76 free parameters g_{ij} that minimize the expression

$$\sum_{\alpha_h > 1} \sum_{z_k} \left(\frac{\hat{s}_{\alpha_h}(x_{hk}) - s_{\alpha_h}(x_{hk})}{s_{\alpha_h}(x_{hk})} \right)^2 \Delta S_{hk} \tag{4.3}$$

where

$$x_{hk} = x_{\alpha_h}(z_k),$$

$$\Delta S_{hk} = S_{\alpha_h}(z_{k+1}) - S_{\alpha_h}(z_{k-1}),$$

s and S represent "true" calibration values, and $\hat{s}$ (and $\hat{S}$ below) indicate the fitted spline approximation values.

The values of g_{ij} that minimize (4.3) are tabulated in Table 1. All the above calculations were performed in FORTRAN REAL *8, or approximately 16 significant digits. However, it was found that truncating to 11 digits as in Table 1 did not impair the precision of the approximation.

5. Precision

In the important range $1 \leq \alpha \leq 2$, it was found that

$$\max_{\alpha,x} \mid \hat{S}_\alpha(x) - S_\alpha(x) \mid = 2.2 \times 10^{-5}$$

and

$$\max_{\alpha,x} \mid \hat{s}_\alpha(x) - s_\alpha(x) \mid = 6.6 \times 10^{-5},$$

so that our approximation dominates four-place tables in terms of absolute precision. These precisions continued down to $\alpha = 0.92$ for both the distribution and the density.

The expected computational error in a log likelihood calculated from our density approximation is N times the expected value of $\log(\hat{s}/s)$, where N is the sample size. We found that

$$\max_{\alpha,x} \mid \log(\hat{s}_\alpha(x)/s_\alpha(x)) \mid = .014,$$

with the worst case occurring near normality and far out in the tail, at $\alpha = 1.98$ with $z = .94$. However, the expected relative error is far less than this, because of the low density in the problem regions. We have

$$\log\left(\frac{\hat{s}}{s}\right) = \frac{\Delta s}{s} - \frac{1}{2}\left(\frac{\Delta s}{s}\right)^2 + \frac{1}{3}\left(\frac{\Delta s}{s}\right)^3 + \ldots,$$

where $\Delta s = \hat{s} - s$. The expectation of the first term in this expansion is identically zero, since by construction the integral of our $\hat{s}$ is unity. Therefore

$$\mid E \log(\hat{s}/s) \mid = \frac{1}{2} E\left(\frac{\Delta s}{s}\right)^2 + O[E\left(\frac{\Delta s}{s}\right)^3]. \tag{5.1}$$

Table 1				
Spline Coefficients g_{ij}				
$j \backslash i$	1	2	3	4
1	1.8514190959e+02	-4.6769332663e+02	4.8424720302e+02	-1.7639153404e+02
2	-3.0236552164e+02	7.6351931975e+02	-7.8560342101e+02	2.8426313374e+02
3	4.4078923600e+02	-1.1181138121e+03	1.1548311335e+03	-4.1969666223e+02
4	-5.2448142165e+02	1.3224487717e+03	-1.3555648053e+03	4.8834079950e+02
5	5.3530435018e+02	-1.3374570340e+03	1.3660140118e+03	-4.9286099583e+02
6	-4.8988957866e+02	1.2091418165e+03	-1.2285872257e+03	4.4063174114e+02
7	3.2905528742e+02	-7.3211767697e+02	6.8183641829e+02	-2.2824291084e+02
8	-2.1495402244e+02	3.9694906604e+02	-3.3695710692e+02	1.0905855709e+02
9	2.1112581866e+02	-2.7921107017e+02	1.1717966020e+02	3.4394664342e+00
10	-2.6486798043e+02	1.1999093707e+02	2.1044841328e+02	-1.5110881541e+02
11	9.4105784123e+02	-1.7221988478e+03	1.4087544698e+03	-4.2472511892e+02
12	-2.1990475933e+03	4.2637720422e+03	-3.4723981786e+03	1.0174373627e+03
13	3.1047490290e+03	-5.4204210990e+03	4.2221052925e+03	-1.2345971177e+03
14	-5.1408260668e+03	1.1090264364e+04	-1.0270337246e+04	3.4243449595e+03
15	1.1215157876e+04	-2.4243529825e+04	2.1536057267e+04	-6.8490996103e+03
16	-1.8120631586e+04	3.1430132257e+04	-2.4164285641e+04	6.9126862826e+03
17	1.7388413126e+04	-2.2108397686e+04	1.3397999271e+04	-3.1246611987e+03
18	-7.2435775303e+03	4.3545399418e+03	2.3616155949e+02	-7.6571653073e+02
19	-8.7376725439e+03	1.5510852129e+04	-1.3789764138e+04	4.6387417712e+03

The second order term above may be roughly estimated by half the internal sum in (4.3). The maximum of this over the interval $1 \leq \alpha \leq 2$ was only 3.1×10^{-7}, while the estimated expectation of the third order term was never greater than 2×10^{-10}. The binding restriction on the expected relative precision of our density approximation is therefore the relative precision of our density calibration values themselves, namely 10^{-6}.[1]

Differences in the log likelihood of 0.01 or less are rarely of any consequence for statistical tests based on the likelihood ratio. Indeed, tables of critical values of the χ^2 distribution are typically only given to this precision. It would take an extraordinarily large sample, with N in excess of $10,000$, for our approximation to cause an error of such a magnitude to be expected.

We were able to obtain such high expected relative precisions for three reasons: First, we eliminated first order effects altogether by deriving the density by differentiation from a proper distribution. Had we instead fit the log density directly (as we did in an earlier, less satisfactory version), we would have saved having to take a logarithm to compute the log likelihood, but would not have been able to constrain the density to integrate to unity. Second, our transformation $z_\alpha(x)$, together with (2.7), eliminates the relative error entirely in the otherwise difficult region $x \to \infty$ for $\alpha < 2$. Third, the minimized expression (4.3) is essentially the crucial second order term, averaged over α. Although $\mid \log(\hat{s}/s) \mid$ is far greater than 3.1×10^{-7} for many

[1]The internal sum in (4.3) provides a slightly biased estimate of the second order term, because of the familiar degrees of freedom problem: If we had as many free parameters as calibration points, we would obtain a "perfect" fit at the calibration point, that nevertheless swerved violently in between them. We are fitting $4 \times 19 = 76$ parameters to $49 \times 49 = 2401$ points in the region $\alpha > 1$, and so should adjust the second order term by a factor of $2401/(2401 - 76) = 1.0327$. An unbiased estimate of the second order t erm is thus 3.2×10^{-7}.

values of x, our least squares procedure guarantees that these errors will algebraically offset one another across x in a highly desirable fashion. For α near 2, the Paretian behavior of the tail becomes increasingly weak, whence the poor relative density fit for $\alpha = 1.98$ at $z = .94$. However, the rapid dropoff of the tail density near normality compensates for this, with the result that the estimated expected relative precision for $\alpha = 1.98$ is only 7×10^{-8}.

The upper portion of the α range (0, 1) is important, even if we expect to find α in [1, 2], since if an estimate of α is less than 1, we will want to test whether this is significant, or merely due to sampling error. Because our approximation is exact (to within the computational precision of the arctangent) for $\alpha = 1$, and because both the stable distribution and our approximating function are smooth functions of α, we obtain a "free ride" without any loss of precision at the above levels for the distribution and density down to $\alpha = 0.92$. The expected relative imprecision remains below 1×10^{-6} down to $\alpha = 0.84$. This is true despite the fact that α values below 1.00 were given no weight in (4.3).

The expected relative imprecision of the density remains below 10^{-5} down to $\alpha = 0.78$, and below 10^{-4} down to 0.70. For smaller sample sizes, e.g. with $N \leq 1,000$ and 100, respectively, the computed log likelihood may still be useful in these extended ranges.

6. Computational Considerations

This numerical approximation was developed in VS FORTRAN, and later converted to GAUSS for use on a personal computer.

Considerable computation time is saved if all of the calculations down to and including the $p_{\alpha iJ}$ in (3.3) are performed only once for each α value of interest, and these coefficients saved for use with different values of x. In GAUSS, passing multiple x values to the procedure as a vector generates further savings.

Computing the stable density approximation as above for an individual value of x with a new value of α takes 15 milliseconds on a Pentium-100 personal computer with GAUSS-386i (3.2.12). This time is reduced to 870 microseconds on average if the α value remains constant from call to call. And if the x-values are submitted in vectors of 1000 each, the average time per ordinate is only 34 microseconds. The distribution itself, which is not required for ML estimation, was not calculated in this timing exercise.

The accuracy of our approximation depends on that of the exponential, sine, arctangent, cumulative normal and gamma functions used to compute the normal and Cauchy benchmarks and the coefficients on the Bergstrom expansions. The gamma function provided with early versions of GAUSS (3.1-) was accurate to at most 6 decimal places, which interfered unacceptably with the computation. In such a case, the function GAMMLN given by [PTVF] (206ff, after C. Lanczos) is an adequate substitute. Later versions of GAUSS (3.2+) use an improved, acceptable approximation.

A VS FORTRAN program named SYMSTB to compute the symmetric table distribution and density may be downloaded by anonymous FTP from STATLIB:

http://lib.stat.cmu.edu/general, file symstb.

It should be noted that VS FORTRAN supports a generic function GAMMA that works in either double or single precision, depending on context. This function is used SYMSTB, but may not be recognized by non-VS compilers.

GAUSS versions that compute the density by itself, the density together with the distribution, and the density and the distribution together with the numerical derivative with respect to x are available from the American Univ. GAUSS software archive:

http://gurukul.ucc.american.edu/econ/gaussres/GAUSSIDX.HTM

in directory PDFs and RNGs.

7. Conclusion

The symmetric stable density approximation developed in this paper can surely be improved upon, both in terms of precision and compactness, and it is hoped that numerical analysts will at some future date do this. Meanwhile, it provides an adequate tool for statistical work with these distributions.

The present approximation makes no attempt to deal with the important asymmetric stable distributions. These distributions have an additional parameter β (here set to 0) that adds a third dimension to the problem. It is hoped that numerical analysts will take the present note as a challenge to tackle this problem as well. [MP] provide a high precision tabulation of the maximally skewed ($\beta = 1$) stable distribution and density as a first step toward developing such an approximation. [NJ] provides a program to compute the general stable densities, by means of the Zolotarev integrals.

References

[BH] Bergstrom, H. On some expansions of stable distribution functions. *Arkiv fur Matematik* **2** (1952) 375-78.

[DW] DuMouchel, W. H. *Stable Distributions in Statistical Inference*, unpublished doctoral dissertation, Yale University, 1971.

[HC] Holt, D., and Crow, E. L. Tables and graphs of the stable probability density functions, *J. of Research of the National Bureau of Standards* **77B** (1973) 143-98.

[MC] McCulloch, J. H. Financial applications of stable distributions. In G.S. Maddala and C.R. Rao, eds., *Statistical Methods in Finance (Handbook of Statistics, vol. 14)*, Elsevier Science, 393-425.

[MP] McCulloch, J. H., and Panton, D. B. Precise fractiles and fractile densities of the maximally-skewed stable distributions. *Computational Statistics and Data Analysis.* In press for 1996.

[MR] Mittnik, S., and Rachev, S. T. Modeling asset returns with alternative stable distributions, *Econometric Reviews* **12** (1993) 261-330.

[NJ] Nolan, J. P. Program STABLE. American University, Mathematics Dept., 1996. Downloadable from http://cage.cas.american.edu/~jpnolan.

[PTVF] Press, W. H., Teukolsky, S. A., Vetterling, W. T. and Flannery, B. P. *Numerical Recipes in FORTRAN: The Art of Scientific Computing,* 2nd ed. Cambridge Univ. Press, 1992.

[ST] Samorodnitsky, G., and Taqqu, M. S. *Stable Non-Gaussian Random Processes.* Chapman and Hall, New York, 1994.

[ZV] Zolotarev, V. M. *One-Dimensional Stable Laws* American Mathematical Society, Providence, RI, 1986; Russian original 1983.

J. Huston McCulloch
Ohio State Univ.
1945 N. High St.
Columbus, OH 43210

mcculloch.2@osu.edu
(614) 292-0382
FAX 292-3906

Table of the Maximally-Skewed Stable Distributions

J. Huston McCulloch and Don B. Panton

Abstract

Fractiles of the maximally positively skewed stable distributions are given for $p = .0001, .001, .005, .01(.01).99, .995, .999, .9999$, and for $\alpha = .5(.1)2.0$.

1. Introduction

The accompanying table gives selected fractiles of the maximally positively skewed stable distributions $S_{\alpha 1}(x)$. If X is a random variable such that $P(X < x) = S_{\alpha\beta}(x)$, then its log characteristic function is given by

$$\log E e^{iXt} = \psi_{\alpha\beta}(x) = \begin{cases} -\mid t\mid^{\alpha} [1 - i\beta \text{ sign } (t) \tan(\pi\alpha/2)], & \alpha \neq 1, \\ -\mid t \mid [1 + i\beta(2/\pi) \text{ sign } (t) \log \mid t \mid], & \alpha = 1. \end{cases} \tag{1.1}$$

In the present tabulation, the skewness parameter β takes on its maximum permissible value, $+1$. The fractiles are those values of x such that $S_{\alpha,1}(x) = p$, *for* $p = .0001, .001, .005, .01(.01).99, .995, .999$, and for $\alpha = .5(.1)2.0$. The table is excerpted from the more detailed tabulations of the maximally skewed stable fractiles and fractile densities of [MP], q.v. for details of computation. A trailing digit, if present, indicates a power of 10 that must be added. See [ST], p. 597-601, for a comparable tabulation of the fractiles of the standard symmetric stable distributions, $S_{\alpha 0}(x)$.

A location parameter $\delta \in (-\infty, \infty)$ that shifts the distribution to the left or right, and a scale parameter $c \in (0, \infty)$ that expands or contracts it about δ may be added in such a way that the general stable cdf may be written as

$$S(x; \alpha, \beta, c, \delta) = S_{\alpha\beta}((x - \delta)/c). \tag{1.2}$$

The general stable log cf implied by (1.2) is then

$$\log E e^{iXt} = i\delta t + \psi_{\alpha\beta}(ct). \tag{1.3}$$

For further details see [MC1].

In the tabulation, the characteristic exponent α runs from 0.5, which is equivalent to a Lévy (inverse $\chi^2(1)$) distribution, up to its maximum permissible value, 2.0, which corresponds to a normal or Gaussian distribution with variance 2. As α approaches 2.0, the skewness parameter loses its effect, so that the distribution becomes completely symmetrical when α reaches 2.0, regardless of β.

When $\alpha > 1$, the mean of the distribution exists and equals 0 in the standard case tabulated here. Because the distributions are positively skewed with an upper tail that becomes increasingly heavy as α falls towards 1, the median and most other quantiles become increasingly negative as this occurs. The central fractiles, in fact, can be shown to behave very much like $\tan(\pi\alpha/2)$, which is 0 at $\alpha = 2$, and goes to minus infinity as α declines toward 1.

When $\alpha < 1$, $S_{\alpha,1}(x) = 0$ unless $x > 0$. These stable distributions are therefore said to be *positive*. Their mean is undefined, but their location parameter δ (set to 0 in the standard case tabulated here) may be interpreted as their *focus of stability* (see [MC1]). It may be seen that as α approaches 1 from below, the median and other central fractiles become increasingly large. In fact, it can be shown that they are again behaving very much like $\tan(\pi\alpha/2)$, which is positive for $0 < \alpha < 1$ and tends to plus infinity as α increases toward 1.

When $\alpha = 1$, the conventional location parameter δ is defined in such a way that it is relatively near the center of the distribution. When normalized in terms of δ, the fractiles thus all undergo a double discontinuity as α passes 1 unless $\beta = 0$. When $\alpha = 1$ and $\beta \neq 0$, the mean is still undefined, but no focus of stability exists. These unusual stable distributions may thus be characterized as *afocal*.

[ZV], p. 11, has shown that if we define $\zeta_{\alpha\beta}$ by

$$\zeta_{\alpha\beta} = \begin{cases} \delta + \beta\tan(\pi\alpha/2), & \alpha \neq 1, \\ \delta, & \alpha = 1, \end{cases} \tag{1.4}$$

and set $z = x - \zeta_{\alpha\beta}$, then $S_{\alpha,\beta}(z + \zeta_{\alpha\beta})$ undergoes no discontinuity, holding z constant but allowing $\zeta_{\alpha\beta}$ to vary, as α passes unity.

The full table of [MP] is available at http://www.econ.ohio-state.edu/jhm/jhm.html.

Fractiles of Maximally-Skewed Stable Distributions				
	alpha			
	0.5	0.6	0.7	0.8
p				
0.0001	0.0661	0.2064	0.5471	1.4102
0.0010	0.0924	0.2588	0.6337	1.5376
0.0050	0.1269	0.3209	0.7291	1.6704
0.0100	0.1507	0.3607	0.7870	1.7476
0.0200	0.1848	0.4144	0.8620	1.8447
0.0300	0.2123	0.4557	0.9176	1.9146
0.0400	0.2371	0.4915	0.9644	1.9723
0.0500	0.2603	0.5241	1.0061	2.0229
0.0600	0.2827	0.5546	1.0446	2.0689
0.0700	0.3046	0.5839	1.0808	2.1116
0.0800	0.3263	0.6123	1.1153	2.1519
0.0900	0.3479	0.6401	1.1487	2.1904
0.1000	0.3696	0.6675	1.1812	2.2275
0.1500	0.4826	0.8034	1.3370	2.4008
0.2000	0.6089	0.9454	1.4917	2.5663
0.2500	0.7557	1.1008	1.6537	2.7338
0.3000	0.9309	1.2763	1.8292	2.9096
0.3500	1.1449	1.4794	2.0245	3.0992
0.4000	1.4118	1.7200	2.2470	3.3089
0.4500	1.7524	2.0113	2.5064	3.5460
0.5000	2.1981	2.3730	2.8159	3.8204
0.5500	2.7986	2.8344	3.1950	4.1460
0.6000	3.6364	3.4424	3.6738	4.5438
0.6500	4.8567	4.2755	4.3010	5.0466
0.7000	6.7353	5.4758	5.1612	5.7103
0.7500	9.8492	7.3237	6.4151	6.6373
0.8000	1.5580 1	1.0447 1	8.4056	8.0387
0.8500	2.7960 1	1.6532 1	1.2002 1	1.0427 1
0.9000	6.3328 1	3.1715 1	2.0135 1	1.5436 1
0.9100	7.8261 1	3.7601 1	2.3096 1	1.7172 1
0.9200	9.9138 1	4.5505 1	2.6957 1	1.9386 1
0.9300	1.2959 2	5.6526 1	3.2168 1	2.2302 1
0.9400	1.7651 2	7.2658 1	3.9519 1	2.6301 1
0.9500	2.5431 2	9.7867 1	5.0525 1	3.2098 1
0.9600	3.9755 2	1.4108 2	6.8448 1	4.1178 1
0.9700	7.0702 2	2.2642 2	1.0164 2	5.7201 1
0.9800	1.5912 3	4.4213 2	1.7852 2	9.1951 1
0.9900	6.3659 3	1.3942 3	4.7269 2	2.1150 2
0.9950	2.5464 4	4.4111 3	1.2617 3	4.9442 2
0.9990	6.3662 5	6.4311 4	1.2489 4	3.6450 3
0.9999	6.3662 7	2.9832 6	3.3453 5	6.4612 4

Fractiles of Maximally-Skewed Stable Distributions				
	——— alpha ———			
	0.9	1.0	1.1	1.2
p				
0.0001	4.3896	-2.1849	-8.7634	-5.7966
0.0010	4.5630	-1.9613	-8.4861	-5.4627
0.0050	4.7361	-1.7458	-8.2265	-5.1577
0.0100	4.8336	-1.6275	-8.0871	-4.9969
0.0200	4.9530	-1.4855	-7.9226	-4.8097
0.0300	5.0372	-1.3873	-7.8105	-4.6838
0.0400	5.1055	-1.3087	-7.7217	-4.5850
0.0500	5.1646	-1.2413	-7.6463	-4.5017
0.0600	5.2177	-1.1814	-7.5797	-4.4287
0.0700	5.2665	-1.1267	-7.5194	-4.3629
0.0800	5.3122	-1.0759	-7.4638	-4.3025
0.0900	5.3554	-1.0282	-7.4117	-4.2463
0.1000	5.3968	-0.9828	-7.3626	-4.1935
0.1500	5.5860	-0.7789	-7.1447	-3.9621
0.2000	5.7612	-0.5948	-6.9522	-3.7614
0.2500	5.9336	-0.4178	-6.7705	-3.5750
0.3000	6.1099	-0.2405	-6.5918	-3.3946
0.3500	6.2956	-0.0576	-6.4107	-3.2144
0.4000	6.4958	0.1357	-6.2226	-3.0300
0.4500	6.7168	0.3447	-6.0227	-2.8371
0.5000	6.9662	0.5756	-5.8058	-2.6310
0.5500	7.2547	0.8370	-5.5649	-2.4059
0.6000	7.5975	1.1406	-5.2905	-2.1539
0.6500	8.0186	1.5044	-4.9686	-1.8639
0.7000	8.5571	1.9575	-4.5770	-1.5183
0.7500	9.2835	2.5508	-4.0772	-1.0873
0.8000	1.0339 1	3.3843	-3.3956	-0.5151
0.8500	1.2055 1	4.6863	-2.3674	0.3214
0.9000	1.5440 1	7.1287	-0.5213	1.7651
0.9100	1.6568 1	7.9162	0.0572	2.2061
0.9200	1.7982 1	8.8888	0.7629	2.7381
0.9300	1.9808 1	1.0124 1	1.6472	3.3969
0.9400	2.2256 1	1.1752 1	2.7946	4.2405
0.9500	2.5716 1	1.4005 1	4.3550	5.3706
0.9600	3.0974 1	1.7345 1	6.6217	6.9839
0.9700	3.9906 1	2.2848 1	1.0264 1	9.5215
0.9800	5.8311 1	3.3732 1	1.7228 1	1.4239 1
0.9900	1.1662 2	6.6021 1	3.6811 1	2.6929 1
0.9950	2.4179 2	1.3013 2	7.3291 1	4.9350 1
0.9990	1.3972 3	6.4046 2	3.3092 2	1.9284 2
0.9999	1.7905 4	6.3715 3	2.7104 3	1.3201 3

Fractiles of Maximally-Skewed Stable Distributions				
	alpha			
	1.3	1.4	1.5	1.6
p				
0.0001	-4.9565	-4.6522	-4.5665	-4.5950
0.0010	-4.5633	-4.1972	-4.0473	-4.0088
0.0050	-4.2116	-3.7976	-3.5983	-3.5089
0.0100	-4.0291	-3.5930	-3.3711	-3.2585
0.0200	-3.8191	-3.3600	-3.1148	-2.9783
0.0300	-3.6794	-3.2064	-2.9472	-2.7964
0.0400	-3.5706	-3.0877	-2.8184	-2.6574
0.0500	-3.4796	-2.9888	-2.7117	-2.5428
0.0600	-3.4001	-2.9030	-2.6196	-2.4441
0.0700	-3.3290	-2.8265	-2.5376	-2.3566
0.0800	-3.2639	-2.7568	-2.4633	-2.2776
0.0900	-3.2037	-2.6925	-2.3949	-2.2050
0.1000	-3.1472	-2.6325	-2.3312	-2.1377
0.1500	-2.9025	-2.3746	-2.0600	-1.8528
0.2000	-2.6937	-2.1575	-1.8344	-1.6184
0.2500	-2.5025	-1.9612	-1.6328	-1.4110
0.3000	-2.3199	-1.7760	-1.4445	-1.2191
0.3500	-2.1399	-1.5955	-1.2629	-1.0359
0.4000	-1.9581	-1.4154	-1.0836	-0.8567
0.4500	-1.7705	-1.2318	-0.9028	-0.6776
0.5000	-1.5728	-1.0406	-0.7167	-0.4953
0.5500	-1.3601	-0.8376	-0.5215	-0.3062
0.6000	-1.1257	-0.6172	-0.3122	-0.1059
0.6500	-0.8604	-0.3717	-0.0826	0.1109
0.7000	-0.5503	-0.0897	0.1769	0.3521
0.7500	-0.1719	0.2477	0.4815	0.6301
0.8000	0.3184	0.6747	0.8583	0.9664
0.8500	1.0143	1.2637	1.3637	1.4050
0.9000	2.1719	2.2089	2.1457	2.0584
0.9100	2.5171	2.4842	2.3681	2.2392
0.9200	2.9294	2.8099	2.6284	2.4486
0.9300	3.4344	3.2046	2.9405	2.6967
0.9400	4.0734	3.6984	3.3265	2.9995
0.9500	4.9181	4.3431	3.8242	3.3845
0.9600	6.1057	5.2370	4.5049	3.9032
0.9700	7.9396	6.5950	5.5234	4.6670
0.9800	1.1269 1	9.0099	7.3020	5.9779
0.9900	1.9897 1	1.5076 1	1.1654 1	9.1171
0.9950	3.4479 1	2.4954 1	1.8525 1	1.3954 1
0.9990	1.2044 2	7.9294 1	5.4192 1	3.7884 1
0.9999	7.0996 2	4.1128 2	2.5154 2	1.5948 2

Fractiles of Maximally-Skewed Stable Distributions				
——— alpha ———				
	1.7	1.8	1.9	2.0
p				
0.0001	-4.6940	-4.8431	-5.0328	-5.2595
0.0010	-4.0379	-4.1136	-4.2259	-4.3702
0.0050	-3.4850	-3.5056	-3.5599	-3.6428
0.0100	-3.2108	-3.2065	-3.2348	-3.2900
0.0200	-2.9060	-2.8762	-2.8776	-2.9044
0.0300	-2.7094	-2.6644	-2.6499	-2.6598
0.0400	-2.5599	-2.5040	-2.4780	-2.4758
0.0500	-2.4371	-2.3727	-2.3378	-2.3262
0.0600	-2.3317	-2.2603	-2.2182	-2.1988
0.0700	-2.2386	-2.1614	-2.1131	-2.0871
0.0800	-2.1547	-2.0724	-2.0187	-1.9871
0.0900	-2.0778	-1.9911	-1.9328	-1.8961
0.1000	-2.0067	-1.9160	-1.8535	-1.8124
0.1500	-1.7076	-1.6021	-1.5238	-1.4657
0.2000	-1.4641	-1.3487	-1.2599	-1.1902
0.2500	-1.2504	-1.1283	-1.0319	-0.9539
0.3000	-1.0544	-0.9276	-0.8258	-0.7416
0.3500	-0.8689	-0.7390	-0.6336	-0.5449
0.4000	-0.6889	-0.5576	-0.4500	-0.3583
0.4500	-0.5108	-0.3796	-0.2711	-0.1777
0.5000	-0.3311	-0.2015	-0.0938	0.0000
0.5500	-0.1467	-0.0205	0.0851	0.1777
0.6000	0.0464	0.1671	0.2685	0.3583
0.6500	0.2529	0.3652	0.4600	0.5449
0.7000	0.4792	0.5793	0.6644	0.7416
0.7500	0.7354	0.8177	0.8883	0.9539
0.8000	1.0387	1.0941	1.1425	1.1902
0.8500	1.4232	1.4344	1.4470	1.4657
0.9000	1.9733	1.9011	1.8468	1.8124
0.9100	2.1211	2.0224	1.9469	1.8961
0.9200	2.2899	2.1588	2.0577	1.9871
0.9300	2.4871	2.3152	2.1821	2.0871
0.9400	2.7239	2.4992	2.3247	2.1988
0.9500	3.0195	2.7234	2.4932	2.3262
0.9600	3.4101	3.0109	2.7005	2.4758
0.9700	3.9734	3.4114	2.9734	2.6598
0.9800	4.9205	4.0596	3.3801	2.9044
0.9900	7.1465	5.5394	4.2054	3.2900
0.9950	1.0521 1	7.7740	5.3942	3.6428
0.9990	2.6621 1	1.8267 1	1.1324 1	4.3702
0.9999	1.0270 2	6.5046 1	3.7172 1	5.2595

References

[MC1] McCulloch, J. H. (1996). On the parametrization of the afocal stable distributions. *Bull. London Math. Soc.,* **28:** 651-655.

[MC2] McCulloch, J. H. Linear regression with stable disturbances.

[MP] McCulloch, J.H. and Panton, D. B. (1996). Precise fractiles and fractile densities of the maximally-skewed stable distributions. *Computational Statistics and Data Analysis,* in press.

[ST] Samorodnitsky, G. and Taqqu, M. S. (1994). *Stable Non-Gaussian Random Processes.* Chapman & Hall, New York.

[ZV] Zolotarev, V.M. (1986). *One-Dimensional Stable Laws.* Amer. Math. Soc.

J. Huston McCulloch
Economics Dept.
Ohio State Univ.
1945 N. High St.
Columbus, OH 43210
e-mail: mcculloch.2@osu.edu
(614) 292-0382
FAX 292-3906

Don B. Panton
Dept. of Finance and Real Estate
University of Texas at Arlington
Arlington, TX 76019
e-mail: panton@uta.edu

Multivariate Stable Distributions: Approximation, Estimation, Simulation and Identification

John P. Nolan

Abstract

An overview of recent work on multivariate stable distributions is given. Examples are shown of stable density surfaces, simulations of stable random vectors, estimating stable spectral measures from data and identifying the type of a stable process.

1. Introduction

A d-dimensional α-stable random vector is determined by a spectral measure Γ (a finite Borel measure on S_d, the unit sphere in $\mathbf{R}^d$) and a shift vector $\boldsymbol{\mu}^0 \in \mathbf{R}^d$, e.g., Samorodnitsky and Taqqu (1994). The notation $\mathbf{X} \sim S_{\alpha,d}(\Gamma, \boldsymbol{\mu}^0)$ will be used to denote such a stable random vector. Until recently, there has been little understanding of what multivariate stable distributions look like, nor practical methods for working with them. The way that Γ spreads the mass over the sphere S_d determines the joint distribution of $\mathbf{X}$, but this description is indirect (through the characteristic function, see (1.1) below) and is not very useful in applications. Two basic questions in applications are: (1) given a spectral measure, what does the density surface look like? (2) given a sample of multivariate stable data, what does the spectral measure look like?

In this paper we review recent work that makes it possible to answer these questions and to use multivariate stable models for practical problems. Our goal here is make these results available to people who want to understand and use multivariate stable distributions, so we state results without proof; references are given for those who are interested in the mathematical details.

Below we summarize some basic facts about multivariate stable distributions. In Section 2 we describe the class of stable random variables that correspond to discrete spectral measures with a finite number of point masses, and show that they correspond to a dense subset of all stable distributions. Then in Section 3, formulas are given for computing multivariate stable densities and examples of such computations are given. A method of generating stable random vectors is given in Section 4. Section 5 is concerned with the problem of estimating a spectral measure from a data set, and the last section discusses the problem of identifying the type of a stable process by looking at the spectral measure associated with finite dimensional distributions of the process.

The characteristic function (c. f.) of $\mathbf{X} \sim S_{\alpha,d}(\Gamma, \mu^0)$ is

$$\phi_{\mathbf{X}}(\mathbf{t}) = \mathbf{E}\exp\{i < \mathbf{X}, \mathbf{t} >\} = \exp(-I_{\mathbf{X}}(\mathbf{t}) + i < \mu^0, \mathbf{t} >), \tag{1.1}$$

where the function in the exponent is

$$I_{\mathbf{X}}(\mathbf{t}) = \int_{S_d} \psi_\alpha(< \mathbf{t}, \mathbf{s} >)\Gamma(d\mathbf{s}). \tag{1.2}$$

Here $< \mathbf{t}, \mathbf{s} >= t_1 s_1 + \cdots + t_d s_d$ is the inner product, and

$$\psi_\alpha(u) = \begin{cases} |u|^\alpha(1 - i\text{sign }(u)\tan\frac{\pi\alpha}{2}) & \alpha \neq 1 \\ |u|(1 + i\frac{2}{\pi}\text{sign }(u)\ln|u|) & \alpha = 1. \end{cases}$$

Since the c. f. is determined by $I_{\mathbf{X}}(t)$, $t \in \mathbf{R}^d$, the (complex valued) function $I_{\mathbf{X}}(t)$ determines the distribution of X. This idea is used below to estimate Γ from data. A stable distribution is symmetric iff Γ is symmetric ($\Gamma(A) = \Gamma(-A)$) iff $I_{\mathbf{X}}(\mathbf{t})$ is real.

For any $t \in \mathbf{R}^d$, the projection of the r. vector $< \mathbf{t}, \mathbf{X} >$ is a one-dimensional r. variable with c. f. $E\exp(iu < \mathbf{t}, \mathbf{X} >) = \exp(-I_{\mathbf{X}}(u\mathbf{t}) + i < \mu^0, u\mathbf{t} >)$. Hence its scale, skewness and shift are given by (Zolotarev (1986), p. 20, or Samorodnitsky and Taqqu (1994), Example 2.3.4):

$$\begin{aligned}
\sigma^\alpha(\mathbf{t}) &= \Re I_{\mathbf{X}}(\mathbf{t}) = \int_{S_d} | < \mathbf{t}, \mathbf{s} > |^\alpha \Gamma(d\mathbf{s}), \\
\beta(\mathbf{t}) &= \sigma^{-\alpha}(\mathbf{t}) \int_{S_d} \text{sign} < \mathbf{t}, \mathbf{s} > | < \mathbf{t}, \mathbf{s} > |^\alpha \Gamma(d\mathbf{s}) \\
&= \begin{cases} -\Im I_{\mathbf{X}}(\mathbf{t})/(\sigma^\alpha(\mathbf{t})\tan\frac{\pi\alpha}{2}) & \alpha \neq 1 \\ \Im[I_{\mathbf{X}}(2\mathbf{t}) - 2I_{\mathbf{X}}(\mathbf{t})]/(4\sigma(\mathbf{t})\ln 2/\pi) & \alpha = 1 \end{cases} \\
\mu(\mathbf{t}) &= \begin{cases} < \mathbf{t}, \mu^0 > & \alpha \neq 1 \\ < \mathbf{t}, \mu^0 > -\frac{2}{\pi}\int < \mathbf{t}, \mathbf{s} > \ln| < \mathbf{t}, \mathbf{s} > |\Gamma(d\mathbf{s}) & \alpha = 1 \end{cases} \\
&= \begin{cases} < \mathbf{t}, \mu^0 > & \alpha \neq 1 \\ -\Im I_{\mathbf{X}}(\mathbf{t})/\sigma(\mathbf{t}) & \alpha = 1 \end{cases}.
\end{aligned}$$

As noted in Zolotarev (1986), the functions $\sigma(\mathbf{t})$, $\beta(\mathbf{t})$, and $\mu(\mathbf{t})$ determine the distribution. An easy way to see this is to note that they determine $I_{\mathbf{X}}(\cdot)$:

$$I_{\mathbf{X}}(\mathbf{t}) = \begin{cases} \sigma^\alpha(\mathbf{t})(1 - i\beta(\mathbf{t})\tan\frac{\pi\alpha}{2}) & \alpha \neq 1 \\ \sigma(\mathbf{t})(1 - i\mu(\mathbf{t})) & \alpha = 1. \end{cases}$$

By replacing $\mathbf{X}$ with $\mathbf{X} - \mu^0$, we can take $\mu^0 = 0$, which we shall do for the remainder of this paper.

Throughout this paper, we will use the standard parameterization described above. This parameterization is poorly behaved in a neighborhood of $\alpha = 1$: if Γ and μ^0 are fixed and $\alpha \to 1$, then the mode of the density goes to infinity in some direction that depends both on the measure Γ and on whether $\alpha < 1$ or $\alpha > 1$. See Nolan (1998b) for a description of other parameterizations of univariate and multivariate stable distributions which do not have this problem.

2. Discrete spectral measures

In this section we discuss the case when Γ is a discrete spectral measure with a finite number of point masses, i.e.,

$$\Gamma(\cdot) = \sum_{j=1}^{n} \gamma_j \delta_{\mathbf{s}_j}(\cdot), \tag{2.1}$$

where γ_j's are the weights, and $\delta_{\mathbf{s}_j}$'s are point masses (Dirac measures of mass 1) at the points $\mathbf{s}_j \in S_d$, $j = 1, \ldots, n$. Such spectral measures arise naturally in several cases: when the components of $\mathbf{X}$ are independent, when $\mathbf{X}$ arises from the finite dimensional distributions of a stable Ornstein-Uhlenbeck process, when the control measure is discrete or the integrand functions are simple in a stochastic integral representation, when one estimates Γ from data, etc. Discrete spectral measures form a particularly simple class to study, we now explain that they are "dense".

For a discrete spectral measure (2.1), (1.1) takes the form

$$\phi^*(\mathbf{t}) = \exp\left(-\sum_{j=1}^{n} \psi_\alpha(<\mathbf{t}, \mathbf{s}_j>)\, \gamma_j \right). \tag{2.2}$$

This expression is numerically simple, whereas computing an arbitrary $\phi(\mathbf{t})$ is generally difficult because it involves a $(d-1)$ dimensional integral. Let p be the density with characteristic function (1.1) and let p^* be the density corresponding to (2.2).

Theorem 2.1 *(Theorem 1, Byczkowski, Nolan and Rajput (1993)) Given $\epsilon > 0$, there exists an $n = n(\epsilon, d, \alpha, \Gamma)$, and values $\mathbf{s}_1, \ldots, \mathbf{s}_n$ and $\gamma_1, \ldots, \gamma_n$ so that*

$$\sup_{\mathbf{x} \in \mathbf{R}^d} |p(\mathbf{x}) - p^*(\mathbf{x})| < \epsilon. \tag{2.3}$$

This approximation is useful for calculating $p(\mathbf{x})$ pointwise. One can also show that p and p^* are close in any $L^p(\mathbf{R}^d, d\mathbf{x})$, $1 \le p < \infty$ and that the supports of p and p^* can be guaranteed to be the same if the discrete measure is constructed appropriately. Furthermore, the proof generalizes to approximating the probability of arbitrary events, i.e. $\sup |P(A) - P^*(A)| < \epsilon$, where the supremum is over all Borel sets $A \subset \mathbf{R}^d$. Because of this approximation, we will generally restrict ourselves to discrete spectral measures in what follows.

There is a point we wish to stress about discrete spectral measures that is implicit in Theorem 2.1. Even if Γ is absolutely continuous with respect to surface measure, a discrete spectral measure can still approximate the density well. To make this point precise, define the two quantities

$$I_0 = I_0(\mathbf{X}) = \inf\{\Re I_{\mathbf{X}}(\mathbf{t}) : \mathbf{t} \in S_d\} \qquad I_1 = I_1(\mathbf{X}) = \sup\{\Re I_{\mathbf{X}}(\mathbf{t}) : \mathbf{t} \in S_d\}.$$

The first value provides a bound for the decay of the c .f.: $|\phi_{\mathbf{X}}(\mathbf{t})| \leq \exp(-I_0|\mathbf{t}|^\alpha)$ and is used in the proof of Theorem 2.1. Calculating the value of I_0 may be difficult; see Lemmas 6, 7, 8 and 9 of Byczkowski, Nolan and Rajput (1993) and Lemma 1 of Nolan and Rajput (1995) for methods of finding a lower bound for I_0.

The number of point masses needed in Theorem 2.1 is bounded by an expression that depends on ϵ, d, α and two functionals of Γ: the total mass $\| \Gamma \| = \Gamma(S_d)$ and $I_0 = I_0(X)$. Page 23 of Byczkowski, Nolan and Rajput (1993) gives an explicit formula for this number. For example, in two dimensions when $\alpha > 1$,

$$n(\epsilon, 2, \alpha, I_0, \| \Gamma \|) = \left\lceil \frac{k_1 \| \Gamma \|}{[1-(1+k_2 \epsilon I_0^{2/\alpha})^{-\alpha/2}] I_0} \right\rceil, \tag{2.4}$$

where $k_1 = (\pi\alpha/2)(1+|\tan \frac{\pi\alpha}{2}|)$ and $k_2 = 2\pi\alpha/\Gamma(2/\alpha)$ (Γ here is the standard gamma function, not the spectral measure) depend only on α. The constant k_1 in (2.4) tends to infinity as $\alpha \to 1$. We suspect that this difficulty can be eliminated by using the alternate parameterization mentioned at the end of Section 1. Partitioning the unit circle into n equal length arcs $A_1, \ldots, A_n$, letting $\mathbf{s}_j$= midpoint of arc A_j and $\gamma_j = \Gamma(A_j)$ will guarantee that (2.3) holds.

The expression for n increases as I_0 decreases. In fact, it can be shown that, for the total mass $\| \Gamma \|$ fixed, I_0 is maximized when Γ is a uniform measure on S_d. Thus a uniform spectral measure will generally require fewer point masses to accurately approximate than any other spectral measure.

The point that discrete spectral measures are good approximations can be made in a different way. Given two stable r. vectors $\mathbf{X}_1$ and $\mathbf{X}_2$ with respective spectral measures Γ_1 and Γ_2, one measure of the distance between $\mathbf{X}_1$ and $\mathbf{X}_2$ is the total variation norm $\| \Gamma_1 - \Gamma_2 \|$. In Nolan and Panorska (1997), we argue that this metric is too strong and that the metric

$$\| \mathbf{X}_1 - \mathbf{X}_2 \| = \sup_{\mathbf{t} \in S_d} |I_{\mathbf{X}_1}(\mathbf{t}) - I_{\mathbf{X}_2}(\mathbf{t})|$$

is more meaningful. The proof of Theorem 2.1 shows that if this distance is small, then (2.3) will hold. This distance can be small even if Γ_1 and Γ_2 are far apart in the total variation norm, e.g., one is absolutely continuous and the other is discrete.

3. Stable densities

For a set $A \subset S_d$, define the cone generated by A to be $\text{Cone}(A) = \{\mathbf{x} \in \mathbf{R}^d : |\mathbf{x}| > 0, \mathbf{x}/|\mathbf{x}| \in A\} = \{r\mathbf{a} : r > 0, \mathbf{a} \in A\}$. In general, the support of a stable density is the span of the support of its spectral measure. In the totally skewed case when $\alpha < 1$, the support of a stable density is the positive span of the support of its spectral measure.

An important result in inderstanding the connection between the spectral measure Γ and the distribution of $\mathbf{X}$ is the following theorem on the tails of the distribution due to Araujo and Giné (1980).

Theorem 3.1 *(Corollary 6.20, Araujo and Giné, 1980)*

$$\lim_{r\to\infty} \frac{P(\mathbf{X} \in \text{Cone}(A), |\mathbf{X}| > r)}{P(|\mathbf{X}| > r)} = \frac{\Gamma(A)}{\Gamma(S_d)}.$$

In words, the mass that Γ assigns to A determines the tail behavior of $\mathbf{X}$ in the "direction" A. In contrast, the local behavior of the distribution near the mode is bounded by functions of $I_0(\mathbf{X})$ and $I_1(\mathbf{X})$. In particular, $I_0(\mathbf{X})$ and $I_1(\mathbf{X})$ are invariant under rotations of the spectral measure, so local behavior is very different from the directional tail behavior in Theorem 3.1. The justification of these last remarks are contained in Abdul-Hamid (1996).

To understand more about multivariate stable densities, numerical methods can be used. The inversion formula for characteristic functions shows that

$$\begin{aligned} p(\mathbf{x}) &= (2\pi)^{-d} \int_{\mathbf{R}^d} e^{-i<\mathbf{x},\mathbf{t}>} e^{-I_{\mathbf{X}}(\mathbf{t})} d\mathbf{t} \\ &= (2\pi)^{-d} \int_{\mathbf{R}^d} e^{-i[<\mathbf{x},\mathbf{t}> + \Im I_{\mathbf{X}}(\mathbf{t})] - \Re I_{\mathbf{X}}(\mathbf{t})} d\mathbf{t} \\ &= (2\pi)^{-d} \int_{\mathbf{R}^d} \cos[<\mathbf{x},\mathbf{t}> + \Im I_{\mathbf{X}}(\mathbf{t})] e^{-\Re I_{\mathbf{X}}(\mathbf{t})} d\mathbf{t}. \end{aligned}$$

In Nolan and Rajput (1995), the problem of numerically evaluating this integral is addressed. Here we present two examples of such calculations in two dimensions with a discrete spectral measure. More examples can be found in Byczkowski, Nolan and Rajput (1993) and in Nolan and Rajput (1994).

Figure 1 shows a symmetric stable density with $\alpha = 0.9$ and $n = 6$ point masses, Figure 2 shows a nonsymmetric stable density with $\alpha = 1.6$ and $n = 5$ point masses; the location and strength of the masses are given in Table 1. Note that the level curves in Figure 2 are quite smooth, even though the measure only has five point masses.

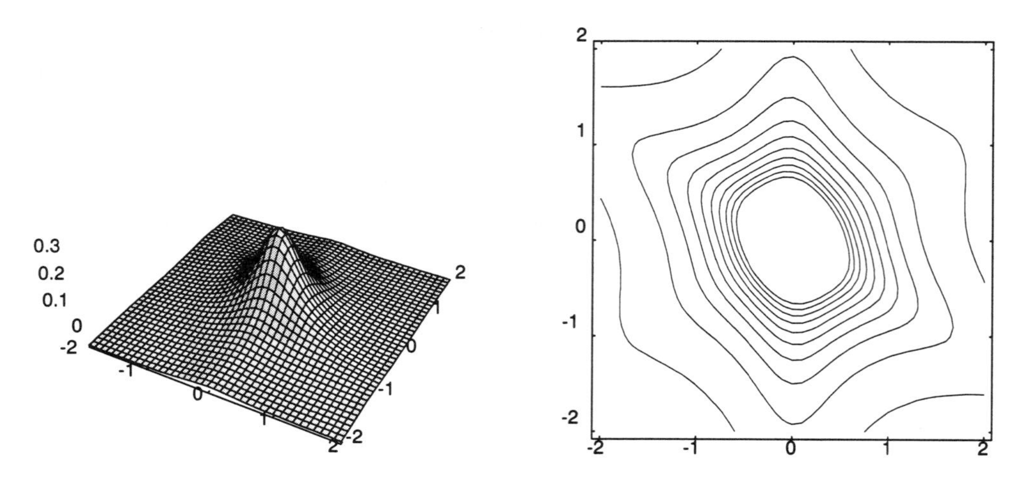

Figure 1. Two dimensional symmetric stable density surface and level curves with $\alpha = 0.9$ and $n = 6$ point masses.

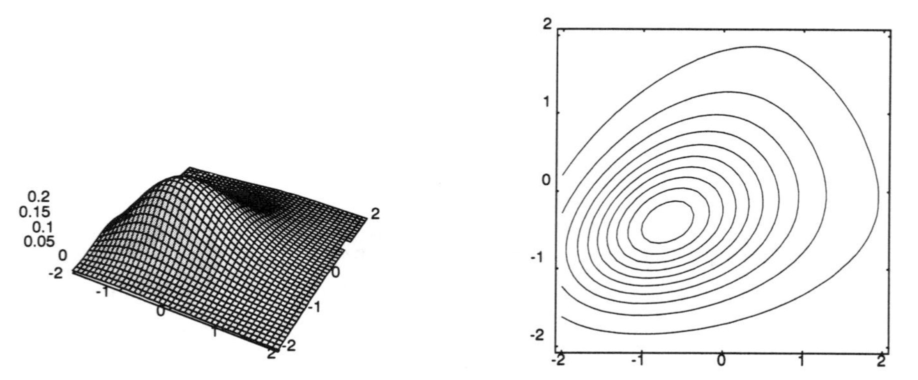

Figure 2. Two dimensional stable density surface and level curves with $\alpha = 1.6$ and $n = 5$ point masses.

Figure 1 $\alpha = 0.9, n = 6$		Figure 2 $\alpha = 1.6, n = 5$	
γ_i	$\mathbf{s}_i$	γ_i	$\mathbf{s}_i$
0.250	(1,0)	0.1	(1,0)
0.125	$(1/2, \sqrt{3}/2)$	0.3	$(\sqrt{3}/2, 1/2)$
0.205	$(-1/2, \sqrt{3}/2)$	0.1	$(1/2, \sqrt{3}/2)$
0.250	(-1,0)	0.3	(0,1)
0.125	$(-1/2, -\sqrt{3}/2)$	0.1	$(-1/2, \sqrt{3}/2)$
0.250	$(1/2, -\sqrt{3}/2)$		

Table 1: Description of the discrete spectral measures used to compute Figures 1 and 2.

While the densities are unimodal, their fine behavior can be very different from a Gaussian distribution. In the symmetric case, the mode is at the origin, in the nonsymmetric case the mode is shifted. The exact location of the mode is not known in general; a necessary step in finding the mode is to find the mode of a one dimensional stable density, which is still not exactly known (see Nolan (1998b) for accurate numerical location of univariate modes). Applied to a discrete measure, Theorem 3.1 says that the tails of the density will have "creases" along rays starting at the origin and passing through the point masses. The level curves show this difference more distinctly than the density surface. The level curves are star-like in Figure 1, while they are egg-shaped in Figure 2. Generally, as α gets closer to 2 and as the mass of Γ is more uniformly spread, the level curves get rounder.

The following formula expresses $p(\mathbf{x})$ in terms of the functions $\sigma(\mathbf{t})$, $\beta(\mathbf{t})$ and $\mu(\mathbf{t})$, $\mathbf{t} \in S_d$. The formula makes density calculations more practical and gives insight into the structure of multivariate stable laws.

Theorem 3.2 *(Abdul-Hamid and Nolan (1996)) Let* $\mathbf{X} \sim S_{\alpha,d}(\Gamma, \mu^0)$ *be a nondegenerate stable random vector with* $d \geq 1$.
(a) When $\alpha \neq 1$*, the density of* $\mathbf{X}$ *is given by*

$$p(\mathbf{x}) = \int_{S_d} g_{\alpha,d}\left(\frac{< \mathbf{x} - \mu^0, \mathbf{s} >}{\sigma(\mathbf{s})}, \beta(\mathbf{s})\right) \sigma^{-d}(\mathbf{s}) d\mathbf{s},$$

where

$$g_{\alpha,d}(v, \beta) = \frac{1}{(2\pi)^d} \int_0^\infty \cos(vu - (\beta \tan \tfrac{\pi\alpha}{2}) u^\alpha) u^{d-1} e^{-u^\alpha} du.$$

(b) When $\alpha = 1$*, the density of* $\mathbf{X}$ *is given by*

$$p(\mathbf{x}) = \int_{S_d} g_{1,d}\left(\frac{< \mathbf{x} - \mu^0, \mathbf{s} > - \mu(\mathbf{s}) + \frac{2}{\pi}\beta(\mathbf{s})\sigma(\mathbf{s}) \ln \sigma(\mathbf{s})}{\sigma(\mathbf{s})}, \beta(\mathbf{s})\right) \sigma^{-d}(\mathbf{s}) d\mathbf{s},$$

where

$$g_{1,d}(v,\beta) = \frac{1}{(2\pi)^d}\int_0^\infty \cos(vu - \frac{2}{\pi}\beta u \ln u)u^{d-1}e^{-u}du.$$

In part (a), the quantity $v(\mathbf{s}) =< \mathbf{x} - \boldsymbol{\mu}^0, \mathbf{s} > /\sigma(\mathbf{s})$ is the projection of $(\mathbf{x}-\boldsymbol{\mu}^0)$ onto $\mathbf{s}$, scaled by $\sigma(\mathbf{s})$. Thus the formula expresses the density in terms of a weighted average over the compact set S_d of $g_{\alpha,d}(v(\mathbf{s}), \beta(\mathbf{s}))$. This representation for the density is much more amenable to numerical calculations than the numerical inversion described above. Note that $g_{\alpha,d}$ is a function of two real variables no matter what the dimension d is, and that this function is the same for every α-stable r. vector $\mathbf{X}$. A numerical evaluation of $g_{\alpha,d}(v, 0)$ has been implemented in Abdul-Hamid (1996).

4. Simulation of stable random vectors

There are several reasons to simulate stable random vectors. One is to have a source of multivariate noise with heavy tails to use in evaluating the robustness of multivariate statistical methods. A second reason is to be able to use Monte Carlo techniques. The numerical computations discussed in the previous section are computationally expensive. The availability of a fast method to generate stable random vectors makes it possible to empirically estimate a stable density or the probability of an event. Currently, it is only feasible to numerically calculate two dimensional densities, and the computation time is a complicated function of α and the number, mass and spread of the points. In any dimension, the computation time for simulation is essentially a linear function of the number of point masses times the number of vectors simulated. (Of course in higher dimensions, many simulations are necessary to get accurate estimates.)

We do not know how to simulate a general stable random vector directly. But as we discussed in Section 2, one can get arbitrarily close to a general stable distribution by using ones with discrete spectral measures. We now describe how to simulate a stable random vector with a discrete spectral measure. The method is based on the following result from Modarres and Nolan (1994), see also Example 2.3.6 of Samorodnitsky and Taqqu (1994). If $\mathbf{X}$ has characteristic function (2.2), then

$$\mathbf{X} \stackrel{D}{=} \begin{cases} \sum_{j=1}^n \gamma_j^{1/\alpha} Z_j \mathbf{s}_j & \alpha \neq 1 \\ \sum_{j=1}^n \gamma_j (Z_j + \frac{2}{\pi}\ln\gamma_j)\mathbf{s}_j & \alpha = 1, \end{cases} \tag{4.1}$$

where $Z_1, \ldots, Z_n$ are iid totally skewed, standardized one dimensional α-stable random variables, i.e. $Z_i \sim S_\alpha(1,1,0)$. (When $\boldsymbol{\mu}^0 \neq 0$, both cases above should be shifted by $+\boldsymbol{\mu}^0$.)

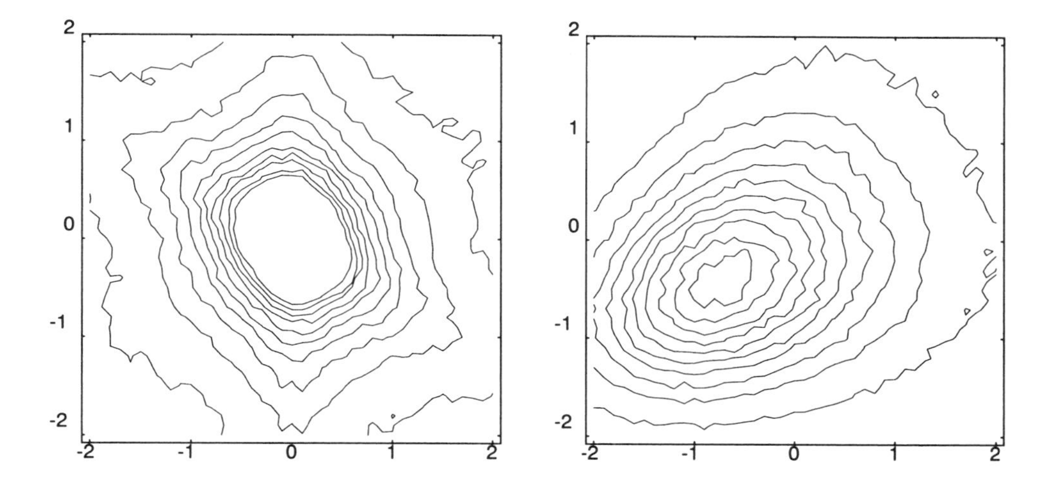

Figure 3. Contour plots of empirical densities from simulation of two dimensional stable distributions. On the left is the distribution shown in Figure 1, on the right is the distribution shown in Figure 2. 1,000,000 simulations were done in each case.

Figure 3 shows results of simulations using the spectral measures described in Figures 1 and 2. For each figure, a grid was set up and the number of simulated vectors that fell into each grid was tabulated. This histogram was regarded as an empirical density and level curves were plotted. Computation times on a 60 MHz Pentium for Figures 1 and 2 were around 30 minutes each, whereas the simulations in Figure 3 took approximately 10 minutes each.

5. Estimation of stable spectral measures

The need for an estimator of a multivariate α-stable spectral measure arises in connection with stochastic modeling of financial portfolios. The spectral measure carries essential information about the vector, in particular about the dependence structure between the individual stocks that make up the portfolio. Other applications involve modeling the relation between exchange rates of different currencies. In Tsakalides and Nikios (1995), it is shown that modeling data from an array of radar sensors with a two dimensional (radially symmetric) Cauchy distribution yields significant improvement over the standard Gaussian model. Analysis of such experimental data may give a better model for the radar data that could lead to further improvements in signal processing. Finally, combined with Section 6, methods of estimating a spectral measure give a way of identifying the type of a stable process by looking at a time series.

One solution to this estimation problem is due to Rachev and Xin (1993) and Cheng and Rachev (1995). In Nolan, Panorska and McCulloch (1996), two other solutions are given; all 3 methods are implemented and compared on test data. Below we give a brief description of all three methods. Let $\mathbf{X}_1, \ldots, \mathbf{X}_k$ be the sample vectors; for conciseness we assume that $\boldsymbol{\mu}^0 = 0$.

The Rachev-Xin-Cheng method is based on Theorem 3.1. An ad hoc value is picked for r and it is used to estimate the measure of a set $A \subset S_d$ by

$$\widehat{\Gamma}(A) = \text{const.} \frac{\#\{\mathbf{X}_i : |\mathbf{X}_i| > r, \mathbf{X}_i \in \text{Cone}(A)\}}{\#\{\mathbf{X}_i : |X_i| > r\}}.$$

The suggested value of r is the 80th percentile of the lengths of the data vectors: $|\mathbf{X}_1|, |\mathbf{X}_2|, \ldots, |\mathbf{X}_k|$.

The next two estimators are based on using the sample to estimate the c .f. on some grid. More precisely, we will estimate $I_{\mathbf{X}}(\cdot)$ on a grid $\mathbf{t}_1, \ldots, \mathbf{t}_n \in S_d$. Below we show how to recover an estimate of the spectral measure from this.

The empirical characteristic function (ECF) method is straightforward. Given an i.i.d. sample $\mathbf{X}_1, \ldots, \mathbf{X}_k$ of α-stable random vectors with spectral measure Γ, let $\widehat{\phi}_k(\mathbf{t})$ and $\widehat{I}_k$ be the empirical counterparts of ϕ and I, i.e. $\widehat{\phi}_k(\mathbf{t}) = (1/k)\sum_{j=1}^k \exp(i < \mathbf{t}, \mathbf{X}_j >)$ is the sample c .f., and $\widehat{I}_k(\mathbf{t}) = -\ln \widehat{\phi}_k(\mathbf{t})$. Given a grid $\mathbf{t}_1, \ldots, \mathbf{t}_n \in S_d$, $\vec{I}_{ECF,k} = \left[\hat{I}_k(\mathbf{t}_1), \ldots, \hat{I}_k(\mathbf{t}_n)\right]'$ is the ECF estimate of $I_{\mathbf{X}}(\cdot)$.

The third method was suggested by McCulloch (1994). We call it a projection method because it is based on one-dimensional projections of the data set. For each gridpoint $\mathbf{t}_j$, define the one-dimensional data set $< \mathbf{t}_j, \mathbf{X}_1 >, \ldots, < \mathbf{t}_j, \mathbf{X}_k >$. Use some method to estimate the scale $\hat{\sigma}(\mathbf{t}_j)$ and skewness $\hat{\beta}(\mathbf{t}_j)$ (and shift $\hat{\mu}(\mathbf{t}_j)$ when $\alpha = 1$) of this one-dimensional data. The best method for doing this is to use maximum likelihood estimation. While several researchers have used this approach, it is prohibitively expensive (in computational time) in general. Efficient programs exist for calculating symmetric stable densities for $\alpha > 0.85$ and new programs will soon be available for calculating general stable densities, so this method will become practical soon. See McCulloch (1996) for references to ML estimation and the symmetric density program, see Nolan (1997) for the nonsymmetric case. Since we needed a fast method for estimating parameters of general stable distributions, we used the quantile based estimators of McCulloch (1986) in the examples below. Define

$$\hat{I}_k(\mathbf{t}_j) = \begin{cases} \hat{\sigma}^\alpha(\mathbf{t}_j)\left(1 - i\hat{\beta}(\mathbf{t}_j)\tan\frac{\pi\alpha}{2}\right) & \alpha \neq 1 \\ \hat{\sigma}(\mathbf{t}_j)\left(1 - i\hat{\mu}(\mathbf{t}_j)\right) & \alpha = 1. \end{cases}$$

The vector $\vec{I}_{PROJ,k} = \left[\hat{I}_k(\mathbf{t}_1), \ldots, \hat{I}_k(\mathbf{t}_n))\right]'$ is the projection estimator of $I_{\mathbf{X}}(\cdot)$.

In order to obtain an estimate of the spectral measure $\widehat{\Gamma}$, we need to invert the discrete approximations to the characteristic function obtained by the emprical c .f. method and by the projection method. We start with the case when Γ is a discrete spectral measure of form (2.1), in which case $I_{\mathbf{X}}(\mathbf{t}) = \sum_{j=1}^{n} \psi_\alpha(<\mathbf{t}, \mathbf{s}_j>)\gamma_j$. Let $\mathbf{t}_1, \ldots, \mathbf{t}_n \in R^d$ and define the $n \times n$ matrix

$$\Psi = \Psi(\mathbf{t}_1, \ldots, \mathbf{t}_n; \mathbf{s}_1, \ldots, \mathbf{s}_n) = \begin{bmatrix} \psi_\alpha(<\mathbf{t}_1, \mathbf{s}_1>) & \ldots & \psi_\alpha(<\mathbf{t}_1, \mathbf{s}_n>) \\ \vdots & \vdots & \vdots \\ \psi_\alpha(<\mathbf{t}_n, \mathbf{s}_1>) & \ldots & \psi_\alpha(<\mathbf{t}_n, \mathbf{s}_n>) \end{bmatrix}.$$

If $\vec{\gamma} = [\gamma_1, \ldots, \gamma_n]'$, and $\vec{I} = [I_{\mathbf{X}}(\mathbf{t}_1), \ldots, I_{\mathbf{X}}(t_n)]'$, then

$$\vec{I} = \Psi\vec{\gamma}. \tag{5.1}$$

If $\mathbf{t}_1, \ldots, \mathbf{t}_n \in R^d$ are chosen so that Ψ^{-1} exists, then $\vec{\gamma} = \Psi^{-1}\vec{I}$ is the exact solution of (5.1).

For a general spectral measure Γ (not discrete and/or the location of the point masses are unknown), consider a discrete approximation $\Gamma^* = \sum_{j=1}^{n} \gamma_j \delta_{\mathbf{s_j}}$, where $\gamma_j = \Gamma(A_j)$, $i = 1, \ldots, n$ are the weights, and $\delta_{\mathbf{s}_j}$'s are point masses. When $d = 2$, it is natural to take $\mathbf{s}_j = (\cos(2\pi(j-1)/n), \sin(2\pi(j-1)/n)) \in S_d$, and arcs $A_j = (2\pi(j-(3/2))/n, 2\pi(j-(1/2))/n]$, $j = 1, \ldots, n$. In higher dimensions, the A_j's are patches that partition the sphere S_d, with some "center" $\mathbf{s}_j$. In this case, each of the coordinates of $\vec{\gamma} = [\gamma_1, \ldots, \gamma_n]'$ is an approximation to the mass of the patch containing $\mathbf{s}_j$, $j = 1, \ldots, n$.

The principle behind the estimation of Γ is simple: given some grid $\mathbf{t}_j = \mathbf{s}_j$, $j = 1, \ldots, n$ and either estimate ($\vec{I}_{ECF}$ or $\vec{I}_{PROJ}$) of $\vec{I}$, invert (5.1) to get $\vec{\gamma}$. Using these weights and the grid $\mathbf{s}_1, \ldots, \mathbf{s}_n$, define $\hat{\Gamma}$ by (2.1). While the above method is formally correct, it has several numerical problems. Nolan, Panorska and McCulloch (1996) restate this problem as the solution to a real valued constrained quadratic programming problem to deal with the complex values and ill-conditioned nature of the Ψ matrix.

Before giving examples, we will give some guidelines on picking the grid size for these methods in two dimensions. The intended use of the estimated spectral measure has a bearing on grid size. If one wants the spectral measure to get a rough understanding of the dependence structure, e.g., to determine if the components are independent or positively associated, then one only cares about a rough location of the mass and a coarse grid, say $n = 40$, would suffice. The simulations in Nolan and Panorska (1996) indicate that using a coarse grid does not cause either the ECF or PROJ methods to deteriorate. Both put approximately the right mass in the arcs determined by the grid and both estimate the total mass accurately.

On the other hand, if one wants to make precise use of the estimated spectral measure, e.g., to compute the density of the fitted distribution, then more gridpoints may be necessary. For this situation, the required grid size is determined by the number of point masses needed in Theorem 2.1 to approximate a general spectral measure by a discrete one. As (2.4) shows, this depends on $\epsilon, \alpha, d, \| \Gamma \|$, and $I_0(X)$. We choose ϵ and know d, but must estimate α, $\| \Gamma \|$, and $I_0(\mathbf{X})$ from the sample. All three methods provide estimates of α and $\| \Gamma \|$ for any grid size. The value of $I_0(\mathbf{X})$ may be estimated by numerically minimizing $\Re \hat{I}(\mathbf{t})$ (in the ECF method) or $\hat{\sigma}^{\alpha}(\mathbf{t})$ (in the PROJ method). To choose a grid size, we suggest a multi-step procedure: (1) pick moderate grid size, say $n = 40$ and estimate α and $\| \Gamma \|$, (2) estimate I_0 by numerical minimization, (3) compute n by (2.4), and (4) repeat the estimation of Γ with a uniform grid with $4\lceil n/4 \rceil$ grid points. We suggest a multiple of four gridpoints to ensure that the important case of independence (which corresponds to a spectral measure concentrated at the four points $(1, 0), (0, 1), (-1, 0), (0, -1)$) can be distinguished.

5.1 Examples of estimating Γ

We shall consider two examples of estimating the spectral measure for bivariate data. In the first one, the data is simulated using the procedure of Section 4, the second one estimates the spectral measure for a real sample consisting of foreign exchange rates. The empirical characteristic function method and projection methods with quadratic programming solution were used, the procedure of Rachev-Xin-Cheng (RXC) was also used to estimate spectral measures. In both examples, we used 100 evenly spaced gridpoints $\mathbf{t}_j$, $j = 1, \dots, 100$, where we estimated the weights γ_j. Both examples took less than a minute of execution time on a fast PC. We turn to the description of examples now.

Example 1. Data was simulated from the measure $\Gamma_1(\cdot) = \sum_{j=1}^{3} (1/3)\delta_{\mathbf{s}_j}(\cdot)$, where $\mathbf{s}_j = (\cos\theta_j, \sin\theta_j) \in S_d$, $j = 1, \dots, 3$, and $\theta_1 = \pi/3$, $\theta_2 = \pi$, and $\theta_3 = 3\pi/2$. The shape parameter is $\alpha = 1.25$. The estimated shape parameters for the three methods were: $\alpha_{RXC} = 1.383$, $\alpha_{PROJ} = 1.275$, and $\alpha_{ECF} = 1.270$. The results of estimation of the cumulative spectral measure are shown in Figure 4. Both the ECF and PROJ methods do a good job of recovering Γ; both the location and the size of the point masses are accurately determined. The RXC method seriously underestimates the total mass, smooths out the mass instead of identifying point masses and overestimates α.

Example 2. We looked at a portfolio of two assets: German Mark vs US Dollar (DMUS), and Japanese Yen vs US Dollar (JYUS) exchange rates. The data is in the form of daily log-returns, and spans the period from January 1, 1980 to December 7, 1990 (total of 2563 values). The shape parameters for each component were close, suggesting that a jointly stable model is plausible. We plotted the results of estimation for this data in Figure 5. The estimated shape parameters for the three methods were: $\alpha_{RXC} = 1.579$, $\alpha_{PROJ} = 1.531$, and $\alpha_{ECF} = 1.529$.

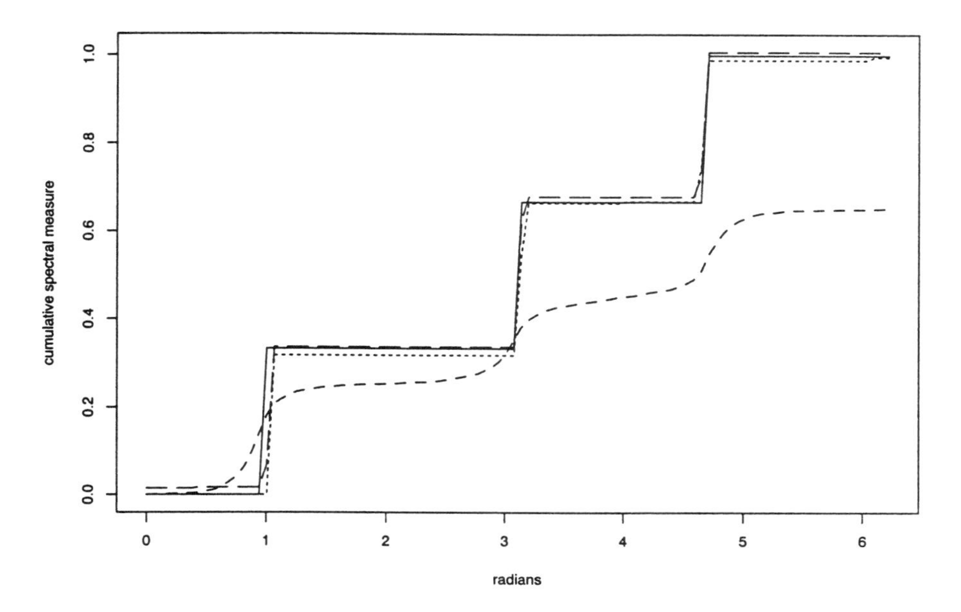

Figure 4. Estimates of the cumulative spectral measure from simulated data. The solid line is the exact cumulative, short dashed line is the Rachev-Xin-Cheng estimator, dotted line is the projection estimator, long dashed line is the characteristic function estimator. (10,000 data vectors)

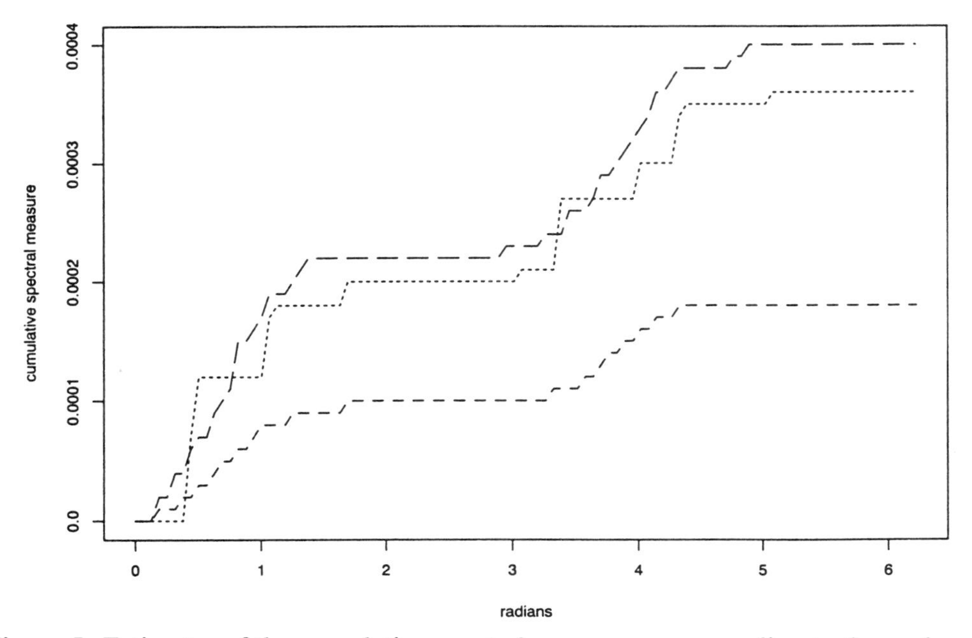

Figure 5. Estimates of the cumulative spectral measure corresponding to the exchange rates data. Line types are as in Figure 4. (2,563 data vectors).

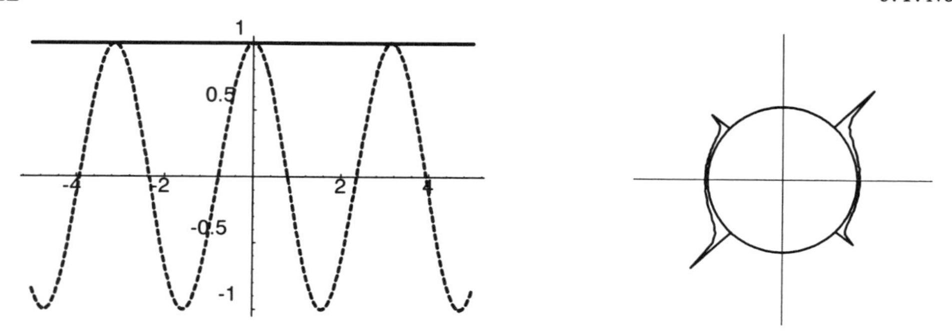

Figure 6. Spectral measure corresponding to $(X(t), X(t+h))$ for a real harmonizable stable process with $\alpha = 0.75$, $h = 2$, and spectral measure $m(dx) = 1/(1+|x|^3)dx$. On the left is the graph of the components of $\mathbf{f}(x) = (1, \cos(hx))$ and on the right is a radial plot of the spectral measure.

6. Identification of stable processes

If a physical process is observed to have stable finite dimensional distributions, it may be of interest to know what type of process is being observed. For example, is it a moving average process, harmonizable, or self-similar? This section will give a brief discussion of how to relate the stochastic integral representation of a stable process to the spectral measure of finite dimensional distributions. By estimating the spectral measure using the preceeding section, we may be able to identify the type of process. Since we will be estimating spectral measures from data, it is unlikely that we can precisely identify which type of process is being observed, rather we can identify plausible models and rule out implausible ones.

An alternative way of representing $\mathbf{X}$ in distribution is the stochastic integral representation, see Theorem 3.5.6 of Samorodnitsky and Taqqu (1994):

$$\mathbf{X} \stackrel{D}{=} \int_M \mathbf{f}(u) W(du) = \left(\int_M f_1(u) W(du), \ldots, \int_M f_d(u) W(du) \right), \tag{6.1}$$

where W is an α-stable measure defined on some measure space $(M, \mathcal{M}, m)$ and $\mathbf{f} = (f_1, \ldots, f_d)$ is a vector of $\mathcal{M}$-measurable funtions. (Again we take $\boldsymbol{\mu}^0 = 0$.)

Theorem 6.1 *(Samorodnitsky and Taqqu (1994), Nolan (1998a)) The spectral measure corresponding to the stochastic integral (6.1) is given by:*

$$\Gamma(A) = \int_{\{x \in M : \mathbf{f}(x) \in \mathrm{Cone}(A)\}} |\mathbf{f}(x)|^\alpha \, m(dx),$$

where A is any Borel subset of S_d.

Note that if the random measure W is symmetric, then $\mathbf{X}$ will always be symmetric, but the measure Γ given by Theorem 6.1 may not be symmetric. It does give the right

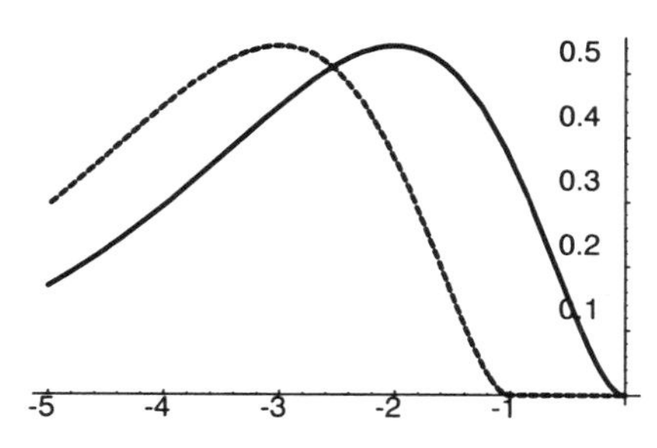

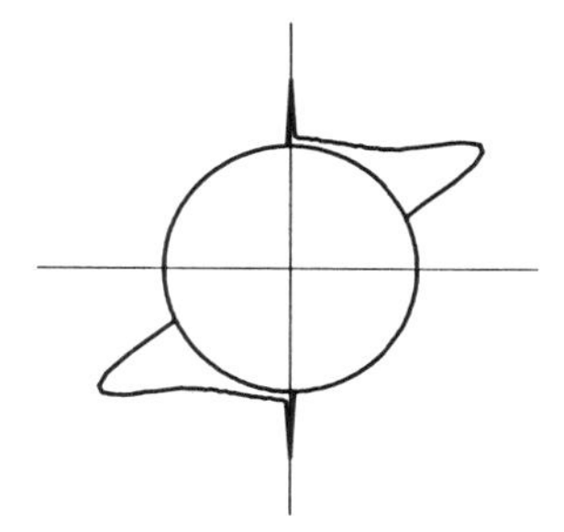

Figure 7. Spectral measure corresponding to $(X(t), X(t+h))$ for a moving average stable process with $\alpha = 1.5$, $h = 1$, and kernel function $g(x) = x^2 e^x 1_{(-\infty,0]}(x)$. On the left is the graph of the components of $\mathbf{f}(x) = (g(x), g(x-h))$ and on the right is a radial plot of the spectral measure.

integral in (1.2) only if we use the symmetric kernel function $\psi_\alpha(u) = |u|^\alpha$. The (unique) symmetric measure that will yield (1.2) is $\Gamma^{sym}(A) \equiv (\Gamma(A) + \Gamma(-A))/2$.

6.1 Examples

Consider the real part of a harmonizable process, i.e., a process of the form $X(t) = \Re \int_{\mathbf{R}} e^{itu} W(du)$, $t \in \mathbf{R}$, where W is a complex valued, rotationally invariant α-stable random measure determined by a finite Borel control measure m on $\mathbf{R}$. The distribution of $X(t)$ is determined completely by m. The rotational symmetry of W means that $X(t)$, the real part of the stochastic integral above, is given by $X(t) = k(\alpha) \int_{\mathbf{R}} \cos(tu) W_1(du)$, where W_1 is a real symmetric random measure with the same control measure m and $k(\alpha) = \left(\frac{1}{2\pi} \int_0^{2\pi} |\cos\theta|^\alpha d\theta\right)^{1/\alpha}$.

The next theorem gives an explicit formula for the spectral measure of the two dimensional distributions of such processes. For the statement, define for $\theta \in (-\pi/4, \pi/4)$ and for $h \in \mathbf{R}$, $B_h(\theta) = \{x \in \mathbf{R} : \cos(hx) \le \tan\theta\}$, and $Sector(\theta_1, \theta_2) = \{(\cos\theta, \sin\theta) : \theta_1 \le \theta \le \theta_2\}$.

Theorem 6.2 *Let Γ be the spectral measure of $(X(t), X(t+h))$ from a real harmonizable process with control measure m. Then support $(\Gamma) \subset Sector(-\pi/4, \pi/4) \bigcup Sector(3\pi/4, 5\pi/4)$ and for any $\theta \ge \pi/4$,*

$$\Gamma\left(Sector(-\pi/4, \theta)\right) = k(\alpha)^\alpha \int_{B_h(\theta)} \left(1 + \cos^2 hx\right)^{\alpha/2} m(dx).$$

We close with some numerical examples. Figure 6 shows the spectral measure of $(X(t), X(t+h))$ where $X(t)$ is a harmonizable process. By stationarity, we can assume $t = 0$, in which case $\mathbf{f}(x) = (1, \cos(hx))$ on $M = \mathbf{R}$. The right graph shows a polar plot of the spectral measure. For reference, the inner circle is the unit circle and the outer curve is a plot of $r(\mathbf{s}) = 1 + c\Gamma(ds)$. (The constant c was chosen for visual display purposes.) The last figure corresponds to a moving average process $X(t) = \int_{-\infty}^{t} g(x-t)W(dx)$, where W is the standard Lévy process (m is Lebesgue measure). Both examples used $m = 200$ points spread uniformly around the unit circle. More theoretical and numerical results are contained in Nolan (1998a).

References

[1] Abdul-Hamid, H. (1996), Approximation of multivariate stable densities, PhD Dissertation, American University.

[2] Abdul-Hamid, H. and Nolan, J. P. (1996), Multivariate stable densities as functions of their one dimensional projections. Submitted.

[3] Araujo, A. and Giné, E. (1980), *The Central Limit Theorem for Real and Banach Valued Random Variables*, Wiley, NY.

[4] Byczkowski, T., Nolan, J. P., and Rajput, B. (1993), Approximation of Multidimensional Stable Densities, *J. of Mult. Anal.*, 46(1), 13-31.

[5] Cheng, B.N., and Rachev, S.T., (1995), Multivariate stable future prices, *Mathematical Finance* 5(2), 133-153.

[6] McCulloch, J. H. (1986), Simple Consistent Estimators of Stable Distribution Parameters, *Communications in Statistics. Simulation and Computation*, 15, 1109-1136.

[7] McCulloch, J. H. (1994), Estimation of bivariate stable spectral densities, Unpublished manuscript, Department of Economics, Ohio State University.

[8] McCulloch, J. H. (1996), Financial Applications of Stable Distributions, *Statistical Methods in Finance, Handbook of Statistics, Volume 14*, Edited by Maddala, G. S. and Rao, C. R., North-Holland, NY, In press.

[9] Modarres, R., and Nolan, J. P., (1994), A Method for Simulating Stable Random Vectors, *Computational Statistics*, **9**, 11-19.

[10] Nolan, J. P. and Rajput, B., (1995) Calculation of multidimensional stable densities, *Commun. Statist. - Simula.*, **24**: 551-556.

[11] Nolan, J. P. (1997), Numerical computation of stable densities and distribution functions, *Stochastic Models* **13**, 759–774.

[12] Nolan, J. P. (1998a), Identification of stable measures, In preparation.

[13] Nolan, J. P. (1998b), Parameterizations and modes of stable distributions, To appear in *Stat. and Prob. Letters*.

[14] Nolan, J. P., Panorska, A. K. (1997), Data analysis for heavy tailed multivariate samples, *Stochastic Models* **13**, 687-702.

[15] Nolan, J. P., Panorska, A. K., McCulloch, J. H. (1996). "Estimation of stable spectral measures", Submitted.

[16] Rachev S.T., and Xin, H., (1993), Test for Association of Random Variables in the Domain of Attraction of Multivariate Stable Law, *Probability and Mathematical Statistics*, **14** (I) 125 - 141.

[17] Samorodnitsky, G. and Taqqu, M. S. (1994) *Stable Non-Gaussian Random Processes*. Chapman and Hall, NY, NY.

[18] Tsakalides, P. and Nikias, C. (1995) Maximum likelihood localization of sources in noise modeled as a stable processes. *IEEE Trans. on Signal Proc.* **43**.

[19] Zolotarev, V. M. (1986) *One-dimensional Stable Distributions, Amer. Math. Soc. Transl. ofMath. Monographs, Vol. 65*. Amer. Math. Soc., Providence, RI. (Transl. of the original 1983 Russian)

American University
Department of Mathematics and Statistics
Washington, DC 20016

Univariate Stable Distributions: Parameterizations and Software

John P. Nolan

Abstract

This note announces the availability of the STABLE software package - a collection of programs that compute densities, distribution functions and quantiles for general stable distributions. That package uses three parameterizations of stable laws, which we describe briefly.

1. Introduction

One obstacle to the use of stable distributions in applications has been the lack of effective means of calculating densities, distribution functions and quantiles of stable distributions. Except for a few special cases - Gaussian ($\alpha = 2$), Cauchy ($\alpha = 1, \beta = 0$) and Lévy ($\alpha = 1/2$, $\beta = \pm 1$) - there are no known closed formula for stable densities or distribution functions. In Nolan (1997), we describe efficient programs for computing these quantities for a general stable law. Our purpose here is to announce that package, called STABLE, and to describe the three parameterizations used by the program.

It takes four parameters to define a general stable distribution. One parameter is the index of stability α, which occurs naturally in the definition of stable laws, i.e. the norming constants for the sum of n i.i.d. stable random variables must be of the form $n^{1/\alpha}$ (Theorem VI.1 of Feller (1971)) if the limit is nondegenerate. Some kind of skewness parameter is needed and using $\beta \in [-1, 1]$ seems as good as any other. Some kind of scale and location parameters are needed, but we believe the ones most often used (equation (1.1) below) are not the most appropriate ones for applications. Below we will give two other parameterizations that are more intuitively meaningful. While the expressions for the characteristic function are more complicated, the connection between the parameters and the qualitative behavior of the densities is more intuitive. We suggest that users of stable distributions will find these parameterizations more practical than the standard ones.

Section 2 describes the three parameterizations and Section 3 gives brief information on how the program functions and how to obtain it.

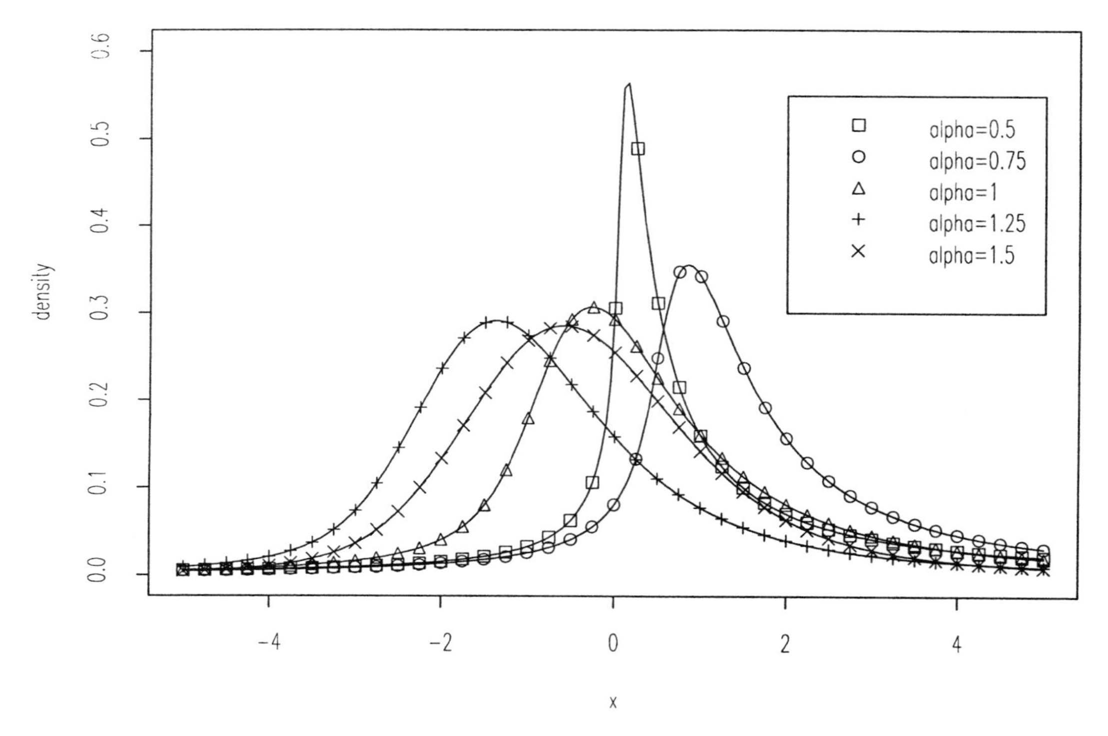

Figure 1: Stable densities in the $S_\alpha(\sigma, \beta, \mu)$ parameterization. $\beta = 0.5, \sigma = 1, \mu = 0$, and α as indicated.

1. Parameterizations

The parameterization most often used now (see Samorodnitsky and Taqqu (1994)) is the following: $X \sim S_\alpha(\sigma, \beta, \mu)$ if the characteristic function of X is given by

$$E \exp(itX) = \begin{cases} \exp\{-\sigma^\alpha |t|^\alpha \left[1 - i\beta(\tan \frac{\pi\alpha}{2})(\text{sign } t)\right] + i\mu t\} & \alpha \neq 1 \\ \exp\{-\sigma |t| \left[1 + i\beta \frac{2}{\pi}(\text{sign } t) \ln |t|\right] + i\mu t\} & \alpha = 1. \end{cases} \tag{1.1}$$

The range of parameters are $0 < \alpha \leq 2, -1 \leq \beta \leq 1$, scale $\sigma > 0$, and location $\mu \in R$. Equation (1.1) is a slight variation of the (A) parameterization of Zolotarev (1986). See Figure 1 for graphs of stable densities of standardized stable random variables ($\sigma = 1$, $\mu = 0$).

We define a second parameterization that is a variation of Zolotarev's (M) parame-

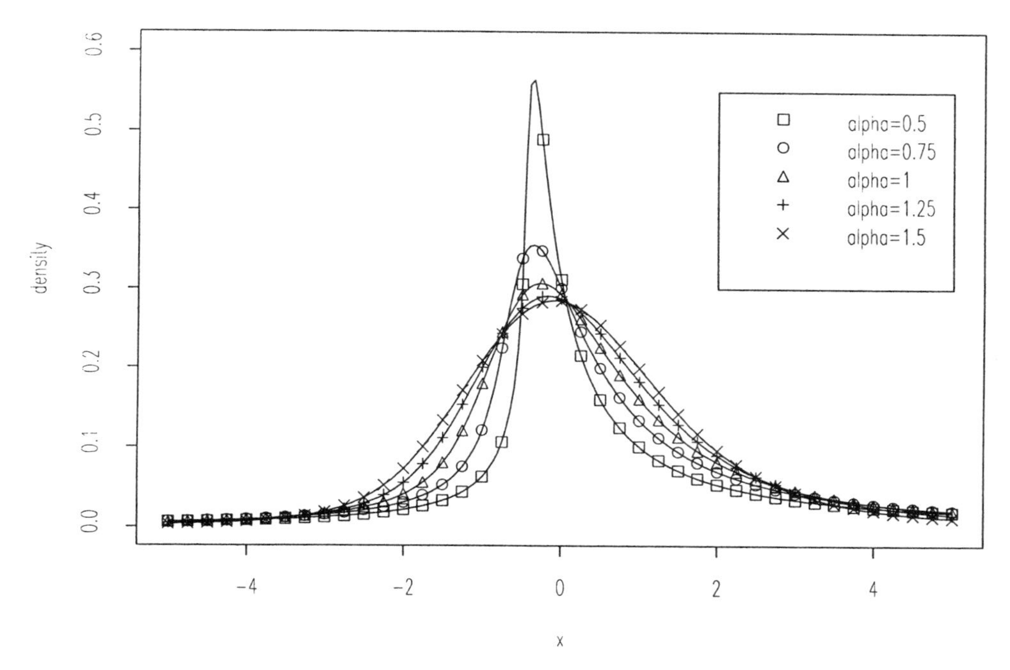

Figure 2: Stable densities in the $S_\alpha^0(\sigma, \beta, \mu^0)$ parameterization. $\beta = 0.5$, $\sigma = 1$, $\mu^0 = 0$, and α as indicated.

terization: $X^0 \sim S_\alpha^0(\sigma, \beta, \mu^0)$ if the characteristic function of X^0 is given by

$$E \exp(itX^0) = \begin{cases} \exp\{-\sigma^\alpha |t|^\alpha \left[1 + i\beta(\tan \frac{\pi\alpha}{2})(\text{sign } t)((\sigma|t|)^{1-\alpha} - 1)\right] + i\mu^0 t\} \\ \qquad\qquad\qquad\qquad\qquad\qquad\qquad \alpha \neq 1 \\ \exp\{-\sigma|t|\left[1 + \beta\frac{2}{\pi}(\text{sign } t)(\ln|t| + \ln\sigma)\right] + i\mu^0 t\} \\ \qquad\qquad\qquad\qquad\qquad\qquad\qquad \alpha = 1 \end{cases}$$

The value of this representation is that the characteristic functions (and hence the corresponding densities and distribution functions) are jointly continuous in all four parameters. Because of this property, it is better conditioned numerically to compute the density and distribution function of a standardized $S_\alpha^0(1, \beta, 0)$ distribution and base other calculations on it. Furthermore, accurate numerical calculations of the corresponding densities show that in this representation α and β have a clear meaning as measures of the heaviness of the tails and skewness parameters, compare Figure 1 to Figure 2. The numerical advantages of this parameterization was recognized early by Chambers, Mallows and Stuck (1976). Their program for simulating stable random variables generates an $S_\alpha^0(1, \beta, 0)$ random variate, not an $S_\alpha(1, \beta, 0)$ random variate.

The parameters α, β and σ have the same meaning for the S and the S^0 parameterizations, while the location parameters of the two representations are related by

$$\mu = \begin{cases} \mu^0 - \beta(\tan \frac{\pi\alpha}{2})\sigma & \alpha \neq 1 \\ \mu^0 - \beta\frac{2}{\pi}\sigma \ln \sigma & \alpha = 1. \end{cases}$$

The particular form of the characteristic function was chosen to make the S^0 parameterization a location and scale family: if $Y \sim S_\alpha^0(\sigma, \beta, \mu^0)$, then for any $a \neq 0$, b, $aY + b \sim S_\alpha^0(|a|\sigma, (\text{sign } a)\beta, a\mu^0 + b)$.

It is known that all stable densities are unimodal, see Zolotarev (1986). In the $S_\alpha^0(\sigma, \beta, \mu^0)$ parameterization, modes are reasonably behaved because of the joint continuity. Denote the mode of an $S_\alpha^0(1, \beta, 0)$ density by $m(\alpha, \beta)$. Symmetry arguments show that $m(\alpha, 0) = 0$ and $m(\alpha, -\beta) = -m(\alpha, \beta)$, so it suffices to locate the mode for $0 \leq \beta \leq 1$. In Nolan (1998), we have numerically computed $m(\alpha, \beta)$; the results are plotted in Figure 3. The estimated accuracy in the values of $m(\alpha, \beta)$ is 0.0001.

The third and final parameterization is motivated by the desire to make the scale and location parameters even more meaningful. An annoying issue with the standard scale is that as $\alpha \uparrow 2$, an $S_\alpha(\sigma, \beta, \mu)$ r.v. converges in distribution to a normal distribution with mean μ and standard deviation $\sigma/\sqrt{2}$, not standard deviation σ. This scale mismatch is an artifact of the way the characteristic function is specified. The definition below is one way to make the scale agree with the standard deviation in the normal case. As for the location parameter, the shift of $\beta \tan \frac{\pi\alpha}{2}$ from the (A) to the (M) parameterization makes things continuous in all parameters, but so does any shift $\beta \tan \frac{\pi\alpha}{2} + h(\alpha, \beta)$, where h is any continuous function. Thus the location parameter was somewhat arbitrary in the $S_\alpha^0(\sigma, \beta, \mu^0)$ parameterization. Modes are easily understood, every stable distribution has a unique mode, and every user of the normal distribution is used to thinking of the location parameter as the mode. We suggest doing the same for stable distributions.

These two points motivate the following definition: we will say $X^* \sim S_\alpha^*(\sigma^*, \beta, \mu^*)$ if the characteristic function of X^* is

$$E \exp(itX^*) = \begin{cases} \exp\{-\frac{1}{\alpha}(\sigma^*|t|)^\alpha \left[1 - i\beta(\tan \frac{\pi\alpha}{2})(\text{sign } t)\right] \\ \qquad + i\left[\mu^* - \alpha^{-1/\alpha}\sigma^*(\beta \tan \frac{\pi\alpha}{2} - m(\alpha, \beta))\right] t\} & \alpha \neq 1 \\ \exp\{-\sigma^*|t| \left[1 + i\beta\frac{2}{\pi}(\text{sign } t) \ln |t|\right] \\ \qquad + i\left[\mu^* - \sigma^*(\beta\frac{2}{\pi} \ln \sigma^* - m(1, \beta))\right] t\} & \alpha = 1. \end{cases}$$

It is straightforward that α and β are the same in all three parameterizations, while the scales and locations are related by:

$$\begin{aligned} \sigma^* &= \alpha^{1/\alpha}\sigma \\ \mu^* &= \mu^0 - \sigma m(\alpha, \beta) = \begin{cases} \mu + \sigma(\beta \tan \frac{\pi\alpha}{2} - m(\alpha, \beta)) & \alpha \neq 1 \\ \mu + \sigma(\beta\frac{2}{\pi} \ln \sigma - m(1, \beta)) & \alpha = 1 \end{cases} \end{aligned}$$

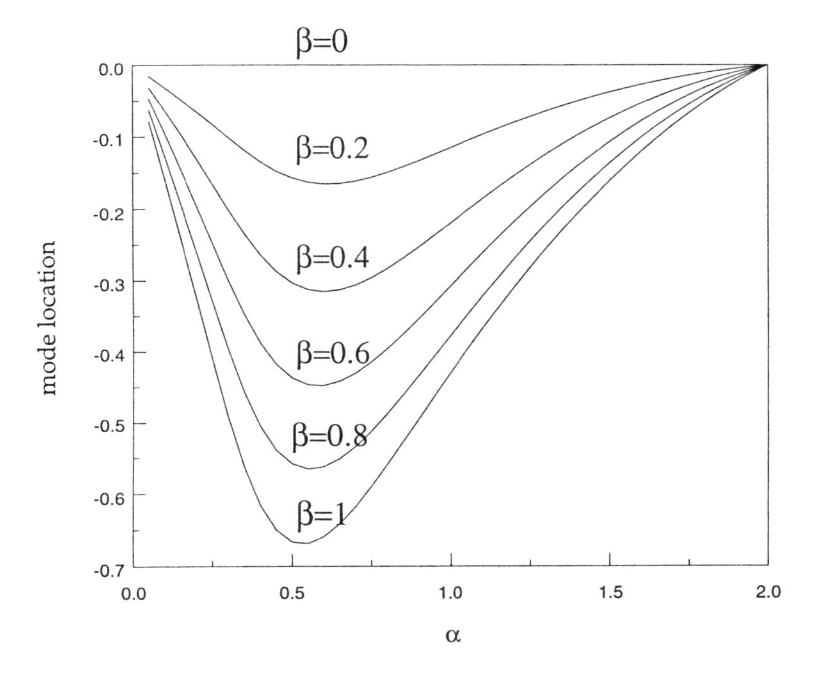

Figure 3: Plot of the mode $m(\alpha, \beta)$ of a standardized stable random variable $Z \sim S_\alpha^0(1, \beta, 0)$.

Basic properties of the S^* parameterization are: (a) The mode of an $S_\alpha^*(\sigma^*, \beta, \mu^*)$ density is at μ^*, (b) In the Gaussian case ($\alpha = 2$), σ^* is the standard deviation and in the Cauchy case ($\alpha = 1, \beta = 0$), σ^* is the standard scale parameter, (c) σ^* and μ^* are true scale and location parameters: if $Y \sim S_\alpha^*(\sigma^*, \beta, \mu^*)$, then for any $a \neq 0$, b, $aY + b \sim S_\alpha^*(|a|\sigma^*, (\text{sign } a)\beta, a\mu^* + b)$, (d) The characteristic function (and hence the density and distribution functions) are jointly continuous in all four parameters $(\alpha, \sigma^*, \beta, \mu^*)$. See Figure 4 for a plot of selected stable densities in the $S_\alpha^*(\sigma^*, \beta, \mu^*)$ parameterization. In addition to centering at the mode, this parameterization emphasizes the fact that stable distributions with larger α are more concentrated around the mode, while those with lower α have heavier tails - this is not as clear in the other parameterizations.

More information on these parameterizations and generalizations to multivariate stable laws can be found in Nolan (1996b).

3. The STABLE program

The program STABLE calculates general stable densities, distribution functions and quantiles. An executable PC version of the program can be downloaded from the World Wide Web. This is done by opening `http://www.cas.american.edu/~jpnolan`

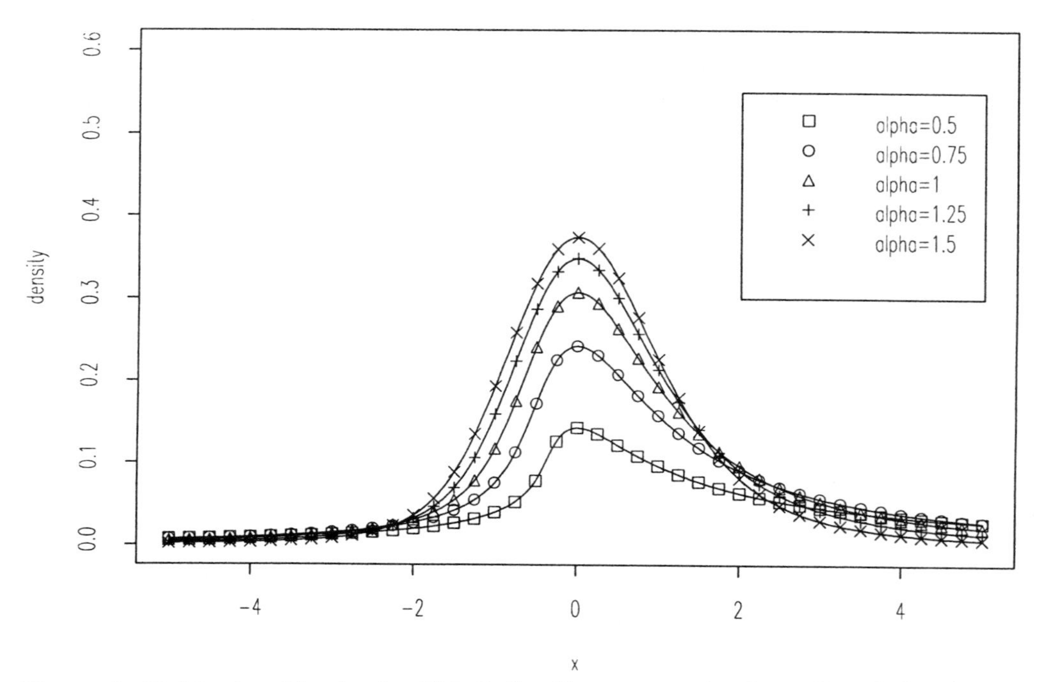

Figure 4: Stable densities in the $S^*_\alpha(\sigma^*, \beta, \mu^*)$ parameterization. $\beta = 0.5$, $\sigma^* = 1$, $\mu^* = 0$, and α as indicated.

and clicking on the link to stable software.

The central parts of the program STABLE are FORTRAN routines that calculate the density $f(x; \alpha, \beta)$ and distribution function $F(x; \alpha, \beta)$ of an $S^0_\alpha(1, \beta, 0)$ distribution. These algorithms are based on a numerical evaluation of integral formulas for f and F, see Nolan (1997). The program has built in transformations for a general stable distribution in any of the three parameterizations: $S_\alpha(\sigma, \beta, \mu)$, $S^0_\alpha(\sigma, \beta, \mu^0)$, or $S^*_\alpha(\sigma^*, \beta, \mu^*)$.

STABLE works for most values of the (α, β) parameter space. The output has been compared to existing tables of densities and distribution function and agrees to the printed accuracy. For α very close to 1, i.e. $|\alpha - 1| < 0.02$, or α close to zero, there are some numerical difficulties. When α is close to 1, the integrands in the formulas for the density and distribution function change very rapidly and it is difficult to accurately approximate the integrals. When α is close to zero, the densities have a steep spike at the mode $m(\alpha, \beta)$. We are not aware of tables for these parameter values and it is hard to assess the accuracy of STABLE. For example, when $\alpha = 0.1$ and $\beta = 0$, $f(m(\alpha, \beta); \alpha, \beta) = f(0; 0.1, 0) = 1.155 \times 10^6$, while $f(0.01; 0.1, 0) = 1.666$ - a change of six orders of magnitude when x changes by 0.01! Further calculations show $P(|X| \leq 0.01) = .2244$ and $P(|X| > 5) = .5527$, which is very different from a standardized Gaussian (or Cauchy) distribution: most of the mass is on the tails and much of what's left is concentrated very near the mode. It would be interesting to know of any real data sets with this property.

The relative accuracy of the calculations is 10^{-6}, i.e. the program tries to guarantee that the computed density $\hat{f}$ and computed distribution function $\hat{F}$ satisfy $|f(x) - \hat{f}(x)| \leq 10^{-6} f(x)$ and $|F(x) - \hat{F}(x)| \leq 10^{-6} F(x)$. The accuracy can be improved by recompiling the program, but this will generally slow down the program. Approximately 200 density or cumulative distribution function calculations per second can be done on a 60 MHz Pentium PC.

The current version of the program runs on a PC computer under DOS. STABLE is currently being used in a prototype program that computes maximum likelihood estimators of stable parameters from data. Planned future developments include machine independent source code, an interface to S-Plus, and programs for working with multivariate stable distributions.

References

[1] Chambers, J. M., Mallows, C. and Stuck, B. W. (1976) A method for simulating stable random variables. *JASA* **71**, 340-344.

[2] Nolan, J. P. (1997) Numerical calculation of stable densities and distribution functions, *Stochastic Models* **13**, 759-774.

[3] Nolan, J. P. (1996b) Parameterizations and modes of stable distributions. To appear in *Stat. and Prob. Letters.*

[4] Samorodnitsky, G. and Taqqu, M. S. (1994) *Stable Non-Gaussian Random Processes.* Chapman and Hall, NY, NY.

[5] Zolotarev, V. M. (1986) *One-dimensional Stable Distributions, Amer. Math. Soc. Transl. of Math. Monographs, Vol. 65.* Amer. Math. Soc., Providence, RI. (Transl. of the original 1983 Russian)

Department of Mathematics and Statistics
American University
Washington, DC 20016